Springer Series in
SOLID-STATE SCIENCES
151

Springer Series in
SOLID-STATE SCIENCES

Series Editors:

M. Cardona P. Fulde K. von Klitzing R. Merlin H.-J. Queisser H. Störmer

The Springer Series in Solid-State Sciences consists of fundamental scientific books prepared by leading researchers in the field. They strive to communicate, in a systematic and comprehensive way, the basic principles as well as new developments in theoretical and experimental solid-state physics.

Volumes 92–138 are listed at the end of the book.

Peter Sigmund

Particle Penetration and Radiation Effects

General Aspects
and Stopping of Swift Point Charges

With 131 Figures

 Springer

Dr. Peter Sigmund
University of Southern Denmark
Department of Physics and Chemistry
Campusvej 55
5230 Odense, Denmark
psi@dou.dk

Series Editors:

Professor Dr., Dres. h. c. Manuel Cardona
Professor Dr., Dres. h. c. Peter Fulde*
Professor Dr., Dres. h. c. Klaus von Klitzing
Professor Dr., Dres. h. c. Hans-Joachim Queisser

Max-Planck-Institut für Festkörperforschung, Heisenbergstrasse 1, 70569 Stuttgart, Germany
* Max-Planck-Institut für Physik komplexer Systeme, Nöthnitzer Straße 38
  01187 Dresden, Germany

Professor Dr. Roberto Merlin
Department of Physics, 5000 East University, University of Michigan
Ann Arbor, MI 48109-1120, USA

Professor Dr. Horst Störmer
Dept. Phys. and Dept. Appl. Physics, Columbia University, New York, NY 10027 and
Bell Labs., Lucent Technologies, Murray Hill, NJ 07974, USA

Library of Congress Control Number: 2007934058

ISSN 0171-1873

ISBN 978-3-540-72622-7 Springer Berlin Heidelberg New York
ISBN 978-3-540-31713-5 Springer Berlin Heidelberg New York

Springer is a part of Springer Science+Business Media

springer.com

Typesetting by the author and LE-TEX Jelonek, Schmidt & Vöckler GbR, Leipzig, Germany
Production: LE-TEX Jelonek, Schmidt & Vöckler GbR, Leipzig, Germany
Cover design: eStudio Calamar S.L., F. Steinen-Broo, Girona, Spain

SPIN 12025138   57/3180/YL – 5 4 3 2 1 0   Printed on acid-free paper

To Pia

Preface

This book has emerged from lectures given to physics students at the University of Copenhagen and the University of Southern Denmark in Odense. Part of the material was first compiled in connection with lectures at international summer schools in Predeal (Romania), Split (Croatia), Porto Vecchio (Corsica, France), Haifa (Israel), Viana do Castelo (Portugal) and Alicante (Spain).

The book addresses mainly two different groups of readers. Chapters 1–5 of this volume serve as the main text of a standard course for graduate students in physics. Here I have tried to focus on general-physics aspects which should be of interest even to someone who never gets in touch with a charged-particle beam. General principles are introduced ab initio, and important physical arguments and mathematical derivations are presented in extenso. Consultation of the original literature should only rarely be necessary during the first reading, but pertinent references have been supplied generously on all major items for the reader who wants to dig deeper.

The second group of prospective readers is scientists, engineers and medical doctors who apply charged-particle beams, in particular ion beams, in a wide variety of contexts and who need an introduction to the underlying physical principles in order to understand and quantify some of their activities. Many of those readers are not physicists and have typically only a minor background – if any at all – in collision theory. Therefore also this readership may appreciate a reasonably self-contained introduction into the field. On the other hand, those readers, as well as students specializing in the field, need to go further. Chapters 6–9 are intended to guide them up to the level of current research in the area.

I shall assume the reader to be familiar with classical dynamics of one- and two-particle systems and elements of electromagnetic theory. Special relativity will enter in numerous instances, but the reader uninterested in relativistic beam energies will be able to jump over those passages without loss of essential information for what follows. Quantal concepts enter throughout the book, but mainly in the common Schrödinger picture. Yet quantal collision theory, which is frequently omitted from introductory courses in quantum theory, will be introduced ab initio. Some experience with elements of probability theory will be beneficial. A generous amount of supporting material that is – more or less – standard in a physics curriculum but not necessarily familiar to a reader with a different background has been collected in an appendix.

The main text contains passages which a novice in the field might prefer to jump over in a first reading. I have tried to mark such passages by a star ($\star$), even though there might be an occasional reference to such passages in unstarred sections.

Problems added to all chapters are intended to serve four purposes,

- to provide order-of-magnitude estimates of important experimental parameters,
- to illustrate the significance of a general result by adopting specific input,
- to train the student in casting a physics question into a form tractable by quantitative calculation, and
- to complete formal derivations quoted in the text where appropriate. This type of problem is dominating in the second half of the volume.

Some help has been provided whenever judged necessary, although I do not by any means want to discourage the reader from trying to take a different and, possibly, more efficient route.

A first draft of this book was written at IBM Watson Research Laboratory many years ago. That version focused on energy loss of ion beams and did not get finished because I noticed a lack of knowledge on key features of the theory. I could, in principle, have published a status report, but instead I initiated research programs addressing topics that I felt needed attention. This process continued over the years, interrupted by an extended period of administrative duties but compensated by a sabbatical at Argonne National Laboratory, short visiting appointments at Université Paris-Sud, École Polytechnique and the University of Pretoria, and a month at San Cataldo monastery. While the style of the book is still very much like that of the first draft, I find that the subject matter, or at least my personal grasp of it, has developed to the point that this book can serve a useful purpose.

This first volume presents general concepts and, moreover, focuses on the stopping of swift point charges. The first part provides an overview of the field and its application areas, a chapter on elementary penetration theory inspired by the 'Danish school' of Bohr and Lindhard, and elements of classical and quantal scattering theory as far as needed in context. Chapters 4 and 5 outline classical and quantal theory of electronic stopping of swift point charges. Taken together, Chapters 1–5 provide ample material for a semester course in particle penetration. Of the remaining four chapters, only a fraction is likely to be found suitable for presentation in a university course, but I hope that they will be useful to those who want to specialize in the field.

A second volume is planned to extend the treatment to heavy ions, including molecule and cluster ions, stopping at lower velocities, as well as multiple scattering, ion ranges and channeling. Associated radiation effects are planned to be treated in a third volume.

My own introduction to the field has been inspired indirectly by the late Niels Bohr whom I never got a chance to meet or listen to, and more directly by his close collaborator, the late Jens Lindhard whose lectures and papers as well as numerous discussions greatly stimulated my approach to the field.

This book would never have come along without the excellent working conditions provided by the University of Southern Denmark, as well as my hosts at various sabbatical stays, most notably J.F. Ziegler at IBM, D.S. Gemmell and H.G. Berry at Argonne, Y. Lebeyec at Orsay, Y. Quéré and Annie Dunlop at Palaiseau, and E. Friedland and J.B. Malherbe in Pretoria.

Numerous colleagues, junior and senior ones, have contributed to develop my knowledge of and insight into the field over the years. In connection with the present volume I like to mention in particular H.H. Andersen, J.U. Andersen, N.R. Arista, G. Basbas, A. Belkaçem, F. Besenbacher, E. Bonderup, H. Esbensen, D.J. Fu, L.G. Glazov, A. Gras-Martí, U. Haagerup, P. Hvelplund, M. Inokuti, J. Jensen, K. Johannessen, H. Knudsen, E. Merzbacher, H.H. Mikkelsen, E.H. Mortensen, S.P. Møller, J. Oddershede, H. Paul, R.H. Ritchie, A. Schinner, A.H. Sørensen, A. Tofterup, S.M. Tougaard, K.B. Winterbon, and J.F. Ziegler.

Drafts of the first five chapters have been read by numerous students who returned useful corrections, comments and questions. Lev Glazov and Nestor Arista have spent much time, care and thought in reading the whole volume and have provided invaluable feedback. Nevertheless, the blame for any serious omissions and errors is the author's and not theirs.

Financial support has been received from the Danish Natural Science Research Council (FNU) which has enabled the author to travel, to conduct an extensive visitors' program, and to support research students and postdoctoral fellows. Support has also been received from NORDITA, the Carlsberg Foundation, the Ib Henriksen Foundation, the NATO Collaborative Research Programme, the IBM World Trade Program and the Institution San Cataldo. It was a particular honor to spend a year at Argonne National Laboratory as an Argonne Fellow.

Odense, October 2005   *Peter Sigmund*

Contents

Part III Straggling

Part IV Appendices

General Notations

- Sections marked by a star ($\star$) can be jumped over in a first reading.
- Problems marked by a star ($\star$) are considerably more difficult and/or time-consuming than average.
- Notations $< \cdots >$ or $\overline{\cdots}$ are utilized synonymously to indicate averages, dependent on readability.
- The symbol $(T, \mathrm{d}T)$ indicates the interval limited by T and $T + \mathrm{d}T$. Similarly, $(\boldsymbol{\Omega}, \mathrm{d}^2\boldsymbol{\Omega})$ indicates a solid angle $\mathrm{d}^2\boldsymbol{\Omega}$ around the unit vector $\boldsymbol{\Omega}$, and $(\mathbf{r}, \mathrm{d}^3\mathbf{r})$ indicates a volume element $\mathrm{d}^3\mathbf{r} = \mathrm{d}x\mathrm{d}y\mathrm{d}z$ located at a vector distance $\mathbf{r}$ from the origin.

a	screening radius
a_0	Bohr radius, 0.529177 Å
a_ad	adiabatic radius, v/ω
a_TF	Thomas-Fermi radius, $0.8853a_0 Z^{-1/3}$
$\boldsymbol{a}_0, \boldsymbol{b}_0$	oscillation amplitudes in Bohr theory
A_1	mass number of projectile ion
A_n	Fourier coefficient
$\mathrm{Ai}(z)$	Airy function
$\boldsymbol{A}(\boldsymbol{r}, t)$	vector potential
$\arg(z)$	phase of complex number, ϕ in $z = A\mathrm{e}^{\mathrm{i}\phi}$
Å	Ångström, 0.1 nm
α	fine structure constant, 1/137.0360
α_0	shifted resonance frequency, $\sqrt{\omega_0^2 + \omega_\mathrm{P}^2}$
$\boldsymbol{\alpha}, \alpha$	multiple-scattering angle, vectorial, polar
b	barn, 10^{-28} m^2
b	collision diameter, $2e_1 e_2/m_0 v^2$
$\boldsymbol{b}$	projection of molecular axis on impact plane
B	Bethe parameter, $2mv^2/\hbar\omega$

continued on next page

continued from previous page

B'	Barkas-Andersen parameter, $Z_1 e^2 \omega / m v^3$
$\boldsymbol{B}(\boldsymbol{r},t)$	magnetic field
β	v/c
c	speed of light, $2.99792 \cdot 10^8$ m/s
$c_j(t)$	expansion coefficient in wave function
$c^{(\nu)}$	atomic fraction, N_ν / N
χ	parameter in Lindhard-Scharff model
$\chi(k,\omega)$	electric susceptibility, $\big(\epsilon_l(k,\omega) - 1\big)/4\pi$
C	coefficient in Bohr stopping formula, $2\exp(-\gamma) = 1.1229$
$-\mathrm{d}E/\mathrm{d}x$	stopping force, stopping power
$\mathrm{d}\sigma(T)$	differential energy-loss cross section, $\big(\mathrm{d}\sigma(T)/\mathrm{d}T\big)\mathrm{d}T$
$\mathrm{d}^2\boldsymbol{\Omega}$	element of solid angle
$\mathrm{d}^3\boldsymbol{r}$	volume element
δ_ℓ	phase shift
$\delta(\xi)$	Dirac function
$\delta_{j\ell}$	Kronecker symbol, 1 for $j = \ell$, 0 for $j \neq \ell$
ΔE	total energy loss
$\langle \Delta E \rangle$	mean energy loss
ΔE_p	most probable energy loss
$\langle \Delta E \rangle_\mathrm{e}$	mean electronic energy loss
$\langle \Delta E \rangle_\mathrm{n}$	mean nuclear energy loss
Δx	penetration depth or pathlength element
e	elementary charge, $1.602176 \cdot 10^{-19}$ Coulomb
e_1	projectile charge, $Z_1 e$ or $\pm e$
e_2	target charge, $Z_2 e$ or $-e$
esu	electrostatic unit
eV	electron volt, $1.602176 \cdot 10^{-19}$ J
E, E'	kinetic energy
E_+, E_-	energies in Dirac theory
E_i, E_tot	particle energy (relativistic)
E_k	relativistic energy, $\sqrt{(\hbar k c)^2 + (mc^2)^2}$
$E_n(z)$	exponential integral, $\int_1^\infty \mathrm{e}^{-zt}/t^n \mathrm{d}t$
$\mathcal{E}$	total energy
$\boldsymbol{E}(\boldsymbol{r},t)$	electric field
ε	Lindhard dimensionless energy, $M_2 E a / \big[(M_1 + M_2) Z_1 Z_2 e^2\big]$
ϵ_0	vacuum permittivity, $8.85419 \cdot 10^{-12}$ C/(Vm)
$\epsilon_l(k,\omega)$	longitudinal dielectric function
$\epsilon_t(k,\omega)$	transverse dielectric function
ϵ_F	Fermi energy
ϵ_j	energy level

continued on next page

continued from previous page

$f(\boldsymbol{\Omega})$	scattering amplitude
$f(v)\mathrm{d}^3 v$	orbital velocity spectrum
$f_{\mathrm{dist}}(p), f_{\mathrm{close}}(p)$	factors entering energy loss vs. p in Bohr theory
f_{j0}	dipole oscillator strength
$f_{j0}(Q)$	generalized oscillator strength
f_ν	oscillator strength of shell or subshell
$\boldsymbol{F}$	force
$F(\Delta E, x)$	energy-loss spectrum at pathlength x
$F(E, E', x)$	energy spectrum at pathlength x
$F_{j\ell}$	transition matrix element, $\langle j | \sum_\nu \mathrm{e}^{\mathrm{i}\boldsymbol{k}\cdot\boldsymbol{r}_\nu} | \ell \rangle$
$g_2(r)$	pair distribution function
$g_2(r) - 1$	pair correlation function
$G(x, x')$	Green function
γ	Euler's constant, 0.577216
γ	energy transfer factor, $4m_1 m_2/(m_1 + m_2)^2$
γ_v, γ_i	relativistic mass factor, $1/\sqrt{1 - v^2/c^2}, 1/\sqrt{1 - v_i^2/c^2}$,
Γ	infinitesimal damping constant
h	Planck's constant, $6.62607\cdot10^{-34}$ J·s
$\hbar$	$h/2\pi$, $6.58212\cdot10^{-16}$ eV s
H, H_ν	Hamiltonian
i	imaginary unit
I	'I-value', mean logarithmic excitation energy
I_1	I-value for straggling
Im	imaginary part
j	type of event
j_ℓ	spherical Bessel function
$|j\rangle$	state vector in Hilbert space
J_0	Bessel function
$\boldsymbol{J}, J$	particle current density
$\boldsymbol{J}_e$	electric current density
k	absorption coefficient
keV	10^3 eV
$\boldsymbol{k}, \boldsymbol{k}', k, k'$	wave vector, wave number
$k_\pm$	$k_x \pm \mathrm{i}k_y$
k_F	Fermi wave number
K_0, K_1	modified Bessel functions
$K(\phi)$	differential scattering cross section, $\mathrm{d}\sigma(\phi)/(2\pi\phi\mathrm{d}\phi)$

continued on next page

continued from previous page

κ	Bohr kappa parameter, $2|e_1 e_2|/\hbar v$
$\varkappa$	angular-momentum quantum number in Dirac theory
L	stopping number
L_1	Z_1^3 correction to stopping number
L_{e}	electronic stopping number
L_{n}	nuclear stopping number
λ	de Broglie wavelength, h/mv
$\lambdabar$	$\lambda/2\pi$, $\hbar/mv$
Λ	scaled energy loss in Landau theory
m	electron mass, $9.10938{\cdot}10^{-31}$ kg
m_0	reduced mass, $m_1 m_2/(m_1 + m_2)$
m_1	projectile mass, M_1 or m
m_2	target mass, M_2 or m
M	particle mass
M_1	mass of projectile ion
M_2	mass of target atom
M_p	proton mass, $1.672621{\cdot}10^{-27}$kg
$M(1, x, y)$	Kummer's function
$\boldsymbol{M}$	Runge-Lenz vector, $\boldsymbol{v} \times \boldsymbol{L}$
MeV	10^6 eV
μm	micro meter, 10^{-6} m
n, n_j	number of events
n	number of electrons per volume, NZ_2
n	index of refraction
$n_{\boldsymbol{k}_0}$	occupation number number of electrons/state
nm	nanometer 10^{-9} m
N	number of atoms per volume
N_A	Avogadro's number, $6.02214{\cdot}10^{23}$/mol
N	total number of electrons
∇	nabla operator, $(\partial/\partial x, \partial/\partial y, \partial/\partial z)$
ω	angular frequency
ω_{P}	plasma frequency, $\sqrt{4\pi n e^2/m}$
$\omega, \omega_0, \omega_{j0}$	resonance frequency, $(\epsilon_j - \epsilon_0)/\hbar$
ω_k	$\hbar k^2/2m$
$\boldsymbol{\Omega}$	unit vector
Ω^2	energy loss straggling, $\left\langle (\Delta E - \langle \Delta E \rangle)^2 \right\rangle$
Ω_R^2	range straggling

continued on next page

continued from previous page

$\boldsymbol{p}, p$	impact parameter, vectorial or scalar
$\boldsymbol{P}, \boldsymbol{P}_i, \boldsymbol{P}'_i$	momentum
P	Cauchy principal value
$\mathsf{P_n}$	probability for n events
P_j	transition probability
P_J	ionic charge fraction
P_ℓ	Legendre polynomial
P_ℓ^μ	associated Legendre polynomial
ϕ, ϕ_1	single-scattering angle, laboratory system
ϕ_2	recoil angle, laboratory system
$\phi_0(x)$	screening function of neutral Thomas-Fermi atom
Φ	electric potential
$\Phi(r/a)$	screening function
φ	azimuthal angle
$\psi(\boldsymbol{r}), \Psi(\boldsymbol{r}, t)$	wave function
$\psi(x)$	digamma function, $\mathrm{d}\ln\Gamma(x)/\mathrm{d}x$
q	electric charge
$\boldsymbol{q}$	wave vector
$\langle q_1 \rangle$	mean ionic charge number
q_J	ionic charge state
Q	$\hbar^2 q^2/2m$
Q_ν	νth moment over energy-loss cross section, $\int T^\nu \mathrm{d}\sigma(T)$
$\boldsymbol{r}$	position vector of classical electron
$\boldsymbol{r}_0(t)$	orbital motion in Bohr theory
$\boldsymbol{r}, \boldsymbol{r}_\nu$	position operator for electron(s)
r_m	distance of closest approach
r_s	Wigner-Seitz radius, $(3/4\pi na_0^3)^{1/3}$
r_0	classical electron radius, $2.81794{\cdot}10^{-15}$ m
R	path length or range
R	Rydberg energy
$R(r), R_\ell(r)$	radial wave function
$\boldsymbol{R}$	position vector, projectile or center of mass
Re	real part
ρ_m	mass density, mass/volume
ρ_e	charge density, charge/volume
ρ	Lindhard dimensionless pathlength variable, $N\pi a^2 \gamma x$
S	stopping cross section, $\sum_j T_j \sigma_j$ or $\int T \mathrm{d}\sigma(T)$
S_0	stopping cross section in rest frame of target
S_e	electronic stopping cross section
S_n	nuclear stopping cross section

continued on next page

continued from previous page

σ	cross section
$\sigma_x, \sigma_y, \sigma_z$	Pauli spin matrices
σ_j	excitation cross section
$\sigma^{(1)}, \sigma^{(2)}$	transport cross sections
σ_l, σ_t	longitudinal and transverse conductivity
$\sigma(k), \sigma_1(k), \sigma_2(k)$	transport cross sections
t	time
T, T_j	energy transfer per collision, continuum or discrete
T_{max}	maximum energy transfer
τ	collision time
θ	polar angle
Θ	center-of-mass scattering angle
u	atomic mass unit, $1.660538 \cdot 10^{-27}$ kg
$\boldsymbol{u}_i, \boldsymbol{u}_i'$	relative velocity
$u^{(\nu)}(\boldsymbol{k})$	Dirac spinor
U	ionization or binding energy
v	volume
v_0	Bohr velocity, $e^2/4\pi\epsilon_0\hbar = c/137.0360$
$\boldsymbol{v}, \boldsymbol{v}_i, v, v_i$	velocity, speed
$\boldsymbol{v}_e$	orbital electron velocity
v_{F}	Fermi speed
v_{M}	Møller speed
v_{TF}	Thomas-Fermi velocity
$\boldsymbol{V}$	center-of-mass velocity
V	volume
$\mathcal{V}(r)$	potential energy
$\boldsymbol{w}_e$	relative electron velocity
W	straggling parameter, $\sum_j T_j^2 \sigma_j$ or $\int T^2 \mathrm{d}\sigma(T)$
W_{B}	Bohr straggling parameter, $4\pi Z_1^2 Z_2 e^4$
x	depth or pathlength
ξ	$mv^3/Z_1 e^2 \omega$, Bohr parameter
y_ℓ	spherical Bessel function
$Y_{\ell\mu}(\boldsymbol{\Omega})$	spherical harmonic
Z_1	atomic number of projectile nucleus
Z_2	atomic number of target nucleus
ζ	scaled impact parameter, $\omega p/v$

Part I

General

1

Introduction

1.1 Brief Historical Survey

1.1.1 Particle Penetration

Research in the field of interaction of charged particles with matter has its roots in two distinctly different disciplines.

Early experimental observations were made around 1850 in gas-discharge tubes. Most spectacular was the light emitted from a discharge which provided much of the stimulation behind the development of atomic spectroscopy. Moreover, solid cathode material was observed to erode, a phenomenon now called sputtering. The atomistic nature of the processes going on in a discharge was largely unknown at the time, and only gradually did it become clear that the current in a tube was carried by electrons and positive ions. Now we know that light emitted from a discharge tube originates in the excitation of gas atoms and ions by electrons accelerated to sufficient energy to cause inelastic collisions. Similarly we know that sputtering of cathode material is caused by disruptive collisions between accelerated gas ions and metal atoms in a shallow layer of the cathode surface.

Beams of electrons ('cathode rays') and positive ions ('canal rays') were identified by Thomson in 1897 and Goldstein in 1902, respectively, and rapid progress was achieved in the extraction and manipulation of such beams. Experiments under reasonably controlled conditions became eventually possible. Numerous phenomena were observed early in the previous century such as reflection of impinging beam particles, emission of secondary electrons or ions and of light from the target material, as well as various kinds of surface modification. An important step in the development of atomic physics, involving the interaction of electrons with a gaseous target, was the famous Franck & Hertz experiment in 1913 which demonstrated a relation between the excitation levels of gas atoms in a discharge tube and the acceleration voltage for the electron current.

The second important source of information on particle penetration became available with the discovery of radioactivity. With the identification of alpha and beta rays as being made up of heavy and light charged particles, respectively, well-defined beams became available with energies several orders of magnitude higher than what was achievable in gas discharges. This implied greater penetration depth and, therefore, more well-defined quantitative measurements. A key experiment addressed the range of alpha particles in a cloud chamber (Fig. 1.1). It is seen that at atmospheric pressure, the range is of the order of a few centimeters and therefore measurable with good accuracy.

Fig. 1.1. Tracks of alpha particles in a Wilson cloud chamber. The scale is about 1:2. From Meitner and Freitag (1926)

A most important set of measurements addressed the scattering of a beam of alpha particles in gaseous and solid matter (Rutherford et al., 1910). This prompted a famous analysis (Rutherford, 1911) which established the structure of the atom as being made up of a heavy nucleus surrounded by electrons. At the same time, this analysis established the *scattering cross section* as a key concept in the statistical description of the interaction of fast particles with matter. Only after this concept had become available was it feasible to develop a theory of particle penetration from first principles. Central papers by Thomson (1912) and Bohr (1913, 1915) appeared almost immediately and were followed by numerous others in rapid sequence. Early studies had to be based entirely on classical-mechanics concepts, augmented by relativistic extensions when appropriate. Gradually quantal concepts were incorporated as they developed. A key step here was the appearance of first-order perturbation theory of quantal scattering by Born (1926) and the theory of charged-particle stopping based on this approximation (Bethe, 1930). Another important step was the statistical theory of multiple collisions (Bothe, 1921b). Although the principles underlying this theory were rediscovered repeatedly, we now know that Bothe provided a universally applicable tool in the statistical theory of particle penetration.

Subsequent research brought about a widening of the energy range accessible to experiments. With the discovery of cosmic rays, penetration properties of particle beams in the extreme relativistic regime became of interest. The trajectories of such particles are characterized by wide tracks, i.e., large numbers of target atoms interacting simultaneously with a penetrating beam particle. As a consequence, macroscopic polarization phenomena had to be expected (Fermi, 1940), and collective effects like Cherenkov radiation were observed and interpreted (Cherenkov, 1934).

With the appearance (Cockcroft and Walton, 1932) and gradual development of high-energy accelerators, detection devices were developed such as photographic emulsion, bubble chambers, and many more. All aspects of particle penetration became key input into the development of this instrumentation.

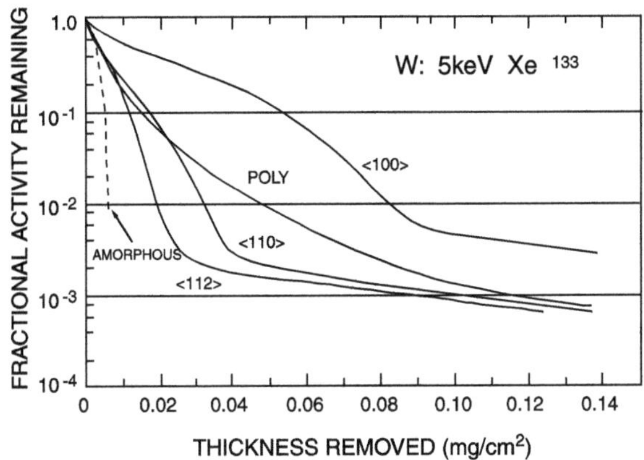

Fig. 1.2. Penetration of keV xenon ions into crystalline tungsten. Plotted is the integrated range profile, i.e., the amount of xenon still buried in the target after a quantity of material corresponding to the value of the abscissa has been etched away. Etching was performed by anodic oxidation and subsequent dissolution. Implantation was performed with a radioactive beam. This enabled the amount of xenon buried in the target material to be determined by monitoring the beta activity. See also problem 1.3. From Kornelsen et al. (1964)

The discovery of nuclear fission in 1938 brought about a substantial extension of the range of accessible particle masses. Beams of fission fragments provided the main source of medium-mass fast particles in the MeV energy range until the 1960s when such beams could finally be generated in tandem van de Graaf accelerators. Swift heavy ions are typically not point charges, and even though theoretical studies were initiated immediately (Bohr, 1940, 1941), the understanding of penetration phenomena involving these beams, in particular in solids and liquids, still poses challenges (Sigmund, 2004).

Precision measurements of the ranges of products of nuclear reactions were a major tool in the development of nuclear and particle physics since such measurements provide an estimate of the energy liberated during a reaction.

A very wide range of particles and energies was of interest, covering recoiling nuclei with energies in the keV range at the one end and mesons and other light particles with much higher energies at the other. This technique has proved very successful in high-energy physics, but ingenious ideas were conceived also to get it to work in low-energy nuclear reaction physics (Davies et al., 1960).

A comparatively recent development is the option of generating high-energy molecular beams. While beams of small molecules like H_2^+ and low hydrocarbons have long been available, present-day technology allows to accelerate metal clusters or fullerene (C_{60}) ions up to energies in the MeV range (Della-Negra et al., 1993). This has opened up another exciting period of exploration of hitherto unknown collision and penetration phenomena.

Once experimental techniques were at hand that allowed penetration depths in the submicron regime to be measured accurately (Fig. 1.2), quantitative studies became finally possible of phenomena that had been discovered in gas discharges more than a hundred years earlier. A major stimulus around 1965 was the desire to produce integrated electronic circuits by means of ion implantation, i.e., controlled doping of semiconducting materials by ion beams. Several steps in this highly successful development will be described in detail in Volumes II and III.

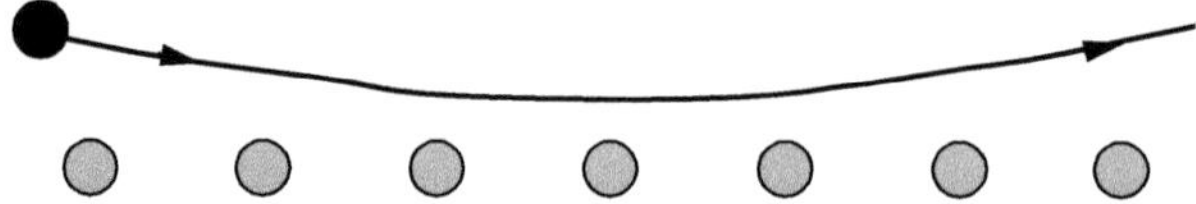

Fig. 1.3. Channeling: Reflection of a channeled ion from a string or plane of atoms in a crystal by a sequence of soft collisions

Early theoretical research on particle penetration was directed at gaseous scattering media, i.e., media composed of random scatterers. Even for crystalline solids, the regular structure was thought to be of minor significance since free paths for significant scattering events frequently exceed interatomic distances in solid matter by several orders of magnitude. After the discovery of electron diffraction by Davisson and Germer (1927), this argument was no longer tenable for the elastic interaction between electrons and the ionic cores of a penetrated crystal. However, random penetration theory was still considered approximately valid for inelastic interactions with the electron gas and has been successful in many applications such as the quantitative interpretation of electron microscope images.

De Broglie wavelengths of heavy charged particles in the energy range from a few eV upward are small enough to discourage the consideration of diffraction effects during passage through crystalline matter. It came, therefore, as a genuine surprise to the scientific community that even the classical trajectory of a particle through a regular crystal may be strongly governed by

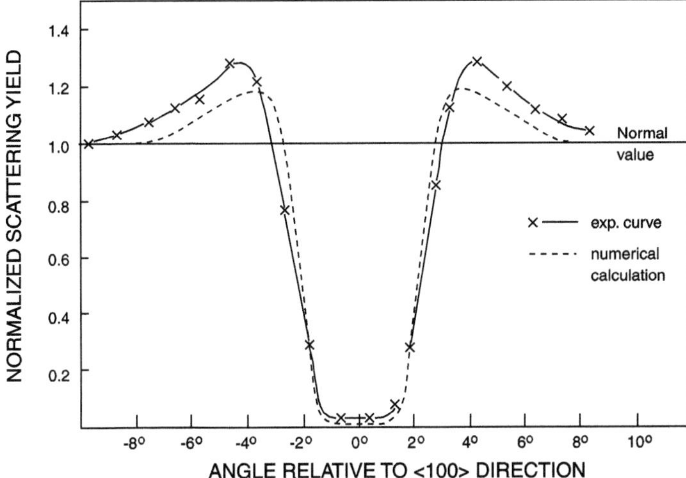

Fig. 1.4. Channeling: Rutherford scattering yield of 480 keV protons reflected under a wide angle from a tungsten monocrystal. Because of the dominance of the process shown in Figure 1.3, the yield is reduced dramatically at angles of incidence around a channeling direction. From Andersen (1967)

crystal structure. Two simple prototype cases were identified here. A particle moving at a small angle to a string or plane of crystal atoms may undergo a sequence of weak scattering events, with the individual momentum transfers all directed into one direction (Fig. 1.3). As a result, the particle will be reflected almost specularly from the string or plane. This differs drastically from random multiple scattering, which would result in a broadened directional profile, roughly gaussian and centered around the direction of incidence. More important, the chance for wide-angle deflection is reduced dramatically (Fig. 1.4). First indications of this effect were extracted from computer simulations of particle penetration through crystals (Robinson and Oen, 1963). Experimental evidence was found in the form of deeply penetrating tails in the range profiles of ions in single crystals by Piercy et al. (1963), as you have seen in Fig. 1.2, and a comprehensive theory outlining numerous consequences appeared shortly after (Lindhard, 1965).

Another effect, predicted theoretically and discovered in independent experiments, was found to be complementary to channeling. If an atom located at a regular crystal site is set in motion by some process such as a recoil from a nuclear reaction or decay, its motion will be 'blocked' in certain directions by the presence of rows and planes of lattice atoms (Fig. 1.5). As a result the angular distribution of moving atoms – which may be observed as an emission pattern from a surface – may be anisotropic with pronounced minima near close-packed crystal directions and planes. No such anisotropies would be expected if the crystal acted as a random scattering medium.

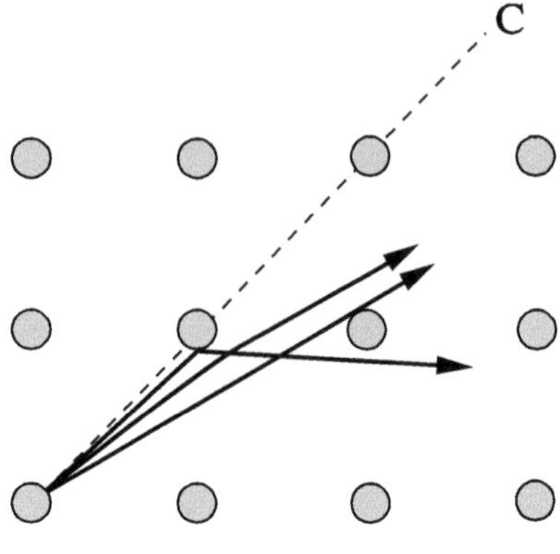

(a) Blocking, schematically

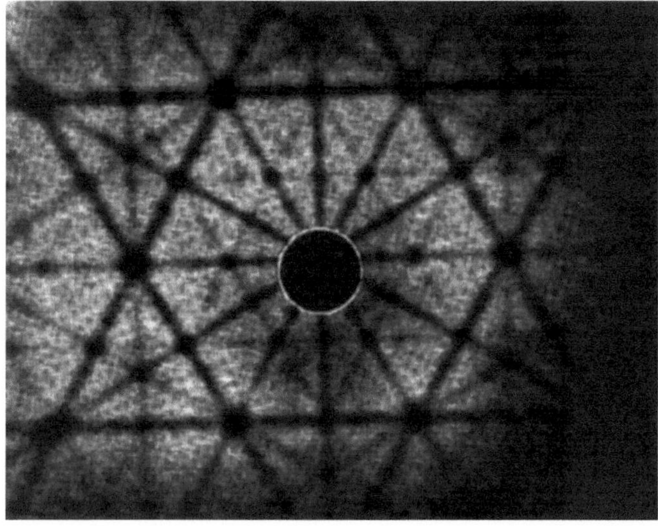

(b) Experimental blocking pattern. From Barrett et al.
(1968)

Fig. 1.5. Blocking: a) Blocking effect on the trajectory of a particle set in motion
from a regular lattice site in a crystal: Particles moving initially in the direction
marked **C** will be steered away by nuclei located along a string. b) Experimental
blocking pattern of 100 keV protons backscattered from cobalt single crystal

The discovery of channeling and blocking generated renewed interest in
the interaction of charged particles with solids and surfaces and caused much
rethinking on phenomena that had been seemingly well understood. Most
important was the classification of directions of motion into 'random' and
'channeling' directions and the splitting of an incident beam into a channeled

and a random beam, where the latter is supposed to follow the statistical laws of random penetration theory.

1.1.2 Radiation Effects

Intimately connected to the fate of penetrating particles are the radiation effects caused by them. Indeed, within conventional terminology, penetration properties can roughly be grouped into *scattering* (i.e., angular deflection) and *stopping* (i.e., loss of kinetic energy). A scattering event leading to a significant change in momentum of a penetrating particle will leave kinetic energy to the

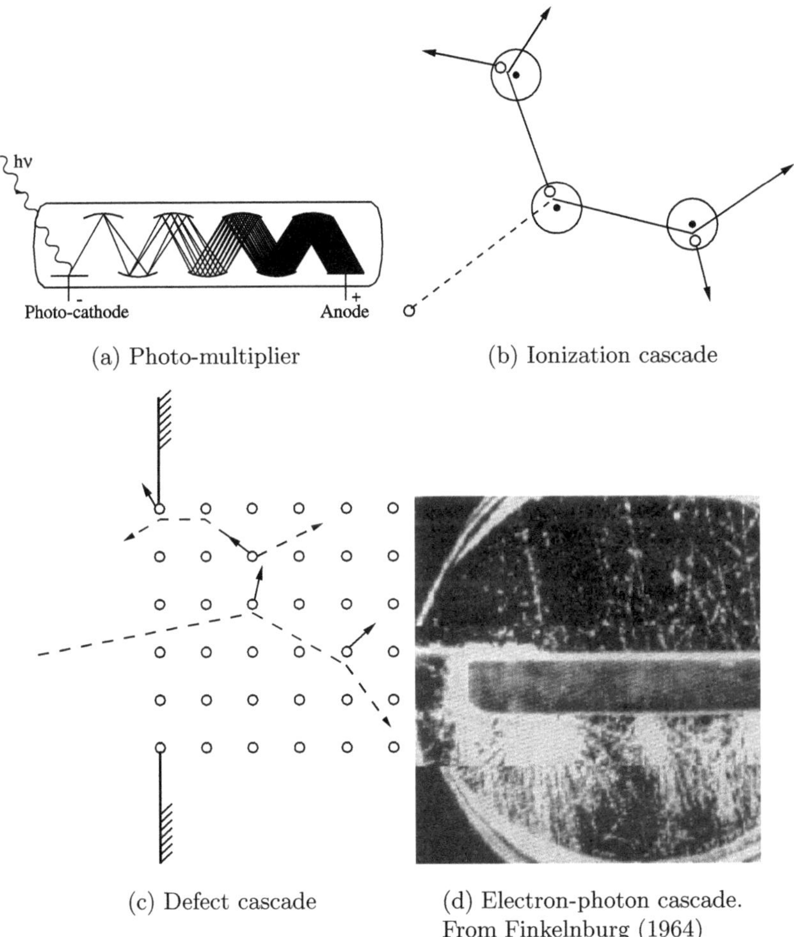

(a) Photo-multiplier   (b) Ionization cascade

(c) Defect cascade   (d) Electron-photon cascade.
                              From Finkelnburg (1964)

Fig. 1.6. Cascade processes: a) Amplification in photo-multiplier; b) Ionization cascade in a gas; c) Defect cascade in a crystal; d) Electron-photon cascade collision partner which is typically a recoiling atom of the medium. Similarly, energy lost to the medium by a penetrating particle must leave the medium in an excited state. Subsequent relaxation processes will typically end up with the medium being in a metastable state characterized by electronic and structural defects and with energy being emitted in the form of photons, electrons, atomic particles, heat, and possibly sound.

Early studies of radiation effects addressed ionization phenomena in gas and scintillation counters, photo-multipliers, radiation shields and the like. The study of structural defects, 'radiation damage', came into focus with the development of nuclear reactors in the Manhattan Project during World War II. While primary particles in that context were neutrons, it was evident that both fast and slow neutrons could generate fast recoil atoms, the former by elastic collisions and the latter by nuclear reactions. Therefore the field of neutron-generated radiation damage has been closely connected to the field of charged-particle penetration since the early work of Seitz and Koehler (1956).

A universal feature of radiation effects induced by high-energy particles is the possibility of cascade processes, i.e., sequences of events involving several generations of secondary particles. Cascade processes may be field-enhanced like in a photo-multiplier (Fig. 1.6a) but may also develop in the absence of an external field if sufficient energy is available. Figure 1.6b shows an ionization cascade initiated by an electron with an energy a few orders of magnitude above the ionization threshold of the atoms in the penetrated medium. This is relevant to gas or semiconductor counters for all kinds of ionizing radiation. Figure 1.6c shows a defect cascade initiated by a heavy particle with a kinetic energy a few orders of magnitude above the minimum energy for formation of a defect in a crystal lattice. This is relevant to radiation effects caused by fast recoil atoms in e.g., fission and fusion reactor materials. Figure 1.6d shows a cascade shower generated by a relativistic charged particle, involving photons generated by bremsstrahlung and electron-positron pairs generated by those photons.

Radiation effects in inorganic and organic including living matter have been studied intensely since the end of World War II. Virtually all sources of high-energy radiation cause damage to matter, sometimes desired, frequently not, as will be discussed in Sect. 1.2. Special interest is attached to those radiation effects which may serve as radiation detectors such as heat development, light emission, electric signals and others.

1.2 Applications

The potential for applications of charged-particle beams seems truly unlimited. Conferences on applications of accelerators in research and industry collect easily over a thousand active participants. Not all of those applications offer themselves for detailed analysis in terms of fundamental physics, but almost everywhere, knowledge of some basic physical facts is required. The present survey is by no means complete. In particular, the number of applications of ion beams mentioned here exceeds by far that of electron or neutron irradiation. This is at least partly caused by the author's horizon which is biased toward ion-beam physics.

1.2.1 Fundamental Physics Research

It has already been mentioned that penetration properties of recoil atoms in the keV and MeV range are of interest in the analysis of nuclear reactions. Penetration depths serve as a measure of reaction energies. Measurement of the Doppler shift of gamma rays emitted from a short-lived nucleus set in motion by a nuclear reaction may provide information on nuclear lifetimes if the rate of energy loss is known. Nuclear lifetimes may also be extracted from blocking patterns in crystals: If a nucleus recoiling from a crystal lattice site decays while still near its origin, the angular distribution of the reaction products will exhibit a blocking pattern. This will not be the case if the lifetime is long enough to allow the recoil to move into an interstitial site. Fast ions moving in a dense medium may create transient magnetic fields which, in turn, may affect the magnetic moment of the moving nucleus and thus provide insight into the latter.

Penetration properties form basic input into all aspects of high-energy physics instrumentation such as ion sources, beam handling, beam interaction with residual gas and the walls of accelerator tubes, slits and apertures, detectors, radiation shields as well as environmental aspects. Similar statements may be made about modern atomic-collision physics, with the pertinent energy range being typically six orders of magnitude lower. An interesting, fairly new aspect is the growing field of research involving highly charged ions.

1.2.2 Astrophysics and Space Science

In addition to cosmic rays, other types of slow and fast particles penetrate inner and outer space. Prominent representatives in our solar system are the solar wind and solar flare consisting of keV protons and MeV helium ions, respectively. These particles interact with planetary atmospheres where those exist, and with planetary or lunar surfaces where not. Polar light is presumably the most well-known radiation effect caused by these particles. The analysis of meteorites and lunar samples provides evidence on truly high-dose exposure of materials to radiation and allows modelling of solar wind and intergalactic radiation. Exposure of space vehicles to local radiation fields may severely affect materials properties of a spacecraft, the function of the instruments carried including computers, and the health and well-being of astronauts.

1.2.3 Plasma Physics and Fusion Research

Space and laboratory plasmas have much in common except for pertinent length scales. Plasmas of interest in fusion research contain isotopes of hydrogen with energies of tens of keV. The interaction between a fusion plasma and its environment, in particular the walls of the reactor vessel, is an aspect of primary importance in all fusion research, whether based on magnetic or inertial confinement. Moreover, neutrons are the prime reaction product and carry the energy which is to be converted into electricity in one way or another. Therefore, radiation effects caused by neutrons in any accessible part of a fusion reactor such as superconducting magnets, first wall, converter or blanket as well as diagnostic tools need to come under control.

Fusion plasmas may be fed by neutral-beam injection of hydrogen or by injection of pellets of solid hydrogen. In either case it is the interaction between the plasma particles and the neutral gas or solid that determines whether or not the injected fuel ends up in an ionized state at the right position in a fusion plasma.

One of the concepts studied in inertial-confinement fusion involves the fact that a dense ion beam efficiently deposits energy into a dense medium and thus rapidly may heat a plasma to temperatures sufficient to trigger fusion. Beams of light and heavy ions are considered as relevant although for different reasons, and even beams of large clusters have come into consideration. In all cases the stopping of an ion beam in a dense (solid or gaseous) hydrogen target is an issue of prime importance.

Finally one may mention the field of muon-catalyzed fusion, where the rate of spontaneous fusion events in a material containing hydrogen isotopes is enhanced by the capture of a muon into an atomic orbit. The time dependence and the process of slowing-down and capture of a muon is a typical problem of charged-particle penetration.

1.2.4 Materials Research and Engineering

Analysis by Ion and Electron Beams

Beams of light and heavy charged particles are playing an ever increasing role in materials analysis. Well-known examples are the electron microscope and the electron microprobe. Understanding the imaging properties of a microscope is in part a penetration problem as far as elastic and inelastic scattering of electrons is concerned. Conversely, imaging in a microprobe is frequently done on the basis of emitted x-rays or secondary electrons, i.e., a radiation effect is utilized as the signal.

The penetration properties of low- to medium-energy electrons ($\sim$100 eV to a few keV) are of primary interest in surface analysis by Auger and x-ray photoelectron spectroscopy. It is the mean free path for inelastic scattering that determines the information depth of these and other surface-sensitive techniques, and it is the significance of multiple elastic scattering events that determines their respective ranges of applicability.

A variety of techniques exists for application of positron beams in materials analysis. Knowledge of the penetration properties of these beams is a prerequisite for any quantitative interpretation.

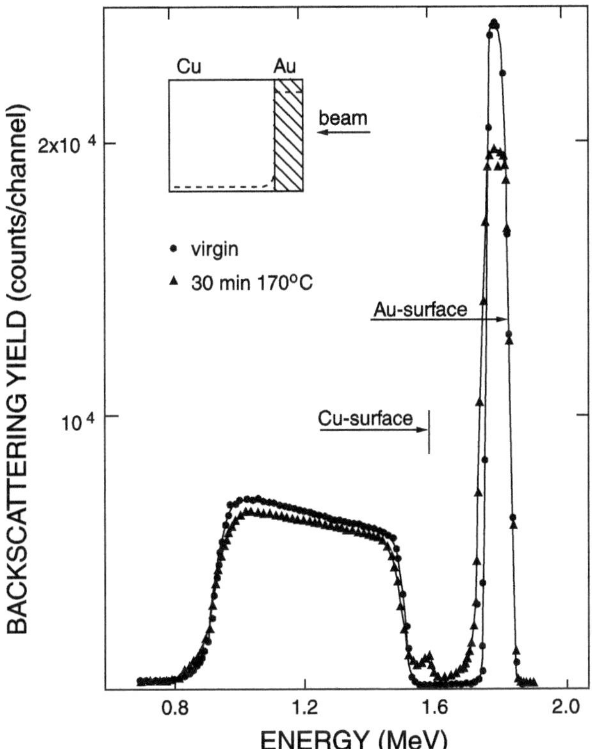

Fig. 1.7. Ion beam analysis by Rutherford backscattering: See text. From Campisano et al. (1973)

Materials analysis by ion beams – briefly although somewhat misleading: 'ion beam analysis' – has developed into a powerful tool based on a variety of physical principles. Several techniques are built on binary-collision dynamics: Conservation laws of momentum and energy determine uniquely the energy of an ion scattered into a given angle from a particular scatterer. Measurement of that energy, therefore, delivers the mass of the scatterer. Hence, measuring an energy spectrum of ions scattered at a definite angle provides information on the composition of the scattering medium (Fig. 1.7). The information depth depends on the energy of the incident beam.

Rutherford Backscattering (RBS) utilizes hydrogen or helium ions in the upper-keV or lower-MeV range or heavier ions at even higher energies. Characteristic depths range from a few nm to $\sim$ 1μm. Depth resolutions down to a monolayer can be achieved. Figure 1.7 shows the spectrum of He$^+$ ions backscattered at a fixed angle to the incident beam from a target consisting of a gold layer evaporated onto a copper film. The high-energy branch of the spectrum represents ions reflected from gold atoms. Gold is heavier than copper and therefore takes up less recoil momentum from He during scattering. The area under this part of the spectrum is a measure of the amount of gold per unit area. The left portion of the spectrum shows ions scattered on Cu atoms. Those particles with the highest energies were scattered at the interface. Their energy is lower than what would be expected from

a consideration of the recoil energy given to a copper atom because these ions also lose energy while travelling through the gold film. A second curve, found after heat treatment, shows that the two components have interpenetrated each other by diffusion.

Ion-surface scattering (ISS) utilizes light ions (noble-gas or alkali) in the upper-eV or lower-keV range for analysis of one to two top surface monolayers. Numerous modifications of these prototypes exist. Recoiling target atoms may be analysed instead of scattered beam particles such as in elastic-recoil detection analysis (called ERDA at MeV energies). Photons may be the signal instead of atomic particles such as in proton-induced x-ray emission (PIXE) or surface composition analysis by ion-induced radiation (SCANIIR). Signals of high sensitivity and resolution may be gained by charged-particle-induced nuclear reaction analysis (NRA). Finally, high-sensitivity surface analytic techniques may utilize atoms or ions emitted by sputter processes as a signal such as secondary-ion or secondary-neutral mass spectrometry (SIMS, SNMS).

A particular role is being played by ion beams in the analysis of crystal lattice structure and the spatial configuration of lattice defects and implanted atoms. Rutherford backscattering patterns observed on crystalline targets show direct evidence of channeling (Fig. 1.4 on page 7) and/or blocking (Fig. 1.5). Since the signal is sensitive to the target mass it is possible to record separate patterns for each species present. If one species occupies regular lattice positions and another one is located interstitially, the respective emission patterns will be radically different. Therefore, RBS and equivalent signals provide information on the defect state of a crystal and the lattice location of dopant atoms both before and after annealing treatment. This is very important in the analysis of the behavior of dopants in micro and nano electronics.

Ion-Beam Modification of Materials

Unlike ion beam analysis which is to be conducted at low beam intensities, ion-beam modification of materials may involve very large implant doses.

Ion implantation is a technique to produce dilute alloys with good control of the composition: The integrated beam intensity determines the number of implanted ions and the beam energy determines the implant depth. The combination of these two independent experimental parameters makes ion implantation superior to competing techniques such as diffusion doping where density and penetration depth are not independent of each other. Dilute alloys are of great interest in microelectronics: In fact, ion implantation is a crucial step in the industrial production of microprocessors. Moreover, surface properties like corrosion resistance or compatibility of medical implants with living tissue can be controlled by ion implantation doping. An important feature of ion implantation is the option to produce alloys that are thermodynamically metastable and therefore cannot be generated by equilibrium processes.

Another technique, called ion beam mixing, is used to produce non-dilute metastable alloys. Here the starting point is a material containing layers of different pure materials. Bombardment with an ion beam generates cascades of atomic collisions as a result of which atoms from one layer recoil into a neighboring layer. The result resembles very much that of an interdiffusion, but in view of the high energies involved the process also takes place if the resulting alloy is metastable just as in ion implantation.

Several techniques utilize ion beams as tools in thin-film deposition. It was recognized almost 150 years ago that the flux of atoms sputtered from an ion-bombarded sample could be collected on a substrate where a smooth film would form. Simultaneous bombardment of different materials generates sputtered fluxes which, when collected on one substrate, may form alloy films of reasonably controlled composition. This is now one of the standard techniques to produce high-T_c superconducting films. Films may also be formed by collection of the fluxes of (low-energy) ion beams themselves on a substrate. In 'ion-assisted film deposition', ion bombardment serves as an energizer to stabilize a film that is produced by a more conventional technique such as evaporation.

It follows from the above that ion beams are prime tools in 'surface engineering': In addition to ion implantation, mixing, and deposition, that term comprises etching and polishing as well as roughening. By proper use of masks and/or focused beams, one-, two-, and three-dimensional structures may be generated on a submicron scale and utilized in numerous contexts.

Radiation Damage

Radiation effects on a larger scale may more readily be generated by neutron or high-energy light-ion irradiation. The penetration depth of a fast-neutron beam may be up to ~ 1 m. Therefore, changes brought about in a material by atoms recoiling from collisions with fast neutrons tend to be distributed homogeneously throughout a macroscopic sample. This dilution in space implies, on the other hand, fairly high minimum irradiation doses for radiation effects to be visible at all. Typical irradiations in ion implantation may last seconds or minutes while equivalent effects in a fission reactor may need weeks or months.

Neutron-generated radiation effects have been studied for more than half a century. Radiation damage in fission-reactor materials such as uranium, graphite, light and heavy water, cadmium and many others is of prime interest from the point of view of safety, stability of supply, and lifetime. Similar concerns exist with regard to radiation effects in fusion-reactor materials such as refractory metals, various ceramics, stainless steel, magnetic and superconducting materials and the like. To this adds the item of radiation effects in materials pertinent to nuclear-waste treatment and deposition, i.e., rock salt and minerals.

While damage effects to superconductors may be detrimental to the function of a fusion reactor it has long been recognized that defects generated by irradiation may affect the function of a superconductor also in a positive direction, in particular by raising the transition temperature. Indeed, introduction of defects may turn a 'soft' superconductor into a 'hard' one by introduction of trapping centers for flux lines.

1.2.5 Analytical Chemistry

Several analytic techniques mentioned above have enhanced the arsenal of available tools in analytical chemistry. Desorption techniques in conjunction with mass spectrometry are of prime importance here. Ion beams serve in two distinct functions,

- as a tool to promote *intact* molecules from the liquid or solid phase into the gas phase by desorption, and
- to enable an adequate fraction of the desorbed molecules to appear ionized and, therefore, identifiable by time-of-flight or conventional mass spectrometry.

The fact that ion beams with energies far in excess of what is needed to break a molecule into pieces can desorb even very big intact molecules at all was one of the great surprises in ion beam physics. Considerable attention will be paid to pertinent physical processes in Volume III.

While the SIMS technique mentioned above has been used successfully to mass-analyse large molecules like proteins and insulin, several modifications of the technique have proved even more efficient. In one technique, called 'liquid SIMS' or 'fast atom bombardment' (FAB), the clue is to embed the sample to be analyzed into a liquid matrix like glycerol or nitrocellulose: This seems to enhance both the desorption yield and the ionization probability of emitted molecules, thus allowing for high counting rates at low bombardment doses. It was found that desorption processes may be ascribable both to ionization and collision cascades (Fig. 1.6 on page 9). Thus, analytical techniques may be based not only on low-energy beams (SIMS) but also on beams in the fission fragment regime. In the latter case the somewhat confusing notion of 'plasma desorption mass spectrometry' (PDMS) has become common.

An important technique, called 'accelerator mass spectrometry' (AMS), serves to determine isotopic compositions of materials, e.g., with the aim of determining the age of archeologic or geologic samples. The sample is inserted as a sputter source into a tandem accelerator, the beam optics of which is utilized as a mass spectrometer. The high beam energy allows single-atom detection. This leads to great sensitivity and thus allows determination of very small percentages of the isotope in question. In this manner the age regime covered by ^{14}C dating has been extended significantly beyond what is achievable by radioactivity measurements.

1.2.6 Biomedical Research

As in materials research, charged-particle beams serve two distinct purposes in biomedical research: They may be utilized for analysis and diagnostics on the one hand and for tailoring materials properties and for therapy on the other. Undesired radiation effects on living tissue have been a prime issue in radiation research ever since the discovery of x rays and radioactivity, and this research has been intensified after the events generated by atomic bomb explosions during and after World War II.

Radiation protection and dosimetry and the development of standards involves measurements with charged-particle beams at all stages, and highly advanced theoretical calculations enter here. Indeed, laboratories involved in this type of research usually carry out a certain amount of basic atomic-physics research because it is found necessary to base radiation standards on the best available theoretical foundation.

Numerous applications of high-energy electron beams include radiation sterilization of medical tools, radiation chemistry, and radiation treatment of plants. Amongst more recent applications of charged-particle beams, diagnostics by proton and heavy-charged particle beams as well as charged-particle therapy need to be mentioned. In diagnostics the energy loss is utilized as an indicator: A beam particle penetrating a tumor may suffer another energy loss than one penetrating healthy tissue. Heavy-particle beams suffer little angular scattering and are detected easily. Therefore efficient diagnostics can be performed at low irradiation doses.

Much more in use is charged-particle therapy, which is based on the fact that the rate of energy loss of swift charged particles increases with decreasing beam energy (for details cf. Chapter 2). Therefore a properly chosen beam will deposit its energy efficiently in a given depth range where a tumor has been located, and comparatively little damage will be generated in the overlying healthy tissue. This is very different from conventional radiation therapy involving x or gamma rays where the density of energy deposition decreases with increasing depth. Charged-particle therapy is being carried out both with electrons, mesons, protons, and heavy ions. The variety of tumors that can be treated successfully by this technique is continuously increasing. It is a major challenge to construct reasonably priced machines that can be made available to general health care.

1.3 Measurements and Experimental Tools

This section brings nothing new to a routined experimentalist but is intended to serve as a rough orientation on experimental options for students of theoretical physics.

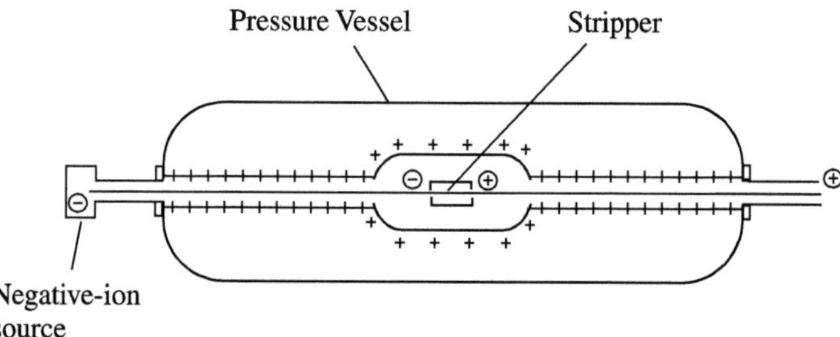

Fig. 1.8. Principle of a Tandem electrostatic accelerator. The point at the highest potential is located near the middle of the acceleration line. At the left end, negative ions are injected at ground potential. These ions are accelerated toward the high-voltage terminal; at this point they pass through a thin foil or a gas cell where they get stripped off some of their electrons and, consequently, get positively charged. In continuing their motion toward the right, they get accelerated further until arriving at ground potential again. From Andersen (1991)

1.3.1 Sources of Energetic Charged Particles

Not all experimentation with charged-particle beams requires a big accelerator. Radioactive alpha emitters provide beams of well-defined energy, and by proper use of apertures a reasonably collimated beam may be obtained. Beta emitters may be utilized similarly but are less attractive because of the wide spread of the kinetic energies of emitted electrons. Substances undergoing spontaneous fission such as the californium isotope ^{252}Cf are ideally suited as sources of heavy-ion beams in a small laboratory, provided that a certain spread in projectile mass and energy is tolerable.

Nuclear reactions induced by neutrons or fast particles may serve as a supply of a variety of light and heavy charged particles with more or less well defined energies. Moreover, elastic recoils from alpha-particle bombardment form an internal source as discussed in Sect. 1.2.

Beams of charged particles in the energy range up to a few keV require only modest investments in high-voltage equipment regardless of whether the interest is in ions or electrons. For positrons, mesons, antiprotons and other exotic particles, the situation is different since those particles are generated at rather high energies. Here an additional investment may have to be made to *decelerate* particles to some well-defined energy if the initial energy is too high.

Electrons of well defined energy can be generated by ultraviolet or x-ray photoemission with the primary radiation coming from an x-ray tube or a synchrotron.

A very wide arsenal of accelerators is available. These may be classified into electrostatic machines with acceleration voltages from a few volts up to 10-20 million volts. In the upper voltage range many machines work on the basis of the tandem principle (Fig. 1.8). Accelerators for higher energies utilize various types of electromagnetic fields. The well-known cyclotron employs a static magnetic field to force the particles into circular orbits, and acceleration is achieved by alternating electric fields. All other high-energy accelerators, whether circular or linear, require time-dependent fields and provide bunched beams. These machines used to be developed with electrons or protons in mind. Subsequently heavy-ion beams were generated, and several dedicated heavy-ion accelerators as well as storage rings have been constructed during the past quarter of a century.

Next to the acceleration stage an important part of any accelerator is the source of particles. Electrons are conventionally generated by heating of a filament. Ion sources may be constructed on the basis of a variety of principles. Gas ions are conveniently generated in a discharge. Metal ions may be generated either directly by heating and evaporation, by evaporation of suitable compounds, or by exposing the metal to an ion or laser beam and extracting sputtered positive or negative ions. While sputtered atoms are typically neutral, suitable surface treatment of the target may enhance the emission of either positive or negative ions.

By and large, experimental techniques have developed to the point where ions of any element may be generated in any stable charge state and accelerated to almost any energy. Ion optics is well developed to a stage where beams of lateral dimensions in the micron and submicron regime have become available. In addition, a large variety of molecular beams has become available.

1.3.2 Targets and Detecting Devices

Targets to be exposed to particle bombardment include potentially all inorganic and organic solids, crystalline or not, liquids and gases including plasmas over a wide range of temperatures. Control of target parameters ensuring target homogeneity and stability is crucial and quite often more difficult to achieve than a high-quality beam. Experimenters tell of painful experiences with thin foils carrying invisible holes or polycrystalline materials with pronounced texture. If the existence of such features is not expected from the beginning, peculiar experimental results will inevitably be found and much time is spent on finding exotic explanations of seemingly new effects.

Enormous progress has been achieved in the area of radiation detection. Sophisticated techniques in high-energy physics were gradually extended down into the MeV and keV range. It is now possible to simultaneously record the positions of a large number of particles with a very high space and time resolution. I do not wish to provide numbers here since whatever I write will be obsolete by the time this book appears in print, but this development is truly astonishing. Keywords are wire detectors, charge-coupled devices and channel plates but numerous other principles may be applied. Conventional electrostatic or magnetic analysers are still excellent tools in precision measurements of particle energies, but semiconductor detectors and time-of-flight techniques are more convenient for many purposes. Almost unlimited options are available for performing coincidence measurements involving several particles including photons.

Finally, almost any material property may be employed as an indicator of radiation effects. Most frequent are heat, light, and sound emission, but changes in electrical or thermal conductivity, mass density, or specific heat may serve just as well. Direct observation of radiation effects by electron or ion microscopy, by scanning tunneling microscopy or other surface analytic techniques is helpful, as are chemical analysis including ion beam analysis and all kinds of spectroscopy.

1.4 General-Physics and Related Aspects

While the selection of items covered in this book is highly topic-oriented, i.e., dominated entirely by the needs of scientists and engineers working with charged particles in basic and applied research, this does not preclude that many of the topics covered have a much broader significance. For example, the statistical theory of particle penetration is a specific application area of the theory of stochastic processes in physics. In the opinion of the author, experiments involving beams of charged particles constitute one of the most beautiful illustrations of the theory of stochastic processes, even though this feature is not generally appreciated in the statistical-physics community.

The theory of individual scattering processes plays a major role in particle penetration. This does not only refer to the classical or quantal theory of two-body scattering but, even more important, to inelastic collisions involving excitation and ionization of the projectile or target particle and, finally, rearrangement processes like electron capture from a target atom or molecule by a projectile ion. While the theory of such processes belongs under the heading of electronic and atomic collisions, it is prime input and as such needs to be treated in a monograph of the present character.

Due to the wide variety of available experimental tools and techniques as well as a large number of experienced research teams there is a well-established tradition for close cooperation between experimentalists and theorists in the field. This may be a very refreshing aspect for a student of either theoretical or experimental physics, dependent on the local curriculum.

Numerous computer simulation codes have been developed that allow experimentalists to make their own theoretical explorations in connection with experimental studies. Studying the trajectories of fast particles and their reaction products on a computer screen offers the fascination of computer games to those with an appreciation of this type of entertainment. Making science out of this, however, requires a critical attitude and great caution with regard to the pitfalls of statistics.

1.5 Literature

A comprehensive treatment with the present scope has not appeared in print since the monograph of Bohr (1948) more than half a century ago. That work treated all aspects of particle penetration from a basic-physics point of view, focusing on qualitative understanding but making full reference to the theoretical literature available at the time. Applied aspects as well as explicit treatments of radiation effects were suppressed.

The general standard of this field has very much been set by Lindhard and his lecture course at Aarhus University which has been followed by many workers in the field at one time or another. Lindhard never published his course, but an extract has been written up by Bonderup (1981).

Amongst a few treatments covering narrower topics I like to mention an article on light-ion stopping by Fano (1963) which generated a considerable stimulus to both experimentalists and theorists through its admirable rigor both in the general presentation and the formulation of unsolved problems. More recent surveys of ion stopping and ranges are due to Kumakhov and Komarov (1981) and the present author (Sigmund, 2004). Several international summer schools have been held that covered aspects of charged-particle penetration and radiation effects. Closest to the scope of the present book is the one held in Alicante, Spain (Gras-Marti et al., 1991), but some important items were not covered and several central topics treated in the oral presentations did not make it into the monograph.

Several monographs are available on radiation effects in solids, although most of them fairly old (Dienes and Vineyard, 1957, Billington and Crawford, 1961, Strumane et al., 1964, Leibfried, 1965, Dupuy, 1975, Lehmann, 1977, Thompson, 1969), and quite a few books on ion implantation and ion-beam processing have appeared more recently (Mayer et al., 1970, Williams and Poate, 1984, Nastasi et al., 1996).

Summaries on electron interaction with matter may be found in conjunction with medical radiation physics (Inokuti, 1983) as well as surface analysis by electron spectroscopy (Tougaard, 1988).

1.6 Nomenclature

A general problem faced by any textbook author is the limited number of symbols in the alphabet. I have tried to use the same symbol for the same type of quantity and to add indices only when this was found absolutely necessary. Thus, E denotes consistently the kinetic energy of a moving particle, and if there is more than one type of projectile the energies have been called E, E', E'' and the like. An electric field, on the other hand, has been denoted by $\boldsymbol{E}$ but its magnitude is called $|\boldsymbol{E}|$ rather than E to minimize confusion.

In general I have tried to minimize the number of indices in order to keep mathematical equations readable. This, however, implies that the symbol F, which consistently denotes a statistical distribution function, may have very different physical meanings in different chapters and sections: Not only may the number of statistical variables in the function be different from case to case. The function may also refer to penetrating projectiles in one case and recoiling target atoms in another.

For the convenience of the reader a list of general notations has been compiled on page XIX–XXIV which covers symbols and abbreviations which occur at several places in the book.

Problems

1.1. Projectile speeds are conveniently expressed in terms of the Bohr velocity $v_0 = e^2/4\pi\epsilon_0\hbar = c/137$, where $c = 3 \cdot 10^8$ m/s is the speed of light in vacuum. Determine the kinetic energy (in electron volts) of an electron, a proton, an alpha particle, and a uranium nucleus, respectively, moving a) at speed v_0, b) at speed $10\,v_0$.

1.2. Projectile speeds are frequently expressed in terms of kinetic energy per atomic mass unit u, where $u = 1.66054 \times 10^{-27}$ kg is close to the proton mass. Find the velocity in multiples of the Bohr velocity v_0 of a projectile moving with a) 100 keV/u and b) 100 MeV/u and determine the kinetic energy if the particle is a proton or a uranium nucleus, respectively.

1.3. Penetration depths are frequently expressed in terms of weight per area. Find conversion formulas between thickness and weight per area, utilizing the mass density ρ_m of the respective material, expressing length in nm or μm and weight per area either in μg/cm^2 or mg/cm^2, respectively.

1.4. Determine the thickness of a layer of

- 1 mg/cm^2 helium gas at atmospheric pressure,
- 1 mg/cm^2 aluminium foil and
- 1 mg/cm^2 gold foil.

1.5. For a face-centered cubic crystal lattice you may define a monolayer of material as a slab of thickness $d/2$ with a density of $2/d^2$ atoms/area, where d is the cubic lattice constant. Look up or determine values of d for aluminium and gold and determine the weight/area (in $\mu g/c^2$) for a monolayer of aluminium and gold, respectively.

1.6. Determine the De Broglie wavelength $\lambda = \hbar/Mv$ for

a) an electron with energy 13.6 eV and 100 keV, respectively,
b) a neutron with energy 0.025 eV ('thermal'), 1 eV ('epithermal'), and 1 MeV ('fast'), respectively, and
c) an argon ion at 1 keV and 1 MeV, respectively.

Discuss relevant application areas of the cases mentioned.

1.7. Consider a simple model for the slowing-down of a fast neutron in solid matter: Assume a scattering cross section $\sigma = 1 \cdot 10^{-24}$ cm^2 ($= 1$ b) for interaction with the nuclei in the material which are assumed distributed uniformly at a number density $N = 0.05/\text{Å}^3$. The neutron is assumed in the average to lose half its instantaneous energy per collision. Derive an expression for the average energy loss per travelled pathlength, dE/dR, and estimate the total penetration depth of a neutron from an initial energy 1 MeV down to thermal energy. Compare the penetration depth with the mean free path for a scattering event.

1.8. Consider the following, heavily oversimplified model for the slowing down of an alpha particle in solid matter: Assume a cross section $\sigma = 1$ Å^2 for electronic excitation of target atoms and an average energy loss $\langle T \rangle = 20$ eV in every collision. Target atoms are assumed to be distributed uniformly at a number density $N = 0.05/\text{Å}^3$. Find an expression for the average energy loss per pathlength, dE/dR, and estimate the total penetration depth of an alpha particle with initial energy 1 MeV. Compare the total penetration depth with the mean free path for a scattering event. Try to locate several major errors in this model. Make a comparison with the model discussed in problem 1.7.

1.9. Assume that you are allowed to treat a penetration problem on the basis of nonrelativistic mechanics as long as the kinetic energy of a projectile with mass M does not exceed $0.1 Mc^2$. Evaluate the limiting acceleration voltages for a) an electron, b) a proton, c) an Ar$^+$ ion, d) an Ar^{18+} ion, and e) a U^{90+} ion.

References

Andersen H.H. (1991): Accelerators and stopping power experiments. In Gras-Marti et al. (1991), 145–192

Andersen J.U. (1967): Axial and planar dips in reaction yield for energetic ions in crystals lattice. Mat Fys Medd Dan Vid Selsk **36 no. 7**, 1–26

Barrett C.S., Mueller R.M. and White W. (1968): Proton blocking in cubic crystals. J Appl Phys **39**, 4695–4700

Bethe H. (1930): Zur Theorie des Durchgangs schneller Korpuskularstrahlen durch Materie. Ann Physik **5**, 324–400

Billington D.S. and Crawford J.H. (1961): *Radiation damage in solids*. University Press, Princeton

Bohr N. (1913): On the theory of the decrease of velocity of moving electrified particles on passing through matter. Philos Mag **25**, 10–31

Bohr N. (1915): On the decrease of velocity of swiftly moving electrified particles in passing through matter. Philos Mag **30**, 581–612

Bohr N. (1940): Scattering and stopping of fission fragments. Phys Rev **58**, 654–655

Bohr N. (1941): Velocity-range relation for fission fragments. Phys Rev **59**, 270–275

Bohr N. (1948): The penetration of atomic particles through matter. Mat Fys Medd Dan Vid Selsk **18 no. 8**, 1–144

Bonderup E. (1981): *Interaction of charged particles with matter*. Institute of Physics, Aarhus

Born M. (1926): Quantenmechanik der Stoßvorgänge. Z Physik **38**, 803–827

Bothe W. (1921a): Das allgemeine Fehlergesetz, die Schwankungen der Feldstärke in einem Dielektrikum und die Zerstreuung der alpha-Strahlen. Z Physik **5**, 63–69

Bothe W. (1921b): Die Gültigkeitsgrenzen des Gaußschen Fehlergesetzes für unabhängige Elementarfehlerquellen. Z Physik **4**, 161–177

Bothe W. (1921c): Theorie der Zerstreuung der α-Strahlen über kleine Winkel. Z Physik **4**, 300–314

Campisano S.U., Foti G., Grasso F. and Rimini E. (1973): Low temperature interdiffusion in copper-gold thin films analysed by helium backscattering. Thin Solid Films **19**, 339–348

Cherenkov P. (1934): Sichtbares Leuchten von reinen Flüssigkeiten unter der Einwirkung von Gammastrahlen. Doklady Akad NAUK **2**, 451–457

Cockcroft J.D. and Walton E.T.S. (1932): Artificial production of fast protons. Nature **129**, 242

Davies J.A., McIntyre J.D., Cushing R.L. and Lounsbury M. (1960): The range of alkali metal ions of kiloelectron volt energies in aluminum. Can J Chem **38**, 1535–1546

Davisson C. and Germer L.H. (1927): Diffraction of electrons by a crystal of nickel. Phys Rev **30**, 705–740

Della-Negra S., Brunelle A., Lebeyec Y., Curaudeau J.M., Mouffron J.P., Waast B., Håkansson P., Sundqvist B.U.R. and Parilis E. (1993): Acceleration of C_{60}^{n+} molecules to high energy. Nucl Instrum Methods B **74**, 453–456

Dienes G.J. and Vineyard G.H. (1957): *Radiation effects in solids.* Interscience Publishers, New York

Dupuy C.H.S., editor (1975): *Radiation damage processes in materials.* Noordhoff, Leiden

Fano U. (1963): Penetration of protons, alpha particles, and mesons. Ann Rev Nucl Sci **13**, 1–66

Fermi E. (1940): The ionization loss of energy in gases and in condensed materials. Phys Rev **57**, 485–493

Finkelnburg W. (1964): *Structure of matter.* Springer, Berlin

Gras-Marti A., Urbassek H.M., Arista N.R. and Flores F., editors (1991): *Interaction of charged particles with solids and surfaces,* vol. B 271 of *NATO ASI Series.* Plenum Press, New York

Inokuti M. (1983): Radiation physics as a basis of radiation chemistry and biology. Appl Atomic Collision Phys **4**, 179–236

Kornelsen E.V., Brown F., Davies J.A., Domeij B. and Piercy G.R. (1964): Penetration of heavy ions of keV energies into monocrystalline tungsten. Phys Rev **136**, A849–858

Kumakhov M.A. and Komarov F.F. (1981): *Energy loss and ion ranges in solids.* Gordon and Breach, New York

Lehmann C. (1977): *Interaction of radiation with solids and elementary defect production.* North Holland, Amsterdam

Leibfried G. (1965): *Bestrahlungseffekte in Festkörpern.* Teubner, Stuttgart

Lindhard J. (1965): Influence of crystal lattice on motion of energetic charged particles. Mat Fys Medd Dan Vid Selsk **34 no. 14**, 1–64

Mayer J.W., Eriksson L. and Davies J.A. (1970): *Ion implantation in semiconductors. Silicon and germanium.* Academic Press, New York

Meitner L. and Freitag K. (1926): Über die Alpha-Strahlen des ThC+C' und ihr Verhalten beim Durchgang durch verschiedene Gase. Z Physik **37**, 481–517

Nastasi M., Hirvonen J.K. and Mayer J.W. (1996): *Ion-solid interactions: Fundamentals and applications.* Cambridge University Press, Cambridge

Piercy G.R., Brown F., Davies J.A. and McCargo M. (1963): Experimental evidence for the increase of heavy ion ranges by channeling in crystalline structure. Phys Rev Lett **10**, 399–400

Robinson M.T. and Oen O.S. (1963): Computer studies of the slowing down of atoms in crystals. Phys Rev **132**, 2385–2398

Rutherford E. (1911): The scattering of alpha and beta particles by matter and the structure of the atom. Philos Mag **21**, 669–688

Rutherford E., Geiger H. and Bateman H. (1910): The probability variations in the distributions of alpha particles. Philos Mag **20**, 698–707

Seitz F. and Koehler J.S. (1956): Displacement of atoms during irradiation. Solid State Phys **2**, 307–448

Sigmund P. (2004): *Stopping of heavy ions*, vol. 204 of *Springer Tracts of Modern Physics*. Springer, Berlin

Strumane G., Nihoul J., Gevers R. and Amelinckx S., editors (1964): *The interaction of radiation with solids*. North Holland, Amsterdam

Thompson M.W. (1969): *Defects and radiation damage in metals*. University Press, Cambridge

Thomson J.J. (1912): Ionization by moving electrified particles. Philos Mag **23**, 449–457

Tougaard S. (1988): Quantitative analysis of surface electron spectra: Importance of electron transport. Surf Interf Anal **11**, 453–472

Williams J.S. and Poate J.M. (1984): *Ion implantation and beam processing*. Academic Press, Sydney

2

Elementary Penetration Theory

2.1 Introductory Comments

This chapter is intended to provide a qualitative understanding of the stopping and scattering of charged particles in matter and to acquaint the reader with some of the problems to be solved and suitable tools for their solution. Most of the actual estimates given for specific quantities will be improved or generalized in later chapters. This should not keep the novice from going rather carefully through the material.

The ideas described here have been developed early in the past century to describe the penetration of alpha and beta particles through matter. They have since then been applied to a much wider variety of particles at higher and lower energies than those accessible with the products of natural radioactive decay.

In order to appreciate the validity of the approach taken, have a look at a classical cloud-chamber photograph of the trajectories of alpha particles in air, Fig. 1.1 on page 4. The trajectories are essentially straight lines of almost equal length, of the order of a few centimeters in a gas at atmospheric pressure. Once in a while a trajectory is observed to be bent. Although the process of stopping and scattering is the result of the interaction of the alpha particle with a great number of atoms and therefore must be a statistical process, statistical fluctuations seem to be small and angular deflections rare in the case depicted in Fig. 1.1.

In this chapter, unless otherwise stated, general theoretical considerations are illustrated on an alpha particle moving through a layer of gas that is much thinner than the total length of the trajectory, essentially along a straight line and with very small variation in velocity. Basic concepts introduced include cross section, stopping force and straggling, range, single and multiple scattering, and others. Estimates of these quantities will be based on classical mechanics, much in the way performed in early studies in this field (Rutherford, 1911, Thomson, 1912, Darwin, 1912, Bohr, 1913, 1915).

2.2 Collision Statistics

2.2.1 Definition of Cross Section

The concept of a cross section is of paramount importance in all penetration theory. It is appropriate, therefore, to spend some time on ways of defining and determining this parameter.

Macroscopically the cross section of a target is the area within which it can be hit by some bullet. For example, a spherical target with a radius a offers a cross section πa^2 to a point projectile.

Microscopically we have to come to an agreement on what to mean by saying that a target has been hit by a projectile. In view of the great variety of possible projectiles and targets it is desirable to find a broad definition. Let us say that a target has been hit if the interaction of the projectile with the target has had some specific measurable effect. This means that the magnitude of a given cross section does not only depend on the target, the projectile, and their relative velocity, but also on the physical effect that we decided to monitor. Consequently we talk about scattering cross sections, absorption cross sections, energy-loss cross sections, ionization cross sections, cross sections for specific nuclear reactions, and many others.

Most often in penetration phenomena, the target is an atom or a molecule. In the present chapter we shall also refer to nuclei and/or electrons when talking about targets.

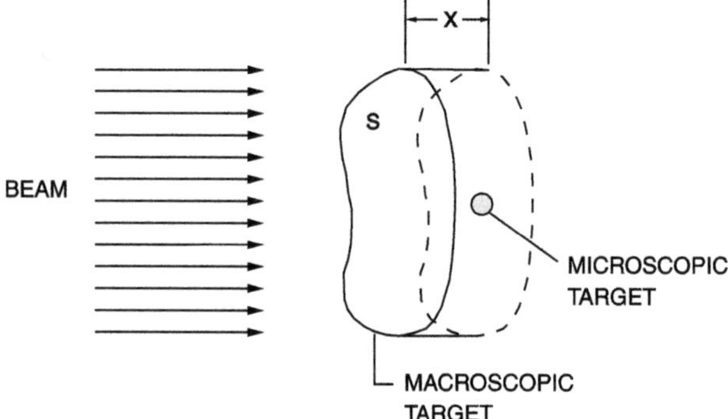

Fig. 2.1. Statistical definition of a cross section: One microscopic target bombarded by a beam

For fundamental and practical reasons there is no way of experimentally determining microscopic cross sections by bombarding one atom with one projectile only. Instead we utilize the statistical information extracted from a large number of bombardments. In that context, we talk about the *stopping medium* and the *beam*. A stopping medium consists of a macroscopic number of target particles in some arbitrary configuration, such as a random assembly of molecules in an ideal gas or a regular structure of atoms in a crystal. A beam consists of a large number of projectiles. Ideally, all projectiles have the same initial state and velocity, but this is not a necessary requirement. We may also talk about a beam when dealing with alpha or beta particles emitted isotropically from a radioactive source. More important is the initial requirement that *individual projectiles only interact with the stopping medium and not with each other*; this can be achieved by letting individual bombardments be well separated in time. In other words, we consider penetration phenomena in the limit of low beam current here.

Let some microscopic target be bombarded by a beam of projectiles spread homogeneously over an area S (Fig. 2.1). If σ_A is the cross section for some process A – such as ionization of the projectile particle – then, σ_A/S is the fraction of all projectiles that undergo the process A by interacting with the target particle, provided that the number of projectiles is large enough to make statistical fluctuations vanishingly small.

If the beam has a current density J [projectiles/time/area], then

$$JS \times \frac{\sigma_A}{S} = J\sigma_A \tag{2.1}$$

is the number of events A induced by the beam per unit time.

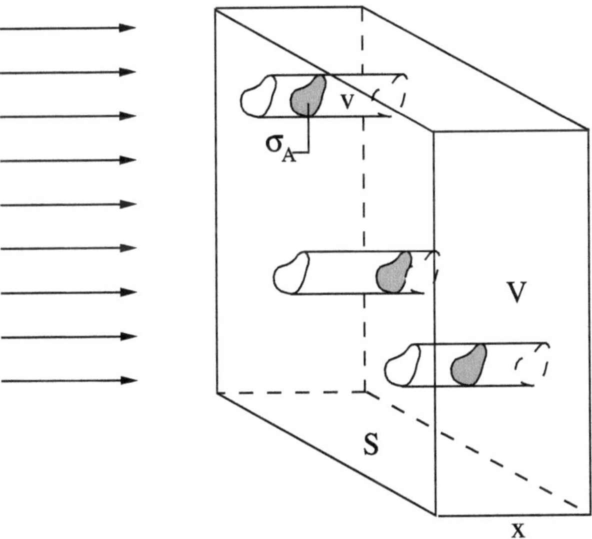

Fig. 2.2. Macroscopic target consisting of randomly placed microscopic targets at number density N, bombarded by a beam

Consider next a stopping medium with a number density N [targets per volume] within a volume $V = \mathsf{S}x$ (Fig. 2.2). According to (2.1) the number of events A induced per unit time by the beam is given by

$$N \times \mathsf{S}x \times J\sigma_A = J\mathsf{S} \times Nx\sigma_A. \tag{2.2}$$

Here, $J\mathsf{S}$ is the number of projectiles per unit time; hence, $Nx\sigma_A$ is the mean number of events per projectile. If the target is thin enough so that $Nx\sigma_A \ll 1$, that number becomes identical with the probability P_A for a projectile to undergo an event A while interacting with the stopping medium, i.e.,

$$\mathsf{P}_A = Nx\sigma_A \qquad \text{for} \qquad Nx\sigma_A \ll 1. \tag{2.3}$$

Thus, cross sections are measured most conveniently with gas targets or thin foils where either N or x is small, respectively.

Eqs. (2.1–2.3) are roughly equivalent, and each of them serves an important purpose. Eq. (2.1) is the most basic one, dealing with one target particle only; it will be employed in theoretical determinations of cross sections from the equations of motion. Eq. (2.2) is most closely related to measurable quantities and is therefore utilized when cross sections are determined experimentally. Eq. (2.3) applies to probability statements about what happens to a projectile during passage through some small path element x; it serves as the starting point of penetration theory.

We shall encounter situations where (2.2) and (2.3) are not valid: It has been assumed in the step from from (2.1) to (2.2) that a scattering event is not influenced by the presence of other target particles. That assumption need not be fulfilled.

There is one major difference between the conventional macroscopic concept of a cross section and the present, statistical one. When you fire a bullet at some macroscopic target you will usually hit it if you have aimed at the right area and you won't if you have not. This is not so in the microscopic world governed by quantum effects. When defining a microscopic cross section as above you may be dealing with a product of some geometric cross section σ_{g} and a probability p_A,

$$\sigma_A = \sigma_{\mathrm{g}} \times \mathsf{p}_A. \tag{2.4}$$

The problem with (2.4) is that it may be hard to obtain independent information on σ_{g} and p_A. All you obtain from (2.2) is the product of the two quantities, and σ_{g} may not even be defined. If you have an idea about the magnitude of σ_{g}, you may, however, determine p_A from (2.4), and that way you get some understanding of whether or not σ_A corresponds to a well-defined area within which the projectile has to aim in order to initiate an event A with a reasonable degree of certainty.

There is a wide range of penetration phenomena where we do not need independent information on either σ_{g} or p_A. In cases where such information is important, a more precise separation of the geometric factor than in (2.4) needs to be made. This point will come up in particular in Chapter 8.

2.2.2 Multiple Collisions; Poisson's Formula

In the preceding paragraph the probability for a collision event during a passage was assumed to be small. In penetration phenomena we deal most often with projectiles undergoing many collisions in an individual passage. Therefore we need some statistical information on the number of events in case P_A as given by (2.3) is not small.

Let us illustrate the situation geometrically (Fig. 2.2) by associating a 'black area' σ_A to each target particle. The number of events A is equal to the number of times a projectile hits such a black area. Then, (2.3) can be written in the form

$$P_A = \frac{NSx \times \sigma_A}{S} = \frac{\text{total black area}}{\text{total area}} \tag{2.5}$$

provided that $P_A \ll 1$ and that the target particles do not shadow each other systematically as, e.g., in a crystal. If the number of target particles NSx increases, e.g., due to increasing thickness x at constant number density N, the total black area will increase to the point where individual black areas overlap appreciably; ultimately, for $Nx\sigma_A \gg 1$, the entire area appears black. This implies that the probability for a projectile to undergo at least one event A is essentially equal to 1, but the actual number of events may be substantially larger.

Let us ask for the probability P_n for the projectile to initiate precisely n events A during its passage through the medium. In our geometrical picture this is the same as asking for the probability for n individual black areas to overlap with the trajectory of a given beam particle. Alternatively, we may associate a cylindrical volume

$$v = x\sigma_A \tag{2.6}$$

with each trajectory and try to find the probability for n target particles to be located within one such volume. The latter question is a standard problem in kinetic gas theory: Given an ideal gas of average density N, what is the probability to find precisely n gas molecules within some specified volume v at any instant of time? The answer is given by the Poisson distribution,

$$P_n = \frac{(Nv)^n}{n!}\,e^{-Nv}. \quad n = 0, 1, 2 \ldots \tag{2.7}$$

If you are unfamiliar with Poisson's formula you may wish to consult Appendix A.2.1 for an elementary derivation. The main assumptions entering into (2.7) are the following:

- the positions of any two or more gas molecules are uncorrelated, and
- the sample volume v must be small compared with the total volume filled with gas.

The latter assumption allows to consider the number of available atoms as practically unlimited; then one easily verifies the relationships:

$$\sum_{n=0}^{\infty} \mathsf{P}_n = 1 \tag{2.8}$$

and

$$\sum_{n=0}^{\infty} n\mathsf{P}_n = \langle n \rangle = N\mathsf{v}, \tag{2.9}$$

the brackets indicating an average. For the variance one finds[1]

$$\overline{(n - \langle n \rangle)^2} = \langle n \rangle. \tag{2.10}$$

Eq. (2.10) is a central property of the Poisson distribution. It implies that the relative fluctuation

$$\frac{\overline{(n - \langle n \rangle)^2}}{\langle n \rangle^2} = \frac{1}{\langle n \rangle} \tag{2.11}$$

goes to zero in the limit of large $\langle n \rangle$.

Let us now go back to the statistics of collision events. In order that (2.7) be applicable we must require that the positions of target particles be uncorrelated, i.e., that target particles act as if they were the constituents of an ideal gas. This may be an essential restriction; we shall come back to this point in Chapter 8. The requirement that v be small compared with the size of the container implies that $\sigma_A \ll \mathsf{S}$, cf. (2.6); this is generously satisfied in case of a macroscopic area S.

By combining (2.6) with (2.9) we find the average number of events A to be given by

$$\langle n \rangle = Nx\sigma_A \tag{2.12}$$

and, from (2.10), the fluctuation

$$\overline{(n - \langle n \rangle)^2} = \langle n \rangle = Nx\sigma_A. \tag{2.13}$$

Eq. (2.12) generalizes (2.2) to the case where $\langle n \rangle$ is not small. As an additional benefit we also have found an expression for the fluctuation in the number of events, (2.13).

---

[1] Notations $< \cdots >$ or $\overline{\cdots}$ are utilized synonymously to indicate averages, dependent on readability.

The probability P_0 for no event at all follows from (2.7) and (2.12) for $n = 0$,

$$\mathsf{P}_0 = e^{-Nx\sigma_A}. \tag{2.14}$$

This is called Lambert & Beer's law and governs absorption phenomena.

In case of $Nx\sigma_A \ll 1$ we find that

$$\mathsf{P}_n \simeq \begin{cases} 1 - Nx\sigma_A & \text{for } n = 0 \\ Nx\sigma_A & \text{for } n = 1 \\ 0 & \text{for } n \geq 2 \end{cases} \tag{2.15}$$

up to first order in $Nx\sigma_A$. Thus (2.7) goes over into (2.3) in the limit where the probability for double events is vanishingly small, cf. (A.19) and (A.20) in Appendix A.2.1.

The Poisson distribution is discussed in mathematical terms in standard texts on probability theory (Feller, 1968). Its significance to kinetic gas theory was pointed out by v. Smoluchowski (1904), and the connection to penetration theory was drawn by Bohr (1915).

2.2.3 Energy Loss

Consider now specifically the process of energy loss by a charged particle moving through a stopping medium; in order to be sure that the collision events be distributed according to Poisson's formula, assume the medium to be a gas for the time being.

On account of the conclusions drawn from Fig. 1.1, ignore initially all angular scattering of the projectile. In colliding with the atoms or molecules of the gas a projectile may transfer part of its kinetic energy to those atoms and thus suffer a decrease in velocity. The observation of nearly equal track lengths (Fig. 1.1) indicates that the typical energy lost in a single encounter is small compared with the projectile energy.

Assume that the projectile can lose energy in discrete bits of T_j, with $j = 1, 2 \ldots$, and that $T_j \ll E$ for all j, where E is the projectile energy. T_j might represent the excitation levels above the ground state of a target atom or molecule.

While penetrating a layer of thickness Δx which is assumed small compared with the total penetration depth, a projectile loses an amount of energy ΔE given by

$$\Delta E = \sum_j n_j T_j, \tag{2.16}$$

where n_j is the number of collisions of type j, each leading to an energy transfer T_j.

In order to find the average energy loss $\langle \Delta E \rangle$ and its fluctuation for a great number of projectiles, let us employ the statistical arguments outlined in the previous section with the one addition that we now deal with a spectrum of energy transfers T_j rather than one single possible event A.

For the average energy loss $\langle \Delta E \rangle$ we find from (2.16) that

$$\langle \Delta E \rangle = \sum_j \langle n_j \rangle T_j. \tag{2.17}$$

Introducing the *energy-loss cross section* σ_j for a quantum T_j according to (2.1) we find from (2.12)

$$\langle n_j \rangle = N \Delta x \, \sigma_j \tag{2.18}$$

and hence,

$$\langle \Delta E \rangle = N \Delta x \sum_j T_j \sigma_j. \tag{2.19}$$

Here,

$$S = \sum_j T_j \sigma_j \tag{2.20}$$

is the *stopping cross section*, and the ratio

$$\frac{\langle \Delta E \rangle}{\Delta x} = NS = N \sum_j T_j \sigma_j \tag{2.21}$$

is called the *stopping force* or *stopping power*[2]. While the stopping force is a property of the stopping medium, the stopping cross section is a microscopic quantity. In the literature one frequently finds the symbol S used for the stopping force rather than the stopping cross section. This should cause little confusion since the former has the dimension of [energy/length] while the latter is an [energy×area].

As defined by (2.21) the stopping force is a positive quantity. This is not a universal convention in the literature but reasonable to the extent that energy *loss* rather than *gain* is in focus. However, the function dE/dx to be introduced below in (2.34) must be taken negative whenever the projectile energy decreases with time or pathlength.

2.2.4 Energy-Loss Straggling

Consider now the mean-square fluctuation Ω^2 in energy loss ΔE. According to (2.16) and (2.17) we have

---

[2] Evidently, the term 'stopping force' is in agreement with common nomenclature, while 'stopping power' would be the correct term for the energy loss per unit time. Nevertheless, the latter term has been in use for almost a century and is only slowly disappearing from the literature. Cf. Sigmund (2000).

$$\Delta E - \langle \Delta E \rangle = \sum_j (n_j - \langle n_j \rangle) T_j \tag{2.22}$$

and therefore

$$\Omega^2 = \overline{(\Delta E - \langle \Delta E \rangle)^2} = \sum_{j,l} \overline{(n_j - \langle n_j \rangle)(n_l - \langle n_l \rangle)} T_j T_l; \tag{2.23}$$

Now take the terms $j = l$ and $j \neq l$ separately. For $j = l$,

$$\overline{(n_j - \langle n_j \rangle)^2} = \langle n_j \rangle = N \Delta x \sigma_j \tag{2.24}$$

when the Poisson relation (2.13) applies and (2.18) is inserted. For $j \neq l$, split the average of the product into the product of averages

$$\overline{(n_j - \langle n_j \rangle)(n_l - \langle n_l \rangle)} = \overline{n_j - \langle n_j \rangle} \times \overline{n_l - \langle n_l \rangle} \tag{2.25}$$

because of the statistical independence of different types of collision events; since $\overline{n_j - \langle n_j \rangle} = 0$, drop all terms with $j \neq l$ in (2.23) and find

$$\Omega^2 = \sum_j \langle n_j \rangle T_j^2 = N \Delta x \sum_j T_j^2 \sigma_j. \tag{2.26}$$

In analogy with the stopping cross section, introduce the *straggling parameter*

$$W = \sum_j T_j^2 \sigma_j, \tag{2.27}$$

which has the dimension of [energy$^2 \times$area]. Just as the stopping cross section it is a microscopic property.

A word of caution is indicated with respect to the validity of the relations (2.19) and (2.26). Although these expressions are formally similar, (2.19) is much more general than (2.26). Little can go wrong with the derivation of (2.19); we shall see later that the stopping force is to some extent independent of the structure of the stopping medium. Conversely, not only was explicit use made of the Poisson relationship (2.13) in the derivation of (2.26), but also were collisions leading to different energy transfers T_j and T_l assumed statistically independent. The latter assumption is readily justified in a dilute stopping medium provided that the ion has no memory, i.e., does not undergo changes in state that may influence the stopping cross section. On the other hand, if the medium were closely packed like a solid or a liquid, there would be a more or less pronounced anticorrelation between collisions leading to different energy transfers; such effects will be analyzed in Chapter 8 and it will be found that straggling is sensitive to the structure of the stopping medium.

2.2.5 Differential Cross Section

Let us finally go over to the case of a continuous spectrum of energy loss in individual encounters; such a spectrum applies, e.g., to ionizing collisions with an atom, collisions with a molecule leading to dissociation, etc. You may then apply the present description in a heuristic sense, i.e., make the replacement

$$\sigma_j \rightarrow \frac{\mathrm{d}\sigma}{\mathrm{d}T} \Delta T_j \tag{2.28}$$

and let the interval size ΔT_j be sufficiently small to replace the sums in (2.20) and (2.27) by integrals. In short-hand notation this yields

$$S = \int T \mathrm{d}\sigma, \tag{2.29}$$

$$W = \int T^2 \mathrm{d}\sigma, \tag{2.30}$$

where

$$\mathrm{d}\sigma = \frac{\mathrm{d}\sigma(T)}{\mathrm{d}T} \mathrm{d}T \tag{2.31}$$

is called the *differential energy-loss cross section* and the integrations extend over the spectrum of possible energy transfers. According to (2.3) the quantity

$$\mathrm{d}P = Nx\mathrm{d}\sigma \tag{2.32}$$

is the probability for a projectile to undergo a collision with energy loss[3] $(T, \mathrm{d}T)$ when interacting with the stopping medium under single-collision conditions, i.e., sufficiently small N and/or x.

2.2.6 Range

Up till now the layer thickness was assumed small compared with the penetration depth of the projectile. This made it possible to assume the projectile energy E to be essentially constant. In general the differential cross section and hence the microscopic parameters S and W will depend on energy. This energy dependence is essential for the understanding of the stopping of a projectile down to zero energy.

Consider first the case where the fluctuation in energy loss is negligibly small, as appears to be the case in Fig. 1.1. This implies that the projectile energy is a well-defined function of the penetration depth x,

---

[3] The symbol $(T, \mathrm{d}T)$ indicates the interval limited by T and $T + \mathrm{d}T$. Similarly, $(\Omega, \mathrm{d}^2\Omega)$ indicates a solid angle $\mathrm{d}^2\Omega$ around the unit vector Ω, and $(\mathbf{r}, \mathrm{d}^3\mathbf{r})$ indicates a volume element $\mathrm{d}^3\mathbf{r} = \mathrm{d}x\mathrm{d}y\mathrm{d}z$ located at a vector distance $\mathbf{r}$ from the origin.

$$E = E(x) \tag{2.33}$$

which obeys the differential equation

$$\frac{\mathrm{d}E}{\mathrm{d}x} = -NS(E) \tag{2.34}$$

that follows from (2.21). The minus sign accounts for the decrease in projectile energy. Equation (2.34) has the solution

$$x = \int_{E(x)}^{E_0} \frac{\mathrm{d}E'}{NS(E')} \tag{2.35}$$

in implicit form, where $E_0 = E(0)$ is the initial energy. In particular, the total path length or range R is found by setting $E(x) = 0$, i.e.,

$$R = \int_0^{E_0} \frac{\mathrm{d}E'}{NS(E')}. \tag{2.36}$$

This estimate of the range, based on the *continuous-slowing-down-approximation* (csda) (2.34), is valid only in the case of negligible straggling; (2.36) is not strictly identical with the average range when statistical fluctuations become significant.

An estimate of the fluctuation Ω_R^2 in range – valid for small Ω_R^2 – can be found as follows. When slowing down from E to some energy $E - \Delta E$, the projectile travels a path length $\Delta x \simeq \Delta E / NS(E)$ on the average; the corresponding fluctuation is of the order of $\Omega_x^2 \simeq \Omega^2/(\mathrm{d}E/\mathrm{d}x)^2$; this follows most easily by dimensional arguments. Insertion of Ω^2 from (2.26) and (2.27) as well as (2.34) and Ω_x yields

$$\Omega_x^2 \simeq \frac{NW(E)\Delta E}{\left[NS(E)\right]^3} \tag{2.37}$$

for the *fluctuation in projectile path length* during slowing down from E to $E - \Delta E$. Consequently, the *fluctuation in total range* Ω_R^2 is found by integration down to zero energy,

$$\Omega_R^2 \simeq \int_0^{E_0} \mathrm{d}E' \frac{NW(E')}{\left[NS(E')\right]^3}. \tag{2.38}$$

If a precise meaning is to be assigned to (2.38) some detailed information on the shape of the statistical distributions of energy loss and range is needed. This point will be considered in Chapter 8 and Volume II.

There is little in the substance of the present section that cannot be found in the review by Bohr (1948). Historically, the most important step was the experience that alpha and beta rays, unlike gamma rays, lose energy gradually rather than getting absorbed. That observation appears to date back to Sklodowska-Curie (1900), while the notion of stopping power may be found with Bragg and Kleeman (1905). At the same time and independently, Leithäuser (1904) demonstrated that cathode rays lose energy in penetrating thin metal foils. The first attempt to relate the stopping force to an atomic stopping cross section dates back to Thomson (1912). Extensive range calculations were made by Darwin (1912). The treatment of fluctuations in energy loss and range dates back to Bohr (1915).

2.3 Electronic and Nuclear Stopping

2.3.1 General Considerations

The present section serves as a first qualitative orientation on the dominating mechanisms of energy loss. This problem will have to be treated again and again as we dig deeper into the field.

Let us keep to the case of a charged particle penetrating through a gaseous stopping medium. At moderate velocities a projectile may experience a change in speed in a collision with an individual gas atom or molecule by means of the following processes:

- excitation or ionization of target particles,
- transfer of energy to center-of-mass motion of target atoms,
- changes in the internal state of the projectile, and
- emission of radiation.

In a rough manner the first process may be characterized as a loss of projectile energy into kinetic and potential energy of target electrons while the second process deals essentially with energy transfer to target nuclei; note that the mass, and therefore the kinetic energy of target atoms or molecules, is essentially contained in the nuclei. Therefore the first process is usually called 'electronic energy loss' and the second 'nuclear energy loss'. One may also refer to electronic energy loss as the transfer of internal energy to the target particle as opposed to nuclear energy loss as the transfer of center-of-mass energy. Therefore one may find the notions of *inelastic* and *elastic* energy loss, although the reader may be inclined to assign a different meaning to those concepts, depending on background. Those familiar with neutron or photon scattering may call a scattering event elastic if the incident particle does not suffer energy loss; the present notion of elastic collisions implies only that *energy is not transferred into internal degrees of freedom of the target or projectile.* Moreover, one may be inclined to call a collision event inelastic whenever it is not elastic, and consequently split the collision *cross section* into an elastic and an inelastic contribution. That is certainly a most reasonable concept. Nevertheless, in the stopping literature, many authors tend to split the *energy loss in an individual event* into an elastic and an inelastic contribution, thereby calling the nuclear energy loss elastic and the electronic one inelastic, but it may take some time to realize that this is what those authors do. In order to minimize confusion I shall avoid the notion of an inelastic collision as much as possible and most often talk about nuclear and electronic energy loss or stopping.

Coming back to the above classification of energy-loss processes, ignore changes in the projectile state for the time being. The estimates put forward presently refer to a penetrating point charge. Emission of radiation, especially bremsstrahlung, is a rather different process which becomes important at projectile speeds approaching the velocity of light.

In principle, various processes at the nuclear or subnuclear level would have to be incorporated into the above classification scheme. Whether or not this is important does not only depend on the velocity range and the projectile-target combination in question but also on the reader's motivation to study penetration phenomena: If the interest is in nuclear or high-energy physics, this book may be consulted for information on atomic phenomena that influence penetration properties important in the analysis of nuclear or high-energy processes. If, on the other hand, the interest is in atomic, solid-state, or biological phenomena, nuclear processes may have to be included as an energy sink (or source) if important, along with all other pertinent effects. Initially we shall concentrate on electronic and nuclear energy loss.

In case of doubt, you may do well by defining energy loss of a charged particle as the loss in *kinetic energy of its center-of-mass* in the laboratory frame of reference. For an ion carrying electrons, you may, alternatively, operate with *the loss of kinetic energy of the nucleus*. The difference between the two definitions is less than 0.1 %, i.e., below the accuracy of available experimental techniques.

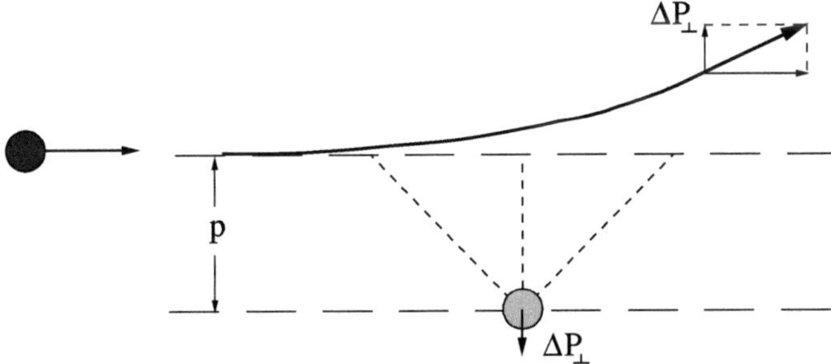

Fig. 2.3. A 'soft' scattering event with momentum transfer $P_\perp$ perpendicular to the beam direction

2.3.2 Momentum and Energy Transfer in Free-Coulomb Collision

For a first qualitative orientation, oversimplify a gaseous stopping medium and treat it as an ideal-gas mixture of free nuclei and electrons. The elementary collision event is, then, the interaction of the projectile, i.e., a point charge e_1 with mass m_1 moving with velocity v, with a target particle of mass m_2 and charge e_2; the target particle can be either a nucleus or an electron. m_2 will turn into a capital M_2 for a nucleon and into m for an electron.

In accordance with Fig. 1.1 on page 4, ignore events leading to a substantial change in projectile velocity; the projectile will be considered to cause an external force that imparts momentum to a target particle that is initially at rest. The collision event is sketched in Fig. 2.3. If the target particle receives only a small momentum it can be considered stationary for the duration of the collision and the momentum transfer is given by

$$\Delta \mathbf{P} = \int_{-\infty}^{\infty} \mathrm{d}t\, \mathbf{F}(t), \tag{2.39}$$

where $\mathbf{F}(t)$ is the Coulomb force[4]

$$F(t) = \frac{e_1 e_2}{p^2 + (vt)^2} \tag{2.40}$$

between the two point charges as a function of time t. Splitting the force into components parallel and normal to the projectile velocity $\mathbf{v}$ (Figure 2.4) we find

$$\Delta P_{\parallel} = e_1 e_2 \int_{-\infty}^{\infty} \mathrm{d}t\, \frac{vt}{(p^2 + v^2 t^2)^{3/2}} = 0 \tag{2.41}$$

and

$$\Delta P_{\perp} = e_1 e_2 \int_{-\infty}^{\infty} \mathrm{d}t\, \frac{p}{(p^2 + v^2 t^2)^{3/2}} = \frac{2|e_1 e_2|}{pv}, \tag{2.42}$$

where p, the *impact parameter*, is the distance between the straight-line trajectory and the initial position of the target. It has been assumed for convenience that this distance is reached at time $t = 0$.

Equations (2.41) and (2.42) constitute a special case of the *momentum approximation* in classical scattering theory. It can be understood as the first term in a perturbation expansion in powers of the interaction between two particles[5]. Because of the symmetry of the Coulomb interaction the longitudinal momentum transfer $\Delta P_{\parallel}$ vanishes to first order, cf. (2.41). Hence momentum is only transferred normal to the projectile velocity in that approximation.

We may find an estimate for the effective collision time τ from the expression

$$\Delta P_{\perp} \simeq F_{\mathrm{max}}\, \tau \tag{2.43}$$

---

[4] From now on electromagnetic quantities will be taken in Gaussian units. Pertinent relationships have been collected in Appendix A.1.1.

[5] This expansion will be discussed in more detail in Appendix A.3.1.

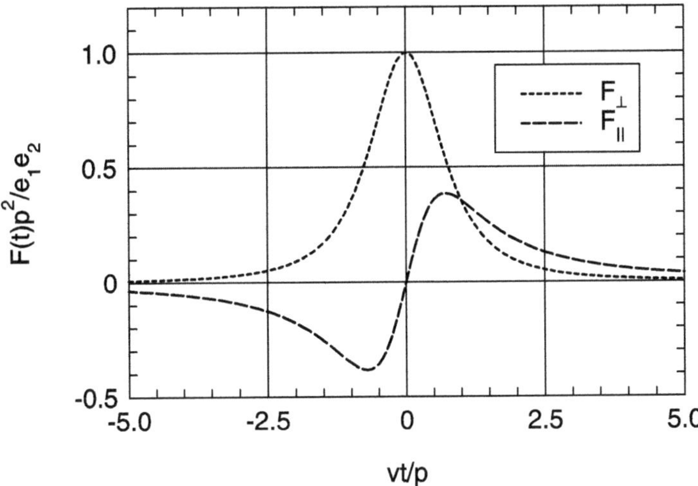

Fig. 2.4. Time dependence of the force parallel and perpendicular to the beam direction according to (2.41) and (2.42)

where F_{max} is the force at closest approach ($t = 0$) and directed normal to the initial velocity. With $F_{\mathrm{max}} = |e_1 e_2|/p^2$ and by comparison to (2.42) one finds

$$\tau \simeq \frac{2p}{v},\tag{2.44}$$

i.e., the two particles interact effectively over a length $\simeq 2p$ of the incoming trajectory. This has been indicated by the stipled lines in Fig. 2.3. Note that the Coulomb force has half its maximum value at a distance $p\sqrt{2}$.

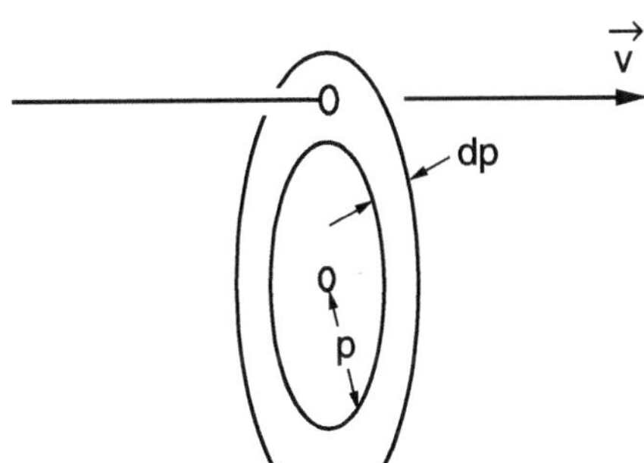

Fig. 2.5. Differential cross section and impact parameter

From (2.42) follows

$$T = \frac{\Delta P_\perp^2}{2m_2} \simeq \frac{2e_1^2 e_2^2}{m_2 v^2 p^2} \tag{2.45}$$

for the energy T lost to particle 2 as a function of impact parameter p. In order that such a collision lead to an energy transfer in an interval $(T, \mathrm{d}T)$, the projectile must aim at a cross-sectional area

$$\mathrm{d}\sigma = 2\pi p \mathrm{d}p = \left|\frac{\mathrm{d}(\pi p^2)}{\mathrm{d}T}\right| \mathrm{d}T \tag{2.46}$$

around the target (Fig. 2.5). By differentiation of (2.45) one finds

$$\mathrm{d}\sigma \simeq 2\pi \frac{e_1^2 e_2^2}{m_2 v^2} \frac{\mathrm{d}T}{T^2}. \tag{2.47}$$

Eq. (2.47) is more accurate than (2.45) from which it was derived. In fact it is an exact version of Rutherford's cross section for Coulomb scattering except for the fact that no upper limit for T is specified by (2.45). A more rigorous derivation of Rutherford's law for classical nonrelativistic scattering will be given in Chapter 3; at present, remember that for a head-on collision ($p = 0$) between a projectile of initial velocity v and a target of initial velocity zero, the conservation laws of energy and momentum, applied to one-dimensional motion, require the final velocity of the target to be

$$v_{\mathrm{max}} = \frac{2m_1}{m_1 + m_2} v. \tag{2.48}$$

This yields a maximum energy transfer

$$T_{\mathrm{max}} = m_2 v_{\mathrm{max}}^2/2 = \gamma E \tag{2.49}$$

with

$$\gamma = \frac{4m_1 m_2}{(m_1 + m_2)^2}, \tag{2.50}$$

where $E = m_1 v^2/2$.

2.3.3 Stopping and Straggling: Preliminary Estimates

When (2.47) is inserted into (2.29) and (2.30) on p. 36 one finds

$$S \simeq 2\pi \frac{e_1^2 e_2^2}{m_2 v^2} \ln \frac{T_{\mathrm{max}}}{T_{\mathrm{min}}}; \tag{2.51}$$

$$W \simeq 2\pi \frac{e_1^2 e_2^2}{m_2 v^2} (T_{\mathrm{max}} - T_{\mathrm{min}}). \tag{2.52}$$

A lower integration limit corresponding to a truncation of the interaction (2.45) at large impact parameters p was introduced in order to remove an apparent divergence from the stopping cross section. Ways of specifying such a cutoff will be discussed in sections 2.3.4 and 2.3.7.

Let us first consider a situation where all energy loss to electrons is ignored. Take N to be the number density of nuclei of charge $e_2 = Z_2 e$ and mass $m_2 = M_2$. e is the elementary charge and Z_2 the atomic number. Then, according to (2.19) and (2.26),

$$\langle \Delta E \rangle_\mathrm{n} \simeq N \Delta x \times \frac{4\pi e_1^2 Z_2^2 e^2}{M_2 v^2} L_\mathrm{n}, \tag{2.53}$$

$$\langle (\Delta E - \langle \Delta E \rangle)^2 \rangle_\mathrm{n} \simeq N \Delta x \times \frac{4\pi e_1^2 Z_2^2 e^2 m_1^2}{(m_1 + M_2)^2}, \tag{2.54}$$

where the subscript 'n' indicates nuclear energy loss. In (2.53), the abbreviation

$$L_\mathrm{n} = \frac{1}{2} \ln \left(\frac{T_\mathrm{max}}{T_\mathrm{min}} \right)_\mathrm{n} \tag{2.55}$$

has been introduced while in (2.54), T_min has been dropped.

Next, consider the opposite case where all energy loss to nuclei is ignored. With a number density $N Z_2$ of electrons (mass m, charge $-e$) one finds correspondingly

$$\langle \Delta E \rangle_\mathrm{e} \simeq N \Delta x \times \frac{4\pi e_1^2 e^2 Z_2}{m v^2} L_\mathrm{e}, \tag{2.56}$$

$$\langle (\Delta E - \langle \Delta E \rangle)^2 \rangle_\mathrm{e} \simeq N \Delta x \times \frac{4\pi e_1^2 e^2 Z_2 m_1^2}{(m_1 + m)^2}, \tag{2.57}$$

where the subscript 'e' indicates electronic energy loss.

Finally, take the ratios of equivalent quantities for nuclear and electronic stopping,

$$\frac{\langle \Delta E \rangle_\mathrm{n}}{\langle \Delta E \rangle_\mathrm{e}} \simeq \frac{m}{M_2} Z_2 \frac{L_\mathrm{n}}{L_\mathrm{e}}, \tag{2.58}$$

$$\frac{\langle (\Delta E - \langle \Delta E \rangle)^2 \rangle_\mathrm{n}}{\langle (\Delta E - \langle \Delta E \rangle)^2 \rangle_\mathrm{e}} \simeq \left(\frac{m_1 + m}{m_1 + M_2} \right)^2 Z_2. \tag{2.59}$$

The ratio of the mean energy losses, (2.58), is obviously dominated by the factor $m Z_2 / M_2$ which is less than 10^{-3}. Regardless of the accurate values of T_max and T_min the ratio of logarithms will not compensate for this large difference unless $L_{\rm e}$ is close to zero. That case will show up only at low projectile velocities where target electrons cannot be considered free.

Thus, under the present somewhat oversimplified assumptions, the *mean energy loss* is heavily dominated by the electronic contribution for both light and heavy point charges, despite the fact that the *maximum energy transfer* from a heavy projectile to a nucleus is many times larger than that to an electron. In order to appreciate the physical origin of this very central conclusion, note that the mean energy loss receives a substantial contribution from rather gentle collisions of the type sketched in Fig. 2.3 on page 39. While the average *momentum* transferred to target electrons does not differ dramatically from the average momentum transferred to target nuclei because forces are comparable, a pronounced difference occurs in transferred *energy*, cf. (2.45), where the mass enters into the denominator.

Next, consider the ratio of fluctuations, (2.59). If the projectile is an electron ($m_1 = m$), that ratio becomes $4Z_2(m/M_2)^2$ which is $\sim 10^{-7}$ or less. Thus, electronic processes dominate even more strongly than in case of the mean energy loss. For a heavy projectile on the other hand, $m_1 = M_1$, the ratio is less than 1 for $M_1 \ll M_2$ but may exceed 1 for $M_1 \geq M_2$.

The physical reason for the significance of nuclear energy losses in straggling is to be found in the high-energy-loss tail of the Rutherford spectrum (2.47). Evidently that tail is more important in the integral $\int T^2 {\rm d}\sigma$ than in $\int T {\rm d}\sigma$. The presence of this tail makes it a rather delicate task to estimate the actual shape of an energy loss profile. Substantial effort will be devoted to this task in Chapter 9.

The dominating role of the electronic stopping force in the penetration of alpha and beta rays was recognized in the earliest investigations into the field and became the basis of Thomson's stopping model where stopping and ionization were treated essentially synonymously (Thomson, 1912). With regard to stopping force and straggling, the present discussion does not go beyond what can be found in the monographs by Bohr (1948) and especially Bonderup (1981).

2.3.4 Adiabatic Limit to Electronic Stopping

According to (2.58) the stopping force on a point charge is essentially electronic. From (2.56) we obtain the electronic stopping cross section

$$S_{\rm e} = \frac{4\pi Z_2 e_1^2 e^2}{mv^2} L_{\rm e} \tag{2.60}$$

where $L_{\rm e} = \frac{1}{2} \ln(T_{\rm max}/T_{\rm min})_{\rm e}$. A quantity L defined by (2.60) is called a *stopping number*. Consequently, $L_{\rm e}$ denotes the *electronic stopping number*. For a heavy projectile, $m_1 \gg m$, (2.49) and (2.50) yield

$$(T_{\rm max})_{\rm e} = 2mv^2. \tag{2.61}$$

In this section an estimate of the lower limit $T_{\min}$ of energy transfer will be derived.

Note first that $T_{\min}$ is strictly zero in the case of a free, isolated target electron. Thus, the long-range Coulomb interaction causes a logarithmic divergence of the stopping cross section. There are two most obvious ways of removing this artifact, taking into account the *binding* of electrons to the atoms or molecules in a neutral stopping medium, or to consider *screening* of the Coulomb interaction. Let us consider binding here.

Offhand one might be inclined to incorporate electronic binding forces by merely inserting the lowest electronic excitation energy of a target atom or molecule for $(T_{\min})_e$ or, as suggested by Thomson (1912), the lowest ionization energy. While this is justified for a qualitative estimate, it turned out that the argument of Bohr (1913, 1915), based on the simple picture of a spring force between electron and nucleus does not only lead to a different quantitative result but also has a very close quantal analog.

Bohr treated the individual target electron as a classical harmonic oscillator with a resonance angular frequency of, say, ω_0. When an external force F acts on such an oscillator during a limited period of time τ, the exchange of momentum depends essentially on the magnitude of τ compared with the oscillation period $2\pi/\omega_0$. For $\tau \ll 2\pi/\omega_0$, the oscillator takes up an impulse $\sim F \times \tau$ just as if it were a free particle; this effect may be experienced by giving a push with a hammer to the steel ball of a pendulum. Conversely, for $\tau \gg 2\pi/\omega_0$, the oscillator tends to respond adiabatically to the external force and it will tend to calm down as the disturbance vanishes even in the absence of damping forces. Thus, the takeup of momentum will be much smaller than that experienced by an otherwise equivalent free particle.

The duration of the force acting on a target electron from a moving projectile is given approximately by (2.44) for a soft collision; hence an effective adiabatic cutoff occurs at an impact parameter where $2p/v \ll 2\pi/\omega_0$, i.e., $p_{\max}$ will be of the order of

$$p_{\max} \sim \frac{v}{\omega_0} \tag{2.62}$$

apart from a numerical constant that has to be determined by a more accurate calculation. Eq. (2.62) specifies *Bohr's adiabatic radius*.

Combining (2.62) with (2.45) we obtain[6]

$$T_{\min} \sim \frac{2e_1^2 e^2 \omega_0^2}{mv^4} \tag{2.63}$$

or

$$L = \frac{1}{2} \ln \frac{T_{\max}}{T_{\min}} = \ln \frac{Cmv^3}{|e_1 e| \omega_0} \tag{2.64}$$

---

[6] The subscript 'e' will be dropped when there is no doubt that we deal with electronic stopping.

with a yet undetermined constant C which is ~ 1 but will be evaluated in Sect. 4.5.1.

2.3.5 Relativistic Extension

While relativistic effects play a role in particle penetration on all levels, most of the phenomena discussed in this book are observable equally well at non-relativistic and relativistic velocities. Hence, relativity may affect quantitative details but – with very few exceptions – is not the essential aspect. Therefore the strategy will be followed to present nonrelativistic treatments and add pertinent relativistic corrections where appropriate.

The present section serves the purpose to extend the simple treatment of Coulomb scattering discussed in Sect. 2.3.2 to relativistic velocities. Appendix A.3.2 recapitulates pertinent formulas from special relativity.

On the basis of the Lorentz transformation for electromagnetic fields (cf. Appendix A.3.2) one finds the following relativistic extensions of (2.41) and (2.42)[7],

$$\Delta P_{\parallel} = e_1 e_2 \int_{-\infty}^{\infty} dt \, \frac{\gamma_v vt}{(p^2 + (\gamma_v vt)^2)^{3/2}} = 0 \tag{2.65}$$

and

$$\Delta P_{\perp} = e_1 e_2 \int_{-\infty}^{\infty} dt \, \frac{\gamma_v p}{(p^2 + (\gamma_v vt)^2)^{3/2}} = \frac{2|e_1 e_2|}{pv}, \tag{2.66}$$

where

$$\gamma_v = \frac{1}{\sqrt{1 - v^2/c^2}}. \tag{2.67}$$

It is seen that there is no change in the momentum transfer at a given impact parameter, even though the maximum force has been enhanced by a factor γ_v. This implies that the collision time in (2.43) has become smaller by the same factor, i.e.,

$$\tau \simeq \frac{2p}{\gamma_v v} \tag{2.68}$$

instead of (2.44). This leads to a change in the adiabatic radius (2.62)

$$p_{\max} \sim \frac{\gamma_v v}{\omega_0} \tag{2.69}$$

---

[7] An attempt has been made to use a notation that should prevent the reader from mixing up the symbol γ_v in (2.65 – 2.67) with the quantity γ defined in eq. (2.50).

which in turn affects the minimum energy transfer (2.63) in an electronic collision,

$$T_{\min} \sim \frac{2e_1^2e^2\omega_0^2}{m\gamma_v^2v^4}. \tag{2.70}$$

In order to find the maximum energy transfer in a binary collision between a heavy particle and an electron, look at a central collision in a reference frame in which the projectile is at rest. Here the electron moves with a velocity $-\boldsymbol{v}$ before and $\boldsymbol{v}$ after the collision. This implies a momentum $\boldsymbol{P}' = m\gamma_v\boldsymbol{v}$ and a total energy $E'_{\text{tot}} = m\gamma_vc^2$. Lorentz transformation into the laboratory system yields the energy in the laboratory frame,

$$E_{\text{tot}} = \gamma_v(E'_{\text{tot}} + vP') \tag{2.71}$$

and, hence, an energy transfer

$$T_{\max} = E_{\text{tot}} - mc^2 = 2m\gamma_v^2v^2. \tag{2.72}$$

This results in a stopping number

$$L = \frac{1}{2}\ln\frac{T_{\max}}{T_{\min}} = \ln\frac{Cm\gamma_1^2v^3}{|e_1e|\omega_0} \tag{2.73}$$

instead of (2.64). It is a preliminary estimate which will be modified in Sect. 4.2.3.

2.3.6 Validity of Classical-Orbit Picture

So far, moving particles have been assigned classical orbits, and the limitations imposed by quantum mechanics were barely mentioned. The tacit justification of this procedure lies in the fact that we have been dealing with Coulomb interaction only; indeed, the differential cross section for scattering of two point charges on each other is known to be identical with Rutherford's law when calculated on the basis of nonrelativistic quantum mechanics (Gordon, 1928). This will be shown in Chapter 3. Hence, even though the actual state of motion will differ from classical Kepler orbits, the difference may be argued to be immaterial since the energy-loss spectrum in an individual collision event is unaffected by quantal corrections.

   This simplifying feature need no longer be true

- when the force between the particles is not Coulomb-like,
- in the presence of binding forces, and
- at relativistic velocities.

Any of these three situations may occur in penetration phenomena. In addition we shall also need to consider spatial correlations in collision problems.

For all these reasons an estimate of the range of validity of the classical-orbit picture is needed.

From the point of view of penetration theory for swift charged particles it is most often the gentle collisions leading to small energy transfers that are considered most representative. Not only are those events by far the most frequent ones – as is seen from (2.47) – they also offer a number of challenges due to their long-range nature that suggests the possibility of collective effects in dense stopping media. Nevertheless close collisions, although rare, do occur and are not insignificant as is evidenced by the occurrence of $T_{\max}$ in the expressions for stopping force and straggling.

In the present context gentle collisions are considered mainly since we know already that binding corrections play a role in the determination of the lower limit $T_{\min}$ of the energy-loss spectrum in (2.60), and hence that the assumption of free-Coulomb scattering becomes invalid in that limit.

Go back once again to Fig. 2.5 on page 41. In order to estimate the limitations of a classical orbit we may try to find the lateral spreading of a wave packet centered around the straight projectile path depicted there.

According to Williams (1945) and Bohr (1948) the physical situation is as follows. One may try to pin down an impact parameter p with an uncertainty δp by constructing a Gaussian wave packet with a lateral spread δp, thereby introducing a transverse momentum δP_1 of the order of

$$\delta P_1 \sim \frac{\hbar}{2\delta p} \tag{2.74}$$

according to the Heisenberg uncertainty principle[8]. At the same time an uncertainty in the impact parameter results in an uncertainty in momentum transfer, δP_2 according to (2.42),

$$\delta P_2 \sim \delta p \left| \frac{\mathrm{d}(\Delta P_\perp)}{\mathrm{d}p} \right| = \frac{2|e_1 e_2|}{p^2 v} \times \delta p, \tag{2.75}$$

where $\Delta P_\perp$ follows from (2.42). The uncertainty principle implies that these two uncertainties are uncorrelated. We may estimate the total uncertainty from the sum of squares,

$$\delta P^2 \sim \delta P_1^2 + \delta P_2^2. \tag{2.76}$$

This quantity takes on its minimum value when δp^2 is chosen to be

$$\delta p^2_{\min} = \frac{\hbar p^2 v}{4|e_1 e_2|} \tag{2.77}$$

and $\delta P_{\min}$ becomes, then,

$$\delta P_{\min} \sim \sqrt{\frac{2|e_1 e_2|\hbar}{p^2 v}}. \tag{2.78}$$

---

[8] Appendix A.4.1 reviews Gaussian wave packets for the interested reader.

In order that the momentum transfer at a given impact parameter $p \pm \delta p_{\min}$ be well-defined we have to require that

$$\frac{\delta P_{\min}}{\Delta P_{\perp}} \sim \sqrt{\frac{\hbar v}{2|e_1 e_2|}} \ll 1 \tag{2.79}$$

In the notation of Bohr (1948) this reads

$$\kappa = \frac{2|e_1 e_2|}{\hbar v} \gg 1 \tag{2.80}$$

or

$$v \ll 2v_0 \left| \frac{e_1 e_2}{e^2} \right|, \tag{2.81}$$

where the Bohr velocity $v_0 = e^2/\hbar = c/137$ is the orbital speed of an electron in the ground state of a hydrogen atom, c being the speed of light in vacuum (Appendix A.1.2).

If the target particle is an electron, (2.81) reads

$$v \ll 2Z_1 v_0. \tag{2.82}$$

For low charge numbers Z_1 the limit imposed by (2.82) is rather severe. For alpha particles it lies at 1.6 MeV.

Now, the whole description of a projectile interacting with a target electron at rest can only be meaningful provided that

$$v \gg v_0, \tag{2.83}$$

i.e., if the projectile speed substantially exceeds some characteristic orbital velocity of the target electrons which has been set to be the Bohr velocity v_0. For low charge numbers, (2.82) is in direct conflict with (2.83). We may therefore conclude that the theory of electronic stopping for electrons and positrons is intrinsically quantal. Conversely, for heavier particles, and in particular for heavy ions with $e_1 \gg e$, (2.82) specifies a velocity range within which the assumptions underlying Bohr's classical stopping formula (2.64) are approximately valid.

2.3.7 Screening in Nuclear Stopping

There were good reasons to ignore nuclear stopping for a while. After all, electronic energy loss appears to dominate the stopping force on point charges by more than three orders of magnitude. This estimate, however, is based on the assumption that the nuclear stopping force does not diverge, i.e., that some mechanism is going to limit the Coulomb interaction between projectile and target nucleus to a finite range.

The keyword is *screening*. Indeed, if the stopping medium consists of neutral atoms or molecules, the Coulomb force between a point charge and a nucleus comes into action only when the point charge penetrates the electron shells. Thus, an effective maximum impact parameter $(p_{\max})_n$ for nuclear collisions is equivalent to a representative radius a of the atomic or molecular electron cloud. Let us be satisfied for the moment with the statement that a is of the order of or less than the Bohr radius a_0 (cf. Appendix A.1.2).

Why don't we use Bohr's adiabaticity limit to find $(p_{\max})_n$ as was done in electronic stopping? After all, nuclei are bound to each other in molecules and solids, approximately by oscillator forces. Well, in order that such an adiabatic limit be significant, it has to be smaller than a_0, or

$$v \leq a_0 \omega = v_0 \frac{\hbar\omega}{2R} \tag{2.84}$$

according to (2.62), where ω is an effective binding frequency. Here, R is the Rydberg energy, R $= 13.6$ eV. (Appendix A.1.2). For typical vibrational frequencies of atoms in molecules, $\hbar\omega$ lies in the range around or below 0.1 eV; hence, nuclear collisions become adiabatic at projectile speeds that are at least two orders of magnitude lower than what we considered so far. Note that an alpha particle at a velocity $v = v_0$ carries a kinetic energy of about 100 keV (cf. problem 1.1 on page 22).

Incidentally, for nuclear stopping the condition for validity of the classical-orbit picture, (2.81) reads

$$v \ll 2Z_1 Z_2 v_0. \tag{2.85}$$

This is less stringent than (2.82) because of the occurrence of $Z_2 \geq 1$. We may conclude that nuclear collisions may follow the laws of classical mechanics over a considerably wider range of projectiles and speeds than what was found for electronic collisions.

Let us get back to screening and insert $(p_{\max})_n = a$ into (2.45). This yields an effective minimum energy loss in nuclear collisions,

$$(T_{\min})_n = \frac{2}{M_2 v^2} \frac{e_1^2 Z_2^2 e^2}{a^2} \tag{2.86}$$

or

$$L_n = \frac{1}{2} \ln \left(\frac{T_{\max}}{T_{\min}} \right)_n \simeq \ln 2\varepsilon, \tag{2.87}$$

where

$$\varepsilon = \frac{M_2}{M_1 + M_2} \frac{Ea}{e_1 Z_2 e} \tag{2.88}$$

and M_2 is the nuclear mass. ε is a dimensionless measure of the projectile energy that is of central significance in nuclear stopping (Lindhard et al., 1968).

2.4 Multiple Scattering

2.4.1 Small-Angle Approximation

Stopping and scattering are closely related phenomena in particle penetration. Strictly speaking you cannot deal with one and ignore the other. From a formal point of view it might look appealing to treat both phenomena at the same time, representing longitudinal and transverse changes in velocity, respectively. Yet the dominating processes will turn out to differ in the two cases. This discourages a joint treatment from a physical point of view. Instead one may come a long way by treating the change in projectile speed as independent of the changes in direction of motion.

Switch over to scattering and ignore all energy loss for a while. As in the case of stopping we normally deal with a sequence of events in particle penetration. Thus, the equivalent of stopping theory is the theory of *multiple scattering*. The formal treatment will largely turn out to be a generalization of the statistical theory of stopping into two dimensions.

Let us again take our starting point at the experimental fact demonstrated in Fig. 1.1 on page 4 that the trajectories of alpha particles are essentially straight lines. This means that small-angle scattering events appear to dominate.

One may at different stages ascribe different meanings to the notion of a 'small' angle. The term *small-angle approximation*, however, will consistently imply that the respective angle, say, θ, is small in an *absolute* sense so that the relations

$$\sin\theta \simeq \theta; \ \cos\theta \simeq 1 \tag{2.89}$$

hold within some prescribed accuracy.

Let us then have a look at the trajectory of an alpha particle, sketched schematically in Fig. 2.6. If all energy loss is ignored, the motion under a series of small-angle scattering events can be characterized by the projection on a plane perpendicular to the initial direction of motion. Introducing spherical polar coordinates v, α and χ and applying the small-angle approximation we find the expression

$$\mathbf{v} = (v\cos\alpha, v\sin\alpha\cos\chi, v\sin\alpha\sin\chi)$$
$$\approx v(1, \alpha\cos\chi, \alpha\sin\chi) \tag{2.90}$$

for the velocity vector of the projectile if the initial direction of motion is directed along the x-axis. Thus, pertinent information on the direction of motion is contained in a two-dimensional vector $\boldsymbol{\alpha}$,

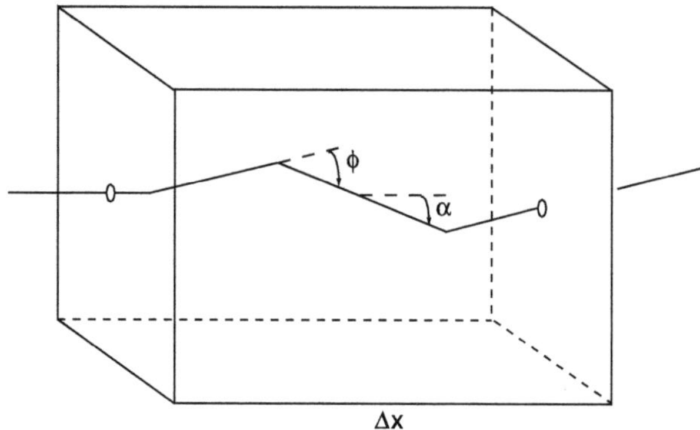

Fig. 2.6. Trajectory of an alpha particle during multiple scattering (schematically). α denotes the angle between the scattered particle and the initial beam direction and ϕ the scattering angle in an individual event

$$\boldsymbol{\alpha} = (\alpha \cos \chi, \alpha \sin \chi) \tag{2.91}$$

which represents the lateral component of a unit vector pointing in the instantaneous direction of motion of the projectile.

2.4.2 Statistics

Now let the projectile be able to undergo a discrete spectrum of scatterings at such vectorial angles $\boldsymbol{\phi}_j$, $j = 1, 2 \ldots$, and let the cross sections for these events be σ_j. If, during an individual passage, the projectile undergoes n_j deflections at angle $\boldsymbol{\phi}_j$, the final direction of motion is determined by the vector

$$\boldsymbol{\alpha} = \sum_j n_j \boldsymbol{\phi}_j \tag{2.92}$$

perpendicular to the initial velocity. This relation is a two-dimensional analog of (2.16). Consequently, averages can be found in the same way as in the treatment of energy loss.

In particular, the average deflection follows by analogy with (2.19),

$$\langle \boldsymbol{\alpha} \rangle = N \Delta x \sum_j \boldsymbol{\phi}_j \sigma_j. \tag{2.93}$$

This quantity will most often be zero, namely whenever the scattering centers look azimuthally symmetric to the projectile; then σ_j will be independent of the azimuth of the scattering angle $\boldsymbol{\phi}_j$, and the sum (or integral) (2.93) vanishes according to (2.91) since $\int_0^{2\pi} d\chi \cos \chi = \int_0^{2\pi} d\chi \sin \chi = 0$.

Next, the mean-square deflection angle follows by analogy to (2.23) and (2.26),

$$\overline{(\alpha - \langle\alpha\rangle)^2} = N\Delta x \sum_i \phi_i^2 \sigma_i = \langle\alpha^2\rangle, \tag{2.94}$$

and this quantity is nonzero if only one $\sigma_i \neq 0$. Here the first identity is general while the second implies azimuthal symmetry.

As in Sect. 2.2.3 we go over to continuum notation and write (2.94) in the form

$$\langle\alpha^2\rangle = N\Delta x \int \phi^2 d\sigma \tag{2.95}$$

where $d\sigma$ as given by

$$d\sigma = K(\phi)d^2\phi \tag{2.96}$$

is the differential cross section for scattering into the solid angle $(\phi, d^2\phi)$. In the considered case of azimuthal symmetry we have $K(\phi) \equiv K(\phi)$, where ϕ is the polar scattering angle.

While the overall number of scattering events undergone by a projectile during a passage is typically a large number, the situation becomes different when attention is limited to large scattering angles. Indeed, take the probability

$$P(\alpha^*) = Nx \int_{\phi>\alpha^*} d\sigma \tag{2.97}$$

for a scattering event by an angle exceeding some lower limit α^*; (2.97) is a straight application of (2.3). Figure 1.1 on page 4 indicates that for alpha particles, $P(\alpha^*)$ must be a small number ($\ll 1$) if α^* exceeds a few degrees. Therefore, in situations like this the distribution $F(\alpha) \times 2\pi\alpha d\alpha$ of projectiles after passage through a layer Δx must approach the single-event limit

$$F(\alpha) \Rightarrow N\Delta x K(\alpha) \tag{2.98}$$

for $\alpha \gg \alpha^*$. A rough estimate of the limiting angle α^* is given by the angle which makes the probability (2.97) equal to unity, i.e.,

$$N\Delta x \int_{\phi>\alpha^*} d\sigma = 1. \tag{2.99}$$

This may be taken to set a limit between single and multiple scattering; it is seen that α^* as given by (2.99) depends on the thickness Δx: The larger Δx the larger α^*.

As a first approximation the following qualitative shape (Fig. 2.7) arises for the multiple-scattering distribution $F(\alpha)$. At angles $\alpha \ll \alpha^*$, where many

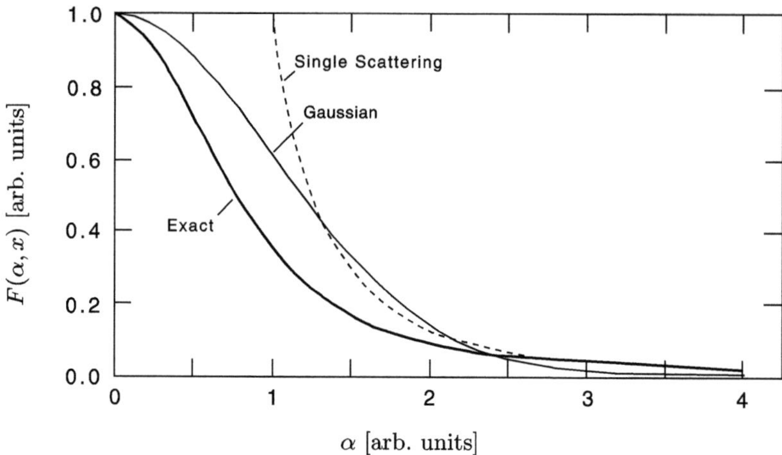

Fig. 2.7. Multiple scattering distribution for heavy projectiles in the model of Williams (1940). Units refer to a quantitative model to be discussed in Volume II. See text

individual deflections contribute to the final value of $F(\alpha)$, one may assume a Gaussian profile with a width given by (2.95)

$$\langle \alpha^2 \rangle = N \Delta x \int_{\phi < \alpha^*} \phi^2 d\sigma; \tag{2.100}$$

in the opposite end, at angles $\alpha \gg \alpha^*$, (2.98) holds.

The above treatment which dates back to Williams (1940) gives a first qualitative orientation. It is often quite reliable. However, a more comprehensive treatment is available and will be discussed in Volume II.

2.4.3 Nuclear and Electronic Scattering

The present section serves the purpose of providing a qualitative estimate of the relative significance of nuclear and electronic collisions in multiple scattering. The treatment will follow that of energy loss in Sect. 2.2.3 on page 33.

The angle of deflection ϕ at impact parameter p follows from the momentum transfer

$$\phi \sim \frac{\Delta P_\perp}{m_1 v} = \sqrt{\frac{m_2 T}{m_1 E}} \tag{2.101}$$

by means of (2.45). Therefore (2.95) reads

$$\langle \alpha^2 \rangle = N \Delta x \int \frac{m_2 T}{m_1 E} \, d\sigma = \frac{m_2}{m_1 E} \langle \Delta E \rangle, \tag{2.102}$$

where $\langle \Delta E \rangle$ is given by (2.19), or $\langle \Delta E \rangle = N \Delta x S$. Equation (2.102) holds for one target species at a time, i.e., the nuclear and electronic contributions read

$$\langle \alpha^2 \rangle_{\mathrm{n}} = \frac{M_2}{m_1 E} \langle \Delta E \rangle_{\mathrm{n}} \tag{2.103}$$

and

$$\langle \alpha^2 \rangle_{\mathrm{e}} = \frac{m}{m_1 E} \langle \Delta E \rangle_{\mathrm{e}}, \tag{2.104}$$

respectively. These rough estimates – which ignore the upper limit α^* that was introduced in (2.97) – show that the ratio of nuclear to electronic multiple-scattering widths is of the order of

$$\frac{\langle \alpha^2 \rangle_{\mathrm{n}}}{\langle \alpha^2 \rangle_{\mathrm{e}}} \sim Z_2 \frac{L_{\mathrm{n}}}{L_{\mathrm{e}}} \tag{2.105}$$

by means of (2.58).

For heavy target atoms where $Z_2 \gg 1$, the *nuclear* contribution dominates clearly. Moreover, the nuclear stopping number L_{n} normally exceeds the electronic one when the projectile is an ion or another heavy particle – because of the large difference between $(T_{\mathrm{max}})_{\mathrm{n}}$ and $(T_{\mathrm{max}})_{\mathrm{e}}$. This results in a predominantly nuclear multiple-scattering distribution even when the target is light.

The situation is even clearer in the limit of single collisions, $\alpha \gg \alpha^*$. Since the momentum transferred to an electron in a single collision cannot exceed $2mv$ a heavy projectile can be scattered from an electron at most by an angle $\sim 2m/M_1$. This angle – less than 0.1 degree – is frequently within the multiple-scattering limit α^*. Hence, single scattering of heavy projectiles ($m_1 \gg m$) is determined by *nuclear* contributions with the exception of extremely small scattering angles. Another exception is the case of channeling (Fig. 1.3 on page 6) where nuclear scattering events are suppressed.

Let us finally write down the scattering cross section in the small-angle approximation. According to (2.42) and (2.91) we have

$$\phi \sim \frac{|e_1 e_2|}{pE} \tag{2.106}$$

and therefore

$$\mathrm{d}\sigma \sim \left| \frac{\mathrm{d}(\pi p^2)}{\mathrm{d}\phi} \right| \sim \frac{e_1^2 e_2^2}{E^2} \frac{2\pi \phi \mathrm{d}\phi}{\phi^4}. \tag{2.107}$$

In the single-collision limit the probability for a scattering event $(\alpha, \mathrm{d}\alpha)$ is then determined by (2.98), i.e.,

$$F(\alpha_1) \, 2\pi\alpha \mathrm{d}\alpha \simeq N \Delta x \, 2\pi \frac{e_1^2 e^2}{E^2} Z_2^2 \frac{\mathrm{d}\alpha}{\alpha^3} \tag{2.108}$$

for $2m/m_1 \leq \alpha \ll 1$, where $Z_2 e$ is the nuclear charge. For $\alpha \leq 2m/M_1$, target electrons contribute to the scattering cross section. In that case the factor Z_2^2 due to the target charge is to be replaced by Z_2 since there are Z_2 electrons per nucleus. Hence, the effective scattering cross section due to both nuclei and electrons reads

$$\mathrm{d}\sigma \simeq \frac{e_1^2 e^2 Z_2(Z_2+1)}{E^2} \frac{2\pi\phi\,\mathrm{d}\phi}{\phi^4} \tag{2.109}$$

for $\alpha < 2m/m_1$.

The treatment presented here applies to the nonrelativistic regime as it stands. Extension to the relativistic regime is straightforward since the central quantity, the transverse momentum transfer (2.42) is unaffected by relativity. Therefore, the only noticeable change at this level is the replacement of the projectile momentum $m_1 v$ in (2.101) by $m_1 \gamma_v v$, i.e.,

$$m_1 \to \gamma_v m_1 \tag{2.110}$$

in all relations on small-angle multiple scattering.

2.5 Estimates

2.5.1 Alpha Particles

Now let us interpret the qualitative features of the cloud-chamber photograph shown in Fig. 1.1 on page 4. Within the spirit of this chapter we do not aim at high accuracy but rather try to find simple order-of-magnitude estimates.

Start with the range. It is found by insertion of the electronic stopping cross section (2.60) into (2.36) and integrating. This yields very roughly

$$R \sim \frac{mE^2}{M_1\, 4\pi N Z_2 e_1^2 e^2} \left\{\frac{1}{L_\mathrm{e}}\right\}, \tag{2.111}$$

where the brackets indicate an average over the trajectory. Note that L_e varies rather slowly with E.

Next get an order-of-magnitude estimate of the range straggling from (2.38) which reads

$$\frac{\Omega_R^2}{R^2} \sim \frac{2m}{M_1} \frac{\{L_\mathrm{e}^{-3}\}}{\{L_\mathrm{e}^{-1}\}^2}; \tag{2.112}$$

This shows that $\Omega_R/R \sim 10^{-2}$ for alpha particles, a result that is consistent with Fig. 1.1 and shows that the continuous-slowing-down-approximation is quite accurate in this case.

Next get an order-of-magnitude estimate of the multiple-scattering angle by setting $\Delta x = R$ in (2.103), i.e., look at the angular spread of an initially well-collimated beam over the entire trajectory,

$$\langle \alpha^2 \rangle = \frac{mZ_2}{2M_1} \left\{ \frac{L_n}{L_e} \right\}. \tag{2.113}$$

It is seen that whatever the accurate value of the weighted average $\{L_n/L_e\}$, the factor $mZ_2/2M_1 \sim Z_2/15000$ will make sure that $\langle \alpha^2 \rangle$ is a small quantity for an alpha particle. This explains why visible portions of the particle tracks in Fig. 1.1 are essentially straight lines.

Finally, according to (2.108) for $\Delta x = R$, the probability for a large-angle scattering event over the entire trajectory is of the order of

$$NR\frac{\pi e_1^2 e^2 Z_2^2}{E^2} \sim \frac{mZ_2}{16M_1} \left\{ \frac{1}{L_e} \right\}. \tag{2.114}$$

This shows that even for heavy target nuclei only a minute fraction, less than 10^{-2} of the alpha particles, will undergo a major deflection over the main part of their range.

2.5.2 Preview: Energy and Z_1 dependence

Figure 2.8 gives you an impression of what we have learned so far and where we are going. Theoretical stopping cross sections for H, Ar and U ions in Si, found from reliable sources, have been plotted over a wide range of beam energies. All stopping cross sections show the characteristic v^{-2} dependence, modified by the logarithmic variation of the stopping number at high but nonrelativistic energies. In that energy regime, S_n lies more than three orders of magnitude below S_e.

However, both S_e and S_n go through maxima at lower energies and then decrease according to some power-law dependence on the beam energy. Evidently, such maxima must exist, because a particle cannot lose more than its total kinetic energy, but their locations differ dramatically, and no easy estimate of their location and height emerges from what we have learned up till now.

For both Ar and U, there is a low-energy regime where S_n dominates, while this is not expected for hydrogen ions. The height of the maximum in S_e varies over almost three orders of magnitude from H to U, but note that the factor is signifcantly smaller than 92^2, the ratio of the respective values of Z_1^2.

Also, note that S_e starts to increase again at relativistic energies, ~ 1 GeV. The importance of relativistic collision kinematics was recognized right from the beginning (Bohr, 1915).

Finally, experimental data from ~ 25 different sources have been included in case of H-Si, a comparatively well-studied ion-target combination. You may note that at energies above the peak position there is excellent agreement both between different sets of experimental data and between experiment and theory. Discrepancies occur at lower speeds.

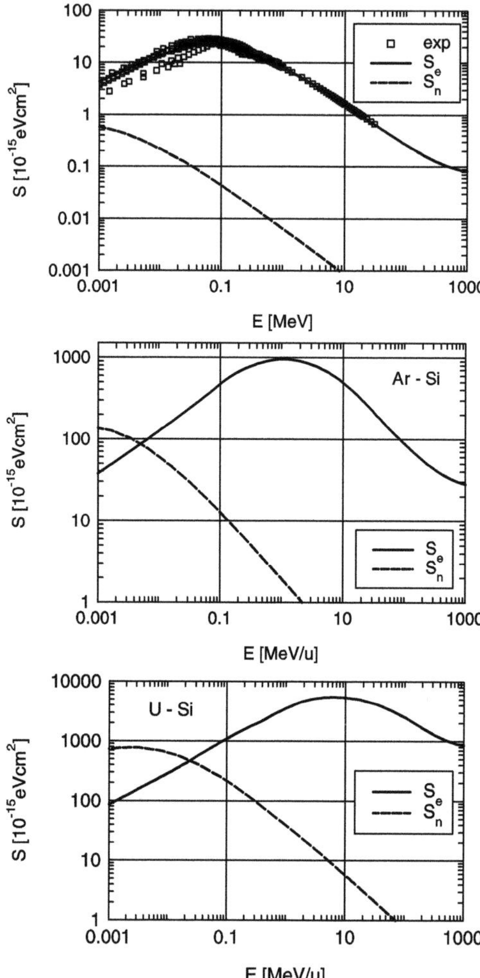

Fig. 2.8. Electronic and nuclear stopping cross sections for H in Si, Ar in Si and U in Si versus beam energy per nucleon. S_n evaluated according to Lindhard et al. (1968). S_e from ICRU (1993) for H-Si, from ICRU (2005) for Ar-Si and for U-Si computed from binary theory (Sigmund and Schinner, 2002). Experimental data from numerous sources compiled by Paul (2005). Note the different abscissa scales

2.6 Electron and Positron Penetration

Several of the explicit results quoted above for stopping and scattering parameters have been derived for projectile masses exceeding the electron mass, i.e., for ions, mesons etc. This implies that the maximum energy transfer to a target electron is close to $2mv^2$ according to (2.61). This does not apply to a positron where the maximum energy transfer is

$$T_{\max} = \frac{m}{2}v^2. \tag{2.115}$$

However, it does not make much sense to insert this into (2.64), because the classical scheme has a very limited range of validity. Instead, we just note that the stopping force on a positron is smaller than that on a proton at the same projectile speed.

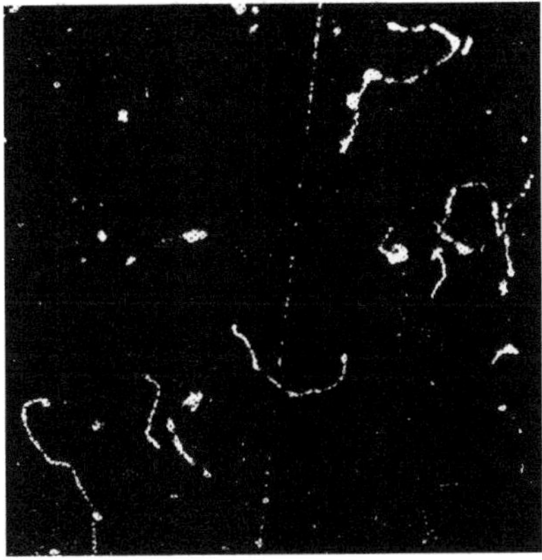

Fig. 2.9. Trajectories of beta particles. From Gentner et al. (1954)

For penetrating electrons a further complication is caused by the indistinguishability of a scattered projectile electron from an ejected target electron in case of substantial energy transfer. In applications of particle penetration theory it is then necessary to consider the possible effects of *both* electrons emerging from a collision. The stopping force is, then, no longer defined in a strict sense. On a less rigorous basis, one may define the mean energy loss as the difference between the initial electron energy and the mean energy of an electron emerging with energy greater than $E/2 = mv^2/4$ from an interaction, i.e.,

$$S = \int_{T_{\min}}^{E/2} T \mathrm{d}\sigma(T) + \int_{E/2}^{E} (E - T) \mathrm{d}\sigma(T). \tag{2.116}$$

This may be rewritten in the form

$$S = \int_{T_{\min}}^{E} T \mathrm{d}\sigma(T) + \int_{E/2}^{E} (E - 2T) \mathrm{d}\sigma(T) \tag{2.117}$$

or, for Coulomb scattering,

$$S_{\text{electron}} = \frac{4\pi Z_2 e^4}{mv^2} \left(L_{\text{positron}} + \frac{1}{2} - \ln 2 \right) \tag{2.118}$$

Electrons and positrons behave dramatically differently from heavy particles with regard to angular deflection. This is most easily seen from (2.113) and (2.114) where replacement of the heavy projectile mass M_1 by m in the denominators implies an increase by a factor of $\gtrsim 2000$ in both $\langle \alpha^2 \rangle$ and the mean number of large-angle scattering events over the length R of the trajectory. This means that the slowing down of electrons and positrons is much like a diffusive motion. This is illustrated in Fig. 2.9 which differs qualitatively from Fig. 1.1 on page 4. The origin of this difference is the long range of an electron or positron compared with that of a heavy particle *at the same energy*, while the scattering probabilities over a given track length are comparable. If this appears puzzling, it might help to note that the stopping force depends on the projectile *speed* while the cross section for angular deflection depends on the *energy*.

Finally, note that collisions with electrons may give rise to substantial angular deflection, unlike what was found for heavy projectiles. Therefore the expression (2.109), valid only at very small scattering angles for heavy projectiles, has a much wider range of validity for electrons or positrons. It is, however, still a small-angle formula.

2.7 Discussion and Outlook

It may be appropriate at this stage to summarize points where the theory as outlined up till now needs improvement.

Most of all, the arguments presented in Sects. 2.3.4 and 2.3.6 indicate that an evaluation of the electronic stopping force on the basis of a classical electron theory is questionable, at least at high velocities and for low-charge projectiles. Bethe's theory, which will be discussed in Chapter 4, avoids several oversimplifications that entered the estimates discussed above. In particular the incorporation of electronic binding, discrete electron states, zero-point motion of electrons, and the interaction between electrons within a target atom or molecule are all effects that at least in principle are taken fully and properly into account in a quantum theory of the interaction between a moving point charge and an atom or molecule.

Next, from the treatment in Sect. 2.3.4 it follows that a penetrating charged particle effectively interacts with target electrons up to a distance equal to Bohr's adiabatic radius, (2.62). Thus, in condensed matter, even at moderately high speed, the projectile may simultaneously interact with a large number of electrons, i.e., the medium may experience substantial polarization. It is to be expected that a proper theory of charged-particle stopping has to account for collective excitations, in particular so for high-speed particles and in dense matter such as solids and liquids.

With increasing projectile speed various relativistic effects cannot be ignored. Although the basic theory is a straight extension of what has been described in Sects. 2.3.3 and 2.3.4 (Bohr, 1915), the topic has been postponed to Chapters 5 and 6 for stopping and Volume II for scattering.

More complex are those effects that originate in the fact that moving particles need not be point charges, especially ions at moderate and low velocities. Indeed, ions with high atomic numbers are usually not fully stripped, and the interaction of an ion carrying an electron cloud with the atoms of the stopping medium can be quite complex, in particular at velocities near or below the orbital velocities of target electrons. The quantum state of a moving ion may undergo changes during a passage, and electron capture and loss processes add to the complexity of the problem. An attempt to discuss some of these problems will be postponed to Volume II when a number of prerequisites will be available.

Nuclear collisions have been treated in a very rough manner in this chapter and mainly from the point of view of its implications to multiple scattering. The reader may have recognized from the discussion in Sect. 2.3.7 that the truncation of nuclear stopping becomes effective at velocities far below those where electronic stopping drops off. This indicates that nuclear stopping dominates at the lowest velocities, but even an order-of-magnitude estimate of this effect requires a more careful discussion of atomic screening. Nuclear scattering and stopping will be studied in Volume II.

Some of the statistical aspects of particle penetration need clarification and specification. The practical limits to Poisson statistics are not clear at this point. In particular it is not obvious to what extent a disordered solid or a liquid can be regarded as a gas at high pressure with respect to the stopping and scattering of ions. Information is needed on the shape of energy-loss and multiple-scattering distributions, and some statistical information on the combined effect of stopping and scattering is desirable. The conditions for separation of electronic from nuclear stopping and scattering are not clear. Finally, the statistics of ion ranges needs to be discussed.

Problems

2.1. The mean number $\langle n \rangle$ of ionizations produced by an alpha particle slowing down to rest from an initial energy E is given with good accuracy by the expression[9] $\langle n \rangle = E/W$, where W is a measure of the energy spent in the creation of an ion pair, i.e., a free electron and a positive ion. W is always greater than the first ionization potential (why?). For atmospheric air, it is empirically given as $W = 29.6$ eV (ICRU, 1979). Find $\langle n \rangle = E/W$ for the case depicted in Fig. 1.1 on page 4 and give a rough estimate of the number of ions produced per micrometer travelled pathlength.

2.2. Repeat the derivation of (2.1–2.3) in the reverse order by starting at (2.3).

2.3. Consult your favored textbooks in classical and quantum mechanics on how they define cross sections and try to reconcile this with the arguments that lead to (2.1–2.3). Most discussions in textbooks refer to particular processes and lead to only one of the three relations.

2.4. Verify the validity of (2.8–2.10) and derive corresponding relations for the cumulants $\overline{(n - \langle n \rangle)^{\nu}}$ for $\nu = 3, 4, 5$.

---

[9] This formula will be derived in Volume III.

2.5. The expression

$$\frac{\left\langle (n - \langle n \rangle)^2 \right\rangle^{1/2}}{\langle n \rangle} \tag{2.119}$$

is sometimes called the relative width of a distribution. The ratio

$$\frac{\left\langle (n - \langle n \rangle)^3 \right\rangle}{\left\langle (n - \langle n \rangle)^2 \right\rangle^{3/2}} \tag{2.120}$$

is called the skewness, and the ratio

$$\frac{\left\langle (n - \langle n \rangle)^4 \right\rangle}{\left\langle (n - \langle n \rangle)^2 \right\rangle^2} \tag{2.121}$$

is called the curtosis. Determine all three quantities for the Poisson distribution.

2.6. Apply a graphics program to generate plots of the Poisson distribution for $0.01 < \langle n \rangle < 100$ at reasonable intervals. Also include gaussian distributions with mean value and variance both given by $\langle n \rangle$ and draw conclusions from the degree of agreement.

2.7. Try to estimate the variance in the number of ionizations made by an alpha particle, referring to Fig. 1.1 and Problem 2.1, assuming Poisson-like distribution. Your result *overestimates* the fluctuation. We shall see in volume III that the distribution is non-poissonian and that the variance is reduced by a numerical factor, the 'Fano factor'.

2.8. Extract an order-of-magnitude estimate for the stopping cross section of an alpha particle from Fig. 1.1 by assuming S to be energy-independent.

2.9. Use (2.36) in conjunction with Fig. 1.1 to estimate the range of an alpha particle in liquid air.

2.10. Extract an order-of-magnitude estimate of Ω_R^2 from Fig. 1.1, use the result of problem 2.9 to extract an order-of-magnitude value of W, and compare your result with the appropriate predictions given in Sect. 2.3.3.

2.11. Determine values of the following quantities in gaussian units,

- an electric field of 1 V/m,
- a voltage of 1 V,
- a magnetic field of 1 T,
- a charge of 1 C and
- a current of 1 A,

2.12. Write down Rutherford's law for the scattering of two point charges on each other in

- SI units,
- gaussian units and
- atomic units.

2.13. Derive a criterion for the range of validity of (2.42) by estimating the pathlength travelled by the hit particle during the collision time τ, assuming constant acceleration in accordance with F_{max}. Require that pathlength to be small compared to the impact parameter.

2.14. You will most likely be unable to find (2.47) in your classical-mechanics text[10]. Instead of the energy transfer T, the running variable is usually the center-of-mass scattering angle Θ. Find the relation between the two variables and the relation between the two corresponding differential cross sections[11].

2.15. Equation (2.45) is an approximate form of an exact relation which, in terms of the center-of-mass scattering angle Θ instead of T, is derived in standard textbooks of classical mechanics. Locate the exact relation in your favored text and use the result of problem 2.14 to transform it into $T = T(p)$. Find out which quantity you have to assume small in order to arrive at (2.45) as a limit, and verify that this is consistent with what has been assumed in the derivation of (2.45).

2.16. Derive (2.47) from the exact relation $T = T(p)$ which you found in problem 2.15 and identify the reason why both the approximate and the exact relation lead to the same differential cross section.

2.17. Verify the validity of each of (2.48 - 2.50) either by consulting a classical-mechanics text or by writing down conservation laws of energy and momentum for a linear collision.

2.18. Referring to Fig. 2.4, try to analyse how the projectile can slow down, i.e., lose momentum in the beam direction, despite the fact that momentum is transferred to the target electron perpendicular to its velocity.

2.19. Make a model to illustrate the adiabatic limit by explicitly evaluating the energy transferred to a linear classical harmonic oscillator by a force $F(t)$ of some adopted shape. Try a) a simple pulse with $F = \mathrm{const}$ over some time interval τ, b) a triangular pulse, c) a lorentzian, and d) a gaussian. Use the Green function of the classical harmonic oscillator. If you are unfamiliar with this concept, consult Appendix A.2.5 first or check (4.6) in chapter 4. Express all results in terms of the transferred momentum $\int_{-\infty}^{\infty} F(t)dt$ to a free electron, define appropriate effective collision times, and compare the results.

---

[10] A notable exception is the book of Landau and Lifshitz (1960).

[11] The reader who has difficulties solving problems 2.14–2.17 may wish to study Chapter 3 first.

2.20. Go carefully through the various steps leading to (2.79). Try to generalize the estimate by adopting an arbitrary relation between the scattering angle $\phi(p)$ of the projectile in the laboratory system and the impact parameter.

2.21. Use three-dimensional graphics to illustrate (2.80). Include the limit expressed by (2.83).

2.22. Check a few texts on quantum theory and atomic physics for what they write about screening. Find expressions for the electron density as a function of the distance from the nucleus. If necessary, take an average over the angular variables. Extract representative values of the screening radius a.

2.23. Devise a simple estimate showing that $\hbar\omega_0$ for typical vibrational frequencies in molecules lies in the range around or below 0.1 eV, as mentioned in the beginning of Sect. 2.3.7.

2.24. Derive (2.85) from the results of Sect. 2.3.6.

2.25. Go explicitly through the steps leading to (2.93) and (2.94) and make sure that no other complications arise from the vectorial nature of the scattering angles than those mentioned in the text.

2.26. The cross sections listed in the end of Sect. 2.4.3 are all divergent at small angles. Verify that these divergencies are equivalent with the $1/T^2$ divergence of the Rutherford cross section expressed as a function of energy transfer. What is it that causes an integral like (2.100) to converge at small angles?

2.27. Verify (2.115).

2.28. Discuss qualitatively possible differences between electrons and positrons regarding scattering and stopping.

References

Bohr N. (1913): On the theory of the decrease of velocity of moving electrified particles on passing through matter. Philos Mag **25**, 10–31

Bohr N. (1915): On the decrease of velocity of swiftly moving electrified particles in passing through matter. Philos Mag **30**, 581–612

Bohr N. (1948): The penetration of atomic particles through matter. Mat Fys Medd Dan Vid Selsk **18 no. 8**, 1–144

Bonderup E. (1981): *Interaction of charged particles with matter.* Institute of Physics, Aarhus

Bragg W.H. and Kleeman R. (1905): On the alpha particles of radium, and their loss of range in passing through various atoms and molecules. Philos Mag **10**, 318–341

Darwin C.G. (1912): A theory of the absorption and scattering of the α rays. Phil Mag (6) **23**, 901–921

Feller W. (1968): *An introduction to probability theory and its applications*, vol. I. J. Wiley and Sons, London

Gentner W., Mayer-Leibnitz H. and Bothe W. (1954): *Atlas of typical expansion chamber photographs*. Pergamon Press, London

Gordon W. (1928): Über den Stoß zweier Punktladungen nach der Wellenmechanik. Z Physik **48**, 180–191

ICRU (1979): *Average energy to produce an ion pair*, vol. 31 of *ICRU Report*. International Commission of Radiation Units and Measurements, Bethesda, Maryland

ICRU (1993): *Stopping powers and ranges for protons and alpha particles*, vol. 49 of *ICRU Report*. International Commission of Radiation Units and Measurements, Bethesda, Maryland

ICRU (2005): *Stopping of ions heavier than helium*, vol. 73 of *ICRU Report*. Oxford University Press, Oxford

Landau L. and Lifshitz E.M. (1960): *Mechanics*, vol. 1 of *Course of theoretical physics*. Pergamon Press, Oxford

Leithäuser G.E. (1904): Über den Geschwindingkeitsverlust, welchen die Kathodenstrahlen beim Durchgang durch dünne Metallschichten erleiden. Ann Physik **15**, 283–306

Lindhard J., Nielsen V. and Scharff M. (1968): Approximation method in classical scattering by screened Coulomb fields. Mat Fys Medd Dan Vid Selsk **36 no. 10**, 1–32

Paul H. (2005): Stopping power graphs. URL `www.exphys.uni-linz.ac.at/stopping/`

Rutherford E. (1911): The scattering of alpha and beta particles by matter and the structure of the atom. Philos Mag **21**, 669–688

Sigmund P. (2000): Stopping power: Wrong terminology? ICRU News **1**, 5–6

Sigmund P. and Schinner A. (2002): Binary theory of electronic stopping. Nucl Instrum Methods B **195**, 64–90

Sklodowska-Curie M. (1900): Sur la penetration des rayons de Becquerel non deviables par le champ magnetique. Compt Rend Acad Sci **130**, 76–79

v. Smoluchowski M. (1904): Über Unregelmäßigkeiten in der Verteilung von Gasmolekülen und deren Einfluß auf Entropie und Zustandssume. In *Festschrift Ludwig Boltzmann*, 626–641

Thomson J.J. (1912): Ionization by moving electrified particles. Philos Mag **23**, 449–457

Williams E.J. (1940): Multiple scattering of fast electrons and alpha-particles, and "curvature" of cloud tracks due to scattering. Phys Rev **58**, 292–306

Williams E.J. (1945): Application of ordinary space-time concepts in collision problems and relation of classical theory to Born's approximation. Rev Mod Phys **17**, 217–226

3

Elastic Scattering

3.1 Introductory Comments

The present chapter serves to summarize some essentials of the theory of binary elastic scattering. Much of this is standard material in courses and textbooks on classical (Goldstein, 1953, Landau and Lifshitz, 1960a, Symon, 1960) and quantum (Schiff, 1981, Merzbacher, 1970, Bransden and Joachain, 2000) mechanics. Therefore the presentation is kept brief but, hopefully, self-contained apart from some mathematical tools. A sizable fraction of the material will not be utilized until much later[1]. The chapter has been placed here because the quantum theory of elastic scattering forms a useful introduction to the treatment of electronic stopping which is to follow in Chapter 4. Also, the theory of elastic scattering is well developed because of its importance in several branches of physics and related fields. A particularly attractive feature is the option to present descriptions in terms of classical and quantal concepts in parallel.

The treatment of scattering and energy loss in the previous chapter was kept simple via the underlying assumption that all interactions between particles were weak. This made it possible to determine the momentum transferred from a projectile to a target particle by time-integrating the force between a projectile in uniform motion and a target particle at rest. This scheme, when valid, allows for a correct description of the simultaneous interaction between a projectile and a number of target particles, i.e., it incorporates features of a many-body treatment.

The scheme breaks down when a projectile interacts strongly with a target particle, e.g., in a wide-angle scattering event of an alpha particle hitting a nucleus, or in an ionization event where an electron is ejected with high kinetic energy. Such violent collisions are typically close encounters. They are comparatively rare because of the small pertinent area (cross section) and therefore do not typically involve more than two particles at a time. This justifies a description in terms of binary collisions.

---

[1] Sections marked with an asterisk ($\star$) can be jumped over during a first reading.

In the absence of forces from other particles the two-body problem may be decomposed into the relative motion of the particles and the motion of their common center of mass, the latter being uniform. Since this decomposition rests on the conservation law of momentum it is equally valid in classical and quantum mechanics. Therefore the relations between relative and center-of-mass velocities on the one hand and velocities in the laboratory system on the other are likewise valid in both classical and quantum mechanics. Conversely, the cross section for a scattering event is governed by the interaction force between the collision partners. Here quantum theory may lead to different predictions.

The question remains whether forces from other particles – which are always present in particle penetration – may be neglected because of lacking significance. A qualitative answer may frequently be found from a consideration of characteristic impact parameters. Another option is a systematic comparison between predictions of binary-scattering theory and either a more comprehensive theory or experimental findings. If the effect of forces due to other particles is not negligible but small, incorporation of their action via perturbation theory may be justified. When such forces cannot be expected to be either negligible or small, numerical simulation is often the most constructive way of theoretical attack.

3.2 Conservation Laws

3.2.1 Laboratory and Center-of-Mass Variables: Nonrelativistic Regime

In elastic scattering the identity of the collision partners and their total kinetic energy is conserved during interaction. For binary scattering the interaction is characterized by a conservative force which vanishes in the limit of infinite distance between the collision partners. This force can be expressed by a potential energy $\mathcal{V}$.

Let two particles with masses m_1, m_2 collide with initial velocities $\boldsymbol{v}_1, \boldsymbol{v}_2$. It is convenient to describe the interaction in a reference frame moving with the center-of-mass velocity

$$\boldsymbol{V} = \frac{m_1}{m_1 + m_2}\boldsymbol{v}_1 + \frac{m_2}{m_1 + m_2}\boldsymbol{v}_2. \tag{3.1}$$

In this frame the collision partners move with individual velocities

$$\boldsymbol{u}_1 = \boldsymbol{v}_1 - \boldsymbol{V} = \frac{m_2}{m_1 + m_2}\boldsymbol{v} \tag{3.2a}$$

$$\boldsymbol{u}_2 = \boldsymbol{v}_2 - \boldsymbol{V} = -\frac{m_1}{m_1 + m_2}\boldsymbol{v}, \tag{3.2b}$$

where $\boldsymbol{v} = \boldsymbol{v}_1 - \boldsymbol{v}_2$ is the relative velocity; their momenta are given by $\pm m_0 \boldsymbol{v}$ where

$$m_0 = \frac{m_1 m_2}{m_1 + m_2} \tag{3.3}$$

is the reduced mass. If the potential energy depends only on the (vectorial or scalar) interparticle distance, i.e., if $V = V(r)$ where $r = r_1 - r_2$, the relative motion is equivalent with the motion of one particle with mass m_0 in a fixed force field $V(r)$.

As a consequence of an elastic scattering event the momentum $m_0 v$ may change in direction but its initial and final magnitude must equal each other. Therefore also the velocities u_1, u_2 of the collision partners in the moving reference frame can only change direction but not magnitude.

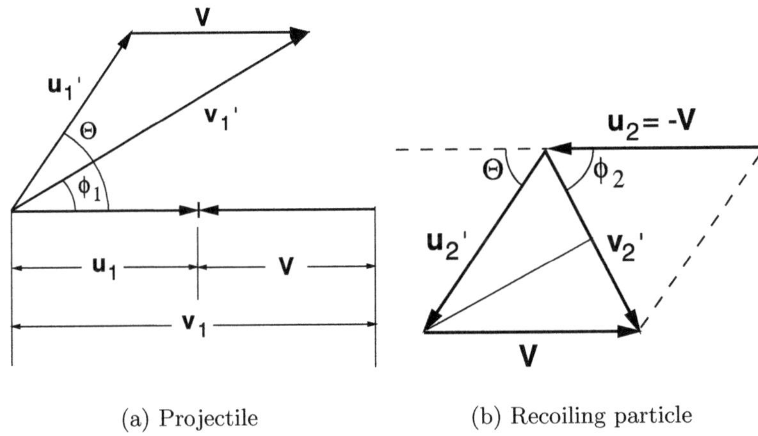

Fig. 3.1. Relation between scattering angles in laboratory and center-of-mass system for target particle initially at rest. For notation see text.

One Collision Partner Initially at Rest

Figure 3.1(a) shows the velocities of particle 1 for the specific case where particle 2 is at rest initially, $v_2 = 0$. A scattering event is characterized by some angle Θ in the center-of-mass frame; in other words, Θ, the *center-of-mass scattering angle*, is the angle between the velocities u_1 and u_1' before and after the collision, respectively. The corresponding velocities in the laboratory frame are given by $v_1 = u_1 + V$ and $v_1' = u_1' + V$, respectively. The angle between v_1 and v_1' is ϕ_1, the scattering angle of particle 1 in the laboratory frame. Figure 3.1(a) shows that

$$\tan \phi_1 = \frac{u_1 \sin \Theta}{u_1 \cos \Theta + V}, \tag{3.4}$$

which reduces to

$$\tan \phi_1 = \frac{\sin \Theta}{\cos \Theta + m_1/m_2} \tag{3.5}$$

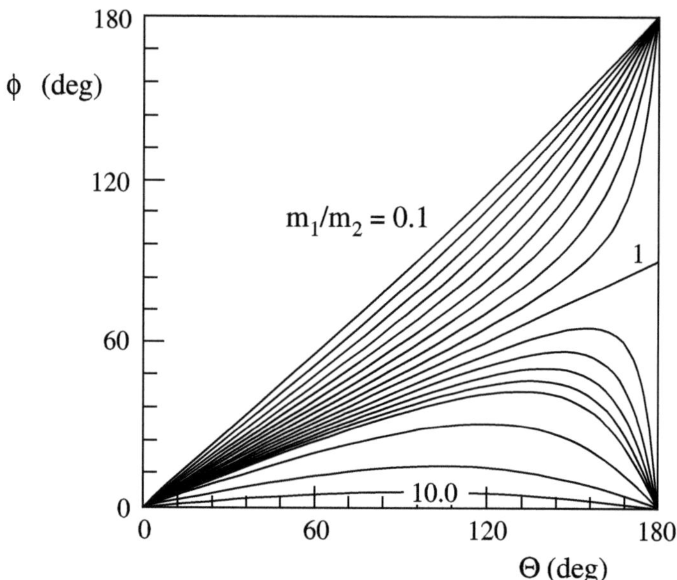

Fig. 3.2. Laboratory scattering angle ϕ_1 *versus* center-of-mass scattering angle Θ. Projectile/target mass ratio m_1/m_2 varying from 0.1 to 1.5 in steps of 0.1. The remaining three curves refer to mass ratios 2, 4, and 10

after insertion of v_1 and V. This relation is illustrated in Fig. 3.2.

It is seen that in the limit of $m_1 \ll m_2$ the difference between laboratory and center-of-mass scattering angle becomes small, $\phi_1 \simeq \Theta$, while for equal masses, $m_1 = m_2$, we get $\phi_1 = \Theta/2$. For $m_1 > m_2$ the scattering angle ϕ_1 has a maximum at $\cos\Theta = -m_2/m_1$ given by

$$\sin\phi_{1,\max} = \frac{m_2}{m_1}. \tag{3.6}$$

For $\phi_1 < \phi_{1,\max}$ two values of Θ belong to every value of ϕ_1.

Figure 3.1(b) shows the corresponding situation for particle 2. Since $u_2 = V$, the two triangles are equilateral and hence $2\phi_2 + \Theta = \pi$ or

$$\phi_2 = \frac{\pi - \Theta}{2}, \tag{3.7}$$

where the *recoil angle* ϕ_2 is the angle between the recoil velocity $\boldsymbol{v}_2'$ and the initial velocity of particle 1.

From Fig. 3.1(b) one deduces that $v_2'/2 = V\sin(\Theta/2)$ or

$$E_2' \equiv T = \frac{m_2}{2}{v_2'}^2 = \gamma E \sin^2\frac{\Theta}{2} \tag{3.8}$$

where $\gamma = 4m_1m_2/(m_1 + m_2)^2$ has been introduced in (2.50).

Laboratory scattering angles can be expressed by the recoil energy T according to

$$\cos \phi_2 = \sqrt{\frac{T}{\gamma E}} \tag{3.9}$$

and

$$\cos \phi_1 = \left(1 - \frac{T}{E}\right)^{-1/2} \left(1 - \frac{m_1 + m_2}{2m_1} \frac{T}{E}\right) \tag{3.10}$$

or

$$\cos \phi_1 = \left(1 - \frac{T}{E}\right)^{1/2} + \frac{1}{2}\left(1 - \frac{m_2}{m_1}\right) \frac{T}{E}\left(1 - \frac{T}{E}\right)^{-1/2}. \tag{3.11}$$

Non-Negligible Target Motion ($\star$)

When the initial motion of the target particle is not negligible, other scattering angles may become of interest such as the angle between the initial and final motion of a target particle and the like. Frequently in this book, the most interesting quantity is the energy transfer T, i.e., the energy lost by particle 1 and gained by particle 2. This can be expressed in terms of transformed velocities,

$$T = \frac{1}{2}m_1(v_1^2 - v_1'^2) = m_1 \boldsymbol{V} \cdot (\boldsymbol{u}_1 - \boldsymbol{u}_1') = m_2 \boldsymbol{V} \cdot (\boldsymbol{u}_2' - \boldsymbol{u}_2), \tag{3.12}$$

from which (3.8) may be recovered for $\mathbf{v}_2 = 0$.

3.2.2 Laboratory and Center-of-Mass Variables: Relativistic Regime

When the velocity of at least one of the collision partners is not small compared to the speed of light it is necessary to generalize the above relationships such as to make them conform with relativistic kinematics. It is then advisable to describe the motion of individual particles in terms of their momenta and energies,

$$\boldsymbol{P}_i = m_i \gamma_i \boldsymbol{v}_i \tag{3.13a}$$
$$E_i = m_i \gamma_i c^2, \qquad (i = 1, 2) \tag{3.13b}$$

where

$$\gamma_i = \frac{1}{\sqrt{1 - v_i^2/c^2}}. \tag{3.14}$$

For elastic collisions the sum of the energies and the vector sum of the momenta are conserved.

Momentum and energy of a particle form a four-vector. This implies that the transformation to a reference frame moving with a velocity $\mathbf{V}$ obeys the Lorentz-transformation

$$\bar{P}_x = \Gamma \left(P_x - \frac{VE}{c^2} \right) \tag{3.15a}$$

$$\bar{P}_y = P_y \tag{3.15b}$$

$$\bar{P}_z = P_z \tag{3.15c}$$

$$\bar{E} = \Gamma \left(E - VP_x \right), \tag{3.15d}$$

where it has been assumed that $\mathbf{V}$ is directed along the x-axis and

$$\Gamma = \frac{1}{\sqrt{1 - V^2/c^2}}. \tag{3.16}$$

This transformation applies to either collision partner.

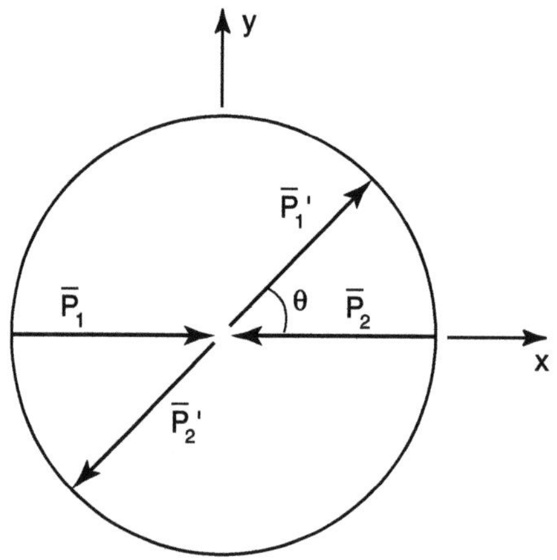

Fig. 3.3. Elastic scattering viewed in the center-of-mass system

One Collision Partner Initially at Rest

Consider now a projectile with rest mass m_1 and velocity $\mathbf{v}_1 = (v_1, 0, 0)$ hitting a target particle with rest mass m_2 and $\mathbf{v}_2 = \mathbf{0}$. The center-of-mass velocity is determined by the requirement that the sum of the transformed momenta, $\bar{P}_{x1} + \bar{P}_{x2}$ be vanishing. This yields

$$V = \frac{P_1 c^2}{E_1 + E_2}. \tag{3.17}$$

After an elastic scattering event the momenta $\bar{P}_{xi}$ and energies $\bar{E}_i$ in the center-of-mass frame have not changed in magnitude but the momenta may have changed direction (Fig. 3.3). Let the scattering plane coincide with the $x - y$ plane and denote post-collision parameters by a prime. Then

$$\bar{P}'_{x1} = \bar{P}_{x1}\cos\Theta\,; \quad \bar{P}'_{y1} = \bar{P}_{x1}\sin\Theta\,; \quad \bar{E}'_1 = \bar{E}_1 \tag{3.18}$$

and

$$\bar{P}'_{x2} = -\bar{P}_{x1}\cos\Theta\,; \quad \bar{P}'_{y2} = -\bar{P}_{x1}\sin\Theta\,; \quad \bar{E}'_2 = \bar{E}_2, \tag{3.19}$$

where Θ is the center-of-mass scattering angle defined in Sect. 3.2.1.

In order to arrive at final velocities in the laboratory frame we need to apply the inverse Lorentz transformation, i.e., (3.15a) with V replaced by $-V$. This yields momenta P'_x, P'_x and an energy E' of the scattered particle from which we find the kinetic energy of the recoiling particle

$$T = E'_2 - E_2 = \frac{2\Gamma^2 P_1^2 c^2 m_2 c^2}{(E_1 + E_2)^2}\sin^2\frac{\Theta}{2} \tag{3.20}$$

as the relativistic extension of (3.8). After elimination of V this reduces to

$$T = T_{\max}\sin^2\frac{\Theta}{2}. \tag{3.21}$$

with

$$T_{\max} = \frac{2m_1^2\gamma_1^2 m_2 v_1^2}{m_1^2 + m_2^2 + 2m_1\gamma_1 m_2} = \frac{2(\gamma_1 + 1)m_1 m_2 E_{\mathrm{kin},1}}{m_1^2 + m_2^2 + 2\gamma_1 m_1 m_2}. \tag{3.22}$$

In the limit of $v_1 \ll c$, this approaches the nonrelativistic limit (2.49), while for $v_1 \sim c$ we find $T_{\max} \sim m_1\gamma_1 c^2$.

The scattering angles may be written in the form

$$\tan\phi_1 = \frac{P'_{y1}}{P'_{x1}} = \frac{\sin\Theta}{\Gamma}\bigg/\left(\cos\Theta + \frac{m_1}{m_2}\frac{m_1 + m_2\gamma_1}{m_1\gamma_1 + m_2}\right) \tag{3.23}$$

and

$$\tan\phi_2 = \frac{|P'_{y2}|}{P'_{x2}} = \frac{1}{\Gamma}\tan\left(\frac{\pi - \Theta}{2}\right) \tag{3.24}$$

which are relativistic extensions of (3.5) and (3.7). After elimination of V the factor Γ eq. (3.16) reads

$$\Gamma = \frac{m_1\gamma_1 + m_2}{\sqrt{m_1^2 + m_2^2 + 2m_1 m_2\gamma_1}}. \tag{3.25}$$

It is seen that Γ becomes large for large γ_1, i.e., v_1 approaching c, and hence both ϕ_1 and ϕ_2 tend to become small in absolute terms at a given center-of-mass scattering angle Θ. This is a characteristic feature of relativistic kinematics.

We shall frequently deal with the case of $m_1 \gg m_2$ where 1 denotes an ion and 2 an electron. Unless γ_1 comes close to or exceeds m_1/m_2, the energy transfer (3.20) reads

$$T = T_{\max} \sin^2 \frac{\Theta}{2} \tag{3.26}$$

with

$$T_{\max} = \frac{2mv_1^2}{1 - v_1^2/c^2}. \tag{3.27}$$

Non-Negligible Target Motion ($\star$)

In case of non-negligible target motion the Lorentz transformation (3.15a) needs to be written in vector form,

$$\bar{\boldsymbol{P}} = \Gamma \left(\boldsymbol{P} - \frac{\boldsymbol{V}E}{c^2} \right) ; \quad \bar{E} = \Gamma (E - \boldsymbol{V} \cdot \boldsymbol{P}) \tag{3.28}$$

with the center-of-mass velocity

$$\boldsymbol{V} = \frac{c^2(\boldsymbol{P}_1 + \boldsymbol{P}_2)}{E_1 + E_2}, \tag{3.29}$$

and the energy transfer reads

$$T = E_2' - E_2 = \Gamma \boldsymbol{V} \cdot (\bar{\boldsymbol{P}}_2' - \bar{\boldsymbol{P}}_2), \tag{3.30}$$

which is the relativistic extension of eq. (3.12).

3.3 Classical Scattering Theory for Central Force

3.3.1 The Scattering Integral

Figure 3.4(a) shows a classical orbit in the center-of-mass frame for the case of repulsive interaction. For a central-force potential $\mathcal{V}(r)$ the angular momentum $\boldsymbol{L} = m_0 \boldsymbol{r} \times \boldsymbol{v}$ is conserved. The magnitude of $\boldsymbol{L}$ is determined by the initial configuration, $|\boldsymbol{L}| = m_0 p v$, where v is the initial speed and p the impact parameter. $\boldsymbol{L}$ is directed into the drawing plane. It follows from angular-momentum conservation that the straight line going through the center of force and the point of closest approach (cf. Fig. 3.4(b)) must be a symmetry axis. That line (or the associated vector) is commonly called the 'apsis'.

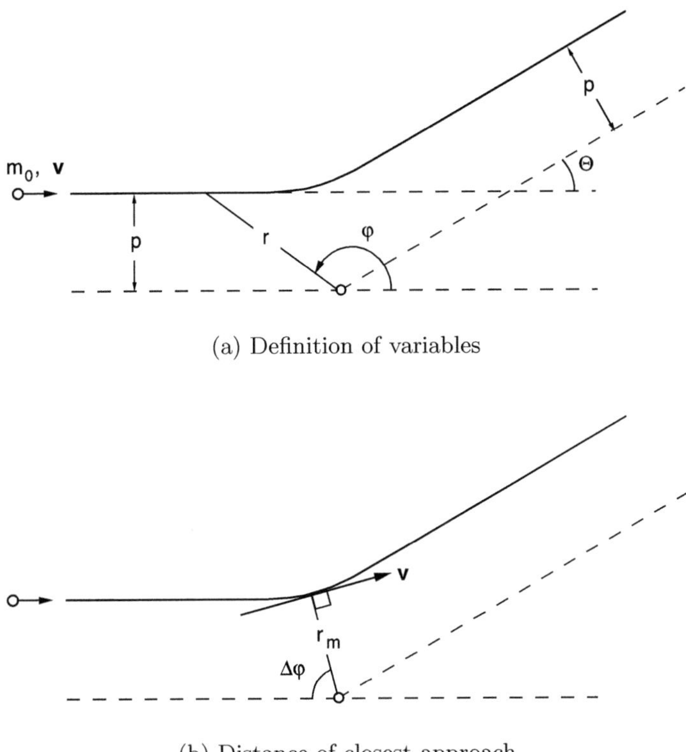

(a) Definition of variables

(b) Distance of closest approach

Fig. 3.4. Classical scattering orbit in center-of-mass system for repulsive interaction (central force)

Introduce polar coordinates $r(t), \varphi(t)$ as functions of time, where r is the distance from the origin (the force center) and φ the angle between the position vector of the moving particle and the initial direction of motion (Fig. 3.4(a)). Then the scattering angle Θ is given by $\Theta = \varphi(\infty)$. In terms of these variables, energy conservation reads

$$\frac{m_0}{2}\left[\left(\frac{\mathrm{d}r}{\mathrm{d}t}\right)^2 + r^2\left(\frac{\mathrm{d}\varphi}{\mathrm{d}t}\right)^2\right] + \mathcal{V}(r) = \frac{m_0}{2}v^2 \equiv E_r \tag{3.31}$$

where m_0 has been defined in (3.3) and E_r, the relative energy, is the initial kinetic energy of the relative motion. Angular momentum conservation reads

$$m_0 r^2 \frac{\mathrm{d}\varphi}{\mathrm{d}t} = -m_0 p v. \tag{3.32}$$

After insertion of $\mathrm{d}\varphi/\mathrm{d}t$ into (3.31) and isolation of $\mathrm{d}r/\mathrm{d}t$ one finds

$$\frac{\mathrm{d}\varphi}{\mathrm{d}r} = \frac{\mathrm{d}\varphi/\mathrm{d}t}{\mathrm{d}r/\mathrm{d}t} = \pm\frac{p}{r^2}\left(1 - \frac{\mathcal{V}(r)}{E_r} - \frac{p^2}{r^2}\right)^{-1/2}. \tag{3.33}$$

The sign can be determined by inspection of Fig. 3.4(a): Initially we have $r = \infty$, $\mathrm{d}\varphi/\mathrm{d}t < 0$ and $\mathrm{d}r/\mathrm{d}t < 0$ and hence $\mathrm{d}r/\mathrm{d}\varphi > 0$, i.e., the positive sign, but after passage of the apsis the sign of $\mathrm{d}r/\mathrm{d}t$ changes and hence $\mathrm{d}r/\mathrm{d}\varphi < 0$. This holds for both repulsive and attractive interaction. Integration yields the angle $\Delta\varphi$ indicated in Fig. 3.4(b),

$$\Delta\varphi = \int_{r_m}^{\infty} \mathrm{d}r \frac{p}{r^2}\left(1 - \frac{\mathcal{V}(r)}{E_r} - \frac{p^2}{r^2}\right)^{-1/2}, \tag{3.34}$$

where r_m, the distance of closest approach, is determined by the zero of the square root, i.e., the point where the radial velocity $\mathrm{d}r/\mathrm{d}t$ vanishes. This yields the scattering angle

$$\Theta = \pi - 2\Delta\varphi \tag{3.35}$$

in the center-of-mass system.

The integral in (3.34) can be evaluated analytically for a number of simple potential functions including the Coulomb interaction. For the latter, however, a more elegant derivation of the scattering law is possible which is given in the next section. In general (3.34) is evaluated numerically for a given potential. Printed tabulations exist (Robinson, 1970), and subroutines evaluating this scattering integral are built into computer codes simulating sequences of binary collisions (Robinson and Torrens, 1974). Approximation methods for analytical evaluation have been summarized by Leibfried (1965). A particularly elegant treatment of scaling relations in scattering from screened-Coulomb potentials was given by Lindhard et al. (1968), to be described in Volume II.

3.3.2 Runge-Lenz Vector and Rutherford's Law

For the specific case of Coulomb interaction (Kepler motion and Rutherford scattering) there exists one more constant of motion in addition to energy and angular momentum. The Runge-Lenz vector[2] $\boldsymbol{M}$ is defined by the relation

$$\boldsymbol{M} = \boldsymbol{v} \times \boldsymbol{L} + e_1 e_2 \frac{\boldsymbol{r}}{r}. \tag{3.36}$$

---

[2] The vector defined by (3.36) is commonly called the Runge-Lenz or Laplace vector. It appears clear that neither Runge nor Lenz claimed any priority on this particular invariant, the history of which has been exposed in two illuminating notes by Goldstein (1975, 1976). The present derivation of Rutherford's law has been published by Basano and Bianchi (1980) but became known to the author already in a lecture by G. Leibfried in 1960.

Its time derivative reads

$$\frac{\mathrm{d}\boldsymbol{M}}{\mathrm{d}t} = \frac{\mathrm{d}\boldsymbol{v}}{\mathrm{d}t} \times \boldsymbol{L} + e_1 e_2 \frac{\boldsymbol{v}}{r} - e_1 e_2 \boldsymbol{r} \frac{\mathrm{d}r/\mathrm{d}t}{r^2} \tag{3.37}$$

since $\mathrm{d}\boldsymbol{L}/\mathrm{d}t = \boldsymbol{0}$. Insertion of the force equation, $\mathrm{d}\boldsymbol{v}/\mathrm{d}t = (1/m_0)e_1 e_2 \boldsymbol{r}/r^3$ and execution of the triple vector product

$$\frac{\mathrm{d}\boldsymbol{v}}{\mathrm{d}t} \times \boldsymbol{L} = \frac{\mathrm{d}\boldsymbol{v}}{\mathrm{d}t} \times (m_0 \boldsymbol{r} \times \boldsymbol{v}) = \frac{e_1 e_2}{r^3}\left[(\boldsymbol{r} \cdot \boldsymbol{v})\boldsymbol{r} - r^2 \boldsymbol{v}\right] \tag{3.38}$$

leads to

$$\frac{\mathrm{d}\boldsymbol{M}}{\mathrm{d}t} = 0, \tag{3.39}$$

i.e., the Runge-Lenz vector is a constant of motion.

In order to utilize this property in Coulomb scattering consider the point of closest approach where

$$\boldsymbol{v} \times \boldsymbol{L} = m_0 \boldsymbol{v} \times (\boldsymbol{r} \times \boldsymbol{v}) = m_0 \left(v^2 \boldsymbol{r} - (\boldsymbol{v} \cdot \boldsymbol{r})\boldsymbol{v} \right). \tag{3.40}$$

The second term in the parentheses vanishes since $\boldsymbol{v} \cdot \boldsymbol{r} = 0$ at closest approach (Fig. 3.4(b)). It follows that the two terms making up $\boldsymbol{M}$ in (3.36) both point in the direction of $\boldsymbol{r}$ which, at closest approach, points in the direction of the apsis. Since $\boldsymbol{M}$ is a constant of motion, this implies that the Runge-Lenz vector must be directed along the apsis at all times. Thus, knowledge of the direction of $\boldsymbol{M}$ is synonymous with knowing the direction of the apsis and, hence, the scattering angle.

In order to find $\boldsymbol{M}$ for a given initial condition we note that at $t = -\infty$ the two terms in (3.36) have components $M_\perp$ and $M_\parallel$ perpendicular and parallel to $\boldsymbol{v}$, respectively. Their magnitudes are

$$M_\perp = m_0 v^2 p; \qquad M_\parallel = -e_1 e_2. \tag{3.41}$$

From this we find the scattering angle utilizing (3.35),

$$\tan \frac{\Theta}{2} = \frac{-M_\parallel}{M_\perp} = \frac{b}{2p}, \tag{3.42}$$

where

$$b = 2e_1 e_2/m_0 v^2 \tag{3.43}$$

is the 'collision diameter'. For repulsive interaction, b is the distance of closest approach in a central collision, as follows from energy conservation, $e_1 e_2/b = m_0 v^2/2$.

Equation (3.42) expresses the functional relation between scattering angle and impact parameter for Coulomb scattering. This relation could also have been found from (3.35) and (3.34) by insertion of the Coulomb potential and integration (cf. problem 3.6).

Inversion of (3.42), $p^2 = (b/2)^2 / \tan^2(\Theta/2)$, and differentiation yields

$$\mathrm{d}\sigma = |\pi dp^2| = \left(\frac{b}{4}\right)^2 \frac{2\pi \sin\Theta \, \mathrm{d}\Theta}{\sin^4(\Theta/2)}, \tag{3.44}$$

where $2\pi \sin\Theta \, \mathrm{d}\Theta \equiv \mathrm{d}\Omega$ is a circular element of solid angle. Because of azimuthal symmetry we may write instead

$$\mathrm{d}\sigma = \left(\frac{b}{4}\right)^2 \frac{\mathrm{d}^2\Omega}{\sin^4(\Theta/2)}, \tag{3.45}$$

where $\mathrm{d}^2\Omega$ is an arbitrary element of solid angle. This is Rutherford's scattering law in its standard form.

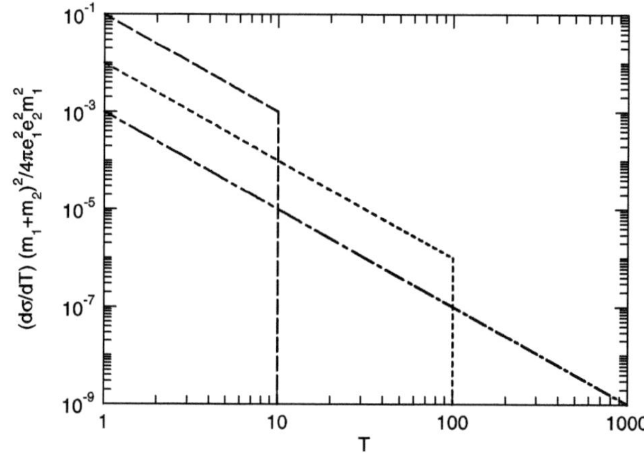

Fig. 3.5. Energy dependence of differential cross section for Coulomb scattering, (3.47). Arbitrary units. For $\gamma E = 1000$, 100, 10 (right to left)

An alternative version, important for stopping problems and radiation effects, may be obtained by introduction of the recoil energy. From (3.8) and (3.42) one arrives at Thomson's formula[3]

$$T = \frac{\gamma E}{1 + (2p/b)^2}. \tag{3.46}$$

---

[3] In relativity the symbol γ is the standard notation for the Lorentz factor $1/\sqrt{1 - v^2/c^2}$, and in nonrelativistic scattering theory it denotes the energy transfer efficiency $4m_1 m_2/(m_1 + m_2)^2$, (2.50). To minimize confusion, the Lorentz factor is here denoted by γ_i, cf. (3.14).

Inversion and differentiation yield

$$\mathrm{d}\sigma = \pi \left(\frac{b}{2}\right)^2 \gamma E \frac{\mathrm{d}T}{T^2} = 2\pi \frac{e_1^2 e_2^2}{m_2 v^2}\frac{\mathrm{d}T}{T^2}, \qquad 0 < T \le \gamma E \tag{3.47}$$

in complete agreement with (2.47). Note that the derivation of (2.47) assumed the scattering angle to be small. That limitation has now evaporated. Figure 3.5 illustrates one of the main features of Rutherford's law: The magnitude of the cross section decreases with increasing beam energy, while the maximun energy transfer increases.

The most outstanding features of Rutherford's law are the strong singularity at small angles (or energy transfers) and the small magnitude of the cross section for violent events such as scattering at an angle $\Theta > \pi/2$ or energy loss $T > \gamma E/2$. This is in pronounced contrast to, e.g., the scattering on an impenetrable sphere of radius a (problem 3.6) where

$$\mathrm{d}\sigma = \left(\frac{a}{2}\right)^2 \mathrm{d}^2\boldsymbol{\Omega} = \pi a^2 \mathrm{d}T/\gamma E. \tag{3.48}$$

3.3.3 Scaling Relations

According to the Bohr criterion, (2.80), classical scattering theory is an appropriate tool in the description of the scattering of heavy particles (ions and/or atoms) on each other. Such particles will typically be taken to interact via some screened-Coulomb potential of the form

$$\mathcal{V}(r) = \frac{Z_1 Z_2 e^2}{r}\Phi(\frac{r}{a}), \tag{3.49}$$

where $\Phi(r/a)$ is a function of the internuclear distance r and a some screening radius. Ways of determining Φ and a will be discussed in Volume II. Here some useful scaling properties will be mentioned, following Lindhard et al. (1968).

Insertion of (3.49) into (3.34) leads to

$$\Theta = \pi - 2p \int_{r_m}^{\infty} \frac{\mathrm{d}r}{r^2}\left[1 - \frac{Z_1 Z_2 e^2}{r E_r}\Phi\left(\frac{r}{a}\right) - \frac{p^2}{r^2}\right]^{-1/2}. \tag{3.50}$$

This can be expressed in terms of the two dimensionless variables p/a and $\varepsilon = a E_r / Z_1 Z_2 e^2$,

$$\Theta = \Theta(\varepsilon, p/a). \tag{3.51}$$

Eq. (3.51) represents a scaling relation for the scattering angle. Note that the parameter ε is identical with the one introduced in (2.88). Inversion of (3.51) in conjunction with (3.8) shows that p/a is a function of $T/\gamma E$ and ε, and hence

$$\mathrm{d}\sigma = \pi |\mathrm{d}\,p^2| = \pi a^2 g\left(\frac{T}{\gamma E}, \varepsilon\right) \mathrm{d}\left(\frac{T}{\gamma E}\right). \tag{3.52}$$

With this we find the following scaling relationships for the stopping cross section and straggling parameter for elastic scattering by screened-Coulomb interaction,

$$S = \int T\mathrm{d}\sigma = \pi a^2 \gamma E\, g_S(\varepsilon) \tag{3.53}$$

and

$$W = \int T^2 \mathrm{d}\sigma = \pi a^2 (\gamma E)^2\, g_W(\varepsilon) \tag{3.54}$$

where the dimensionless functions $g_S(\varepsilon)$ and $g_W(\varepsilon)$ are determined by the functional behavior of $\Phi(r/a)$.

From (3.53) we find the stopping force $dE/dx = NxS(E)$ which may be written in the form

$$\frac{\mathrm{d}\varepsilon}{\mathrm{d}\rho} = \varepsilon g_S(\varepsilon) \tag{3.55}$$

with the dimensionless length variable

$$\rho = N\pi a^2 \gamma x. \tag{3.56}$$

3.3.4 Time Integral ($\star$)

The transformations derived in Sect. 3.2.1 provide all information necessary to relate *velocity* vectors in the center-of mass frame of reference to those in the lab frame. In some applications of scattering theory the need occurs to reconstruct a classical trajectory in the laboratory system. This requires coordination of the clocks governing relative and center-of-mass motion, since the relation between time and pathlength is different in the two reference frames. The center of mass moves uniformly. Complications arise from the relative motion where the velocity does not only change direction but also its magnitude during the interaction.

From (3.31) and (3.32) on page 75 one finds

$$v\mathrm{d}t = \pm \frac{\mathrm{d}r}{\sqrt{1 - \mathcal{V}(r)/E_r - p^2/r^2}} \tag{3.57}$$

or, after integration,

$$v(t_f - t_i) = \left(\int_{r_m}^{r_i} + \int_{r_m}^{r_f} \right) \frac{dr}{\sqrt{1 - \mathcal{V}(r)/E_r - p^2/r^2}}, \tag{3.58}$$

where r_i is the distance between the colliding particles at some time t_i before, and r_f at some time t_f after the interaction, and r_m the distance of closest approach, determined by the zero of the radicand. If r_f and r_i are both large enough so that $\mathcal{V}(r_f)$ and $\mathcal{V}(r_i)$ are effectively zero, the relation may be rewritten in the form

$$v(t_f - t_i) = 2 \int_{r_m}^{\infty} dr \left(\frac{1}{\sqrt{1 - \mathcal{V}(r)/E_r - p^2/r^2}} - \frac{1}{\sqrt{1 - p^2/r^2}} \right)$$

$$+ \left(2 \int_{r_m}^{\infty} - \int_{r_i}^{\infty} - \int_{r_f}^{\infty} \right) \frac{dr}{\sqrt{1 - p^2/r^2}} \tag{3.59}$$

or, after integration,

$$r_f + r_i = v(t_f - t_i) + 2\tau \tag{3.60}$$

where

$$\tau = \sqrt{r_m^2 - p^2}$$

$$- \int_{r_m}^{\infty} dr \left(\frac{1}{\sqrt{1 - \mathcal{V}(r)/E_r - p^2/r^2}} - \frac{1}{\sqrt{1 - p^2/r^2}} \right) \tag{3.61}$$

is called the *time integral.*

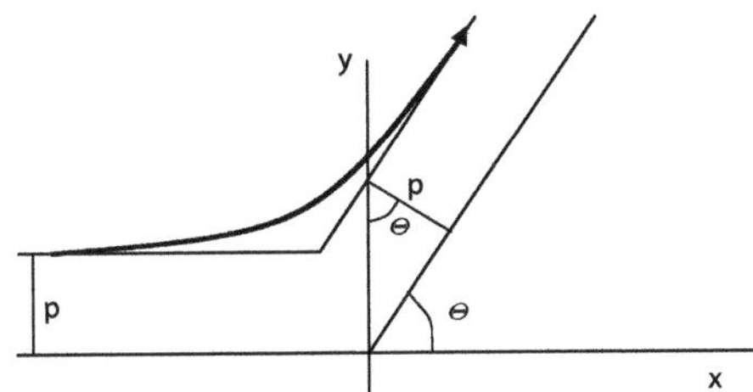

Fig. 3.6. Asymptotic orbits in center-of-mass frame

Now, let the initial motion (for $t = -\infty$) be given by

$$x = vt \tag{3.62a}$$

$$y = p. \tag{3.62b}$$

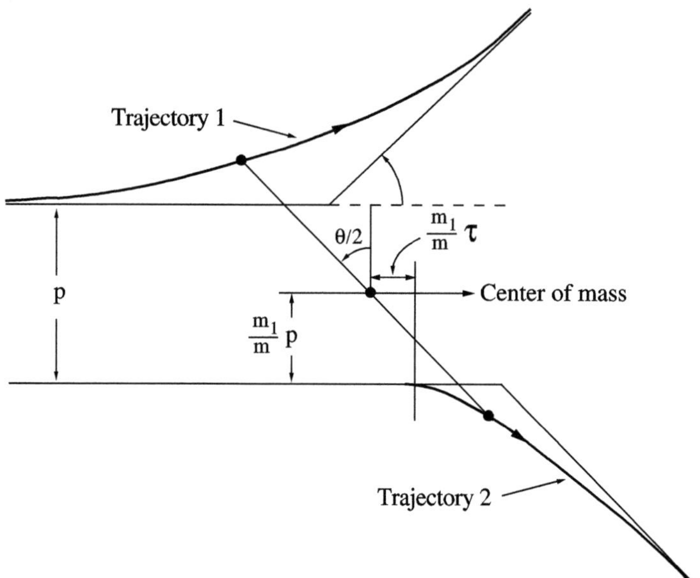

Fig. 3.7. Geometrical significance of the time integral. From Robinson (1970)

Then, (3.60) reduces to

$$r = vt + 2\tau \tag{3.63}$$

if both $t_f = t$ and t_i are far outside the interaction region, so the final trajectory may be written in the form

$$x = (vt + 2\tau)\cos\Theta - p\sin\Theta \tag{3.64a}$$
$$y = (vt + 2\tau)\sin\Theta + p\cos\Theta. \tag{3.64b}$$

Here, the terms containing p enter because the orbit must comply with angular-momentum conservation (Fig. 3.6, problem 3.9). With this, as well as the transformation (3.2a) trajectories may be located in the laboratory system, cf. problem 3.10.

Figure 3.7 illustrates the two trajectories, the positions of the particles being taken at closest approach, i.e., at the time $t_m = -\tau/v$.

3.3.5 Relativistic Scattering Integral ($\star$)

The relativistic extension of the treatment presented in Sect. 3.3.1 starts from the law of energy conservation

$$\sqrt{P^2c^2 + (mc^2)^2} + \mathcal{V}(r) = E, \tag{3.65}$$

where E and m denote the total energy and rest mass of a projectile interacting with an infinitely heavy target particle. Introduction of planar polar coordinates leads to

$$L = m\gamma_1 r^2 \frac{\mathrm{d}\varphi}{\mathrm{d}t} = -m\gamma_1 pv = \text{constant} \tag{3.66}$$

and

$$P^2 = \left(m\gamma_1 \frac{\mathrm{d}r}{\mathrm{d}t}\right)^2 + \frac{L^2}{r^2} \tag{3.67}$$

so that

$$m\gamma_1 \frac{\mathrm{d}r}{\mathrm{d}t} = \pm\sqrt{\frac{1}{c^2}[E - \mathcal{V}(r)]^2 - (mc)^2 - L^2/r^2}. \tag{3.68}$$

Combination of (3.66) with (3.68) yields

$$\frac{\mathrm{d}\varphi}{\mathrm{d}r} = \pm\frac{L}{r^2\sqrt{\frac{1}{c^2}[E - \mathcal{V}(r)]^2 - (mc)^2 - L^2/r^2}}, \tag{3.69}$$

which leads to (3.35) with

$$\Delta\varphi = L \int_{r_m}^{\infty} \frac{\mathrm{d}r}{r^2\sqrt{[E - \mathcal{V}(r)]^2/c^2 - (mc)^2 - L^2/r^2}}. \tag{3.70}$$

The integral can be carried out for Coulomb scattering, $\mathcal{V}(r) = e_1 e_2/r$ with the result (Landau and Lifshitz, 1971)

$$\Theta = \pi - \frac{2Lc}{\sqrt{L^2 c^2 - e_1^2 e_2^2}} \arctan \frac{v\sqrt{L^2 c^2 - e_1^2 e_2^2}}{e_1 e_2 c}. \tag{3.71}$$

Such a classical calculation can only be expected to be meaningful for $L \gg \hbar$, i.e., for

$$\frac{L^2 c^2}{e_1^2 e_2^2} \gg \left(\frac{137 e^2}{e_1 e_2}\right)^2. \tag{3.72}$$

Unless both collision partners have high charges, the square root in eq. (3.71) will be close to Lc and, hence,

$$\Theta \simeq \pi - 2 \arctan \frac{vL}{e_1 e_2} = \pi - 2 \arctan \frac{2p}{b} \tag{3.73}$$

with

$$b = \frac{2 e_1 e_2}{\gamma_1 m v^2}. \tag{3.74}$$

Following the procedure for evaluating (3.42) in Sect. 3.3.2, you readily red-erive (3.45) with the sole change that the collision diameter b now is defined by eq. (3.74). Thus, the dependence on scattering angle of the Rutherford

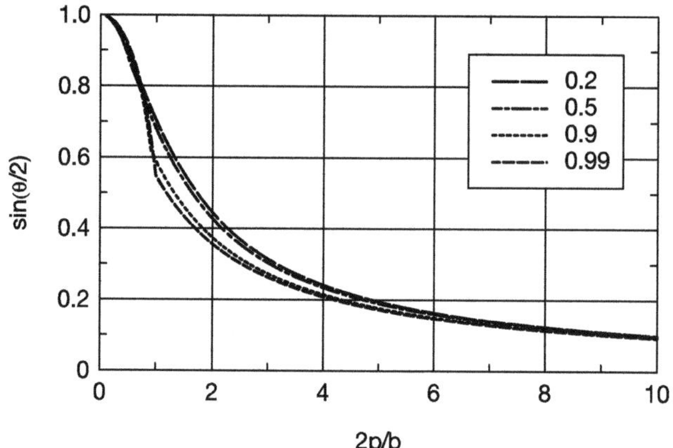

Fig. 3.8. Relativistic Coulomb scattering for $v/c = 0.2, 0.5, 0.9, 0.99$, according to (3.71)

spectrum is not changed by relativistic dynamics, except for deviations at small impact parameters, where the classical calculation cannot be expected to be valid. Figure 3.8 shows a graph for v/c ranging from 0.2 to 0.99. Curves for $v/c < 0.2$ coincide within the resolution of the graph.

In the special case of $m_1 \gg m_2$, you can easily verify (problem 3.13) that

$$d\sigma(T) = \frac{2\pi e_1^2 e_2^2}{mv^2} \frac{\mathrm{d}T}{T^2}; \quad T \leq T_{\max} \tag{3.75}$$

with $T_{\max}$ given by (3.27).

3.3.6 Perturbation Theory ($\star$)

The principle of classical perturbation theory has been illustrated in Sect. 2.3.2: Solve the equation of motion in the absence of the perturbation, evaluate the force resulting from the perturbation, assuming unperturbed motion, and find the momentum transfer as a function of time. This scheme, which provides a first-order estimate of the effect of a perturbing force, is applicable to a wide variety of single- and many-body systems. In scattering theory it goes under the name of impulse or momentum approximation.

The scheme can be extended to higher orders by iteration. This is not attractive in general, since such iteration needs to be done numerically, in which case a straight numerical solution of the problem is to be preferred. However, a higher-order estimate may be desirable in attempts to determine the range of validity of first-order perturbation theory. Such a scheme was developed for binary scattering by Lehmann and Leibfried (1963). Starting from the scattering integral (3.35) and (3.34) one finds

$$\Theta = -\frac{1}{pE_r} \int_p^\infty \frac{\mathrm{d}r}{\sqrt{1-p^2/r^2}} \frac{\mathrm{d}}{\mathrm{d}r}[r\mathcal{V}(r)]$$

$$-\frac{1}{4pE_r^2} \int_p^\infty \frac{r\mathrm{d}r}{\sqrt{r^2-p^2}} \frac{\mathrm{d}^2}{\mathrm{d}r^2} r^2\mathcal{V}(r)^2 \ldots . \quad (3.76)$$

You may find the derivation in Appendix A.3.1, where also the perturbation expansion of the time integral is described.

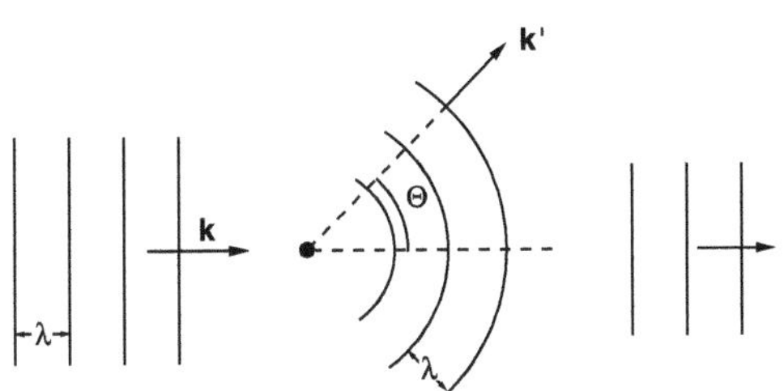

Fig. 3.9. Elastic scattering of an incident plane wave

3.4 Quantum Theory of Elastic Scattering

In quantum mechanics we cannot at the same time operate with a well-defined impact parameter and a well collimated particle current. While a well-defined impact parameter implies the position perpendicular to the beam to be accurately defined, a good collimation implies negligible momentum in the transverse direction. Combining the two requirements must generate a conflict with the uncertainty principle which may be tolerable so long as the Bohr parameter κ, (2.80) is $\gg 1$. If that condition is not satisfied, a classical trajectory will not be a valid starting point in general.

Since the particle current is measurable while the impact parameter is usually not, it is tempting not to introduce an impact parameter at all and instead to operate with an incident plane wave. This ensures a well-defined energy and direction of incidence (Fig. 3.9).

3.4.1 Laboratory and Center-of-Mass Variables

Only the relative motion will be considered in the following sections. The transformation to laboratory variables is achieved by the relations

$$\boldsymbol{r}_1 = \boldsymbol{R} + \frac{m_2}{m}\boldsymbol{r}, \qquad \boldsymbol{r}_2 = \boldsymbol{R} - \frac{m_1}{m}\boldsymbol{r} \qquad (3.77)$$

and

$$\boldsymbol{p}_1 = \frac{m_1}{m}\boldsymbol{P} + \boldsymbol{p}, \qquad \boldsymbol{p}_2 = \frac{m_2}{m}\boldsymbol{P} - \boldsymbol{p}, \tag{3.78}$$

where $\boldsymbol{R}$ and $\boldsymbol{P}$ denote center-of-mass position and momentum operator, respectively, and $\boldsymbol{r}$ and $\boldsymbol{p}$ characterize the relative motion. If the potential depends only on the relative coordinate $\boldsymbol{r}$, Schrödinger's equation can be separated in these variables because the splitting of the initial kinetic energy according to

$$\frac{\boldsymbol{p}_1^2}{2m_1} + \frac{\boldsymbol{p}_2^2}{2m_1} = \frac{\boldsymbol{P}^2}{2m} + \frac{\boldsymbol{p}^2}{2m_0}, \tag{3.79}$$

is equally valid for classical momenta and quantal momentum operators.

3.4.2 Scattering Amplitude and Differential Cross Section

From now on we address the relative motion. The pertinent energy is $E_r = m_0 v^2/2$, where v is the initial relative speed, just as in the classical case discussed in Sect. 3.3.1.

With a well-defined incident energy, quantal scattering may be treated as a stationary problem, i.e., the time-dependent factor in the wave function, $\exp(-iE_r t/\hbar)$, need not be written up explicitly. An incident plane wave is then described by the spatial part of the wave function,

$$\psi^{(0)}(\boldsymbol{r}) = A e^{i\boldsymbol{k}\cdot\boldsymbol{r}} \equiv A e^{ikx} \tag{3.80}$$

in a coordinate system where the x-axis coincides with the direction of the incident beam. This represents an incident density of particle current $J^{(\mathrm{in})} = |A|^2 \hbar k/m_0$, and the beam energy per particle is given by $E_r = \hbar^2 k^2/2m_0$.

The problem is, then, to construct a solution of Schrödinger's equation for a given potential with the boundary condition that at $x = -\infty$ that solution should coincide with the incident plane wave (Fig. 3.9). The behavior of that solution in an arbitrary direction and far away from the scattering center will then reflect the scattered intensity and thus allow determination of the scattering cross section.

A general solution of Schrödinger's equation representing the scattering of a beam from a potential is presented in Appendix A.2.5 on page 394. Following (A.104) we write it in the form

$$\psi(\boldsymbol{r}) = \psi^{(0)}(\boldsymbol{r}) - \frac{m_0}{2\pi\hbar^2} \int \mathrm{d}^3 r' \frac{e^{\pm ik|\boldsymbol{r} - \boldsymbol{r}'|}}{|\boldsymbol{r} - \boldsymbol{r}'|} \mathcal{V}(\boldsymbol{r}')\psi(\boldsymbol{r}'). \tag{3.81}$$

This is often called the *Lippmann-Schwinger equation*. Assume the scattering potential $\mathcal{V}(\boldsymbol{r})$ to be vanishing outside some sphere of radius R around the center of force, and consider (3.81) in the limit of $r \gg R$, i.e., $r \gg r'$. Then, in that region, we may expand

$$|\boldsymbol{r} - \boldsymbol{r'}| \simeq r - \frac{\boldsymbol{r} \cdot \boldsymbol{r'}}{r} \tag{3.82}$$

and obtain

$$\psi(\boldsymbol{r}) = \psi^{(0)}(\boldsymbol{r}) - \frac{m_0}{2\pi\hbar^2} \frac{e^{\pm ikr}}{r} \int d^3 r' e^{\mp ik\boldsymbol{\Omega} \cdot \boldsymbol{r'}} \mathcal{V}(\boldsymbol{r'})\psi(\boldsymbol{r'}), \tag{3.83}$$

to the leading order, where $\boldsymbol{\Omega} = \boldsymbol{r}/r$ is a unit vector pointing in the direction of observation. Details of this expansion may be studied in problem 3.15.

For clarity include the time dependence for a moment which was omitted above. Then (3.83) may be written in the form

$$\psi(\boldsymbol{r},t) = A e^{i(kx - E_r t/\hbar)} + A f(\boldsymbol{\Omega}) \frac{e^{i(\pm kr - E_r t/\hbar)}}{r} \tag{3.84}$$

The second term represents a quasi-spherical wave with an amplitude $A f(\boldsymbol{\Omega})$ dependent on direction,

$$A f(\boldsymbol{\Omega}) = -\frac{m_0}{2\pi\hbar^2} \int d^3 r' e^{\mp i\boldsymbol{k'} \cdot \boldsymbol{r'}} \mathcal{V}(\boldsymbol{r'})\psi(\boldsymbol{r'}), \tag{3.85}$$

where $\boldsymbol{k'} = k\boldsymbol{\Omega}$. In conjunction with the time dependence it is seen that the upper sign represents an outgoing wave in accordance with the physical situation, while the lower sign describes a quasi-spherical wave incoming from all directions. That solution must be rejected in the present context.

The outgoing current density in a direction $\boldsymbol{\Omega}$ is then determined by $\boldsymbol{J}^{(\text{out})} = |Af(\boldsymbol{\Omega})/r|^2 \hbar k \boldsymbol{\Omega}/m_0$. The cross section for scattering of particles into an angular interval $(\boldsymbol{\Omega}, d^2\boldsymbol{\Omega})$ emerges then from

$$d\sigma(\boldsymbol{\Omega}) = \frac{J^{(\text{out})} r^2 d^2\boldsymbol{\Omega}}{J^{(\text{in})}} = |f(\boldsymbol{\Omega})|^2 d^2\boldsymbol{\Omega} \tag{3.86}$$

according to (2.1).

This specifies the main task in quantal scattering for elastic collisions: Find a solution of Schrödinger's equation for the potential under consideration and make sure that the asymptotic behavior at large distances contains the incoming plane wave. Then the asymptotic behavior should also show an outgoing wave of quasi-spherical form. Write the asymptotic wave function in the form of (3.84) and extract the function $f(\boldsymbol{\Omega})$ which determines the differential cross section according to (3.86).

There are some open ends here: Firstly, the particle current following from (3.81) contains three terms, namely $J^{(\text{in})}$, $J^{(\text{out})}$, and an interference term. Secondly, any real scattering experiment deals with a beam of finite dimension and finite duration. The first problem evaporates if the second is solved. Indeed, if an incident plane wave is replaced by a beam of finite width, the outgoing wave at any nonvanishing angle, however small, can be observed sufficiently far out so there is no interference with the incoming wave. The second problem finds its solution in a suitable wave packet treatment. The interested reader is referred to Merzbacher (1970) Chapter 11 for a comprehensive solution.

A third problem concerns the assumption made above that the scattering potential vanishes outside some radius R. This restriction turns out to be too strong in the important case of Coulomb interaction which, therefore, needs separate attention.

3.4.3 Born Approximation

The Born approximation has been designed to describe scattering under conditions of 'weak interaction'. Its basis is a series expansion of the wave function in powers of the scattering potential $\mathcal{V}(\boldsymbol{r})$,

$$\psi(\boldsymbol{r}) = \psi^{(0)}(\boldsymbol{r}) + \psi^{(1)}(\boldsymbol{r}) + \psi^{(2)}(\boldsymbol{r}) \dots, \tag{3.87}$$

where $\psi^{(0)}(\boldsymbol{r})$ characterizes the incident wave which would be present also in the absence of a scattering potential, and $\psi^{(\nu)}(\boldsymbol{r})$ is of ν'th order in $\mathcal{V}$, i.e., $\propto (e_1 e_2)^\nu$ in case of Coulomb interaction. In the theory of elastic scattering the 'Born series' (3.87) is rarely carried on beyond the first or perhaps the second term.

The expansion (3.87) may be inserted into the integrated version (3.83) of the wave equation. Collection of terms linear in the scattering potential yields

$$\psi^{(1)}(\boldsymbol{r}) = -\frac{m_0}{2\pi\hbar^2} \frac{\mathrm{e}^{ikr}}{r} \int \mathrm{d}^3 r' \mathrm{e}^{-ik\boldsymbol{\Omega}\cdot\boldsymbol{r'}} \mathcal{V}(\boldsymbol{r'})\psi^{(0)}(\boldsymbol{r'}), \tag{3.88}$$

where the important point is the replacement of the unknown function $\psi(\boldsymbol{r'})$ by the incident wave $\psi^{(0)}(\boldsymbol{r'})$ in the integrand. At large distances $r \gg R > r'$, this may be written in the form of (3.84) with

$$f(\boldsymbol{\Omega}) = -\frac{m_0}{2\pi\hbar^2} \langle \boldsymbol{k'}|\mathcal{V}|\boldsymbol{k} \rangle \tag{3.89}$$

and

$$\langle \boldsymbol{k'}|\mathcal{V}|\boldsymbol{k} \rangle = \int \mathrm{d}^3 r' \mathrm{e}^{-i\boldsymbol{k'}\cdot\boldsymbol{r'}} \mathcal{V}(\boldsymbol{r'})\mathrm{e}^{i\boldsymbol{k}\cdot\boldsymbol{r'}} \tag{3.90}$$

in Dirac notation, where $\boldsymbol{k'} = k\boldsymbol{\Omega}$ is a wave vector at some chosen direction of observation.

Eq. (3.89) represents the elastic-scattering amplitude in the first Born approximation. In this derivation the scattering potential does not need to be spherically symmetric for the approximation to work.

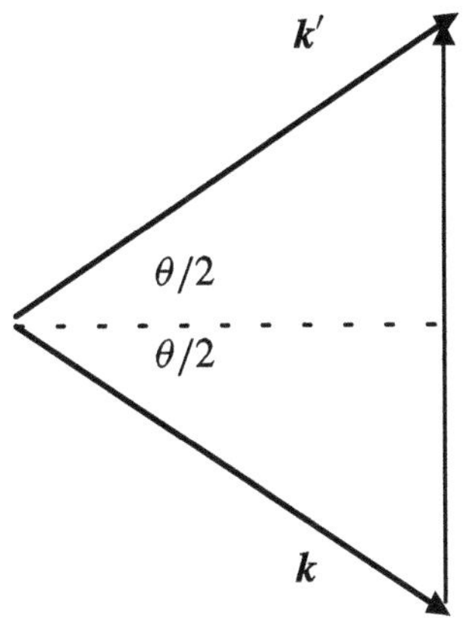

Fig. 3.10. Relation between incoming and outgoing wave vectors $\boldsymbol{k}, \boldsymbol{k}'$ and scattering angle Θ, eq. (3.93)

Screened and Unscreened Coulomb Interaction

The scattering amplitude (3.89) can be evaluated for a potential

$$\mathcal{V}(\boldsymbol{r}) = \frac{e_1 e_2}{r} e^{-r/a}, \tag{3.91}$$

which approaches the Coulomb interaction in the limit of $a = \infty$. This potential keeps coming up under various names in physics, such as Debye, Yukawa, or Bohr potential. The exponential accounts qualitatively for the screening of the nuclear charge by electrons surrounding the nucleus.

After insertion of (3.91), integration of (3.90) in spherical coordinates yields

$$f(\boldsymbol{\Omega}) = -\frac{m_0}{2\pi\hbar^2} \frac{4\pi e_1 e_2}{K^2 + 1/a^2}, \tag{3.92}$$

where

$$K^2 = |\boldsymbol{k}' - \boldsymbol{k}|^2 \equiv 4k^2 \sin^2 \Theta/2. \tag{3.93}$$

Eq. (3.93) follows from the fact that $\boldsymbol{k}$ and $\boldsymbol{k}'$ form an equilateral triangle. The scattering angle Θ is, by definition, the angle between the incident and outgoing wave vector (Fig. 3.10). Eq. (3.86) yields the differential cross section

$$\begin{aligned}
d\sigma(\boldsymbol{\Omega}) &= \left(\frac{2m_0 e_1 e_2}{\hbar^2}\right)^2 \frac{d^2\boldsymbol{\Omega}}{\left[4k^2 \sin^2(\Theta/2) + 1/a^2\right]^2} \\
&\equiv \left(\frac{b}{4}\right)^2 \frac{d^2\boldsymbol{\Omega}}{\left[\sin^2(\Theta/2) + (\lambda/2a)^2\right]^2},
\end{aligned} \tag{3.94}$$

where $\lambda = \hbar/m_0 v$ is the de Broglie wavelength and $b = 2e_1 e_2/m_0 v^2$.

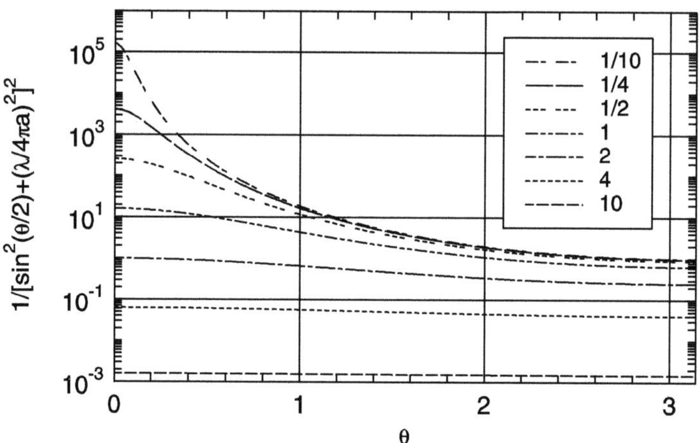

Fig. 3.11. Differential cross section $(4/b)^2 \mathrm{d}\sigma/\mathrm{d}^2\Omega$ for screened Coulomb interaction according to (3.94). Numbers indicate the value of $\lambda/2a$

In the limit of unscreened Coulomb interaction, $a = \infty$, the cross section (3.90) becomes identical with Rutherford's law, (3.45). As will be seen in Sect. 3.5.2, that form of the cross section remains even valid beyond the limit of weak interaction. This, however, does not imply that the Born approximation characterizes Coulomb scattering rigorously already in the first order. Indeed, when carried on to second-order the scheme delivers divergent results as the reader may verify in problem 3.24. It will be seen in Sect. 3.5.2 that this feature is related to the behavior of the phase of the Coulomb wave function at large distances. It is the introduction of a finite interaction radius R in connection with (3.83) which becomes problematic for Coulomb scattering.

For a finite value of the screening radius a these limitations become less severe. The fact that the cross section does not diverge at small angles lends credibility to the assumption of weak interaction. Integration over the angular variables yields the total cross section

$$\sigma_{\text{tot}} = \int \mathrm{d}\sigma(\Omega) = \frac{\pi a^2 \kappa^2}{1 + (\lambda/2a)^2}, \tag{3.95}$$

where $\kappa = 2e_1 e_2 / \hbar v$ is the Bohr parameter. For high-speed projectiles we have $\lambda/2a \ll 1$. Hence, σ_{tot} will be small if the Bohr parameter κ is small. Thus, the condition of weak interaction is equivalent with

$$\kappa \ll 1, \tag{3.96}$$

i.e., a condition complementary to that for the classical-orbit picture to be valid. This is commonly called the Sommerfeld criterion[4]. Conversely, at low speed when $\lambda/2a \gg 1$, the total cross section approaches $\sigma_{\mathrm{tot}} \simeq 16\pi a^4 (e_1 e_2 m_0/\hbar^2)^2$. If this is to be consistent with the assumption of weak interaction, the ratio $\sigma_{\mathrm{tot}}/\pi a^2$ should be $\ll 1$, or

$$2e_1 e_2 a \ll \frac{\hbar^2}{2m_0}. \tag{3.97}$$

You may look more closely into this in problem 3.22

Another way of looking at (3.94) is the deviation from classical Coulomb scattering which occurs at small angles, $\Theta \lesssim \lambda/a$ or, in terms of a classical impact parameter $p = b/\Theta$, at $p \gtrsim \kappa a$, where again κ is the Bohr parameter. The finding in Chapter 2 that for $\kappa > 1$ the essential part of the scattering is described well by the classical-trajectory picture, is thus well confirmed. It is also seen that for $\kappa < 1$ substantial changes may occur as soon as the interaction is not pure-Coulomb like (Fig. 3.11).

Potential Well or Barrier

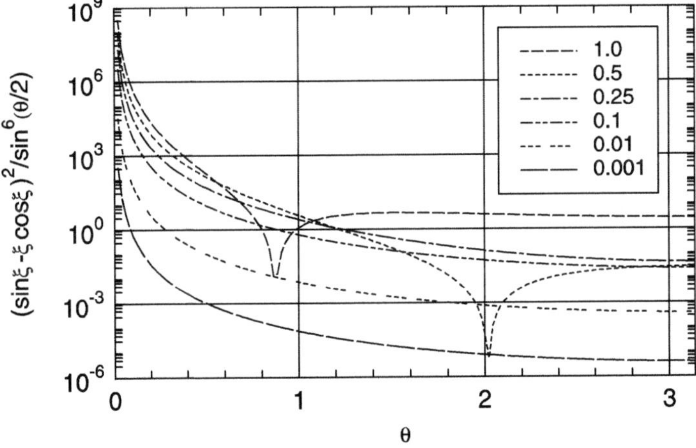

Fig. 3.12. Differential cross section for potential well or barrier in the Born approximation (3.99). Numbers indicate the value of ka

Classical scattering on a perfectly rigid sphere was found to differ drastically from Coulomb scattering, cf. (3.48) and problem 3.6. A quantum analog would be a scattering potential

---

[4] In the literature, the Sommerfeld criterion is commonly written in the form $e_1 e_2/\hbar v \ll 1$, i.e., it differs from the form above by a factor of 2.

$$\mathcal{V}(\mathbf{r}) = \begin{cases} \mathcal{V}_0 & r < a \\ & \text{for} \\ 0 & r > a \end{cases} \qquad (3.98)$$

with a constant $\mathcal{V}_0 > E_r$. In the Born approximation, (3.89) yields the differential cross section

$$\mathrm{d}\sigma(\boldsymbol{\Omega}) = \left(\frac{\mathcal{V}_0\lambda}{8E_r}\right)^2 \left(\frac{\sin\xi - \xi\cos\xi}{\sin^3(\Theta/2)}\right)^2 \mathrm{d}^2\boldsymbol{\Omega} \qquad (3.99)$$

with $\xi = 2ka\sin(\Theta/2)$.

The most striking feature of this result is the smooth dependence of $\mathrm{d}\sigma$ on the height $\mathcal{V}_0$ of the potential barrier: There is no discontinuity at $E_r = \mathcal{V}_0$ as in the classical case. Nor is there evidence for a transition from straight transmission to quantal tunneling. Not even the sign of the interaction enters. Part of this may be an essential difference between the scattering of a wave and a particle, but energy conservation also holds in quantum mechanics: At kinetic energies far below the barrier, the probability density inside the sphere ought to become very small. We are thus justified in suspecting that the Born approximation fails to describe some essentials of scattering at low energies. This need not be a surprise in the light of the initial assumption of weak interaction.

Figure 3.12 shows the angular part of (3.99) for various values of ka. Scattering is weak in general for large wavelengths, while diffraction is observed when λ approaches the radius a.

Relativistic Extension

The Born approximation can be applied also to the Dirac equation, and a cross section can be calculated according to the same principles as in the non-relativistic case. The procedure outlined above involves the Green function for the Dirac equation, often called the Feynman propagator. A derivation for Coulomb interaction will actually emerge in a somewhat indirect way in Sect. 5.6. Therefore, only the result will be given here,

$$d\sigma(\Theta) = \frac{e_1^2 e_2^2}{(2mv^2)^2}\left(1 - \frac{v^2}{c^2}\right)\left(1 - \frac{v^2}{c^2}\sin^2\frac{\Theta}{2}\right)\frac{\mathrm{d}^2\boldsymbol{\Omega}}{\sin^4(\Theta/2)}, \qquad (3.100)$$

which approaches Rutherford's law for $v/c \ll 1$.

3.4.4 Partial-Wave Expansion

A solution of the *central-force scattering problem* in closed form may be found by the partial-wave method which is described in many standard texts on quantum mechanics. It represents an expansion in terms of angular-momentum eigenfunctions that may be familiar to the reader from bound-state problems. An incoming wave, (3.80), can be expressed in the form

$$\psi^{(0)}(\boldsymbol{r}) = A\mathrm{e}^{\mathrm{i}kr\cos\Theta} \equiv A\sum_{\ell=0}^{\infty}(2\ell+1)\mathrm{i}^{\ell}j_{\ell}(kr)P_{\ell}(\cos\Theta) \qquad (3.101)$$

according to Abramowitz and Stegun (1964), (10.1.47), where the $P_{\ell}(\cos\Theta)$ are Legendre polynomials and the $j_{\ell}(kr)$ spherical Bessel functions.

A general solution of Schrödinger's equation at a fixed energy E_r may be written in the form (Schiff, 1981)

$$\psi(\mathbf{r}) = \sum_{\ell=0}^{\infty}\sum_{\mu=-\ell}^{\ell} R_{\ell\mu}(r)Y_{\ell\mu}(\Theta,\phi), \qquad (3.102)$$

where the $Y_{\ell\mu}(\Theta,\phi)$ are spherical harmonics and $R_{\ell\mu}(r)$ solutions of the radial wave equation. Since the incoming wave (3.101) exhibits cylindrical symmetry around the beam direction, we may restrict our attention to azimuthally symmetric solutions which have an azimuthal quantum number $\mu = 0$,

$$\psi(\boldsymbol{r}) = \sum_{\ell=0}^{\infty}(2\ell+1)\mathrm{i}^{\ell}R_{\ell}(r)P_{\ell}(\cos\Theta), \qquad (3.103)$$

where the function $R_{\ell}(r)$ has to be determined from the Schrödinger equation. Numerical coefficients $(2\ell+1)\mathrm{i}^{\ell}$ have been chosen for convenience, suggested by the form (3.101) of the incoming wave.

A scattering calculation then reduces to finding a solution of Schrödinger's equation satisfying all boundary conditions with the incoming wave being described by (3.101).

Scattering Amplitude

Since both the wave equation and the incoming wave have been separated according to angular-momentum quantum number, also the scattering amplitude can be split according to

$$f(\cos\Theta) = \sum_{\ell=0}^{\infty}(2\ell+1)\,\mathrm{i}^{\ell}f_{\ell}\,P_{\ell}(\cos\Theta). \qquad (3.104)$$

Therefore, comparing with eq. (3.84), the elementary scattering event is now characterized by an 'incoming wave' $j_{\ell}(kr)P_{\ell}(\cos\Theta)$ and an outgoing wave of the form $f_{\ell}P_{\ell}(\cos\Theta)\mathrm{e}^{\mathrm{i}kr}/r$, both emerging from the respective limiting behavior of a wave function $R_{\ell}(r)P_{\ell}(\cos\Theta)$.

The quantitative behavior of the radial wave function is governed by the scattering potential $\mathcal{V}(r)$. Its general form may, however, be extracted from the free-particle wave equation if a region can be identified at large distances r where the scattering potential is vanishing or at least insignificant. For $\mathcal{V}=0$ the radial wave equation has the solution

$$R_\ell(r) = a_\ell \, j_\ell(kr) + b_\ell \, y_\ell(kr), \tag{3.105}$$

where both y_ℓ and j_ℓ are modified spherical Bessel functions in the notation of Abramowitz and Stegun (1964), Chapter 10. These functions make up a complete set of solutions for $E_r = \hbar^2 k^2 / 2 m_0$ and angular momentum quantum number ℓ. Both sets characterize damped oscillations and differ from each other most drastically near the origin where j_ℓ is regular while y_ℓ is singular. We may express the arbitrary constants a_ℓ, b_ℓ in terms of an amplitude A_ℓ and a phase δ_ℓ

$$a_\ell = A_\ell \cos \delta_\ell; \quad b_\ell = -A_\ell \sin \delta_\ell, \tag{3.106}$$

so that

$$R_\ell(r) = A_\ell \left(\cos \delta_\ell \, j_\ell(kr) - \sin \delta_\ell \, y_\ell(kr) \right). \tag{3.107}$$

Then the difference between the ℓ'th radial wave function and the incoming ℓ-wave can be written in the form

$$\psi_\ell - \psi_\ell^{(0)} = \left\{ \left[A_\ell \cos \delta_\ell - 1 \right] j_\ell(kr) - A_\ell \sin \delta_\ell \, y_\ell(kr) \right\} P_\ell(\cos \Theta). \tag{3.108}$$

In view of the occurrence of $-\psi_\ell^{(0)}$ on the left-hand side, the function $\psi_\ell - \psi_\ell^{(0)}$ must satisfy the physical requirement that it cannot contain an incoming wave. Incoming and outgoing waves may be identified from the asymptotic behavior which is given by Abramowitz and Stegun (1964)

$$j_\ell(kr) \sim \frac{\cos(kr - \alpha_\ell)}{kr}; \qquad y_\ell(kr) \sim \frac{\sin(kr - \alpha_\ell)}{kr} \tag{3.109}$$

with $\alpha_\ell = (\ell + 1)\pi/2$. These expressions may be inserted into (3.108), and the harmonic functions may be decomposed into terms $\propto e^{\pm ikr}$ representing outgoing and incoming waves, respectively. The requirement that the latter vanish yields

$$A_\ell \cos \delta_\ell - 1 - iA_\ell \sin \delta_\ell = 0, \tag{3.110}$$

i.e., $A_\ell = e^{i\delta_\ell}$, and the sum of the two outgoing waves reduces to

$$f_\ell = \frac{1}{2k} \left(e^{2i\delta_\ell} - 1 \right) (-i)^{\ell+1} \tag{3.111}$$

after elimination of A_ℓ. Thus, (3.104) reads

$$f(\cos \Theta) = \frac{1}{2ik} \sum_{\ell=0}^{\infty} (2\ell + 1) \left(e^{2i\delta_\ell} - 1 \right) P_\ell(\cos \Theta). \tag{3.112}$$

In this way the scattering amplitude has been expressed by a set of *phase shifts* δ_ℓ. While this representation is compact in general it is most suitable in practice when only a small number of angular-momentum components undergo substantial scattering. One speaks of s-wave and p-wave scattering when the scattering amplitude is dominated by the terms from $\ell = 0$ and 1, respectively, and correspondingly for higher components. For s-wave scattering the differential cross section is isotropic while for p-wave scattering it has broad maxima in the forward as well as the backward direction. Practical phase-shift analysis requires most often a major computational effort.

Total Cross Section

From (3.86) and (3.112) follows the total cross section by integration,

$$\sigma_{\text{tot}} = \frac{4\pi}{k^2} \sum_{\ell} (2\ell + 1) \sin^2 \delta_\ell, \tag{3.113}$$

where use has been made of the orthogonality of Legendre polynomials,

$$\int_{-1}^{1} \mathrm{d}\cos\Theta\, P_\ell(\cos\Theta) P_{\ell'}(\cos\Theta) = 0 \quad \text{for } \ell \neq \ell' \tag{3.114}$$

and the normalizing integral,

$$\int_{-1}^{1} \mathrm{d}\cos\Theta\, \left(P_\ell(\cos\Theta)\right)^2 = \frac{2}{2\ell + 1}. \tag{3.115}$$

Incidentally, by comparison with (3.112), the term on the right-hand side of (3.113) can be written in the form

$$\sigma_{\text{tot}} = \frac{4\pi}{k} \mathrm{Im} f(\cos\Theta)\bigg|_{\Theta=0}, \tag{3.116}$$

where Im indicates the imaginary part. This is the famous optical theorem, here for elastic scattering on a spherically symmetric potential. It relates the loss of particle flux in the forward direction to the scattered intensity.

Stopping Cross Section and Straggling Parameter

The stopping cross section S and the straggling parameter W can be determined similarly from (2.29) and (2.30). In addition to (3.115) also the recursion relation (Abramowitz and Stegun, 1964)

$$(2\ell + 1) \cos\Theta\, P_\ell(\cos\Theta) = (\ell + 1) P_{\ell+1}(\cos\Theta) + \ell P_{\ell-1}(\cos\Theta) \tag{3.117}$$

is needed. This yields useful connections

$$S = \frac{\gamma E}{2} \sigma^{(1)} \qquad W = \left(\frac{\gamma E}{2}\right)^2 \sigma^{(2)} \tag{3.118}$$

to the *transport cross sections*

$$\sigma^{(1)} = \int (1 - \cos\Theta)\mathrm{d}\sigma(\Theta) = \frac{4\pi}{k^2} \sum_{\ell} (\ell+1)\sin^2(\delta_\ell - \delta_{\ell+1}) \tag{3.119}$$

and

$$\sigma^{(2)} = \int (1 - \cos\Theta)^2 \mathrm{d}\sigma(\Theta)$$

$$= \frac{4\pi}{k^2} \sum_{\ell} (\ell+1)\left[2\sin^2(\delta_\ell - \delta_{\ell+1}) - \frac{\ell+2}{2\ell+3}\sin^2(\delta_\ell - \delta_{\ell+2}) \right], \tag{3.120}$$

which you may verify by solving problem 3.28. These relations are well-known in gas dynamics but have also proven useful in the theory of electronic stopping (Lindhard and Sørensen, 1996). They will be employed in Chapters. 6 and 8.

Impenetrable Sphere ($\star$)

Phase shifts may be determined rigorously for scattering on a perfectly rigid sphere where the wave function obeys the free-particle wave equation for $r > a$. Since the wave function must vanish inside the sphere and at the same time must be continuous everywhere we must have

$$R_\ell(r) = 0 \text{ for } r = a \text{ and all } \ell. \tag{3.121}$$

Hence, from (3.107), $\cos\delta_\ell\, j_\ell(ka) - \sin\delta_\ell\, y_\ell(ka) = 0$, or

$$\tan\delta_\ell = \frac{j_\ell(ka)}{y_\ell(ka)}. \tag{3.122}$$

This determines all phase shifts and hence both the differential and total cross section at any velocity. A particularly simple analytical result is obtained in the low-speed limit $ka \ll 1$ where

$$j_\ell(z) \sim \frac{z^\ell}{(2\ell+1)!!}; \qquad y_\ell(z) \sim -\frac{(2\ell-1)!!}{z^{\ell+1}} \tag{3.123}$$

with $(2\ell+1)!! = (2\ell+1)(2\ell-1)\cdots 3\cdot 1$. Then

$$\tan\delta_\ell \simeq \delta_\ell \simeq -\frac{2\ell+1}{\left((2\ell+1)!!\right)^2}(ka)^{2\ell+1}, \tag{3.124}$$

i.e., s-wave scattering dominates with $\delta_0 \sim -ka$. This produces a scattering amplitude $f(\Omega) \simeq -a$ and a total cross section

$$\sigma_{\mathrm{tot}} \simeq 4\pi a^2. \tag{3.125}$$

The reader who is astonished at this result, which is four times the classical cross section following from (3.48) for an impenetrable sphere, might recall that the limit $ka \ll 1$ reflects the scattering of a wave with a wavelength significantly greater than the radius of the sphere, i.e., a situation that resembles more the diffraction of a wave on a small sphere than the geometric-optical scattering of a ray on a large sphere.

3.5 Coulomb Scattering

3.5.1 Phase Shifts

For Coulomb interaction,

$$\mathcal{V}(r) = \frac{e_1 e_2}{r},\tag{3.126}$$

the radial wave equation can be written in the form

$$\frac{\mathrm{d}^2 u_\ell(\rho)}{\mathrm{d}\rho^2} + \left[1 - \frac{\kappa}{\rho} - \frac{\ell(\ell+1)}{\rho^2}\right] u_\ell(\rho) = 0,\tag{3.127}$$

where the radial wave function has been set to

$$R_\ell(r) = \frac{u_\ell(\rho)}{\rho},\tag{3.128}$$

with $\rho = kr$. Here, k is the initial wave number defined by the kinetic energy,

$$E = \frac{\hbar^2 k^2}{2m_0}\tag{3.129}$$

and

$$\kappa = \frac{2e_1 e_2}{\hbar v}\tag{3.130}$$

is the Bohr parameter, with $v = \hbar k/m$ the relative speed.

Properties of this important differential equation have been compiled by Abramowitz and Stegun (1964). Solutions can be expressed in terms of confluent hypergeometric functions, of which there exist two classes, one regular and one singular in the origin. It is the regular solution that is of interest here, which can be expressed in the form (Abramowitz and Stegun (1964), 14.5.1 and subsequent equations),

$$u_\ell(\rho) \sim g(\rho) \cos \Theta_\ell + f(\rho) \sin \Theta_\ell,\tag{3.131}$$

where $f(\rho)$ and $g(\rho)$ are real-valued functions,

$$\Theta_\ell = \rho - \frac{\kappa}{2} \ln(2\rho) - \ell\frac{\pi}{2} + \sigma_\ell\tag{3.132}$$

with

$$\sigma_\ell = \arg \Gamma \left(\ell + 1 + i\frac{\kappa}{2} \right), \tag{3.133}$$

and arg denotes the argument[5]. Asymptotically, for large $\rho = kr$, $f(\rho) \sim 1$ while $g(\rho) \sim 1/\rho$ and hence,

$$u_\ell(\rho) \sim \sin\left(\rho - \frac{\kappa}{2} \ln(2\rho) - \ell\frac{\pi}{2} + \sigma_\ell \right) \tag{3.134}$$

This can be compared with with (3.107) and (3.109), from which we find

$$R_\ell(\rho) \sim \frac{1}{\rho} \cos\left(\rho - \alpha_\ell + \delta_\ell \right), \tag{3.135}$$

and hence,

$$\delta_\ell = \sigma_\ell - \frac{\kappa}{2} \ln(2\rho). \tag{3.136}$$

An important feature of this relationship is that the second term, although dependent on $\rho = kr$, is independent of ℓ. Applications in stopping and straggling, cf. (3.53) and (3.54), involve *differences* between two phase shifts. Hence, we may safely operate with

$$\delta_\ell = \arg \Gamma \left(\ell + 1 + i\frac{\kappa}{2} \right) \tag{3.137}$$

as the *Coulomb phase shift*.

This relation will become useful in Sect. 6.3.2. It could also be utilized to evaluate the Rutherford cross section. For that, a more direct procedure is available which will be discussed now.

3.5.2 Cross Section ($\star$)

The long range of the Coulomb force makes the separation of the incoming from the outgoing wave a bit more delicate a task. However, as in the classical case, an exact way of solution can be found. The derivation given here follows standard texts on quantum mechanics (Schiff, 1981, Merzbacher, 1970).

Schrödinger's equation is separated in parabolic coordinates,

$$\xi = r - x; \quad \eta = r + x; \quad \phi = \arctan\frac{z}{y}, \tag{3.138}$$

in which the Laplacian takes on the form

$$\nabla^2 = \frac{4}{\eta + \xi} \left(\frac{\partial}{\partial\xi}\xi\frac{\partial}{\partial\xi} + \frac{\partial}{\partial\eta}\eta\frac{\partial}{\partial\eta} \right) + \frac{1}{\eta\xi} \frac{\partial^2}{\partial\phi^2}. \tag{3.139}$$

---

[5] For a complex number $z = x + iy$ the argument ϕ is given by $\tan\phi = y/x$.

In these variables an incoming plane wave reads $e^{ikx} = e^{ik(\eta-\xi)/2}$. This implies that all dependence on the azimuth ϕ may be ignored from the beginning. An outgoing wave should look like

$$\frac{e^{ikr}}{r} \sim \frac{e^{ik(\eta+\xi)/2}}{r} \tag{3.140}$$

asymptotically. The factor $e^{ik\eta}$ is common to both the incoming and the outgoing wave: One might suspect that it is common even to the exact solution. This suspicion is strengthened by the observation that the other variable, expressed in spherical coordinates, $\xi = r(1-\cos\Theta) = 2r\sin^2(\Theta/2)$, contains the essential angular variation that occurs in Rutherford's law (3.45).

A reasonably straight route toward a tabulated function goes over the *ansatz*

$$\psi(\boldsymbol{r}) = \psi(\xi,\eta,\phi) = e^{ik(\eta-\xi)/2}f(\xi), \tag{3.141}$$

where the unknown function $f(\xi)$ has to be determined. Then Schrödinger's equation reduces to

$$\xi\frac{d^2 f}{d\xi^2} + (1 - ik\xi)\frac{df}{d\xi} - \frac{m_0 e_1 e_2}{\hbar^2}f = 0, \tag{3.142}$$

where, as in previous instances, the energy has been replaced by $E_r = \hbar^2 k^2/2m_0$.

Eq. (3.142) is equivalent with Kummer's equation for confluent hypergeometric functions (Abramowitz and Stegun (1964), 13.1.1) which has two solutions (Kummer's functions) $M(a,b,z)$ and $U(a,b,z)$ with $z = ik\xi$, $a = -im_0 e_1 e_2/\hbar^2 k \equiv -i\kappa/2$ and $b = 1$, where κ is the Bohr parameter. The M-function is regular for all finite ξ and has the asymptotic behavior

$$M \sim \frac{e^{i\pi a}}{z^a \Gamma(1-a)}\left(1 - \frac{a^2}{z}\right) + \frac{e^z z^{a-1}}{\Gamma(a)} \tag{3.143}$$

up to terms $\sim \xi^{-1}$. With this and (3.141), the wave function takes on the asymptotic form

$$\psi(\boldsymbol{r}) = e^{ik(\eta-\xi)/2}M(-i\kappa/2, 1, ik\xi)$$

$$\sim \frac{e^{\pi\kappa/4}}{\Gamma(1+i\kappa/2)}\left[e^{ikx+(i\kappa/2)\ln\left[k(r-x)\right]}\left(1 + \frac{\kappa^2}{4ik(r-x)}\right)\right.$$

$$\left. + \frac{\Gamma(1+i\kappa/2)}{\Gamma(-i\kappa/2)}\frac{1}{2ik\sin^2(\Theta/2)}\frac{e^{ikr-(i\kappa/2)\ln\left[k(r-x)\right]}}{r}\right]. \tag{3.144}$$

Here the first factor in the brackets describes an incident plane wave, apart from a phase factor and a term that vanishes at large distances. This justifies the extraction of a scattering amplitude by collection of those factors in the second term, the outgoing wave, that determine the magnitude and the directional dependence of the phase,

$$
\begin{aligned}
f(\boldsymbol{\Omega}) &= \frac{\Gamma(1 + i\kappa/2)}{i\Gamma(-i\kappa/2)} \frac{1}{2k\sin^2(\Theta/2)} \exp\left\{-i\kappa \ln \sin\left(\frac{\Theta}{2}\right)\right\} \\
&\equiv -\frac{\kappa}{4k\sin^2(\Theta/2)} \exp\left\{-i\kappa \ln \sin\left(\frac{\Theta}{2}\right) + i\chi\right\},
\end{aligned} \tag{3.145}
$$

where $e^{i\chi} = \Gamma(1 + i\kappa/2)/\Gamma(1 - i\kappa/2)$.

With this one obtains the differential cross section

$$
d\sigma(\boldsymbol{\Omega}) = |f|^2 d^2\boldsymbol{\Omega} = \left(\frac{\kappa}{4k}\right)^2 \frac{1}{\sin^4(\Theta/2)} d^2\boldsymbol{\Omega} = \left(\frac{b}{4}\right)^2 \frac{d^2\boldsymbol{\Omega}}{\sin^4(\Theta/2)} \tag{3.146}
$$

in complete agreement with the classical result, (3.45).

Eq. (3.144) demonstrates a problem connected with the Born series applied to Rutherford scattering. Indeed, to leading order in the perturbation – i.e., the parameter $\kappa/2$ – (3.145) goes over into the first-order perturbation solution (3.92). In higher orders, terms depending logarithmically on distance enter into both the incident and the scattered wave. This makes it essentially impossible to extract a scattering amplitude. Since this situation is not foreseen in the formalism it manifests itself in divergent terms.

3.5.3 Relativistic Extension

The relativistic theory of elastic scattering of electrons on nuclei has been an area of considerable activity and controversy. Even for a central potential, the theory is a complex issue since it is the total angular momentum rather than orbital angular momentum that is conserved in Dirac theory (Bransden and Joachain, 2000). As a consequence, scattering in a central field becomes spin-dependent even in the absence of a spin-dependent force.

Exact calculations by Mott (1929, 1932) are complex, but unlike various approximate evaluations, their validity has apparently not been questioned. Subsequent studies were based on Born-type expansions. A detailed survey was given by Dalitz (1951). An approximation by McKinley and Feshbach (1948) is frequently used in the literature, which makes the replacement

$$
1 - \frac{v^2}{c^2}\sin^2\frac{\Theta}{2} \rightarrow 1 - \frac{v^2}{c^2}\sin^2\frac{\Theta}{2} + \pi\frac{Z}{137}\frac{v}{c}\sin\frac{\Theta}{2}\left(1 - \sin\frac{\Theta}{2}\right) \tag{3.147}
$$

in eq. (3.100). Here, the last term represents a higher-order term in the Born series.

Phase shifts for relativistic Coulomb scattering may be found in the book by Berestetskii et al. (1971).

3.6 Discussion and Outlook

Two important assumptions need to be kept in mind. The assumption of binary elastic collisions has been essential in the derivation of practically all relations discussed in this chapter, in particular the transformation between laboratory and center-of-mass velocities, and central-force motion. This is true for both nonrelativistic and relativistic, classical and quantal mechanics. Secondly, relations derived here governing laboratory parameters mostly rely on the assumption of a projectile hitting a target particle initially at rest, with the exception of (3.12) in Sect. 3.2.1 and (3.30) in Sect. 3.2.2. Although this simplification is relevant in particle penetration, it is by no means universally justified.

Physical arguments of the type discussed in the introduction, Sect. 3.1, may be employed to justify the results derived in this chapter also in situations where the underlying assumptions are not rigorously fulfilled. Examples will be discussed in several chapters to come as well as in problems 3.4 and 3.5 in the end of the present chapter.

In contrast to the presentation in Chapter 2 the emphasis was laid in this chapter on results with a considerable degree of rigor. Applications to the important and comparatively transparent case of Coulomb scattering have been discussed but played a less dominating role. The case of screened-Coulomb scattering – which is especially important for scattering between heavy particles like ions on atoms – has been mentioned only very briefly and will be discussed in detail in volume II. Quantal scattering has been discussed mainly as a prerequisite for the theory of electronic stopping which is going to be the main topic of the following chapter.

Problems

3.1. Derive (3.9) and (3.11) from (3.5) and (3.7).

3.2. Determine the maximum angle for scattering of a proton from a free electron.

3.3. Derive (3.8) from (3.12).

3.4. Derive formulas corresponding to (3.5), (3.7) for an inelastic collision in which momentum is conserved but an energy Q is missing after the collision. Show that this is a reasonable model to describe the scattering between heavy ions when some energy is lost by ejection of electrons from inner shells.

3.5. Devise a strategy to determine the inelastic energy loss Q mentioned in problem 3.4 from measurements of the kinetic energy of an emerging particle at a given angle, or from measurements of the two scattering angles in coincidence.

3.6. Evaluate the scattering integral (3.34) for a) Coulomb interaction, b) $\mathcal{V}(r) = Ar^{-2}$, and c) an impenetrable sphere of radius a. Determine the corresponding differential cross sections.

3.7. Evaluate the cross section for wide-angle Rutherford scattering by defining wide-angle scattering by center-of-mass scattering angles $\pi/2 < \Theta < \pi$. Demonstrate that the probability for wide-angle scattering over the total range is much larger for electrons and positrons than for heavier projectiles at the same initial energy.

3.8. Evaluate the scattering integral in the small-angle limit for a Yukawa potential

$$\mathcal{V}(r) = \frac{e_1 e_2}{r} e^{-r/a}, \tag{3.148}$$

where a is a constant. [Hint: You will arrive at a Bessel function.]

3.9. Derive (3.64a). Find the equation for the asymptotic straight-line orbit $y(x)$ at large t by inspecting Fig. 3.6. Express $r = \sqrt{x^2 + [y(x)]^2}$ as a function of x, linearize for large x and find $x(t)$ by comparison with (3.63). Then find $y(t)$.

3.10. Construct asymptotic orbits in the laboratory frame of reference by means of the time integral. Assume particle 2 initially at rest in the origin and scribe the motion of particle 1 by $x_1 = vt; y_1 = p$. Determine $(x_1(t), y_1(t))$ and $(x_2(t), y_2(t))$ for $t \gg 0$ outside the collision region. The result is

$$x_1 = \frac{m_1}{m} vt + \frac{m_2}{m}(vt + 2\tau) \cos \Theta - \frac{m_2}{m} p \sin \Theta \tag{3.149a}$$

$$y_1 = \frac{m_1}{m} p + \frac{m_2}{m}(vt + 2\tau) \sin \Theta + \frac{m_2}{m} \cos \Theta \tag{3.149b}$$

$$x_2 = \frac{m_1}{m} vt - \frac{m_1}{m}(vt + 2\tau) \cos \Theta + \frac{m_1}{m} p \sin \Theta \tag{3.149c}$$

$$y_2 = \frac{m_1}{m} p - \frac{m_1}{m}(vt + 2\tau) \sin \Theta - \frac{m_1}{m} p \cos \Theta, \tag{3.149d}$$

where $m = m_1 + m_2$.

3.11. Show that the time integral does not exist for Coulomb interaction.

3.12. Evaluate the time integral for $\mathcal{V}(r) = Ar^{-2}$.

3.13. Prove (3.75).

3.14. Show that in relativistic Coulomb scattering as described in section 3.3.5 a projectile may 'fall' into the force center. Discuss the motion as a function of impact parameter.

3.15. Go carefully through the expansion leading from (3.81) to (3.83). Make sure to know why you have to go one order higher in the expansion in terms of r'/r in the exponent than in the denominator.

3.16. Your quantum mechanics text quotes a definition of a particle current that differs from the classical one, density $\times$ velocity. Convince yourself that the proper quantal definition is consistent with the classical one for i) a plane wave, ii) a spherical wave, iii) a quasi-spherical wave $f(\cos\Theta)e^{ikr}/r$.

3.17. Go carefully through the evaluation of the integral

$$\int_{-\infty}^{\infty} d\omega \, \frac{e^{i\omega t}}{\omega^2 + \omega_0^2} \tag{3.150}$$

in the complex plane, and make sure not to miss a nonvanishing contribution.

3.18. Find the Green function for a particle of mass m moving in one dimension, acted upon by a friction force $f_c = -b\,dx/dt$ and an arbitrary force $f(t)$. Apply the result to a particle in a constant gravitational field.

3.19. Solve problem 3.18 for the corresponding three-dimensional model with a friction force $\mathbf{f}_c = -b\,d\mathbf{r}/dt$.

3.20. Define and write down Green functions for the electrostatic potential and the electrostatic field on the basis of what you know about electrostatics.

3.21. Find the cross section within the Born approximation for scattering on a potential $\mathcal{V}(r) = C_{12}\delta(\mathbf{r})$ where C_{12} is a constant. This model may describe the interaction of a particle with a long wavelength on a scatterer with very small spatial dimensions, such as the interaction of a thermal neutron with a nucleus.

3.22. Landau and Lifshitz (1960b) mention the following criteria for the validity of the Born approximation:

$$|\mathcal{V}(r)| \ll \frac{\hbar^2}{m_0 r^2} \tag{3.151}$$

or

$$|\mathcal{V}(r)| \ll \frac{\hbar v}{r}. \tag{3.152}$$

Discuss (3.95) in the light of these criteria.

3.23. Evaluate the contribution from the second term in the Born series to the scattering amplitude. Try to go as far as possible without specifying the interaction potential. Start with the exact expression (3.81) and be sure to take the asymptotic solution (3.83) only when this is justified.

3.24. The result of problem 3.23 can be expressed in the form

$$f^{(2)}(\mathbf{\Omega}) = \text{const} \times \int d^3\mathbf{k}'' \frac{\mathcal{V}^*(\mathbf{k}'' - \mathbf{k})\mathcal{V}(\mathbf{k}'' - \mathbf{k}')}{k''^2 - k^2}, \tag{3.153}$$

where $\mathcal{V}(\mathbf{k})$ is the Fourier transform of the potential. Show that $f^{(2)}(\boldsymbol{\Omega})$ and, hence, the differential cross section diverges in this order in the case of Coulomb interaction, but that the divergence is removed by adoption of a finite screening radius.

3.25. Use a computer algebra program to determine the differential and total cross section for an impenetrable sphere for an arbitrary value of ka. Check your result against the analytical finding for the case of $ka \ll 1$.

3.26. $(\star)^6$ Consider the scattering on a potential barrier of the form of (3.98). Utilize both types of spherical Bessel functions outside the barrier and make use of the requirement that the wave function should be continuous everywhere. Derive equations that determine the phase shifts.

3.27. Derive the relations (3.118).

3.28. Derive (3.119) and (3.120). [Hint: Make use of symmetry considerations.]

3.29. Determine the stopping cross section for an impenetrable sphere of radius a from quantal scattering theory, making use of (3.118) and (3.119). Compare with the classical result.

3.30. Perform the same calculation as in problem 3.29 for the straggling parameter W, making use of (3.120).

3.31. $(\star)$ Try to set up a theoretical scheme for treating the scattering of a plane wave on a body with cylindrical symmetry with the cylinder axis perpendicular to the incident wave vector.

3.32. Find a proper form of the Runge-Lenz vector for the quantal Coulomb-Kepler problem from the requirement that $\mathbf{M}$ must commute with the hamiltonian. [Hint: Replace $\mathbf{v}$ by $\mathbf{P}/m$, where $\mathbf{P}$ is the momentum operator and symmetrize as far as neccessary.]

3.33. $(\star)$ Use computer algebra combined with a graphics program to illustrate the difference between the electron densities for Coulomb scattering, calculated either from the exact solution or from the first Born approximation.

References

Abramowitz M. and Stegun I.A. (1964): *Handbook of mathematical functions.* Dover, New York

Basano L. and Bianchi A. (1980): Rutherford's scattering formula via the Runge-Lenz vector. Am J Phys **48**, 400–401

---

^{6} Problems marked by a star are considerably more difficult than average.

Berestetskii V.B., Lifshitz E.M. and Pitaevskii L.P. (1971): *Relativistic quantum theory*, vol. 4 part 1 of *Course of theoretical physics*. Pergamon Press, Oxford

Bransden B.H. and Joachain C.J. (2000): *Quantum Mechanics*. Prentice Hall, Harlow, 2 edn.

Dalitz R.H. (1951): On higher Born approximations in potential scattering. Proc Roy Soc A **206**, 509–520

Goldstein H. (1953): *Classical mechanics*. Addison-Wesley Press, Cambridge, Mass.

Goldstein H. (1975): Prehistory of the "Runge-Lenz" vector. Am J Phys **43**, 737–738

Goldstein H. (1976): More on the prehistory of the Laplace or Runge-Lenz vector. Am J Phys **44**, 1123–1124

Landau L. and Lifshitz E.M. (1960a): *Mechanics*, vol. 1 of *Course of theoretical physics*. Pergamon Press, Oxford

Landau L. and Lifshitz E.M. (1971): *The classical theory of fields*, vol. 2 of *Course of theoretical physics*. Pergamon Press, Oxford

Landau L.D. and Lifshitz E.M. (1960b): *Quantum mechanics. Non-relativistic theory*, vol. 3 of *Course of theoretical physics*. Pergamon Press, Oxford

Lehmann C. and Leibfried G. (1963): Higher order momentum approximations in classical collision theory. Z Physik **172**, 465–487

Leibfried G. (1965): *Bestrahlungseffekte in Festkörpern*. Teubner, Stuttgart

Lindhard J., Nielsen V. and Scharff M. (1968): Approximation method in classical scattering by screened Coulomb fields. Mat Fys Medd Dan Vid Selsk **36 no. 10**, 1–32

Lindhard J. and Sørensen A.H. (1996): On the relativistic theory of stopping of heavy ions. Phys Rev A **53**, 2443–2456

McKinley W.A. and Feshbach H. (1948): The Coulomb scattering of relativistic electrons by nuclei. Phys Rev **74**, 1759–1763

Merzbacher E. (1970): *Quantum mechanics*. Wiley, New York

Mott N.F. (1929): The scattering of fast electrons by atomic nuclei. Proc Roy Soc London A **124**, 425–442

Mott N.F. (1932): The polarization of electrons by double scattering. Proc Roy Soc London Ser A **135**, 429–458

Robinson M.T. (1970): Table of classical scattering integrals. Tech. Rep. ORNL-4556, Oak Ridge National Laboratory

Robinson M.T. and Torrens I.M. (1974): Computer simulation of atomic-displacement cascades in solids in the binary-collision approximation. Phys Rev B **9**, 5008–5024

Schiff L.I. (1981): *Quantum mechanics*. McGraw-Hill, Auckland

Symon K. (1960): *Mechanics*. Addison-Wesley, Reading, Mass.

Part II

Stopping

Stopping of Swift Point Charge I:
Bohr and Bethe Theory

4.1 Introductory Comments

The present chapter is devoted to inelastic collisions between a swift point charge (speed $v \gg v_0$) and an isolated atom or molecule. Calculations aim at the electronic stopping cross section, but concepts introduced, tools utilized and insight gained on the way are central goals to be kept in mind.

Three approaches will be presented separately. Comparisons between them show interesting similarities and differences. The classical theory outlined by Bohr (1913) contains most of the clues. The quantum stopping theory developed by Bethe (1930) and Bloch (1933b) is built upon similar concepts and the results are not dramatically different. However, each approach has its merits, and the overall result should be a rather firm grasp around an important item in atomic physics.

Central to the treatments in this chapter is the use of perturbation theory, either for the entire theoretical scheme or at least parts of it. The expansion parameter is the projectile charge, i.e., the atomic number Z_1. First a classical and a semiclassical treatment are presented which express the electronic energy loss as a function of an impact parameter p, from which the stopping cross section is found by integration,

$$S = \int T(p)\, 2\pi p \mathrm{d}p \,, \tag{4.1}$$

in accordance with (2.46) on page 42. An important finding is the fact that essential features of the $T(p)$ dependence predicted from classical theory survive in the semiclassical calculation. A fully quantal calculation reproduces inelastic cross sections predicted semiclassically for heavy projectiles, but provides an important modification for light projectiles such as electrons and positrons.

In view of Bohr's kappa criterion, (2.80), the velocity regime $v < 2e_1 e/\hbar$ has the distinguishing feature that classical scattering theory provides a reasonably accurate picture of both elastic and inelastic scattering on electrons.

This implies that for ions with atomic numbers $Z_1 \gg 1$, where $e_1 e/\hbar \gg v_0$, the Bohr velocity, classical scattering theory will be able to provide quite accurate estimates of stopping forces and straggling parameters. Hence, studying Bohr stopping theory is not only of historic and didactic interest but also highly relevant from a quantitative point of view.

Unless stated differently, the impact parameter p will in this chapter be defined relative to the *nucleus* of a target atom rather than to an individual target electron. The latter option was heavily used in Chapter 2, but the validity of the underlying physical picture for electronic collisions is limited by the Bohr criterion. This limitation is much less severe for nuclear interactions. In fact, since the impact parameter to the nucleus is directly related to the scattering angle of the projectile, knowledge of $T(p)$ provides information on electronic energy loss as a function of scattering angle.

4.2 Classical Perturbation Theory

4.2.1 Energy Transfer to Harmonic Oscillator

As mentioned in Sect. 2.3.4, Bohr treated target electrons as being bound harmonically to their atomic sites. Each electron is characterized by a single resonance frequency, say ω_0. For a classical electron bound to the origin by a force $-k\boldsymbol{r} = -m\omega_0^2 \boldsymbol{r}$, the nonrelativistic equation of motion is

$$\frac{\mathrm{d}^2 \boldsymbol{r}}{\mathrm{d}t^2} + \omega_0^2 \boldsymbol{r} = -\frac{e}{m} \boldsymbol{E}(\boldsymbol{r}, t), \tag{4.2}$$

where $\boldsymbol{E}(\boldsymbol{r}, t)$ is the electric field generated by the projectile.

Note that the position vector $\boldsymbol{r} = \boldsymbol{r}(t)$ occurs in the field on the right-hand side. This implies that (4.2) will not be a linear equation unless some simplification is made. Ignoring this feature for a moment we may write the field as a function of time,

$$\boldsymbol{E}(\boldsymbol{r}, t) = \boldsymbol{E}(\boldsymbol{r}(t), t) \equiv \boldsymbol{E}(t) \tag{4.3}$$

and, in Fourier space[1]

$$\boldsymbol{E}(t) = \int_{-\infty}^{\infty} \mathrm{d}\omega \, \boldsymbol{E}(\omega) \, \mathrm{e}^{-\mathrm{i}\omega t} \tag{4.4}$$

$$\boldsymbol{E}(\omega) = \frac{1}{2\pi} \int_{-\infty}^{\infty} \mathrm{d}t \, \boldsymbol{E}(t) \, \mathrm{e}^{\mathrm{i}\omega t}. \tag{4.5}$$

Here a common physicist's notation has been employed: The two quantities $\boldsymbol{E}(t)$ and $\boldsymbol{E}(\omega)$ have evidently different functional forms and dimensions. It is the arguments, time t and angular frequency ω, that distinguish the field and its Fourier transform from each other.

---

[1] A brief introduction into Fourier series and Fourier transforms is given in Appendix A.2.2.

A particular solution of (4.2) reads

$$\boldsymbol{r}(t) = -\frac{e}{m\omega_0} \int_{-\infty}^{t} dt'\, \boldsymbol{E}(t') \sin \omega_0(t - t'), \tag{4.6}$$

as is easily verified by differentiation[2]. Eq. (4.6) represents a retarded solution with the property that $\boldsymbol{r}(-\infty) = 0$. This is reasonable when there is no field at $t = -\infty$. If the oscillator had been in motion prior to the action of the field, a solution to the homogeneous equation, i.e., an undisturbed harmonic oscillation at a frequency ω_0 would have to be added.

The time-dependent electromagnetic field acting on the electron is characterized by a gradual increase from zero to a maximum at closest approach of the projectile, and a subsequent decrease toward zero. Therefore we may be able to find some time t_1 after which the action of the field on the target electron has a negligible effect. Then, for $t > t_1$, (4.6) can be rewritten in the form

$$\boldsymbol{r}(t) = -\frac{e}{m\omega_0} \left(\boldsymbol{C} \sin \omega_0 t - \boldsymbol{S} \cos \omega_0 t\right) \tag{4.7}$$

with

$$\boldsymbol{C} = \int_{-\infty}^{\infty} dt'\, \boldsymbol{E}(t') \cos \omega_0 t' \tag{4.8a}$$

$$\boldsymbol{S} = \int_{-\infty}^{\infty} dt'\, \boldsymbol{E}(t') \sin \omega_0 t', \tag{4.8b}$$

since the contributions to the integrals from the interval $t_1 < t' < \infty$ are negligible.

Eq. (4.7) demonstrates that in the absence of damping the electronic motion must be strictly harmonic after the external force has died away. (4.8a) shows that only field components $\boldsymbol{E}(\omega)$ with $\omega = \pm\omega_0$ contribute to the energy transfer. This is a general property of the harmonic oscillator: When offered a white excitation spectrum it picks only the part that oscillates at the resonant frequency. It is this property that causes the empty G string of a violin to have the pitch of a G, regardless of whether it is plucked or bowed. The speed of the bow has no influence on the pitch. The phenomenon of resonant excitation is an important feature not only of the *classical* oscillator model of the atom. It comes again in the quantal description.

In order to find the energy taken up from the field by the oscillator, determine first the velocity of the electron by differentiation of (4.7),

$$\boldsymbol{v} = -\frac{e}{m} \left(\boldsymbol{C} \cos \omega_0 t + \boldsymbol{S} \sin \omega_0 t\right). \tag{4.9}$$

---

[2] For the interested reader who is not familiar with (4.6) a derivation has been given in Appendix A.2.5. The calculational tools utilized there will be applied frequently throughout this book.

Then the energy reads

$$T = \frac{e^2}{2m} \left(\boldsymbol{C}^2 + \boldsymbol{S}^2 \right) \tag{4.10}$$

or

$$T = \frac{1}{2m} \left| \int_{-\infty}^{\infty} dt \, (-e\boldsymbol{E}(t)) \, e^{i\omega_0 t} \right|^2 \tag{4.11}$$

which clarifies the transition from sudden to adiabatic behavior, i.e., small to large ω_0, respectively. Note that for $\omega_0 = 0$, the energy transfer reduces to $\Delta P^2/2m$, where $\Delta \boldsymbol{P}$ is the momentum transfer. However, keep in mind that (4.11) still contains the unknown function $\boldsymbol{r}(t)$ on the right-hand side via $\boldsymbol{E}(t) \equiv \boldsymbol{E}\left(\boldsymbol{r}(t), t\right).$s

4.2.2 Distant Collisions: Dipole Approximation

Consider now the Coulomb field generated by a fast charged particle in uniform motion,

$$\boldsymbol{E}(\boldsymbol{r}, t) = -\boldsymbol{\nabla}\Phi(\boldsymbol{r}, t) \tag{4.12}$$

with the potential

$$\Phi(\boldsymbol{r}, t) = \frac{e_1}{|\boldsymbol{r} - \boldsymbol{R}(t)|} \tag{4.13}$$

and the trajectory of the projectile

$$\boldsymbol{R}(t) = \boldsymbol{p} + \boldsymbol{v}t, \tag{4.14}$$

where $\boldsymbol{p}$ is the vectorial impact parameter, i.e., the shortest vector connecting the force center 0 to the trajectory (Fig. 4.1). Note that $\boldsymbol{p} \cdot \boldsymbol{v} = 0$.

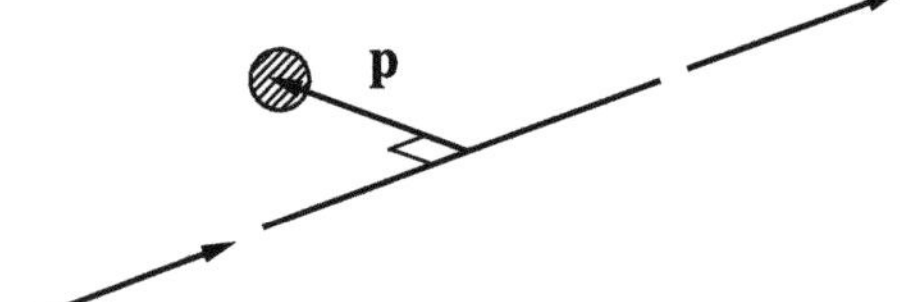

Fig. 4.1. Vectorial impact parameter

Throughout this book we shall need the Fourier transform of the Coulomb potential. It is determined by the relation

$$\frac{1}{r} = \frac{1}{2\pi^2} \int d^3q \, \frac{1}{q^2} \, e^{i\boldsymbol{q} \cdot \boldsymbol{r}}, \tag{4.15}$$

with the integration going over infinite $\boldsymbol{q}$-space. A proof is given in Appendix A.2.2.

Combination of (4.12–4.15) yields

$$\boldsymbol{E}(\boldsymbol{r},t) = -\frac{\mathrm{i}e_1}{2\pi^2} \int \mathrm{d}^3 q \, \frac{\boldsymbol{q}}{q^2} \, \mathrm{e}^{\mathrm{i}\boldsymbol{q}\cdot(\boldsymbol{r}-\boldsymbol{p}-\boldsymbol{v}t)}. \tag{4.16}$$

Equation (4.16) contains $\boldsymbol{r}(t)$ in the exponent. For small displacements from equilibrium, Taylor expansion yields

$$\mathrm{e}^{\mathrm{i}\boldsymbol{q}\cdot\boldsymbol{r}(t)} = 1 + \mathrm{i}\boldsymbol{q}\cdot\boldsymbol{r}\ldots \simeq 1. \tag{4.17}$$

In the second step, all variation of the electric field over a distance of the order of the oscillation amplitude of the target electron has been ignored. This assumption implies 'weak' interaction and, more important, small $\boldsymbol{q}$ or large wavelength. This is the limit that is of interest in classical dispersion theory. Thus, (4.17) can only apply to the interaction at large distances. For close interactions another approach needs to be taken.

Equation (4.17) represents what is called the dipole approximation. It is equivalent with expanding the potential up to first order in $\boldsymbol{r}$ and consistent with the perturbation approach in classical scattering theory discussed in Sect. 3.3.6. In a perturbation treatment the transferred momentum is calculated on the basis of unperturbed trajectories throughout the collision, i.e., uniform trajectory of the projectile and target electron at rest. The same procedure was already applied in Sect. 2.3.2 on page 39.

Once $\boldsymbol{r}(t)$ has been dropped from the exponent in (4.16) by application of (4.17), the Fourier transform of $\boldsymbol{E}(\boldsymbol{r},t)$ emerges from (4.5) and (4.16) and reads

$$\boldsymbol{E}(\omega) = -\frac{\mathrm{i}e_1}{2\pi^2} \int \mathrm{d}^3 q \, \frac{\boldsymbol{q}}{q^2} \, \mathrm{e}^{-\mathrm{i}\boldsymbol{q}\cdot\boldsymbol{p}} \, \delta(\omega - \boldsymbol{q}\cdot\boldsymbol{v}), \tag{4.18}$$

where $\delta(\ldots)$ is the Dirac function. If you are unfamiliar with the Dirac function, have a look at Appendix A.2.4.

The integral is conveniently carried out in cylindrical coordinates with the x-axis chosen along the projectile velocity $\boldsymbol{v}$ and the y-axis along the vectorial impact parameter $\boldsymbol{p}$. With the notation $\boldsymbol{q} = (\omega/v, \boldsymbol{q}_\perp)$, where $\boldsymbol{q}_\perp$ is a vector in the yz plane, one finds

$$\boldsymbol{E}(\omega) = \frac{e_1}{2\pi^2 v} \left(-\mathrm{i}\frac{\omega}{v}, \frac{\partial}{\partial p}, 0 \right) \int \mathrm{d}^2 q_\perp \, \frac{1}{q_\perp^2 + \omega^2/v^2} \, \mathrm{e}^{-\mathrm{i}\boldsymbol{q}_\perp\cdot\boldsymbol{p}}. \tag{4.19}$$

The remaining integral is a modified Bessel function, as follows from

$$\int \mathrm{d}^2 q_\perp \, \frac{1}{q_\perp^2 + \omega^2/v^2} \, \mathrm{e}^{-\mathrm{i}\boldsymbol{q}_\perp\cdot\boldsymbol{p}} = \int_0^\infty \frac{q_\perp \, \mathrm{d}q_\perp}{q_\perp^2 + \omega^2/v^2} \int_0^{2\pi} \mathrm{d}\phi \, \mathrm{e}^{-\mathrm{i}q_\perp p \cos\phi}$$

$$= \int_0^\infty \frac{q_\perp \, \mathrm{d}q_\perp}{q_\perp^2 + \omega^2/v^2} 2\pi J_0(q_\perp p) = 2\pi K_0\left(\frac{\omega p}{v}\right), \tag{4.20}$$

according to Abramowitz and Stegun (1964), p. 488, and hence

$$\boldsymbol{E}(\omega) = -\frac{e_1\omega}{\pi v^2}\left(iK_0\left(\frac{\omega p}{v}\right), K_1\left(\frac{\omega p}{v}\right), 0\right),$$ (4.21)

so that (4.11) reads

$$T = T(p) = \frac{2e_1^2 e^2}{mv^2 p^2}f(p)$$ (4.22)

with

$$f(p) = f_{\mathrm{dist}}(p) = \left[\frac{\omega_0 p}{v}K_0\left(\frac{\omega_0 p}{v}\right)\right]^2 + \left[\frac{\omega_0 p}{v}K_1\left(\frac{\omega_0 p}{v}\right)\right]^2.$$ (4.23)

It follows from (4.21) that the first and second term in the parentheses in (4.23) reflect momentum transferred parallel and perpendicular to the beam direction, respectively.

Both Bessel functions drop off exponentially for $\omega_0 p/v \gg 1$, in agreement with Bohr's adiabatic limit (2.62) discussed on page 45. In the opposite limit we have (Abramowitz and Stegun, 1964)

$$K_0\left(\frac{\omega_0 p}{v}\right) \sim \ln\frac{v}{\omega_0 p}; \quad K_1\left(\frac{\omega_0 p}{v}\right) \sim \frac{v}{\omega_0 p} \text{ for } \frac{\omega_0 p}{v} \ll 1$$ (4.24)

and hence

$$f_{\mathrm{dist}}(p) \simeq 1 \text{ for } \frac{\omega_0 p}{v} \ll 1.$$ (4.25)

This represents the low-p limit of the high-p approximation applied to the Coulomb interaction, in agreement with the qualitative picture arrived at in Sect. 2.3.4.

A convenient analytical representation of this function was found by Grande and Schiwietz (1998),

$$f_{\mathrm{dist}}(p) \simeq \left(1 - 0.174\sqrt{\zeta} + \pi\zeta\right)e^{-2\zeta}, \quad \zeta = \frac{\omega_0 p}{v}$$ (4.26)

the error of which is $\lesssim 1\ \%$ for $0.01 < \zeta < 10$.

4.2.3 Relativistic Extension

The relativistic version of (4.21) may be found on the basis of (2.65) and (2.66)[3],

---

[3] As in Sect. 2 the magnetic interaction is ignored. This is justified in the present perturbational scheme because the Lorentz force, being the product of electron speed and magnetic field which are both proportional to e_1, contributes to terms neglected in the lowest order of the perturbation series.

Another word of caution is indicated if the orbital motion of the target electron is non-negligible. In that case the Lorentz force contributes to the leading term in the perturbation expansion. This may be important for electrons with velocities not too far below the speed of light. However, such effects are most naturally taken into account in relativistic quantum theory.

$$\boldsymbol{E}(\omega) = -\frac{e_1\omega}{\pi\gamma_1 v^2}\left(\frac{\mathrm{i}}{\gamma_1}K_0\left(\frac{\omega p}{\gamma_1 v}\right), K_1\left(\frac{\omega p}{\gamma_1 v}\right), 0\right) \tag{4.27}$$

where $\gamma_1 = 1/\sqrt{1 - v^2/c^2}$. From this one recovers (4.22) with a relativistic version of (4.23),

$$f(p) = f_{\mathrm{dist}}(p) = \frac{1}{\gamma_1^2}\left[\frac{\omega_0 p}{\gamma_1 v}K_0\left(\frac{\omega_0 p}{\gamma_1 v}\right)\right]^2 + \left[\frac{\omega_0 p}{\gamma_1 v}K_1\left(\frac{\omega_0 p}{\gamma_1 v}\right)\right]^2. \tag{4.28}$$

Before drawing more quantitative conclusions we need to have a look at changes brought about by quantum theory.

4.3 Semiclassical Theory

4.3.1 General Considerations

Conventionally a 'semiclassical theory' treats the motion of nuclei by classical mechanics and that of orbiting electrons quantally. In the present context this implies that the physical picture resembles that of the Bohr theory with the sole modification that electrons are no longer treated as classical oscillators but occupy quantum states in a target atom. Fig. 4.1 still specifies the geometry.

In the classical treatment the energy transfer $T(p)$ was uniquely related to the impact parameter by (4.22). In a semiclassical theory we obtain an *excitation probability* $P_j(p)$ for each excitation level j of the target, corresponding to a resonance frequency ω_{j0}, and the quantity to be compared with $T(p)$ from the classical theory is the average energy transfer

$$T_{\mathrm{av}}(p) = \sum_j P_j(p)\hbar\omega_{j0}. \tag{4.29}$$

The stopping cross section following from (4.1) and (4.29) reads

$$S = \sum_j \hbar\omega_{j0} \int 2\pi p\,\mathrm{d}p\, P_j(p), \tag{4.30}$$

and comparison with (2.20) on page 34 shows that

$$\sigma_j = \int 2\pi p\,\mathrm{d}p P_j(p). \tag{4.31}$$

This defines the *excitation cross section* to level j. This important relation – which could have been derived more directly – quantifies the considerations made in the end of Sect. 2.2, in particular the somewhat cryptic relation (2.4): Instead of a 'black area', an atom is viewed by the projectile as a region of varying degrees of 'grey color' expressed by an excitation probability which is typically $\ll 1$ within the range of validity of first-order perturbation theory.

4.3.2 Time-Dependent Perturbation Theory

Consider a target atom or molecule containing Z_2 electrons. It can occupy stationary states $|j\rangle$ with energies ϵ_j, where j stands for a complete set of quantum numbers and $j = 0$ denotes the ground state. Thus, the resonance frequencies for an atom in its ground state are given by

$$\hbar\omega_{j0} = \epsilon_j - \epsilon_0. \tag{4.32}$$

Instead of Newton's second law (4.2), electron motion is governed by Schrödinger's equation

$$(H + \mathcal{V})\Psi(\boldsymbol{r}, t) = \mathrm{i}\hbar \frac{\partial\Psi(\mathbf{r}, t)}{\partial t}, \tag{4.33}$$

where H is the hamiltonian of an isolated target atom or molecule. The potential energy describing the interaction with the projectile is now given by

$$\mathcal{V}(\boldsymbol{r}, t) = \sum_{\nu=1}^{Z_2} (-e)\Phi(\boldsymbol{r}_\nu, t) = \sum_{\nu=1}^{Z_2} \frac{-e_1 e}{|\boldsymbol{r}_\nu - \boldsymbol{R}(t)|}, \tag{4.34}$$

where the vector $\boldsymbol{r}$ stands for $(\boldsymbol{r}_1, \ldots, \boldsymbol{r}_{Z_2})$, $\boldsymbol{r}_\nu$ is the position operator of the ν'th electron and $\boldsymbol{R}(t) = \boldsymbol{p} + \boldsymbol{v}t$ the trajectory of the projectile.

The time-dependent wave function $\Psi(\boldsymbol{r}, t)$ may be expanded in terms of stationary states,

$$\Psi(\boldsymbol{r}, t) = \sum_j c_j(t)\, \mathrm{e}^{-\mathrm{i}\epsilon_j t/\hbar} \, |j\rangle \tag{4.35}$$

with coefficients $c_j(t)$ to be determined from (4.33). In first-order perturbation theory the $c_j(t)$ are expanded in powers of the perturbing potential $\mathcal{V}$. From texts on introductory quantum mechanics or from Appendix A.4.2 you may extract that

$$c_j(t) = \delta_{j0} + c_j^{(1)}(t) + \ldots \tag{4.36}$$

with

$$c_j^{(1)}(t) = \frac{1}{\mathrm{i}\hbar} \int_{-\infty}^{t} \mathrm{d}t' \, \mathrm{e}^{\mathrm{i}\omega_{j0}t'} \langle j|\mathcal{V}(\boldsymbol{r}, t')|0\rangle \tag{4.37}$$

up to first order in $\mathcal{V}(\boldsymbol{r}, t)$. Here δ_{j0} is a Kronecker symbol

$$\delta_{ij} = \begin{cases} 1 & \text{for } i = j \\ 0 & \text{for } i \neq j. \end{cases} \tag{4.38}$$

It is implied in (4.37) that the target atom is in its ground state at $t = -\infty$.

In the classical case, Sect. 4.2.2, only the asymptotic motion at $t = \infty$ was of interest. The same is true here. The coefficients $c_j^{(1)}(\infty)$ represent 'transition amplitudes' which can be evaluated by insertion of (4.34), again expressed as a Fourier transform (4.15). Integration over t' yields

$$c_j^{(1)}(\infty) = -\frac{e_1 e}{\mathrm{i}\pi\hbar} \int \mathrm{d}^3 q \, \frac{\mathrm{e}^{-\mathrm{i}\mathbf{q}\cdot\mathbf{p}}}{q^2} F_{j0}(\mathbf{q})\,\delta(\omega_{j0} - \mathbf{q}\cdot\mathbf{v}), \tag{4.39}$$

where

$$F_{j0}(\mathbf{q}) = \left\langle j \left| \sum_{\nu=1}^{Z_2} \mathrm{e}^{\mathrm{i}\mathbf{q}\cdot\mathbf{r}_\nu} \right| 0 \right\rangle. \tag{4.40}$$

The expectation value of the energy of the oscillator at time $t = \infty$ is given by

$$\langle \Psi(\mathbf{r},t) | H | \Psi(\mathbf{r},t) \rangle_{t=\infty} = \sum_j |c_j(\infty)|^2 \epsilon_j, \tag{4.41}$$

Insertion of (4.36), understood as a series expansion in powers of the ion charge e_1, yields the energy taken up by the atom due to the interaction,

$$T_{\mathrm{av}}(\mathbf{p}) = \langle \Psi(\mathbf{r},t) | H | \Psi(\mathbf{r},\mathbf{t}) \rangle_{t=\infty} - \langle \Psi(\mathbf{r},t) | H | \Psi(\mathbf{r},t) \rangle_{t=-\infty}$$
$$= \sum_j \left| c_j^{(1)}(\infty) \right|^2 (\epsilon_j - \epsilon_0). \tag{4.42}$$

This is an explicit version of (4.29). The transition probabilities are given by

$$P_j(p) = \left| c_j^{(1)}(\infty) \right|^2 \tag{4.43}$$

in first-order perturbation theory.

4.3.3 Distant Collisions

Consider (4.39) at large p, i.e., for distant collisions. In the classical calculation the dipole approximation (4.17) was introduced. After application of the same approximation to (4.40), $F_{j0}(\mathbf{q})$ in (4.39) reduces to

$$F_{j0}(\mathbf{q}) \simeq \mathrm{i}\mathbf{q}\cdot \left\langle j \left| \sum_{\nu=1}^{Z_2} \mathbf{r}_\nu \right| 0 \right\rangle \tag{4.44}$$

to the lowest order in q. The remaining integral in (4.39) is identical with the one in (4.18) and thus leads to

$$c_j^{(1)}(\infty) = -\frac{2e_1 e \omega_{j0}}{i\hbar v^2} \left\langle j \left| \sum_\nu^{Z_2} \boldsymbol{r}_\nu \right| 0 \right\rangle$$

$$\cdot \left(iK_0 \left(\frac{\omega_{j0}p}{v} \right), K_1 \left(\frac{\omega_{j0}p}{v} \right), 0 \right), \quad (4.45)$$

where the plane of incidence has again been assumed to be the xy-plane. With this, the transition probability (4.43) reads

$$P_j(p) = \frac{2e_1^2 e^2 Z_2}{mv^2 p^2 \hbar \omega_{j0}} f_{j0}$$

$$\times \left\{ \left[\frac{\omega_{j0}p}{v} K_0 \left(\frac{\omega_{j0}p}{v} \right) \right]^2 + \left[\frac{\omega_{j0}p}{v} K_1 \left(\frac{\omega_{j0}p}{v} \right) \right]^2 \right\}, \quad (4.46)$$

where the *dipole oscillator strengths*

$$f_{j0} = \frac{2m}{3\hbar^2 Z_2}(\epsilon_j - \epsilon_0) \left| \left\langle j \left| \sum_\nu^{Z_2} \boldsymbol{r}_\nu \right| 0 \right\rangle \right|^2 \quad (4.47)$$

have been introduced. These quantities are familiar from the theory of optical dispersion. Pertinent results have been compiled in Appendix A.5.

The Thomas-Reiche-Kuhn sum rule

$$\sum_j f_{j0} = 1, \quad (4.48)$$

which is proven there, is of prime importance in the present context.

From (4.29) and (4.46) we find the average energy transfer per target electron to be given by (4.22) with

$$f_{\mathrm{dist}}(p) = \sum_j f_{j0} \left\{ \left[\frac{\omega_{j0}p}{v} K_0 \left(\frac{\omega_{j0}p}{v} \right) \right]^2 + \left[\frac{\omega_{j0}p}{v} K_1 \left(\frac{\omega_{j0}p}{v} \right) \right]^2 \right\}. \quad (4.49)$$

Eq. (4.49) was first derived by Bloch (1933b). The similarity to (4.23) is striking. In particular, if there is only one allowed transition, then (4.49) agrees exactly with Bohr's result for a classical harmonic oscillator. The message emerges that in distant interactions the quantum mechanical response of an atom or molecule to the perturbation induced by the projectile is identical with the classical response of an ensemble of harmonic oscillators with the transition frequencies ω_{j0}, weighted by the respective dipole oscillator strengths f_{j0}.

This result is more generally true and well-established in the theory of optical dispersion and absorption. The derivation for that case has been sketched in Appendix A.5.

4.3.4 Excitation Cross Section

From (4.39), (4.43) and (4.31) we find

$$\sigma_j = \frac{e_1^2 e^2}{\pi^2 \hbar^2} \int \frac{\mathrm{d}^3 q}{q^2} \int \frac{\mathrm{d}^3 q'}{q'^2}\, F_{j0}^*(\boldsymbol{q}) F_{j0}(\boldsymbol{q}')$$

$$\times\, \delta\left(\omega_{j0} - \boldsymbol{q}\cdot\boldsymbol{v}\right) \delta\left(\omega_{j0} - \boldsymbol{q}'\cdot\boldsymbol{v}\right) \int \mathrm{d}^2 p\, \mathrm{e}^{\mathrm{i}(\boldsymbol{q}-\boldsymbol{q}')\cdot\boldsymbol{p}}, \quad (4.50)$$

where an asterisk indicates the complex conjugate. The reduction

$$\delta\left(\omega_{j0} - \boldsymbol{q}\cdot\boldsymbol{v}\right) \delta\left(\omega_{j0} - \boldsymbol{q}'\cdot\boldsymbol{v}\right) \int \mathrm{d}^2 p\, \mathrm{e}^{\mathrm{i}(\boldsymbol{q}-\boldsymbol{q}')\cdot\boldsymbol{p}}$$

$$= \frac{(2\pi)^2}{v}\, \delta(\boldsymbol{q}-\boldsymbol{q}')\, \delta\left(\omega_{j0} - \boldsymbol{q}\cdot\mathbf{v}\right) \quad (4.51)$$

emerges in the following way: The integral over $\mathrm{d}^2 p$ yields a two-dimensional Dirac function in $\boldsymbol{q}_\perp - \boldsymbol{q}'_\perp$, i.e., the components of $\boldsymbol{q}$ and $\boldsymbol{q}'$ perpendicular to the velocity. The two Dirac functions on the left-hand side taken together imply that $\boldsymbol{q}\cdot\boldsymbol{v} = \boldsymbol{q}'\cdot\boldsymbol{v}$. Thus, the components of $\boldsymbol{q}$ and $\boldsymbol{q}'$ parallel to the velocity must also be equal. The combined effect is a three-dimensional Dirac function in $\boldsymbol{q} - \boldsymbol{q}'$, and the factor $1/v$ arises from (A.70) in Appendix A.2.4.

With this we arrive at

$$\sigma_j = \frac{4 e_1^2 e^2}{\hbar^2 v} \int \frac{\mathrm{d}^3 q}{q^4}\, |F_{j0}(\boldsymbol{q})|^2\, \delta\left(\omega_{j0} - \boldsymbol{q}\cdot\mathbf{v}\right). \quad (4.52)$$

This represents a general expression for the excitation cross section in the first Born approximation, derived in the semiclassical approximation under the assumption of uniform motion of a projectile much heavier than an electron. It will turn out in Sect. 4.4 that a very similar result emerges from the fully quantal calculation.

Frequently it is of interest to take the orientational average over (4.52). This is appropriate whenever target atoms or molecules are randomly oriented. In practice it is more convenient to average over incoming beam directions, i.e., the angular part of the incident velocity $\mathbf{v}$,

$$\langle \delta\left(\omega_{j0} - \boldsymbol{q}\cdot\mathbf{v}\right)\rangle = \begin{cases} \dfrac{1}{2qv} & \omega_{j0} < qv \\[2mm] & \text{for} \\[1mm] 0 & \text{otherwise.} \end{cases} \quad (4.53)$$

This is easily verified by introduction of spherical coordinates for $\boldsymbol{v}$ with $\boldsymbol{q}$ taken as the polar axis. With this, any explicit orientation dependence of $|F_{j0}(\boldsymbol{q})|^2$ caused by degenerate states must have dropped out. Therefore the angular integration in (4.52) can be carried out. By introduction of the quantity

$$Q = \frac{\hbar^2 q^2}{2m} \tag{4.54}$$

one obtains

$$\sigma_j = \frac{2\pi e_1^2 e^2}{mv^2} \int \frac{\mathrm{d}Q}{Q^2} |F_{j0}(\boldsymbol{q})|^2 \tag{4.55}$$

with the boundary condition

$$2mv^2 Q > (\epsilon_j - \epsilon_0)^2 \tag{4.56}$$

which follows from (4.53).

The derivation presented here is quite direct but has the major weakness that no physical meaning has been attached to the Fourier variables $\boldsymbol{q}$ and Q. In Sect. 4.4, $\hbar\boldsymbol{q}$ will be identified as the transferred momentum.

4.4 Plane-Wave Born Approximation

4.4.1 General Considerations

In (4.31) the excitation cross section was defined as an integral over the impact parameter. The statistical definition of a cross section outlined in Sect. 2.2 did not invoke an impact parameter at all. Therefore, a proper quantal calculation ought to determine σ_j more directly. This requires abandoning the semiclassical picture. Instead, the calculation will be based on the picture underlying the fundamental definition (2.1), where a target particle is exposed to a uniform density J of particle current and σ_j emerges as the mean number of excitations per unit time divided by J. A uniform incident particle current is described as a plane wave, just as in Sect. 3.4 on elastic scattering. The presentation will follow that of the quantum theory of elastic scattering as closely as possible, both in the general outline and in the perturbation expansion (Born approximation). This involves stationary perturbation theory and may be found less elegant than a procedure based on time-dependent perturbation theory, but it illuminates physical processes and is quite explicit. The reader who is familiar with Fermi's golden rule may jump over most of the next section and go directly to (4.77).

4.4.2 Stationary Perturbation Theory

While keeping target definitions as in Sect. 4.3 we now characterize the incoming beam by a plane wave as in (3.80),

$$A\mathrm{e}^{\mathrm{i}\boldsymbol{k}\cdot\boldsymbol{R}} \tag{4.57}$$

with a momentum $\hbar\boldsymbol{k}$ and an amplitude A. The symbol $\boldsymbol{R}$, which denoted the classical trajectory of the projectile in the previous sections, now turns into the position operator of the projectile while r is still reserved to electron coordinates[4]. The interaction with target electrons is characterized by a potential energy $\mathcal{V} = \mathcal{V}(\boldsymbol{R}, \boldsymbol{r})$ similar to (4.34).

For a stationary current the system obeys the time-independent Schrödinger equation,

$$\left(\frac{\boldsymbol{P}^2}{2m_1} + H + \mathcal{V}(\boldsymbol{R}, \boldsymbol{r}) - \mathcal{E} \right) \Psi(\boldsymbol{R}, \boldsymbol{r}) = 0, \tag{4.58}$$

where $\boldsymbol{P} = -i\hbar \boldsymbol{\nabla}_R$ is the momentum operator of the projectile, $\mathcal{E}$ the initial energy of the system,

$$\mathcal{E} = \hbar^2 k^2 / 2m_1 + \epsilon_0 \tag{4.59}$$

and m_1 the projectile mass.

We may expand the wave function Ψ in terms of the eigenstates of the target,

$$\Psi(\boldsymbol{R}, \boldsymbol{r}) = \sum_{\ell} u_{\ell}(\boldsymbol{R}) |\ell\rangle, \tag{4.60}$$

where the $u_{\ell}(\boldsymbol{R})$ are unknown functions of $\boldsymbol{R}$.

After insertion of (4.60) into (4.58), noting that $H|\ell\rangle = \epsilon_{\ell}|\ell\rangle$, and multiplication by $\langle j|$, you find the following system of differential equations governing $u_j(\boldsymbol{R})$,

$$\left(\boldsymbol{\nabla}_R^2 + k_j^2 \right) u_j(\boldsymbol{R}) = \frac{2m_1}{\hbar^2} \sum_{\ell} \mathcal{V}_{j\ell}(\boldsymbol{R}) u_{\ell}(\boldsymbol{R}) \tag{4.61}$$

with

$$\mathcal{V}_{j\ell}(\boldsymbol{R}) = \langle j | \mathcal{V}(\boldsymbol{R}, \boldsymbol{r}) | \ell \rangle \tag{4.62}$$

and a constant k_j defined by

$$k_j^2 = \frac{2m_1}{\hbar^2} \left(\mathcal{E} - \epsilon_j \right). \tag{4.63}$$

At this point it is convenient to apply a perturbation expansion, i.e., to expand the u_j in powers of the interaction potential $\mathcal{V}$,

$$u_j(\boldsymbol{R}) = u_j^{(0)}(\boldsymbol{R}) + u_j^{(1)}(\boldsymbol{R}) + u_j^{(2)}(\boldsymbol{R}) + \cdots , \tag{4.64}$$

where $u_j^{(\nu)}(\boldsymbol{R})$ is proportional to $\mathcal{V}^{\nu}$, i.e., to e_1^{ν}. Separating (4.61) according to equal powers of e_1 one finds

---

[4] It is emphasized that the meaning of the symbols $\boldsymbol{R}$ and $\boldsymbol{r}$ in this chapter is distinctly different from that in Chapter 3.

$$\left(\boldsymbol{\nabla}_R^2 + k_j^2\right)u_j^{(\nu+1)}(\boldsymbol{R}) = \frac{2m_1}{\hbar^2}\sum_\ell \mathcal{V}_{j\ell}(\boldsymbol{R})u_\ell^{(\nu)}(\boldsymbol{R}). \tag{4.65}$$

The zero-order wave function represents a free projectile with the target in the ground state,

$$u_j^{(0)}(\boldsymbol{R}) = \delta_{j0}A\mathrm{e}^{\mathrm{i}\boldsymbol{k}\cdot\boldsymbol{R}}. \tag{4.66}$$

Starting from there one may write down solutions of (4.65) in closed form to arbitrarily high order ν.

Equation (4.65) has the form that was considered in the theory of elastic scattering. Therefore its solution is analogous to (3.81),

$$u_j^{(\nu+1)}(\boldsymbol{R}) = -\frac{m_1}{2\pi\hbar^2}\int \mathrm{d}^3\boldsymbol{R}' \,\frac{\mathrm{e}^{\pm\mathrm{i}k_j|\boldsymbol{R}-\boldsymbol{R}'|}}{|\boldsymbol{R}-\boldsymbol{R}'|}\sum_\ell \mathcal{V}_{j\ell}(\boldsymbol{R}')u_\ell^{(\nu)}(\boldsymbol{R}'). \tag{4.67}$$

as follows from (A.108) in Appendix A.2.5. In principle, a free-particle wave could be added to (4.67), but that aspect is taken care of by (4.66). The $\pm$ sign in (4.67) indicates outgoing and incoming waves. Since there is no incoming wave except the one characterized by eq. (4.66), only the $+$ sign provides a physically valid solution. Here we focus on the lowest order $\nu = 0$. At large distances R from the scattering center that contribution reads

$$u_j^{(1)}(\boldsymbol{R}) = -\frac{m_1}{2\pi\hbar^2}A\frac{\mathrm{e}^{\mathrm{i}k_j R}}{R}\int \mathrm{d}^3\boldsymbol{R}' \,\mathrm{e}^{-\mathrm{i}\boldsymbol{k}_j\cdot\boldsymbol{R}'}\,\mathcal{V}_{j0}(\boldsymbol{R}')\,\mathrm{e}^{\mathrm{i}\boldsymbol{k}\cdot\boldsymbol{R}'} \tag{4.68}$$

with

$$\boldsymbol{k}_j = k_j\frac{\boldsymbol{R}}{R}. \tag{4.69}$$

Combination of (4.60) and (4.68) yields

$$\Psi^{(1)}(\boldsymbol{R},\boldsymbol{r}) = -\frac{m_1}{2\pi\hbar^2}A\sum_j \frac{\mathrm{e}^{\mathrm{i}k_j R}}{R}|j\rangle\langle j,\boldsymbol{k}_j|\mathcal{V}|0,\boldsymbol{k}\rangle, \tag{4.70}$$

where

$$\langle j,\boldsymbol{k}_j|\mathcal{V}|0,\boldsymbol{k}\rangle = \int \mathrm{d}^3\boldsymbol{R}\,\mathrm{e}^{-\mathrm{i}\boldsymbol{k}_j\cdot\boldsymbol{R}}\,\mathcal{V}_{j0}(\boldsymbol{R})\,\mathrm{e}^{\mathrm{i}\boldsymbol{k}\cdot\boldsymbol{R}} \tag{4.71}$$

is the matrix element of the interaction $\mathcal{V}$ between the initial and final states of the system.

4.4.3 Excitation Cross Section

In counting the mean number of events per unit time, an 'event' is the excitation of the target from the ground state $|0\rangle$ to some state $|j\rangle$ *and* the simultaneous deflection of a primary particle from its original direction $\boldsymbol{k}/k$ into some final direction $\boldsymbol{R}/R$.

The coefficient of $|j\rangle$ in (4.70) determines the outgoing transition amplitude and the associated current density is

$$\boldsymbol{J}_j = \frac{\hbar \boldsymbol{k}_j}{m_1} \left(\frac{m_1}{2\pi\hbar^2}\right)^2 \frac{|A|^2}{R^2} \left|\langle j, \boldsymbol{k}_j | \mathcal{V} | 0, \boldsymbol{k}\rangle\right|^2 . \tag{4.72}$$

Now, the number of projectiles passing per unit time through an area at distance R and solid angle $\mathrm{d}^2\boldsymbol{\Omega}$ is

$$\boldsymbol{J}_j \cdot \frac{\boldsymbol{k}_j}{k_j} R^2 \mathrm{d}^2\boldsymbol{\Omega} , \tag{4.73}$$

cf. the corresponding discussion in Chapter 3. Dividing this by the incoming current density $|A|^2 \hbar k/m_1$ and integrating over all directions of the outgoing projectile we obtain the cross section σ_j for excitation to level j according to (2.1),

$$\sigma_j = \frac{k_j}{k} \left(\frac{m_1}{2\pi\hbar^2}\right)^2 \int \mathrm{d}^2\boldsymbol{\Omega} |\langle j, \boldsymbol{k}_j | \mathcal{V} | 0, \boldsymbol{k}\rangle|^2 . \tag{4.74}$$

It is convenient to convert the integration over outgoing directions $\boldsymbol{\Omega}$ into an integration over the outgoing wave vector. This may be achieved by combining (4.59) and (4.63) into

$$\mathcal{E} = \frac{\hbar^2 k_j^2}{2m_1} + \epsilon_j = \frac{\hbar^2 k^2}{2m_1} + \epsilon_0 . \tag{4.75}$$

This reflects energy conservation: If the target has been excited to ϵ_j, the quantity $\hbar^2 k_j^2/2m_1$ is the energy left to the projectile. $\hbar k_j$ is the magnitude of the associated momentum, and hence $\hbar \boldsymbol{k}_j = \hbar k_j \boldsymbol{R}/R$ is the momentum vector of a scattered projectile when the target has been excited to level j. The equality (4.75) may be incorporated into (4.74) by treating k_j as a variable and multiplying the integrand by

$$\delta\left(\frac{\hbar^2 k_j^2}{2m_1} + \epsilon_j - \frac{\hbar^2 k^2}{2m_1} - \epsilon_0\right) \mathrm{d}\left(\frac{\hbar^2 k_j^2}{2m_1}\right) . \tag{4.76}$$

This leads to

$$\sigma_j = \frac{1}{(2\pi)^2 \hbar v} \int \mathrm{d}^3 \boldsymbol{k}' \delta\left(\frac{\hbar^2 k'^2}{2m_1} + \epsilon_j - \frac{\hbar^2 k^2}{2m_1} - \epsilon_0\right) |\langle j, \boldsymbol{k}' | \mathcal{V} | 0, \boldsymbol{k}\rangle|^2 , \tag{4.77}$$

where $\boldsymbol{k}_j$ has been called $\boldsymbol{k}'$ since it has become an integration variable.

Eq. (4.77) is a version of Fermi's golden rule which in general determines the transition rate per unit time to first order of a perturbed system with a quasi-continuous excitation spectrum (Schiff, 1981).

4.4.4 Coulomb Interaction

Now assume Coulomb interaction between the projectile and the target particles. By use of (4.34) and (4.15), (4.71) reduces to

$$\langle j, \boldsymbol{k}'|\mathcal{V}|0, \boldsymbol{k}\rangle = -\frac{4\pi e_1 e}{(\boldsymbol{k} - \boldsymbol{k}')^2} F_{j0}(\boldsymbol{k} - \boldsymbol{k}'), \tag{4.78}$$

where F_{j0} is defined by (4.40). Noting that $\boldsymbol{k}' - \boldsymbol{k}$ is the change in wave vector *of the projectile* we may identify $\hbar\boldsymbol{k} - \hbar\boldsymbol{k}'$ as the momentum transferred from the projectile to the target. Denoting this vector by $\hbar\boldsymbol{q}$, we find from (4.77) and (4.78) that

$$\sigma_j = \frac{4e_1^2 e^2}{\hbar v} \int \frac{\mathrm{d}^3 q}{q^4} \left|F_{j0}(\boldsymbol{q})\right|^2 \delta(\epsilon_j - \epsilon_0 - \hbar\boldsymbol{q} \cdot \boldsymbol{v} + \hbar^2 q^2/2m_1). \tag{4.79}$$

The similarity of this expression to (4.52) is striking. The only significant change is the occurrence of a new term $\hbar^2 q^2/2m_1$ in the argument of the Dirac function. Evidently this term ensures energy conservation. It is essential if the projectile is an electron or positron. For heavy projectiles the term is smaller than the preceding one by a factor of $\hbar q/2m_1 v$, i.e., negligible.

Eq. (4.79) can again be brought into a more handy form under the assumption that the target atom or molecule has no preferred orientation. Then the procedure described in connection with (4.53) yields

$$\sigma_j = \frac{2e_1^2 e^2}{\hbar^2 v^2} \int \frac{\mathrm{d}^3 q}{q^5} |F_{j0}(\boldsymbol{q})|^2, \tag{4.80}$$

where the integral is bounded by

$$\hbar q v > \epsilon_j - \epsilon_0 + \hbar^2 q^2/2m_1. \tag{4.81}$$

With the variable Q defined by (4.54) one finally arrives at

$$\sigma_j = \frac{2\pi e_1^2 e^2}{mv^2} \int \frac{\mathrm{d}Q}{Q^2} |F_{j0}(\boldsymbol{q})|^2 \tag{4.82}$$

with

$$2mv^2 Q > (\epsilon_j - \epsilon_0 + mQ/m_1)^2. \tag{4.83}$$

4.5 The Stopping Cross Section

4.5.1 Bohr Stopping Formula

Before arriving at a stopping cross section by integration of $T(p)$ over the impact parameter following (4.1), we need to remove the logarithmic divergence

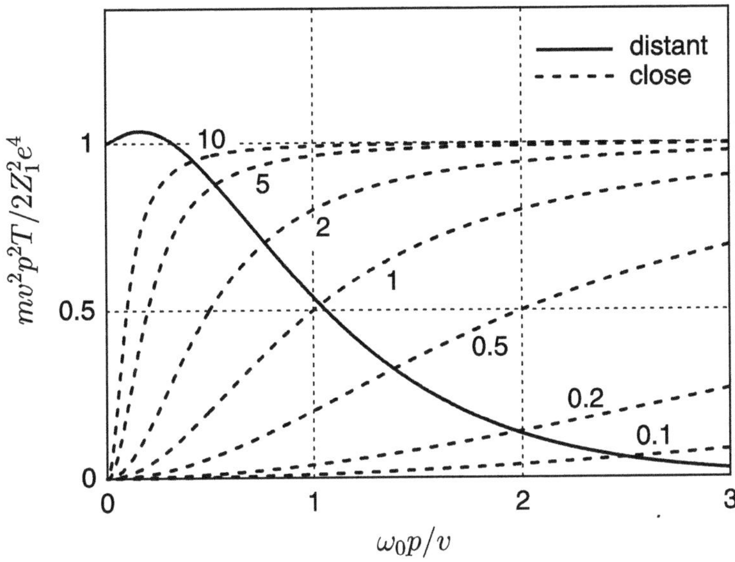

Fig. 4.2. Energy-loss functions T_{dist} and T_{close} *versus* impact parameter p. In the chosen units, T_{dist} (solid line) is a universal curve while T_{close} depends on projectile speed through the parameter $0.1 < mv^3/Z_1e^2\omega_0 < 10$. The crossover defines the limiting impact parameter p_0 (Neufeld, 1954). From Sigmund (1996)

of (4.22) at small p which follows from (4.23). Bohr achieved this by adopting Rutherford's law (3.46) of free-Coulomb scattering for close collisions, which may be cast into the form (4.22) with

$$f(p) = f_{\text{close}}(p) = \frac{1}{1 + (b/2p)^2}, \tag{4.84}$$

where $b = 2Z_1e^2/mv^2$. The two functions $f_{\text{close}}(p)$ and $f_{\text{dist}}(p)$ have been plotted in Fig. 4.2. With the chosen abscissa variable, $f_{\text{dist}}(p)$ becomes a universal function while $f_{\text{close}}(p)$ becomes dependent on the parameter[5]

$$\xi = \frac{2v}{|b|\omega_0} = \frac{mv^3}{Z_1e^2\omega_0}. \tag{4.85}$$

Figure 4.2 suggests to split the integration into two parts at the intersection point p_0. It is seen that for large enough values of ξ the result must be insensitive to the actual choice of p_0 since both functions are $\simeq 1$ over a comfortable interval.

---

[5] Within the context of the Bohr theory we may safely replace $e_1 = Z_1e$, since this theory does not apply to electrons.

With this one finds

$$S_{\text{close}} = \int_0^{p_0} 2\pi p \, \mathrm{d}p \, T_{\text{close}}(p) = \frac{2\pi e_1^2 e^2}{mv^2} \ln\left(1 + \frac{4p_0^2}{b^2}\right) \tag{4.86}$$

or, for large velocities,

$$S_{\text{close}} \simeq \frac{4\pi e_1^2 e^2}{mv^2} \ln \frac{2p_0}{b}. \tag{4.87}$$

In the contribution from distant collisions, note that $K_1(\zeta) = -dK_0(\zeta)/\mathrm{d}\zeta$. By partial integration,

$$S_{\text{dist}} = \int_{p_0}^{\infty} 2\pi p \, \mathrm{d}p \, T_{\text{dist}}(p) = \frac{4\pi Z_1^2 e^4}{mv^2} \times$$

$$\left(\int_{\zeta_0}^{\infty} \mathrm{d}\zeta K_0(\zeta) \left[\zeta K_0(\zeta) + \frac{d}{\mathrm{d}\zeta}\{\zeta K_1(\zeta)\} \right] - \zeta K_0(\zeta) K_1(\zeta) \Big|_{\zeta_0}^{\infty} \right), \tag{4.88}$$

where $\zeta_0 = \omega_0 p_0 / v$. The function in the square brackets is zero (Abramowitz and Stegun (1964), p. 376) and hence,

$$S_{\text{dist}} = \frac{4\pi e_1^2 e^2}{mv^2} \zeta_0 K_0(\zeta_0) K_1(\zeta_0). \tag{4.89}$$

In the range where the curve depicted in Fig. 4.2 is flat, the Bessel functions can be represented by their expansions for small arguments,

$$\zeta_0 K_0(\zeta_0) K_1(\zeta_0) = \ln(2/\zeta_0) - \gamma + O\{\zeta_0^2\}, \tag{4.90}$$

where $\gamma = 0.5772$ is Euler's constant. Hence,

$$S_{\text{dist}} \simeq \frac{4\pi e_1^2 e^2}{mv^2} \ln \frac{2v e^{-\gamma}}{\omega_0 p_0} \tag{4.91}$$

in the limit of high speed. Addition of (4.91) to (4.87) yields

$$S = S_{\text{close}} + S_{\text{dist}} = \frac{4\pi e_1^2 e^2}{mv^2} \ln \frac{2Cv}{b\omega_0}, \tag{4.92}$$

after insertion of b, where $C = 2\exp(-\gamma) = 1.1229$. It is seen that p_0 has dropped out as expected. Thus we have found Bohr's famous formula

$$S = \frac{4\pi e_1^2 e^2}{mv^2} \ln \frac{Cmv^3}{|e_1 e|\omega_0} \qquad \text{with } C = 1.1229 \tag{4.93}$$

for the stopping cross section of an electron bound by a harmonic force hit by uniformly moving point charges. This is identical with (2.64) on page 45, yet the constant C has now been assigned a definite numerical value.

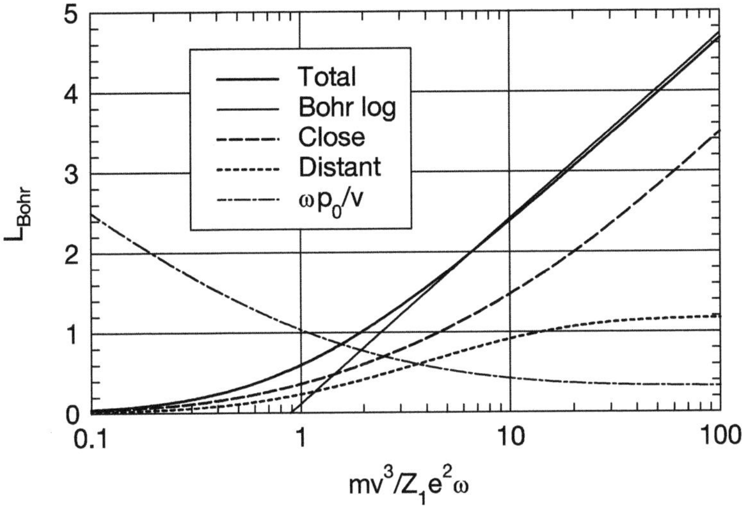

Fig. 4.3. Stopping number L, defined by (2.56), *versus* Bohr parameter $\xi = mv^3/Z_1e^2\omega$. Thick solid line: Straight evaluation of (4.94); contributions from close and distant collisions included separately; thin solid line: Bohr logarithm (4.93); dot-dashed line: Limiting impact parameter p_0 in dimensionless units

Eqs. (4.87) and (4.91) are accurate for sufficiently high projectile speeds. Now try to evaluate S without utilizing the asymptotic relations, i.e.,

$$S = \frac{4\pi e_1^2 e^2}{mv^2}\left\{\frac{1}{2}\ln\left[1+\left(\frac{2v\zeta_0}{b\omega_0}\right)^2\right] + \zeta_0 K_0(\zeta_0)\,K_1(\zeta_0)\right\} \tag{4.94}$$

with ζ_0 denoting the intersection. The resulting expression,

$$S = \frac{4\pi e_1^2 e^2}{mv^2}\,L\left(\frac{2v}{\omega_0 b}\right) = \frac{4\pi e_1^2 e^2}{mv^2}\,L\left(\frac{mv^3}{e_1 e\omega_0}\right) \tag{4.95}$$

is shown in Fig. 4.3 together with Bohr's asymptotic expression (4.93). As was to be expected there is a systematic deviation at low projectile speeds, but otherwise the two curves agree very well.

Basko (2005), in a recent numerical and analytical study, has demonstrated a more complex behavior in the close-collision region. At low projectile speed, when the Coulomb attraction by the projectile competes with the harmonic binding force, the electron may be temporarily bound to the projectile[6], with drastic deviations from Rutherford-like scattering as a direct consequence.

---

[6] In reality, a target electron may actually be captured by the projectile. Such processes are important at sufficiently low projectile speed and will be discussed in considerable detail in Volume II. In the present model, where electrons are bound harmonically to their target sites, such permanent capture is precluded.

This shows up as pronounced oscillations primarily in the energy loss versus impact parameter but is also observed as a deviation from smooth behavior of the stopping cross section as a function of projectile speed. Note, however, that such singular behavior is tied to the neglect of the orbital motion of target electrons. This simplification breaks down at low projectile speeds and needs to be repaired by the *shell correction* which will be discussed in detail in Sect. 6.6.

Relativistic Extension

Evaluation of the distant contribution to the stopping cross section from (4.28) instead of (4.23) yields

$$S_{\mathrm{dist}} = \frac{4\pi e_1^2 e^2}{mv^2} \left(\ln \frac{Cv}{\omega_0 p_0} - \ln \sqrt{1 - \frac{v^2}{c^2}} - \frac{v^2}{2c^2} \right). \tag{4.96}$$

This result, however, only accounts for the longitudinal (Coulomb) interaction. At relativistic speeds also transverse interactions need to be accounted for. This will be done in the context of quantum theory in Sect. 5.6.

From (3.27) we obtain the maximum energy transfer

$$T_{\max} = 2\gamma_1^2 mv^2 = \frac{2mv^2}{1 - v^2/c^2}. \tag{4.97}$$

The energy-loss function in the limit of negligible binding is identical with the nonrelativistic expression when written in the form

$$T(p) = \frac{2e_1^2 e^2}{mv^2 p^2}. \tag{4.98}$$

This may be used to extract an effective lower limit for the impact parameter p_0 by requiring $T(p_0) = T_{\max}$ or

$$p_0 = \frac{e_1 e}{\gamma_1 mv^2}. \tag{4.99}$$

After insertion into (4.96) one finds

$$S = \frac{4\pi e_1^2 e^2}{mv^2} \left[\ln \frac{Cmv^3}{e_1 e \omega_0} - \ln \left(1 - \frac{v^2}{c^2} \right) - \frac{v^2}{2c^2} \right]. \tag{4.100}$$

Comparison with (2.73) reveals the occurrence of a term $-v^2/2c^2$ from distant interactions[7].

---

[7] Bohr (1948) quotes twice the value of this term from a classical calculation. While this is in agreement with the quantal result (cf. below) the actual derivation in Bohr (1915) discusses only the limit of distant collisions. The present result is identical with the one quoted by Jackson (1975).

4.5.2 Semiclassical Theory: Harmonic Oscillator

The range of validity of the semiclassical picture is not restricted to distant collisions. Therefore, $T_{\mathrm{av}}(p)$ can be evaluated and integrated for all impact parameters for any system for which the matrix elements $F_{j0}(\boldsymbol{q})$ occurring in (4.39) are available. While this is the case for atomic hydrogen (Bethe, 1930), an even simpler system will be presented, the quantal harmonic oscillator, which contains the essentials. We have seen that this system is highly relevant in the limit of distant collisions. In the opposite limit, where details of atomic binding are not expected to be of prime importance, this system incorporates an amount of zero-point motion as well as the wave nature of a target electron.

Stationary states of a three-dimensional spherical harmonic oscillator are readily evaluated in cartesian coordinates, where they factorize into one-dimensional states so that $|j\rangle = |j_x, j_y, j_z\rangle = |j_x\rangle|j_y\rangle|j_z\rangle$. Therefore $F_{j0}(\boldsymbol{q})$ can be represented as a product of three factors of the form

$$\langle j_x | e^{i q_x x} | 0 \rangle = \frac{1}{\sqrt{j_x!}} \left(i\sqrt{\frac{\hbar q_x^2}{2m\omega_0}} \right)^{j_x} e^{-\hbar q_x^2/4m\omega_0}. \tag{4.101}$$

A derivation of this relation has been sketched in Appendix A.4.3. Eq. (4.101) supplies the necessary input for an evaluation of $T(p)$ on the basis of (4.39) and (4.42). The results quoted here (Mikkelsen and Sigmund, 1987) were found in practice by an alternate scheme, where the Fourier transform (4.15) of the Coulomb potential has been replaced by another integral representation,

$$\frac{1}{r} = \frac{1}{\sqrt{\pi}} \int_0^\infty \frac{\mathrm{d}\eta}{\sqrt{\eta}}\, e^{-\eta r^2} \tag{4.102}$$

which ensures rapid convergence of numerical integrations.

Figure 4.4 shows $T(p)$ evaluated in this manner, compared with the classical result in the dipole limit (4.23), which is valid in the limit of large p. Results are given for a wide range of velocities, expressed in terms of the 'Bethe parameter' $2mv^2/\hbar\omega_0$. The two sets of curves agree well with each other for $\beta p \gtrsim 1$, i.e., at impact parameters exceeding the oscillator radius $1/\beta = \sqrt{\hbar/m\omega_0}$. At smaller impact parameters an increasing fraction of electronic collisions must be close encounters which are not well characterized by the dipole approximation. At the same time, electrons are distributed in space and have a nonvanishing orbital speed. Specifically, for $p = 0$, Fig. 4.4 suggests a scaling behavior like

$$T(0) = \frac{e_1^2 e^2 \omega_0}{\hbar v^2} g\left(\frac{2mv^2}{\hbar\omega_0} \right), \tag{4.103}$$

where g is an increasing function of the Bethe parameter. Regardless of the detailed form of g, this scaling relationship does not predict the classical result

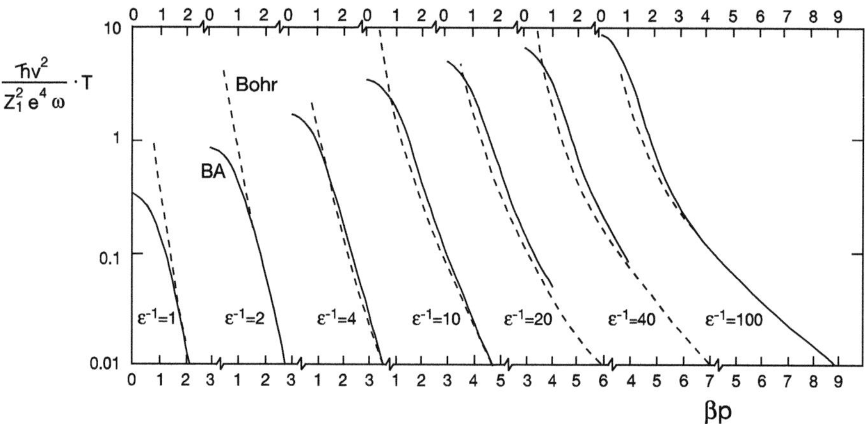

Fig. 4.4. Impact-parameter dependence of mean energy transfer from a projectile in uniform motion to a three-dimensional, spherical harmonic oscillator. Solid lines: First Born approximation. Dashed lines: Classical limit, dipole approximation. Labels indicate the value of the Bethe parameter $2mv^2/\hbar\omega_0$. The abscissa unit is the oscillator radius $(\hbar/m\omega_0)^{1/2}$. From Mikkelsen and Sigmund (1987)

$T(0) = 2mv^2$ for head-on collisions. On the other hand, the functional dependence on impact parameter looks very much like that for classical Coulomb scattering, (4.84). This suggests a behavior like

$$T(p) \simeq \frac{2e_1^2 e^2}{mv^2} \frac{1}{p^2 + (b')^2/4} \tag{4.104}$$

near $p = 0$, which differs from the classical expression (4.84) only in the replacement of the collision diameter b by some parameter b' which is to be specified. If the product $(\hbar v^2/e_1^2 e^2 \omega_0)T(0)$ depends only on the Bethe parameter, then b' must scale as $\hbar/mv$, i.e., as the de Broglie wavelength λ. This is reasonable, since for close collisions an impact parameter cannot be defined to better than some length of the order of the de Broglie wavelength.

Replacement of b in (4.92) by b' yields

$$S \simeq \frac{4\pi e_1^2 e^2}{mv^2} \ln \frac{2mv^2}{\hbar\omega_0} \tag{4.105}$$

as a rough approximation. This is Bethe's formula, derived here for a harmonic oscillator and as an estimate with rough numbers but correct scaling variables. Such a derivation was sketched by Fermi (1950).

4.5.3 Plane-Wave Born Approximation

Generalized Oscillator Strength

Consider the quantity Q defined by (4.54). If target electrons were free and at rest initially, and if all momentum transfer in a collision were to go into one target electron, then Q would be the electron energy after the collision and could be compared with the classical energy transfer T in sect. 2. It is illuminating to write the stopping cross section of an atom in the Born approximation in the form

$$S = \int Q \mathrm{d}\sigma_{\mathrm{R}}(Q) \sum_j f_{j0}(Q), \tag{4.106}$$

where $\mathrm{d}\sigma_{\mathrm{R}}(Q)$ is the free-Coulomb cross section for energy transfer Q. Summation and integration are limited subject to (4.83), and the functions

$$f_{j0}(Q) = \frac{1}{Z_2} \frac{\epsilon_j - \epsilon_0}{Q} \left| F_{j0}(\boldsymbol{q}) \right|^2 \tag{4.107}$$

are called 'generalized oscillator strengths'. In the limit of small $\boldsymbol{q}$ these quantities reduce to

$$f_{j0}(Q) \bigg|_{Q=0} = f_{j0}, \tag{4.108}$$

where f_{j0} is a dipole oscillator strength as specified in (4.47). According to (4.106) the generalized oscillator strengths account for quantum effects on the stopping cross section.

Harmonic Oscillator

As in Sect. 4.5.2 a one-electron atom modelled by a harmonically bound electron will serve as an illustration of more general behavior. From (4.101) or Appendix A.4.3 one finds the oscillator strengths

$$f_{j0}(Q) = \frac{1}{(j-1)!} \left(\frac{Q}{\hbar\omega_0} \right)^{j-1} e^{-Q/\hbar\omega_0} \tag{4.109}$$

for $j = 1, 2, \ldots$. This represents a Poisson distribution.

Figure 4.5 shows a three-dimensional plot of this distribution. It is seen that in the limit of $Q \gg \hbar\omega_0$ it is nonvanishing only on a very thin strip surrounding the straight line $Q = j\hbar\omega_0$, indicating that in hard collisions the energy transfer becomes identical with the free-electron value Q and the binding of the target electron becomes insignificant. The qualitative aspect of this behavior cannot be unique for the oscillator potential: On the contrary, the oscillator potential overestimates binding forces acting on a target electron

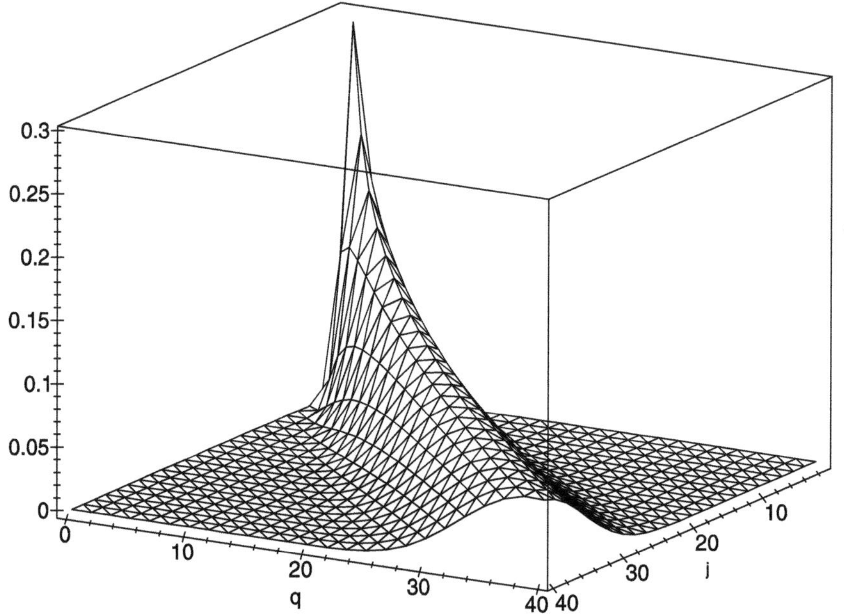

Fig. 4.5. Generalized oscillator strengths for harmonic oscillator; j is the excitation energy in multiples of $\hbar\omega$; $q = Q/\hbar\omega$.

at large momentum transfers. It may, therefore, safely be assumed that this narrowing-in of the oscillator-strength spectrum at high momentum transfers, called the 'Bethe ridge', is a general feature of all target atoms and molecules. In the opposite limit of small Q, the oscillator strength is distributed over a broad range of values of excitation levels j.

Insertion of (4.109) into (4.106) yields

$$S = \frac{2\pi e_1^2 e^2}{mv^2} \sum_{j=1}^{\infty} \frac{1}{(j-1)!} \int_{j^2/B}^{\infty} dt\, t^{j-2} e^{-t}, \tag{4.110}$$

where $B = 2mv^2/\hbar\omega_0$ is the Bethe parameter. Eq. (4.110) has been evaluated by numerical integration (Sigmund and Haagerup, 1986). Figure 4.6 shows the stopping number L, defined in accordance with (2.56) for $Z_2 = 1$ in a semilogarithmic plot *versus* the Bethe parameter. This graph can be compared to the classical result shown in Fig. 4.3.

4.5.4 Bethe Stopping Formula

A general evaluation of (4.106) was presented by Bethe (1930) for a high-speed projectile. The basic idea is very similar to Bohr's classification into close and distant interactions, but now the classification is done in momentum rather than configuration space. Hence distant interactions correspond to low

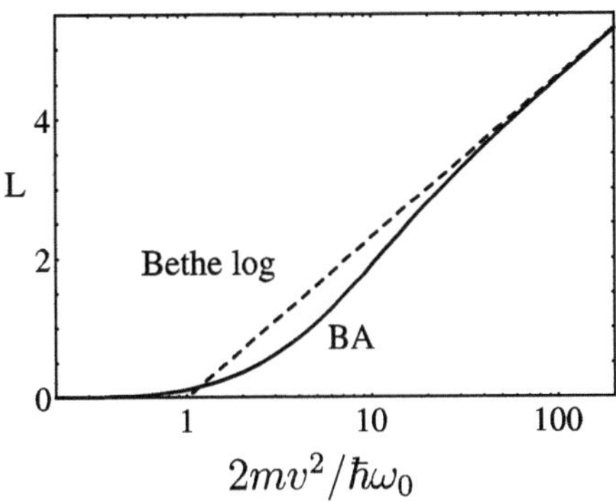

Fig. 4.6. Stopping number for quantal harmonic oscillator in the first Born approximation. Solid line: Exact result; dashed line: Bethe logarithm. From Sigmund and Haagerup (1986)

momentum transfers $\hbar q < \hbar q_0$ and *vice versa*. For $q < q_0$ or $Q < Q_0$ the dipole approximation is assumed to be valid.

The stopping integral then again splits into two parts. The low-Q contribution reduces to

$$S_{\text{dist}} = \sum_j f_{j0} \int_{(\epsilon_j - \epsilon_0)^2/2mv^2}^{Q_0} Q \, d\sigma_{\text{R}}(Q), \tag{4.111}$$

where the lower limit of integration stems from (4.83) for a heavy projectile.

For close interactions, on the other hand, Fig. 4.5 shows that $f_{j0}(Q)$ is nonvanishing only around $Q \simeq \epsilon_j - \epsilon_0$. Insertion of $Q = \epsilon_j - \epsilon_0$ into (4.83) for $m_1 \gg m$ delivers an upper integration limit

$$Q < 2mv^2 \tag{4.112}$$

in agreement with the maximum energy transferrable to a free target electron at rest. Hence,

$$S_{\text{close}} = \int_{Q_0}^{2mv^2} Q \, d\sigma_{\text{R}}(Q) \sum_j f_{j0}(Q). \tag{4.113}$$

Unlike in (4.111) the summation over j is unlimited here. Therefore, 'Bethe's sum rule'

$$\sum_j f_{j0}(Q) = 1, \tag{4.114}$$

which is a straight generalization of the Thomas-Reiche-Kuhn sum rule (4.48) can be applied. A proof of this important relationship is reproduced in Appendix A.4.4.

With this, (4.113) reduces to

$$S_{\text{close}} = \int_{Q_0}^{2mv^2} Q \, d\sigma_{\text{R}}(Q) \equiv \sum_j f_{j0} \int_{Q_0}^{2mv^2} Q \, d\sigma_{\text{R}}(Q), \tag{4.115}$$

where the last step involves (4.48) and has been performed to bring S_{close} into a form compatible with S_{dist}. The two then combine to

$$S = S_{\text{close}} + S_{\text{dist}} = \sum_j f_{j0} \int_{(\epsilon_j - \epsilon_0)^2/2mv^2}^{2mv^2} Q \, d\sigma_{\text{R}}(Q), \qquad (4.116)$$

where Q_0 has dropped out.

Insertion of (2.47) for the free-Coulomb scattering cross section $d\sigma_{\text{R}}(T)$ yields

$$S = \frac{4\pi e_1^2 e^2}{mv^2} Z_2 \sum_j f_{j0} \ln \frac{2mv^2}{\epsilon_j - \epsilon_0}, \qquad (4.117)$$

or

$$S = \frac{4\pi e_1^2 e^2}{mv^2} Z_2 \ln \frac{2mv^2}{I} \qquad (4.118)$$

for $m_1 \gg m$, with the *mean logarithmic excitation energy* I defined by

$$\ln I = \sum_j f_{j0} \ln(\epsilon_j - \epsilon_0). \qquad (4.119)$$

Eq. (4.118) is commonly called the 'Bethe stopping formula', here in the version for an ion or other projectile much heavier than an electron. Try to recollect the assumptions entering this appealing relationship. It was assumed that an interaction at $Q = Q_0$ is distant enough so that the dipole approximation applies. For a harmonic oscillator this implies that Q_0 must be $\ll \hbar\omega_0$ according to (4.109). In addition, for the splitting of the integral to make sense, Q_0 must lie between the upper and lower bound on the integration, $(\hbar\omega_0)^2/2mv^2 < Q_0 < 2mv^2$. For the two conditions to be fulfilled simultaneously we must have

$$2mv^2 \gg \hbar\omega_0. \qquad (4.120)$$

This indicates that Bethe's stopping formula is valid in the limit of high projectile speeds compared to the electron speed in the target. Note that the ion charge does not enter here. Therefore, the condition (4.120) can be only indirectly related to the more fundamental requirement that first-order perturbation theory be valid.

Electrons and Positrons

Equation (4.118) has been derived for a heavy projectile, i.e., for $m_1 \gg m$, the electron mass. Consider now the case of $m_1 = m$, i.e., a positron or electron as a projectile. Disregarding exchange for a moment, go back to (4.83) which, for $m_1 = m$ and $\epsilon_j - \epsilon_0 \ll mv^2$ reads

$$\frac{(\epsilon_j - \epsilon_0)^2}{2mv^2} = Q_{\min} < Q < 2mv^2. \tag{4.121}$$

Here, the lower limit is identical with the one for $m_1 \gg m$. Moreover, recall that according to Fig. 4.5, $Q \simeq \epsilon_j - \epsilon_0$ for large values of Q. Insertion of this relation into (4.83) yields

$$Q < \frac{1}{2}mv^2 = Q_{\max}, \tag{4.122}$$

i.e., the familiar result for the maximum energy transfer between equal masses. This is more restrictive than the upper limit emerging from (4.121).

With this, we find

$$\int_{Q_{\min}}^{Q_{\max}} \frac{dQ}{Q} = \ln \frac{(mv^2)^2}{(\epsilon_j - \epsilon_0)^2} = 2\ln \frac{mv^2}{\epsilon_j - \epsilon_0} \tag{4.123}$$

or

$$S = \frac{4\pi e^4}{mv^2} Z_2 \ln \frac{mv^2}{I}. \tag{4.124}$$

This applies for a positron. If the projectile is an electron, further modification is necessary, cf. Sect. 2.6.

4.5.5 Mean Logarithmic Excitation Energy

Figure 4.7 shows a plot of I/Z_2 for elements *versus* atomic number. Data have been compiled on the basis of stopping measurements and from several types of calculation. They represent the state of the art at the time of appearance of ICRU (1993). Extracting accurate I-values from measurements involves several significant corrections to the Bethe formula (4.118), which will be discussed in Chapter 6. Most useful is the observation that the quantity $I/Z_2 \simeq 10$ eV is roughly constant over the periodic table. Such a scaling relationship was predicted by Bloch (1933a) on the basis of the Thomas-Fermi model of the atom which will be discussed in Chapter 7.

4.6 Discussion and Outlook

Three procedures have been employed in this chapter to estimate the stopping cross section of an atom or molecule for a swift penetrating point charge. Two of these procedures are based on a quantal description of the target atom and differ only in the characterization of projectile motion by means of a classical trajectory or an incident plane wave, respectively. They yield very similar expressions for the stopping cross section[8], and the quantitative

---

[8] This feature is consistent with a more general observation made by Mott (1931).

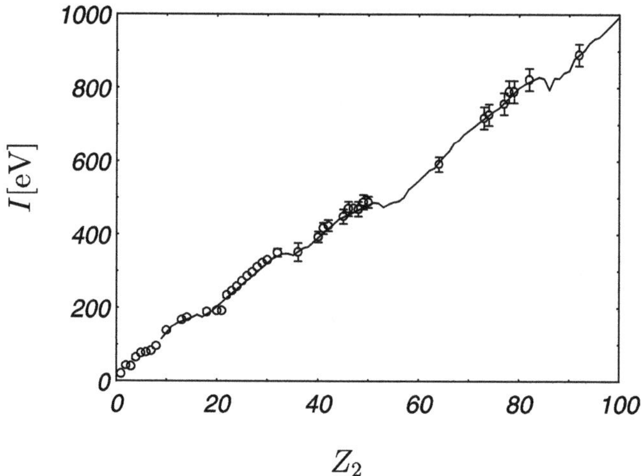

Fig. 4.7. Mean excitation energy of elements *versus* atomic number Z_2. Recommended values from ICRU (1993). Points and error bars based on experimental data. Line based on interpolated values

difference is significant only when the projectile mass is not large compared to the electron mass, i.e., most of all for penetrating electrons or positrons. Even though Bohr's purely classical formula (4.93) differs from Bethe's, (4.118), there are several similar features. Most of all, the energy-loss function at large impact parameters reflects the behavior of an ensemble of classical harmonic oscillators, cf. (4.23) and (4.49), a feature familiar from and related to the theory of optical dispersion and absorption which is to be discussed in the following chapter and Appendix A.5.

The most visible difference between Bohr's and Bethe's expressions is in the arguments of the respective logarithms. It may be instructive to rewrite them in the form

$$L_{\mathrm{Bohr}} = \ln \frac{Cmv^3}{e_1 e \omega_0} = \ln \frac{p_{\max}}{p_{\min}} \tag{4.125}$$

and

$$L_{\mathrm{Bethe}} = \ln \frac{2mv^2}{\hbar \omega_0} = \ln \frac{q_{\max}}{q_{\min}} . \tag{4.126}$$

Remembering that the two treatments coincide at large impact parameters p or small momentum transfers $\hbar q$ we may identify

$$p_{\max} \simeq \frac{v}{\omega_0} \quad \text{and} \quad q_{\min} \simeq \frac{\omega_0}{v} . \tag{4.127}$$

From this follows that

$$p_{\min} \simeq \frac{e_1 e}{mv^2} = \frac{b}{2} \text{ and } \hbar q_{\max} \simeq 2mv.$$ (4.128)

In Chapter 6 we shall have a look at the theory of Bloch (1933b) which combines the two approaches.

A feature common to all treatments is the competition between contributions from close and distant interactions to the stopping force. The logarithmic dependence indicates that neither regime can be neglected in the considered velocity regime, regardless of the quantitative details. In Bethe's estimate, a most notable feature are *two* occurrences of $2mv^2$ under the logarithm, one from the upper limit $Q_{\max}$ and one from the lower limit $Q_{\min} = (\epsilon_j - \epsilon_0)/2mv^2$ in the Q-integration. This suggests the existence of some sort of equipartition in momentum space which can be formulated more quantitatively, as is to be discussed in Chapter 5. Conversely, Fig. 4.3 indicates a tendency for close collisions to be increasingly dominating in the Bohr model with increasing speed.

We may recall that two major approximations were applied in this chapter, both of which geared toward high projectile speeds. One was first-order perturbation theory: The Bethe theory is entirely perturbational, while (classical) perturbation theory was only applied to distant interactions in the Bohr theory. Limitations of this model as well as necessary corrections form one of the subjects of Chapter 6. In addition, the internal motion of target electrons was ignored in the detailed evaluation, even though it is fully allowed for in the expressions like (4.77) or (4.82) and was not ignored in the explicit evaluation of the harmonic oscillator, Fig. 4.6. Deviations from the simple picture are most pronounced for the most rapidly moving electrons, i.e., inner-shell electrons. Necessary *shell corrections* will form another subject of Chapter 6.

The theory has so far addressed the interaction between a point charge and an isolated target atom. To the extent that the range of interaction as expressed by the adiabatic radius (2.62) may amount to many atomic diameters, corrections to this picture must be expected for dense (solid or liquid) matter and for partially stripped ions. Such effects will be considered in Chapter 5 and Volume II, respectively.

Problems

4.1. From the Bohr theory for distant collisions, extract the momentum transferred parallel and perpendicular to the beam direction, and show that after the projectile has passed by, the electron moves on an elliptic trajectory around the force center of the oscillator. Determine the principal axes of this ellipse, the direction of rotation and the angular momentum of the electron.

4.2. In the Bohr model, the motion of the target electron during interaction with and initiated by the projectile is ignored in distant collisions. Devise a simple model to estimate the limitations of this assumption. From the final momentum transfer, determine the distance travelled by the electron during the collision time by assuming uniform acceleration. [Hint: Assume constant acceleration].

4.3. Make a plot of Bohr's stopping formula (4.93) and determine expressions for the position and value of the maximum of this curve.

4.4. Consider the distant interaction between a swift point charge and a classical electron which is bound anisotropically in a potential $V(\mathbf{r}) = m\{\omega_0^2 x^2 + \omega_1^2(y^2 + z^2)\}/2$ with $\omega_0 > \omega_1$. Estimate the difference in the energy transferred to the target electron when the projectile travels in the x and y direction, respectively, but at the same impact parameter.

4.5. Consult a database accessible from your library for current tabulations of atomic dipole oscillator strengths. Make a plot over a wide energy range and identify the absorption edges of the K shell as well as other inner shells, as far as available.

4.6. $(\star)^9$ Evaluate analytical expressions for the dipole oscillator strengths of atomic hydrogen from the ground state to all excited discrete and continuum states.

4.7. $(\star)$ Evaluate analytical expressions for the generalized oscillator strengths between the ground state of atomic hydrogen and the 2s and the 2p level. Check the limiting values in the dipole limit.

4.8. Derive (4.77) by means of time-dependent perturbation theory based on the hamiltonian governing (4.58).

4.9. Repeat problem 4.3 replacing Bohr's formula by Bethe's formula (4.118).

4.10. $(\star)$ Try to evaluate the Thomas-Reiche-Kuhn sum rule on (4.48) the basis of the Dirac equation instead of the Schrödinger equation. You will find out that the result is 0 instead of 1. Try to identify the origin of the problem.

References

Abramowitz M. and Stegun I.A. (1964): *Handbook of mathematical functions.* Dover, New York

Basko M.M. (2005): On the low-velocity limit of the Bohr stopping formula. Europ Phys J D **32**, 9–17

Bethe H. (1930): Zur Theorie des Durchgangs schneller Korpuskularstrahlen durch Materie. Ann Physik **5**, 324–400

---

^{9} Problems marked by a star are considerably more difficult than average.

Bloch F. (1933a): Bremsvermögen von Atomen mit mehreren Elektronen. Z Physik **81**, 363–376

Bloch F. (1933b): Zur Bremsung rasch bewegter Teilchen beim Durchgang durch Materie. Ann Physik **16**, 285–320

Bohr N. (1913): On the theory of the decrease of velocity of moving electrified particles on passing through matter. Philos Mag **25**, 10–31

Bohr N. (1915): On the decrease of velocity of swiftly moving electrified particles in passing through matter. Philos Mag **30**, 581–612

Bohr N. (1948): The penetration of atomic particles through matter. Mat Fys Medd Dan Vid Selsk **18 no. 8**, 1–144

Fermi E. (1950): *Nuclear physics*. Univ. Chicago Press

Grande P.L. and Schiwietz G. (1998): Impact-parameter dependence of the electronic energy loss of fast ions. Phys Rev A **58**, 3796–3801

ICRU (1993): *Stopping powers and ranges for protons and alpha particles*, vol. 49 of *ICRU Report*. International Commission of Radiation Units and Measurements, Bethesda, Maryland

Jackson J.D. (1975): *Classical electrodynamics*. John Wiley & Sons, New York

Mikkelsen H.H. and Sigmund P. (1987): Impact parameter dependence of electronic energy loss: Oscillator model. Nucl Instrum Methods B **27**, 266–275

Mott N.F. (1931): On the theory of excitation by collision with heavy particles. Proc Cambr Philos Soc **27**, 553–560

Neufeld J. (1954): Electron capture and loss by moving ions in dense media. Phys Rev **96**, 1470–1478

Schiff L.I. (1981): *Quantum mechanics*. McGraw-Hill, Auckland

Sigmund P. (1996): Low-velocity limit of Bohr's stopping-power formula. Phys Rev A **54**, 3113–3117

Sigmund P. and Haagerup U. (1986): Bethe stopping theory for a harmonic oscillator and Bohr's oscillator model of atomic stopping. Phys Rev A **34**, 892–910

5

Dielectric Stopping Theory

5.1 Introductory Comments

The theory presented so far considered the interaction between a projectile and an isolated target atom or molecule as the elementary event. The outcome of a sequence of individual events was assumed to be governed by Poisson statistics. This appears appropriate for a dilute stopping medium such as a neutral gas. In a dense medium such as a solid or liquid where target particles interact with each other, complications occur. Firstly, the electric field acting on the electrons in the medium may be affected by the presence of other atoms, i.e., polarization effects may be significant. Secondly, the distribution in space and time of distinct collision events may not follow Poisson statistics. The present chapter addresses the first aspect.

For rough orientation consider the adiabatic radius $a_{\mathrm{ad}} = v/\omega_0$ which defines the range of interaction in electronic collisions and which increases with increasing beam velocity and decreasing ω_0. Thus one expects collective effects to show up most pronouncedly for outer-shell electrons, in particular for conduction electrons in metals. In atomic units we have

$$\frac{a_{\mathrm{ad}}}{a_0} = \frac{v}{\omega_0 a_0} = \frac{v}{v_0}\frac{e^2/a_0}{\hbar\omega_0}, \tag{5.1}$$

where $(e^2/a_0)/(\hbar\omega_0)$ is of the order of ~ 1 for outer electrons. A rough measure of the internuclear distance in condensed matter is ~ 2.5 Å $\sim 5a_0$. Hence, for $v \gg v_0$, collective effects must be expected. This was first recognized by Swann (1938).

A most dramatic manifestation of a collective effect in fast-particle penetration is the Cherenkov radiation emitted by electrons moving with a velocity exceeding the speed of light in the stopping medium. After its discovery by Cherenkov (1934) the phenomenon was explained as resulting from electric polarization of the medium by Frank and Tamm (1937) and Tamm (1939). The connection to stopping theory was established by Fermi (1939, 1940) and clarified by A. Bohr (1948), Halpern and Hall (1948) and others.

Although the intimate connection between optical dispersion and particle stopping has been evident ever since the early studies by Bohr (1913, 1915), treating particle stopping as a polarization phenomenon in terms of Maxwell's equations was a genuinely new development in 1940, and precautions were taken to limit this view to distant interactions where the dipole approximation underlying all dispersion theory was valid.

An even more radical step in the same direction was taken by Lindhard (1954) who demonstrated that *all* collisional stopping could be described in terms of the electromagnetic field equations. The above precaution was avoided by a suitable generalization of the dielectric constant which allowed for fields varying rapidly not only in time but also in space. This involved the introduction of a dielectric function $\epsilon(\boldsymbol{k}, \omega)$ depending on both wave number and frequency, whereas the classical Drude-Lorentz theory, designed to describe the interaction of optical radiation with matter, only allowed for frequency dependence[1].

One purpose of this chapter is to reformulate stopping theory in terms of this important concept and to explore some consequences. These include the nonrelativistic and the relativistic density correction to the stopping force on swift particles. As a byproduct we also obtain the relativistic extension of the Bethe theory for a dilute medium. Other consequences, such as Cherenkov radiation and wake effects, will be discussed in Volumes II and III.

In metals, due to the low binding forces on conduction electrons, collective phenomena must be expected to be noticeable even at moderate projectile speeds. In fact, the term *Lindhard function* denotes the dielectric function of a free-electron gas (Fermi gas). This function has found widespread application as a tool in the description of the response of electrons to an external perturbance (Smith, 1983). It is heavily used in particle stopping, and its use is by no means restricted to metallic conduction electrons.

5.2 Electrodynamics

As long as we are not concerned with fluctuation phenomena, classical electromagnetic theory is an adequate tool to describe phenomena in dense media.

5.2.1 Field Equations in Vacuum

Let us start with Maxwell's equations in vacuum. In gaussian units they read

---

[1] The notion of a dielectric function or a related quantity, dependent not only on frequency but also on wave number, was developed at several places at the same time. Lifshitz and Pitaevskii (1981) do not mention Lindhard but quote Klimontovich and Silin (1952) for work that could well have resulted in the Lindhard function and its application to stopping theory, if that had been one of the goals of their work.

$$\boldsymbol{\nabla} \cdot \boldsymbol{E} = 4\pi \rho_{\mathrm{e}} \tag{5.2a}$$

$$\boldsymbol{\nabla} \times \boldsymbol{E} = -\frac{1}{c}\frac{\partial \boldsymbol{B}}{\partial t} \tag{5.2b}$$

$$\boldsymbol{\nabla} \cdot \boldsymbol{B} = 0 \tag{5.2c}$$

$$\boldsymbol{\nabla} \times \boldsymbol{B} = \frac{1}{c}\frac{\partial \boldsymbol{E}}{\partial t} + \frac{4\pi}{c}\boldsymbol{J}_{\mathrm{e}}, \tag{5.2d}$$

cf. Jackson (1975), where $\rho_{\mathrm{e}}(\boldsymbol{r},t)$ and $\boldsymbol{J}_{\mathrm{e}}(\boldsymbol{r},t)$ denote charge and current density, respectively, which satisfy the continuity equation

$$\boldsymbol{\nabla} \cdot \boldsymbol{J}_{\mathrm{e}} + \frac{\partial \rho_{\mathrm{e}}}{\partial t} = 0. \tag{5.3}$$

If we express the fields in terms of electromagnetic potentials $\boldsymbol{A}$ and $\varPhi$,

$$\boldsymbol{E}(\boldsymbol{r},t) = -\boldsymbol{\nabla}\varPhi(\boldsymbol{r},t) - \frac{1}{c}\frac{\partial \boldsymbol{A}(\boldsymbol{r},t)}{\partial t} \tag{5.4a}$$

$$\boldsymbol{B}(\boldsymbol{r},t) = \boldsymbol{\nabla} \times \boldsymbol{A}(\boldsymbol{r},t) \tag{5.4b}$$

and adopt the *Coulomb gauge*

$$\boldsymbol{\nabla} \cdot \boldsymbol{A} = 0, \tag{5.5}$$

the field equations reduce to

$$\nabla^2 \varPhi = -4\pi \rho_{\mathrm{e}} \tag{5.6a}$$

$$\nabla^2 \boldsymbol{A} - \frac{1}{c^2}\frac{\partial^2 \boldsymbol{A}}{\partial t^2} - \frac{1}{c}\boldsymbol{\nabla}\frac{\partial \varPhi}{\partial t} = -\frac{4\pi}{c}\boldsymbol{J}_{\mathrm{e}}. \tag{5.6b}$$

We may write

$$\rho_{\mathrm{e}}(\boldsymbol{r},t) = \int \mathrm{d}^3k \int \mathrm{d}\omega\, \rho_{\mathrm{e}}(\boldsymbol{k},\omega)\, \mathrm{e}^{\mathrm{i}(\boldsymbol{k}\cdot\boldsymbol{r} - \omega t)}, \tag{5.7}$$

and similarly for $\boldsymbol{J}_{\mathrm{e}}(\boldsymbol{r},t)$ and the fields. This yields field equations in Fourier space

$$k^2 \varPhi(\boldsymbol{k},\omega) = 4\pi \rho_{\mathrm{e}}(\boldsymbol{k},\omega) \tag{5.8a}$$

$$\left(k^2 - \frac{\omega^2}{c^2}\right)\boldsymbol{A}(\boldsymbol{k},\omega) + \frac{\omega}{c}\boldsymbol{k}\,\varPhi(\boldsymbol{k},\omega) = \frac{4\pi}{c}\boldsymbol{J}_{\mathrm{e}}(\boldsymbol{k},\omega). \tag{5.8b}$$

We may split the current density $\boldsymbol{J}_{\mathrm{e}}(\boldsymbol{k},\omega)$ into a longitudinal part $\boldsymbol{J}_{\mathrm{e},\mathrm{l}}$ parallel to $\boldsymbol{k}$ and a transverse part $\boldsymbol{J}_{\mathrm{e},\mathrm{t}}$ perpendicular to $\boldsymbol{k}$. Noting that $\boldsymbol{k}\cdot\boldsymbol{A}(\boldsymbol{k},\omega) = 0$ in view of (5.5), we may split (5.8b) into its longitudinal and transverse components,

$$\frac{\omega}{c}\boldsymbol{k}\,\varPhi(\boldsymbol{k},\omega) = \frac{4\pi}{c}\boldsymbol{J}_{\mathrm{e},\mathrm{l}}(\boldsymbol{k},\omega) \tag{5.9a}$$

$$\left(k^2 - \frac{\omega^2}{c^2}\right)\boldsymbol{A}(\boldsymbol{k},\omega) = \frac{4\pi}{c}\boldsymbol{J}_{\mathrm{e},\mathrm{t}}(\boldsymbol{k},\omega), \tag{5.9b}$$

while the continuity equation (5.3) reduces to

$$\boldsymbol{k} \cdot \boldsymbol{J}_{\mathrm{e,l}}(\boldsymbol{k}, \omega) - \omega \rho_{\mathrm{e}}(\boldsymbol{k}, \omega) = 0. \tag{5.10}$$

5.2.2 Linear Response

In a material medium we may split charge and current densities into external and induced contributions,

$$\rho_{\mathrm{e}} = \rho_{\mathrm{e,ext}} + \rho_{\mathrm{e,ind}} \tag{5.11a}$$
$$\boldsymbol{J}_{\mathrm{e}} = \boldsymbol{J}_{\mathrm{e,ext}} + \boldsymbol{J}_{\mathrm{e,ind}}, \tag{5.11b}$$

where $\rho_{\mathrm{e,ind}}$ and $\boldsymbol{J}_{\mathrm{e,ind}}$ reflect the response of the medium to the field. Assuming linear response we may adopt Ohm's law and write

$$\boldsymbol{J}_{\mathrm{eind,l}}(\boldsymbol{k}, \omega) = \sigma_l(k, \omega) \boldsymbol{E}_l = -\mathrm{i} \boldsymbol{k} \sigma_l(k, \omega) \Phi(\boldsymbol{k}, \omega) \tag{5.12a}$$
$$\boldsymbol{J}_{\mathrm{eind,t}}(\boldsymbol{k}, \omega) = \sigma_t(k, \omega) \boldsymbol{E}_t = \mathrm{i} \frac{\omega}{c} \sigma_t(k, \omega) \boldsymbol{A}(\boldsymbol{k}, \omega), \tag{5.12b}$$

where $\sigma_l(k, \omega)$ and $\sigma_t(k, \omega)$ reflect complex conductivities depending on wave number and frequency, assuming the medium to be isotropic, infinite and homogeneous[2].

From (5.9b) and (5.12b) we find

$$\frac{\omega}{c} \boldsymbol{k} \left(1 + \frac{4\pi \mathrm{i} \sigma_l(k, \omega)}{\omega} \right) \Phi(\boldsymbol{k}, \omega) = \frac{4\pi}{c} \boldsymbol{J}_{\mathrm{eext,l}}(\boldsymbol{k}, \omega) \tag{5.13a}$$
$$\left[k^2 - \frac{\omega^2}{c^2} \left(1 + \frac{4\pi \mathrm{i} \sigma_t(k, \omega)}{\omega} \right) \right] \boldsymbol{A}(\boldsymbol{k}, \omega) = \frac{4\pi}{c} \boldsymbol{J}_{\mathrm{eext,t}}(\boldsymbol{k}, \omega) \tag{5.13b}$$

Multiplying (5.13a) by $\boldsymbol{k}$ and making use of the continuity equation for external charge and current,

$$\boldsymbol{k} \cdot \boldsymbol{J}_{\mathrm{eext,l}} - \omega \rho_{\mathrm{e,ext}} = 0, \tag{5.14}$$

we may rewrite (5.13a) in the form

$$k^2 \left(1 + \frac{4\pi \mathrm{i} \sigma_l(k, \omega)}{\omega} \right) \Phi(\boldsymbol{k}, \omega) = 4\pi \rho_{\mathrm{eext}}(\boldsymbol{k}, \omega). \tag{5.15}$$

After introduction of the dielectric function

$$\epsilon_l(k, \omega) = 1 + \frac{4\pi \mathrm{i} \sigma_l(k, \omega)}{\omega} \tag{5.16}$$

---

[2] The case of anisotropic media, which is of importance in plasma physics, has been discussed in detail by Alexandrov et al. (1984).

this reduces to

$$k^2 \Phi(\boldsymbol{k}, \omega) = 4\pi \left(\rho_{\text{eext}}(\boldsymbol{k}, \omega) - (\epsilon_l(\boldsymbol{k}, \omega) - 1) \frac{k^2}{4\pi} \Phi(\boldsymbol{k}, \omega) \right). \tag{5.17}$$

This relation must be identical with (5.8a). Hence, the second quantity in the brackets must be the induced charge density,

$$\rho_{\text{eind}}(\boldsymbol{k}, \omega) = -\frac{k^2}{4\pi} \left[\epsilon_l(\boldsymbol{k}, \omega) - 1 \right] \Phi(\boldsymbol{k}, \omega), \tag{5.18}$$

at the same time as (5.8a) takes on the familiar form

$$k^2 \epsilon_l(\boldsymbol{k}, \omega) \Phi(\boldsymbol{k}, \omega) = 4\pi \rho_{\text{eext}}(\boldsymbol{k}, \omega). \tag{5.19}$$

In order to unify the notation, also a transverse dielectric function $\epsilon_t(k, \omega)$ is introduced,

$$\epsilon_t(k, \omega) = 1 + \frac{4\pi i \sigma_t(k, \omega)}{\omega}. \tag{5.20}$$

With this, (5.13b) reads

$$\left(k^2 - \frac{\omega^2}{c^2} \epsilon_t(k, \omega) \right) \boldsymbol{A}(\boldsymbol{k}, \omega) = \frac{4\pi}{c} \boldsymbol{J}_{\text{eext,t}}(\boldsymbol{k}, \omega). \tag{5.21}$$

5.2.3 Connection to Stopping Force

Consider a point charge e_1 in uniform motion with a velocity $\boldsymbol{v}$ so that

$$\rho_{\text{e,ext}}(\boldsymbol{r}, t) = e_1 \delta(\boldsymbol{r} - \boldsymbol{v}t). \tag{5.22}$$

In Fourier space this reads

$$\rho_{\text{e,ext}}(\boldsymbol{k}, \omega) = \frac{e_1}{(2\pi)^3} \, \delta(\omega - \boldsymbol{k} \cdot \boldsymbol{v}) \tag{5.23}$$

and, correspondingly,

$$\boldsymbol{J}_{\text{e,ext}}(\boldsymbol{k}, \omega) = \frac{e_1}{(2\pi)^3} \boldsymbol{v} \, \delta(\omega - \boldsymbol{k} \cdot \boldsymbol{v}). \tag{5.24}$$

We may then determine

$$\boldsymbol{J}_{\text{eext,t}} = \boldsymbol{J}_{\text{e,ext}} - \frac{\boldsymbol{k}}{k} \left(\frac{\boldsymbol{k}}{k} \cdot \boldsymbol{J}_{\text{e,ext}} \right). \tag{5.25}$$

The electric field is given by

$$\boldsymbol{E}(\boldsymbol{r}, t) = \int \mathrm{d}^3 k \int \mathrm{d}\omega \, e^{\mathrm{i}(\boldsymbol{k} \cdot \boldsymbol{r} - \omega t)} \left(-\mathrm{i} \boldsymbol{k} \cdot \Phi(\boldsymbol{k}, \omega) + \mathrm{i} \frac{\omega}{c} \boldsymbol{A}(\boldsymbol{k}, \omega) \right). \tag{5.26}$$

The force on the projectile reads

$$\boldsymbol{F} = e_1 \boldsymbol{E}(\boldsymbol{v}t, t). \tag{5.27}$$

In a homogeneous, isotropic and infinite medium, the stopping force is directed opposite to the velocity $\boldsymbol{v}$. Hence, the energy-loss function must be given by

$$-\frac{\mathrm{d}E}{\mathrm{d}x} = -\frac{e_1}{v} \boldsymbol{v} \cdot \boldsymbol{E}(\boldsymbol{v}, t). \tag{5.28}$$

Insertion of the potentials (5.19) and (5.21) and, subsequently, (5.23) and (5.25) yields

$$-\frac{\mathrm{d}E}{\mathrm{d}x} = \frac{ie_1^2}{2\pi^2 v} \int \frac{\mathrm{d}^3\boldsymbol{k}}{k^2} (\boldsymbol{k} \cdot \boldsymbol{v}) \left(\frac{1}{\epsilon_l(k, \omega)} - \frac{v^2}{c^2} \frac{k^2 - (\boldsymbol{k} \cdot \boldsymbol{v})^2/v^2}{k^2 - \epsilon_t(k, \omega)(\boldsymbol{k} \cdot \boldsymbol{v})^2/c^2} \right). \tag{5.29}$$

The integration is conveniently performed in terms of spherical coordinates. Setting $\omega = kv \cos\theta$ (θ being the polar angle) one finds

$$-\frac{\mathrm{d}E}{\mathrm{d}x} = \frac{ie_1^2}{\pi v^2} \int_0^\infty \frac{\mathrm{d}k}{k} \int_{-kv}^{kv} \mathrm{d}\omega\, \omega \left(\frac{1}{\epsilon_l(k, \omega)} - \frac{v^2}{c^2} \frac{k^2 - \omega^2/v^2}{k^2 - \epsilon_t(k, \omega)\omega^2/c^2} \right). \tag{5.30}$$

Evidently, the second term in the parentheses, which originates in the vector potential, i.e., transverse interaction, becomes significant only at relativistic velocities.

In the special case where $\epsilon_t = \epsilon_l \equiv \epsilon$, the longitudinal and transverse terms add up to

$$-\frac{\mathrm{d}E}{\mathrm{d}x} = \frac{ie_1^2}{\pi v^2} \int_0^\infty k\,\mathrm{d}k \int_{-kv}^{kv} \omega\mathrm{d}\omega\, \frac{1/\epsilon - v^2/c^2}{k^2 - \epsilon\omega^2/c^2}. \tag{5.31}$$

5.3 Gaseous Medium

In the present section we shall establish a link between Bethe stopping theory and the dielectric description outlined in the previous section. As a by-product we shall also arrive at the relativistic extension of the Bethe theory.

The stopping material will be considered as being made up by individual atoms whose electron clouds do not overlap significantly. Nevertheless, charges induced by the projectile may cause the local field to deviate from the external Coulomb field and hence modify the stopping force.

5.3.1 Dielectric Function

Consider an atom with a nucleus located in a fixed point $\boldsymbol{R}$ such that its electronic states $|j\rangle$ can be characterized by wave functions $\psi_j(\boldsymbol{r} - \boldsymbol{R})$. We shall need the matrix elements

$$\langle j|\mathrm{e}^{\mathrm{i}\boldsymbol{k}\cdot\boldsymbol{r}}|0\rangle = \int \mathrm{d}^3\mathbf{r}\,\psi_j^\star(\boldsymbol{r} - \boldsymbol{R})\,\mathrm{e}^{\mathrm{i}\boldsymbol{k}\cdot\boldsymbol{r}}\,\psi_0(\boldsymbol{r} - \boldsymbol{R}) = \mathrm{e}^{\mathrm{i}\boldsymbol{k}\cdot\boldsymbol{R}}F_{j0}(\boldsymbol{k}), \qquad (5.32)$$

where

$$F_{j0}(\boldsymbol{k}) = \int \mathrm{d}^3\mathbf{r}\,\psi_j^\star(\boldsymbol{r})\,\mathrm{e}^{\mathrm{i}\boldsymbol{k}\cdot\boldsymbol{r}}\,\psi_0(\boldsymbol{r}). \qquad (5.33)$$

The effect of an arbitrary electric field, specified by a potential $\Phi(\boldsymbol{k}, \omega)$, on an individual electron may be treated within first-order perturbation theory, (4.35-4.37) with

$$\langle j|\mathcal{V}(\mathbf{r}, t)|0\rangle = -e\int \mathrm{d}^3\boldsymbol{k}\int \mathrm{d}\omega\,\Phi(\boldsymbol{k}, \omega)\mathrm{e}^{\mathrm{i}\boldsymbol{k}\cdot\boldsymbol{R}}F_{j0}(\boldsymbol{k})\mathrm{e}^{-\mathrm{i}\omega t} \qquad (5.34)$$

so that

$$c^{(1)} = \frac{e}{\hbar}\int \mathrm{d}^3\boldsymbol{k}\int \mathrm{d}\omega\,\Phi(\boldsymbol{k}, \omega)\,\mathrm{e}^{\mathrm{i}\boldsymbol{k}\cdot\boldsymbol{R}}F_{j0}(\boldsymbol{k})\,\frac{\mathrm{e}^{\mathrm{i}(\omega_{j0}-\omega)t}}{\omega_{j0} - \omega - \mathrm{i}\Gamma}. \qquad (5.35)$$

Here an infinitesimal positive damping constant Γ has been introduced in order to ensure causality or, in other words, to define the behavior near the singularities in the complex ω plane. In principle, this step reflects the physical requirement that the electron be in its ground state at time $t = -\infty$. As discussed in Appendix A.5.1 such a requirement may be imposed either on the electronic states or on the field. In the latter case the interaction potential is amended by a factor $\exp(\Gamma t)$.

The electron density $|\Psi|^2$ may be found from (4.35). We then obtain the induced charge density

$$\rho_{\mathrm{e,ind}}(\boldsymbol{r}, t) = -e\left(\psi^{(0)\star}(\boldsymbol{r}, t)\psi^{(1)}(\boldsymbol{r}, t) + \psi^{(0)}(\boldsymbol{r}, t)\psi^{(1)\star}(\boldsymbol{r}, t)\right) \qquad (5.36)$$

to first order in Φ, or

$$\rho_{\mathrm{e,ind}}(\boldsymbol{r}, t) = -\frac{e^2}{\hbar}\sum_j\left(\psi_0^\star(\boldsymbol{r} - \boldsymbol{R})\psi_j(\boldsymbol{r} - \boldsymbol{R})\int \mathrm{d}^3\boldsymbol{k}\int \mathrm{d}\omega\,\Phi(\boldsymbol{k}, \omega)\right.$$

$$\left.\mathrm{e}^{\mathrm{i}\boldsymbol{k}\cdot\boldsymbol{R}}\,F_{j0}(\boldsymbol{k})\,\frac{\mathrm{e}^{-\mathrm{i}\omega t}}{\omega_{j0} - \omega - \mathrm{i}\Gamma} + \text{complex conjugate}\right) \qquad (5.37)$$

for a single electron.

Consider now a medium with N one-electron atoms per volume distributed randomly in space. There will then be a density of polarization charge

$$\rho_{\mathrm{epol}}(\boldsymbol{r}, t) = N \int \mathrm{d}^3 \boldsymbol{R} \, \rho_{\mathrm{e,ind}}(\boldsymbol{r}, t). \tag{5.38}$$

Noting that

$$\int \mathrm{d}^3 \boldsymbol{R} \, \psi_0^\star(\boldsymbol{r} - \boldsymbol{R}) \, \psi_j(\boldsymbol{r} - \boldsymbol{R}) \, \mathrm{e}^{\mathrm{i}\boldsymbol{k}\cdot\boldsymbol{R}} = \mathrm{e}^{\mathrm{i}\boldsymbol{k}\cdot\boldsymbol{r}} \, F_{0j}(-\boldsymbol{k}) \tag{5.39}$$

similar to (5.32) we find

$$\rho_{\mathrm{epol}}(\boldsymbol{k}, \omega) = -\frac{Ne^2}{\hbar} \, \Phi(\boldsymbol{k}, \omega) \sum_j \frac{F_{0j}(-\boldsymbol{k}) F_{j0}(\boldsymbol{k})}{\omega_{j0} - \omega - \mathrm{i}\Gamma}$$
$$- \frac{Ne^2}{\hbar} \, \Phi(\boldsymbol{k}, \omega) \sum_j \frac{F_{0j}(\boldsymbol{k}) F_{j0}(-\boldsymbol{k})}{\omega_{j0} + \omega + \mathrm{i}\Gamma}, \tag{5.40}$$

where use has been made of the relation

$$\Phi^\star(\boldsymbol{k}, \omega) = \Phi(-\boldsymbol{k}, -\omega), \tag{5.41}$$

which ensures that the potential $\Phi(\boldsymbol{r}, t)$ be real.

For an isotropic medium any explicit directional dependence of $|F_{0j}(\boldsymbol{k})|^2$ drops out upon summation over j. Then (5.18) leads to

$$\epsilon_l(\boldsymbol{k}, \omega) = 1 + \frac{m\omega_{\mathrm{P}}^2}{\hbar k^2} \sum_j |F_{j0}(\boldsymbol{k})|^2$$
$$\times \left(\frac{1}{\omega_{j0} - \omega - \mathrm{i}\Gamma} + \frac{1}{\omega_{j0} + \omega + \mathrm{i}\Gamma} \right), \tag{5.42}$$

where

$$\omega_{\mathrm{P}} = \sqrt{\frac{4\pi Ne^2}{m}} \tag{5.43}$$

is the plasma frequency for a system with N electrons per volume.

5.3.2 Bethe Stopping Formula

The atomic model employed in the above derivation is identical with the one adopted in Bethe theory in Sect. 4.5.4. Therefore, Bethe stopping theory must emerge from the present description in the limit of low target density, i.e., in the limit of low plasma frequency ω_{P}.

At nonrelativistic velocities the second term in (5.30) may be disregarded. Expansion of $1/\epsilon_l(\boldsymbol{k}, \omega)$ up to first order in N yields

$$\frac{1}{\epsilon_l(\boldsymbol{k}, \omega)} \simeq 1 - \frac{m\omega_{\mathrm{P}}^2}{\hbar k^2} \sum_j |F_{j0}(\boldsymbol{k})|^2$$
$$\times \left(\frac{1}{\omega_{j0} - \omega - \mathrm{i}\Gamma} + \frac{1}{\omega_{j0} + \omega + \mathrm{i}\Gamma} \right). \tag{5.44}$$

The integrations can be evaluated by means of the following useful formula, valid for real x,

$$\frac{1}{x - a + i\Gamma}\bigg|_{\Gamma \to 0} = \mathsf{P}\frac{1}{x - a} - i\pi\delta(x - a), \tag{5.45}$$

which the reader may verify by solving problem 5.3. Here, P indicates the Cauchy principal value. According to (5.30) only the imaginary part is of interest in the present connection since $-\mathrm{d}E/\mathrm{d}x$ must be real. Hence, insertion of (5.45) into (5.44) yields

$$\mathrm{Re}\left(\frac{i}{\epsilon_l(k,\omega)}\right) = \frac{\pi m \omega_P^2}{\hbar k^2} \sum_j |F_{j0}(\boldsymbol{k})|^2 \left[\delta(\omega - \omega_{j0}) - \delta(\omega + \omega_{j0})\right] \tag{5.46}$$

and, after insertion into (5.30) and integration over ω,

$$\frac{\mathrm{d}E}{\mathrm{d}x} = -\frac{2e_1^2 m \omega_P^2}{\hbar v^2} \sum_j \omega_{j0} \int_{kv > \omega_{j0}} \frac{\mathrm{d}k}{k^3} |F_{j0}(\boldsymbol{k})|^2. \tag{5.47}$$

Substitution of

$$Q = \hbar^2 k^2 / 2m \tag{5.48}$$

yields

$$-\frac{\mathrm{d}E}{\mathrm{d}x} = \frac{2\pi e_1^2 e^2 N}{mv^2} \sum_j \hbar\omega_{j0} \int_{Q > (\hbar\omega_{j0})^2 / 2mv^2} \frac{\mathrm{d}Q}{Q^2} |F_{j0}(\boldsymbol{k})|^2, \tag{5.49}$$

in agreement with (4.55) and (4.56).

It is left to the reader to get convinced of the equivalence of the physical input in the two derivations of the Bethe formula (problem 5.2). Note in particular that the present derivation did not invoke cross sections.

Eq. (5.47) suggests $\hbar\omega$ to represent the energy transfer in an individual event, but this requires, strictly speaking, a mapping of the integration interval in (5.30) on the interval $0 < \omega < kv$. It also follows from the occurrence of $|F_{j0}(\boldsymbol{k})|^2$ in (5.44) that $\hbar\boldsymbol{k}$ must represent the associated momentum transfer.

5.3.3 Nonrelativistic Density Effect

We may introduce an electric susceptibility $\chi(k,\omega)$ in accordance with the common definition

$$\epsilon_l(k,\omega) = 1 + 4\pi\chi(k,\omega) \tag{5.50}$$

in gaussian units. In the previous section we have approximated

$$\frac{1}{\epsilon_l(k,\omega)} \simeq 1 - 4\pi\chi(k,\omega).$$

$$(5.51)$$

The difference between the exact expression and the expression for a dilute medium,

$$\frac{1}{\epsilon_l(k,\omega)} - \frac{1}{\epsilon_l(k,\omega)}\bigg|_{\text{dilute}} = \frac{(4\pi\chi(k,\omega))^2}{1 + 4\pi\chi(k,\omega)} \equiv \frac{(\epsilon_l(k,\omega) - 1)^2}{\epsilon_l(k,\omega)}$$

$$(5.52)$$

yields the density correction

$$\Delta\left(\frac{\mathrm{d}E}{\mathrm{d}x}\right)_{\text{density}} = \frac{e_1^2}{\pi v^2}\mathrm{i}\int_0^\infty \frac{\mathrm{d}k}{k}\int_{-kv}^{kv}\mathrm{d}\omega\,\omega\,\frac{(\epsilon_l(k,\omega) - 1)^2}{\epsilon_l(k,\omega)}.$$

$$(5.53)$$

We may expect that close collisions are not affected noticeably by the density correction. Therefore, (5.53) should be insensitive to the behavior of the dielectric function at large wave numbers. As a simple example take the standard Drude function for a single resonance frequency ω_0 (Jackson, 1975),

$$\epsilon_l(\omega) = 1 + \frac{\omega_{\mathrm{P}}^2}{\omega_0^2 - \omega^2 - \mathrm{i}\omega\Gamma}.$$

$$(5.54)$$

We now find

$$\mathrm{i}\left(\frac{1}{\epsilon_l(0,\omega)} - 1\right) = -\mathrm{i}\frac{\omega_{\mathrm{P}}^2}{\omega_0^2 + \omega_{\mathrm{P}}^2 - (\omega + \mathrm{i}\Gamma)^2}$$

$$= -\frac{\pi\omega_{\mathrm{P}}^2}{2\alpha_0}\left[\delta(\omega - \alpha_0) - \delta(\omega + \alpha_0)\right] \quad (5.55)$$

with

$$\alpha_0^2 = \omega_0^2 + \omega_{\mathrm{P}}^2$$

$$(5.56)$$

we may carry out the integrations in (5.53), imposing some upper limit $k_{\max}$ in the k-integral. The result is

$$\Delta\left(-\frac{\mathrm{d}E}{\mathrm{d}x}\right) = -\frac{e_1^2\omega_{\mathrm{P}}^2}{v^2}\ln\frac{\sqrt{\omega_0^2 + \omega_{\mathrm{P}}^2}}{\omega_{\mathrm{P}}} = \frac{4\pi e_1^2 e^2 Z_2 N}{mv^2}\ln\frac{\omega_{\mathrm{P}}}{\sqrt{\omega_0^2 + \omega_{\mathrm{P}}^2}},$$

$$(5.57)$$

where the electron density now has been set to $n = Z_2 N$, allowing for Z_2 electrons per target atom[3].

---

[3] In this and the following chapter, the reader will notice an occasional swith between $n = N$ and $n = Z_2 N$ electrons per volume, dependent on whether a calculation is performed for an electron gas, a medium with one-electron atoms or a medium with Z_2-electron atoms. Ambiguities because of this can be avoided by remembering

- that the stopping number L always refers to a single target electron, and
- that the factor in front of the stopping number always refers to the actual number of electrons per volume, $n = NZ_2$.

This is a very plausible result: It implies that ω_0 in the Bohr or Bethe logarithm needs to be replaced by the shifted resonance frequency $\alpha_0 = \sqrt{\omega_0^2 + \omega_P^2}$.

5.4 Static Electron Gas

A particularly instructive example is a static electron gas, i.e., a system of independent Hartree electrons within a periodicity volume V. The Pauli principle being ignored, the ground state of the system is a state for which all electrons are at rest.

5.4.1 Dielectric Function

With the wave functions[4]

$$|j\rangle = \frac{1}{\sqrt{\mathsf{V}}}e^{i\boldsymbol{k}_j\boldsymbol{r}} \tag{5.58}$$

(5.33) reduces to

$$F_{j0}(\boldsymbol{k}) = \delta_{\boldsymbol{k}_j\boldsymbol{k}}, \tag{5.59}$$

where the right-hand side represents a three-dimensional Kronecker symbol and the wave vectors $\boldsymbol{k}_j$ form a discrete set determined by the boundaries of the electron gas enclosed into a volume V. Then, (5.42) reads

$$\epsilon_l(k,\omega) = 1 + \frac{m\omega_P^2}{\hbar k^2}\left(\frac{1}{\omega_k - \omega - i\Gamma} + \frac{1}{\omega_k + \omega + i\Gamma}\right)$$

$$= 1 + \frac{\omega_P^2}{\omega_k^2 - (\omega + i\Gamma)^2}, \tag{5.60}$$

where

$$\omega_k = \frac{\hbar k^2}{2m}. \tag{5.61}$$

This particularly useful model function was first derived by Lindhard (1954). The difference to the well-known Drude-Lorentz function for a harmonic oscillator (Jackson, 1975) is the replacement of the resonance frequency ω_0 of the oscillator by the single-particle excitation frequency ω_k.

---

[4] The use of this particular form of free-electron wave functions will be motivated in sect. 5.7.1.

5.4.2 Relativistic Extension ($\star$)

In the relativistic regime, scalar wave functions need to be replaced by spinors. A brief summary of Dirac theory is presented in appendix A.4.5.

Instead of (5.58), a plane wave representing an electron with a momentum $\hbar\boldsymbol{k}$ may be represented by one of four spinors,

$$|j_\nu\rangle = \frac{1}{\sqrt{V}}\mathrm{e}^{\mathrm{i}\boldsymbol{k}_j\boldsymbol{r}}u^{(\nu)}(\boldsymbol{k}_j), \quad \nu = 1\ldots 4 \tag{5.62}$$

with

$$u^{(1)}(\boldsymbol{k}) = B_k\begin{pmatrix} 1 \\ 0 \\ k_z b_k \\ k_+ b_k \end{pmatrix}; u^{(2)}(\boldsymbol{k}) = B_k\begin{pmatrix} 0 \\ 1 \\ k_- b_k \\ -k_z b_k \end{pmatrix}; \tag{5.63a}$$

$$u^{(3)}(\boldsymbol{k}) = B_k\begin{pmatrix} -k_z b_k \\ -k_+ b_k \\ 1 \\ 0 \end{pmatrix}; u^{(4)}(\boldsymbol{k}) = B_k\begin{pmatrix} -k_- b_k \\ k_z b_k \\ 0 \\ 1 \end{pmatrix}, \tag{5.63b}$$

where

$$k_\pm = k_x \pm \mathrm{i}k_y, \tag{5.64}$$

$$b_k = \frac{\hbar c}{E_k + mc^2}, \tag{5.65}$$

$$E_k = \sqrt{(\hbar kc)^2 + (mc^2)^2}, \tag{5.66}$$

and

$$B_k = \sqrt{\frac{E_k + mc^2}{2E_k}}. \tag{5.67}$$

With the choice

$$|0\rangle = \frac{1}{\sqrt{V}}u^{(1)}(\boldsymbol{0}) \tag{5.68}$$

we find the following generalization of (5.59),

$$F_{j_\nu 0}(\boldsymbol{k}) = \delta_{\boldsymbol{k}_{j_\nu}\boldsymbol{k}}B_k c^{(\nu)}, \tag{5.69}$$

where

$$c^{(1)} = 1; \quad c^{(2)} = 0; \quad c^{(3)} = -b_k k_z; \quad c^{(4)} = -b_k k_-. \tag{5.70}$$

The remaining calculation proceeds in complete analogy to the nonrelativistic case and yields

$$\epsilon_l(k,\omega) = 1 + \frac{m\omega_{\mathrm{P}}^2}{\hbar k^2} B_k^2 \left(\frac{1}{\omega_k^+ - \omega - i\Gamma} + \frac{1}{\omega_k^+ + \omega + i\Gamma} \right.$$
$$\left. + \frac{(b_k k)^2}{\omega_k^- - \omega - i\Gamma} + \frac{(b_k k)^2}{\omega_k^- + \omega + i\Gamma} \right) \quad (5.71)$$

as the extension of (5.60), where

$$\hbar\omega_k^+ = E_k - mc^2; \quad \hbar\omega_k^- = -E_k - mc^2. \quad (5.72)$$

This notation indicates that $\hbar\omega_k^+$ represents the excitation energy of a positive-energy state and approaches $\hbar\omega_k$, (5.61), in the nonrelativistic limit, while $\hbar\omega_k^-$ represents excitation of a negative-energy state. The two terms containing $\hbar\omega_k^-$ vanish in the nonrelativistic limit.

Equation (5.71) can be compacted to

$$\epsilon_l(k,\omega) = 1+$$
$$+ \frac{\omega_{\mathrm{P}}^2}{\omega_k^2 - (1 + \hbar\omega_k/mc^2)(\omega + i\Gamma)^2 + \hbar^2(\omega + i\Gamma)^4/(2mc^2)^2}. \quad (5.73)$$

In this form it was given by Lindhard (1954), who also found the same expression for the transverse dielectric function $\epsilon_t(k,\omega)$.

5.4.3 Stopping Force

In order to appreciate the content of (5.60) consider the expression entering the first part of (5.30),

$$\mathrm{Re}\left(i\frac{1}{\epsilon_l(k,\omega)} \right) = -\mathrm{Re}\left(i\frac{\omega_{\mathrm{P}}^2}{\omega_k^2 + \omega_{\mathrm{P}}^2 - (\omega + i\Gamma)^2} \right)$$
$$= -\frac{\pi\omega_{\mathrm{P}}^2}{2\alpha_k}\left[\delta(\omega - \alpha_k) - \delta(\omega + \alpha_k) \right] \quad (5.74)$$

with

$$\alpha_k^2 = \omega_k^2 + \omega_{\mathrm{P}}^2, \quad (5.75)$$

where the relation (5.45) has been utilized.

The occurrence of the Dirac functions in (5.74) implies that energy is transferred in resonant processes. Indeed, it reveals a unique relationship between energy and momentum transfer,

$$(\hbar\omega)^2 = (\hbar\omega_{\mathrm{P}})^2 + \left(\frac{(\hbar k)^2}{2m} \right)^2. \quad (5.76)$$

This dispersion relation (Fig. 5.1) reflects properties of the medium independent of the external penetrating charge. In the limit of large momentum transfers $\hbar k$ it reduces to the dispersion relation for individual free electrons, $\hbar\omega = \hbar^2 k^2/2m$. In the opposite limit of small k one finds $\hbar\omega \simeq \hbar\omega_\mathrm{P}$, corresponding to a collective excitation similar to a standing wave oscillating with the classical plasma frequency ω_P.

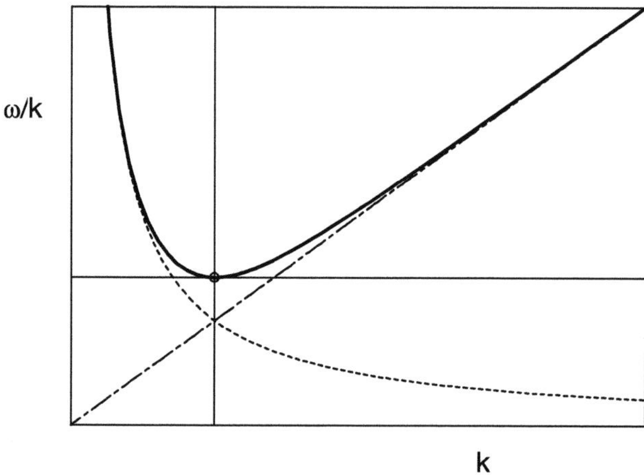

Fig. 5.1. Dispersion for static electron gas, (5.76). Dotted and dot-dashed lines represent the limiting behavior at small and large wave numbers, respectively. The horizontal line represents the minimum phase velocity $(\omega/k)_{\mathrm{min}} = \sqrt{\hbar\omega_\mathrm{P}/m}$. The vertical line separates collective from single-particle-like excitations

Integration of (5.30) yields

$$\frac{\mathrm{d}E}{\mathrm{d}x} = -\frac{e_1^2\omega_\mathrm{P}^2}{v^2}\int_{\alpha_k<kv}\frac{\mathrm{d}k}{k} = -\frac{e_1^2\omega_\mathrm{P}^2}{2v^2}\int_{\omega_k^2+\omega_\mathrm{P}^2<2mv^2\omega_k/\hbar}\frac{\mathrm{d}\omega_k}{\omega_k}$$

$$= -\frac{e_1^2\omega_\mathrm{P}^2}{2v^2}\ln\left(\frac{\zeta+\sqrt{\zeta^2-1}}{\zeta-\sqrt{\zeta^2-1}}\right) = -\frac{4\pi N e_1^2 e^2}{mv^2}\cosh^{-1}\zeta \quad (5.77)$$

where

$$\zeta = \frac{mv^2}{\hbar\omega_\mathrm{P}}. \tag{5.78}$$

For $\zeta \gg 1$ we have $\cosh^{-1}\zeta \sim \ln 2\zeta = \ln(2mv^2/\hbar\omega_\mathrm{P})$. Thus Bethe's asymptotic formula is retained with $\hbar\omega_\mathrm{P}$ taking the role of the mean excitation energy (Kramers, 1947, Lindhard, 1954).

From the asymptotic expansion of the inverse hyperbolic cosine for large arguments (Abramowitz and Stegun, 1964) one finds

$$\frac{\mathrm{d}E}{\mathrm{d}x} = -\frac{4\pi N e_1^2 e^2}{mv^2}\left[\ln\left(\frac{2mv^2}{\hbar\omega_\mathrm{P}}\right) - \left(\frac{\hbar\omega_\mathrm{P}}{2mv^2}\right)^2 - \frac{3}{2}\left(\frac{\hbar\omega_\mathrm{P}}{2mv^2}\right)^4 \cdots\right] \tag{5.79}$$

for $2mv^2 \gg \hbar\omega_\mathrm{P}$. We shall meet expansions of this type again in Chapter 6.

Relativistic Extension ($\star$)

As is obvious from (5.73), the relativistic treatment of the static electron gas lacks the analytical convenience of the nonrelativistic version (5.60).

For the longitudinal interaction, the integrations can be carried out in closed form, but the result is not particularly enlightening and will not be quoted here. However, it will usually make sense to assume that

$$\frac{\hbar\omega_\mathrm{P}}{2mc^2} \ll 1, \tag{5.80}$$

so that a Taylor expansion in terms of this parameter becomes meaningful. When this is done, the longitudinal contribution reduces to

$$\frac{\mathrm{d}E}{\mathrm{d}x} = -\frac{4\pi N e_1^2 e^2}{mv^2}\ln\left(\frac{2mv^2}{\hbar\omega_\mathrm{P}}\right), \tag{5.81}$$

consistent with (5.79).

For the transverse interaction, the denominator in the integrand becomes a third-order polynomial in ω^2, i.e., approximative steps become necessary already in the beginning. One may then treat long- and short-range interactions separately. Results of such a treatment will emerge in more general form in Sect. 5.6.

5.4.4 Oscillator Strength, Equipartition Rule and Differential Cross Section

The integral in (5.77) has the bounds

$$\omega_{k,\min} < \omega_k < \omega_{k,\max} \tag{5.82}$$

with

$$\omega_{k,\max,\min} = \frac{mv^2}{\hbar} \pm \sqrt{\left(\frac{mv^2}{\hbar}\right)^2 - \omega_\mathrm{P}^2}. \tag{5.83}$$

You may verify, by splitting the integration leading to (5.77) into two intervals $(\omega_{k,\min}, \omega_\mathrm{P})$ and $(\omega_\mathrm{P}, \omega_{k,\max})$, that these two intervals contribute equally to the integral. At the same time, the point $\omega_k = \omega_P$ corresponds to the minimum in the dispersion curve shown in Fig. 5.1. Thus, the stopping force of the static electron gas receives exactly equal contributions from the left and right branches of the dispersion curve shown in the figure, representing collective and single-electron-like excitations, respectively. This is a special case of a more general equipartition theorem (Lindhard and Winther, 1964).

Noting the equivalence[5] of $\hbar\omega$ with the excitation energy $\epsilon_j - \epsilon_0$ in the Bethe theory, we may compare the stopping cross section in the Bethe theory with (5.30) to find the quantity equivalent with the generalized oscillator strength defined in (4.107),

$$\sum_j f_j(Q) \cdots \to \frac{1}{2} \int \mathrm{d}\omega \, [\delta(\omega - \alpha_k) + \delta(\omega + \alpha_k)] \ldots \tag{5.84}$$

This shows that the curve depicted in Fig. 5.1 represents the region in k, ω space where the generalized oscillator strength is nonvanishing. Evidently, Bethe's sum rule $\sum_j f_j(Q) = 1$ is satisfied.

Go back for a moment to the evaluation of the stopping force, (5.30) for the dielectric function of the static electron gas, (5.60), but interchange the order of integrations. This can be accomplished by insertion of (5.74) and substitution of α_k for k. The result may be written in the form

$$\frac{\mathrm{d}E}{\mathrm{d}x} = -\frac{e_1^2 \omega_P^2}{2v^2} \int_{\omega_{\min}}^{\omega_{\max}} \mathrm{d}(\hbar\omega)(\hbar\omega) \frac{1}{(\hbar\omega)^2 - (\hbar\omega_P)^2} \tag{5.85}$$

with

$$(\hbar\omega_{\max,\min})^2 = (\hbar\omega_P)^2 + \left(mv^2 \pm \sqrt{(mv^2)^2 - (\hbar\omega_P)^2} \right)^2. \tag{5.86}$$

While the integration over ω leads back to (5.77) as it should, we also have the option of interpreting (5.85) as an integral over a differential cross section equivalent with (2.29), so that

$$\mathrm{d}\sigma(T) = \frac{2\pi e_1^2 e^2}{mv^2} \frac{dT}{T^2 - (\hbar\omega_P)^2} \tag{5.87}$$

for $\hbar\omega_{\min} < T < \hbar\omega_{\max}$. This interpretation is evidently not unique since no assumptions about statistics have entered the theory here. However, $\mathrm{d}\sigma(T)$ is seen to reduce to Rutherford's law in the limit of $\omega_P = 0$. For nonvanishing density N, energy is transferred only if the projectile speed exceeds the threshold defined by $mv^2 \geq \hbar\omega_p$. In the limit of high speed the excitation spectrum is limited by $\hbar\omega_P < T < 2mv^2$.

---

[5] Regarding the sign of ω see the remark on page 149.

5.4.5 Plasmon-Pole Approximation ($\star$)

In the plasmon-pole approximation (Lundqvist, 1967b,a), equation (5.60) is generalized to

$$\epsilon_l(k,\omega) = 1 + \frac{\omega_{\mathrm{P}}^2}{\omega_k^2 + 2\omega_1\omega_k - (\omega + \mathrm{i}\Gamma)^2}, \tag{5.88}$$

where ω_1 is a free parameter that can be chosen to match deviations from the free-electron picture due to band structure or the like.

Evaluation of the stopping force proceeds as in Sect. 5.4.3 and reproduces (5.77), but now with

$$\zeta = \frac{mv^2 - \hbar\omega_1}{\hbar\omega_{\mathrm{P}}}. \tag{5.89}$$

5.5 Assembly of Harmonic Oscillators ($\star$)

As another example of limited complexity consider a dense gas of 'harmonic-oscillator atoms' characterized by a resonance frequency ω_0. This is a specific application of Sect. 5.3. Therefore, keep in mind that possible overlap between individual oscillators has been ignored.

5.5.1 Dielectric Function

Matrix elements may be determined from (4.101), and sums over directionally degenerate states may be carried out by use of the binomial theorem. Then, (5.42) reduces to

$$\epsilon_l(k,\omega) = 1 + \frac{m\omega_{\mathrm{P}}^2}{\hbar k^2} \mathrm{e}^{-\omega_k/\omega_0} \sum_{j=1}^{\infty} \frac{1}{j!} \left(\frac{\omega_k}{\omega_0}\right)^j$$
$$\times \left(\frac{1}{j\omega_0 - \omega - \mathrm{i}\Gamma} + \frac{1}{j\omega_0 + \omega + \mathrm{i}\Gamma}\right) \tag{5.90}$$

or, in terms of Kummer's function (Abramowitz and Stegun, 1964, p. 504),

$$\epsilon_l(k,\omega) = 1 + \frac{m\omega_{\mathrm{P}}^2}{\hbar k^2} \frac{1}{\omega + \mathrm{i}\Gamma}$$
$$\times \left\{ M\left(1, 1 + \frac{\omega + \mathrm{i}\Gamma}{\omega_0}, -\frac{\omega_k}{\omega_0}\right) - M\left(1, 1 - \frac{\omega + \mathrm{i}\Gamma}{\omega_0}, -\frac{\omega_k}{\omega_0}\right) \right\}. \tag{5.91}$$

These results were derived by Belkacem and Sigmund (1990). It is readily verified that $\epsilon_l(k,\omega)$ approaches the free-electron expression (5.60) in the limit of $\omega_0 = 0$ and the classical Drude-Lorentz expression (A.215) in the limit of $\hbar = 0$ or $k = 0$.

It is seen from (5.90) that the dielectric function has simple poles at $\omega = \pm j\omega_0$. This implies that ϵ must vanish at some point in each interval $\nu\omega_0 < \omega < (\nu+1)\omega_0$ with $\nu = -\infty \ldots \infty$, the accurate position depending on density N and wave number k.

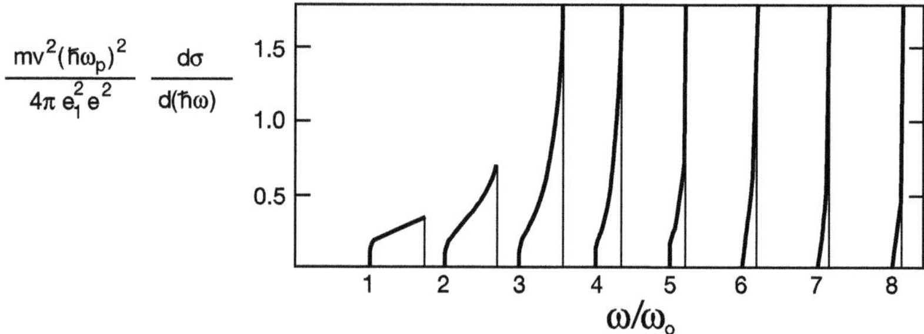

Fig. 5.2. Differential energy loss cross section for gas of harmonic-oscillator atoms. $\omega_P^2/\omega_0^2 = 10$ and $2mv^2/\hbar\alpha_0 = 10$, with $\alpha_0 = (\omega_0^2 + \omega_P^2)^{1/2}$. From Belkacem and Sigmund (1990)

5.5.2 Excitation Spectrum

Following the argument that led to the differential cross section (5.87) above, we may again extract a differential cross section for energy transfer $\hbar\omega_{j0}$ in a single event. Negative values of $\hbar\omega$ need to be mapped on the positive axis by a simple symmetry transformation. This yields

$$\frac{d\sigma(\hbar\omega)}{d(\hbar\omega)} = \frac{ie_1^2}{\pi\hbar^2 nv^2} \int_{\omega/v}^{\infty} \frac{dk}{k} \left(\frac{1}{\epsilon_l(k,\omega)} - \frac{1}{\epsilon_l(k,-\omega)} \right), \qquad (5.92)$$

leaving out the transverse contribution.

Figure 5.2 shows the differential cross section following from (5.91) and (5.92) for an oscillator gas of fairly high density, $\omega_P^2/\omega_0^2 = 10$ at a moderately high projectile speed, $2mv^2/\hbar\sqrt{\omega_0^2 + \omega_P^2} = 10$. At low density n, the cross section would have sharp maxima at integer values of ω/ω_0. It is seen that the excitation levels get somewhat smeared, but despite high electron density the discrete structure of the excitations is still clearly visible. It is also seen that the integral over the broadened oscillator levels $\omega \simeq \nu\omega_0$, $\nu = 1, 2, \ldots$ reaches a maximum at $\nu = 3$ corresponding to $\omega \sim \omega_P$ and decreases for large excitation levels ν.

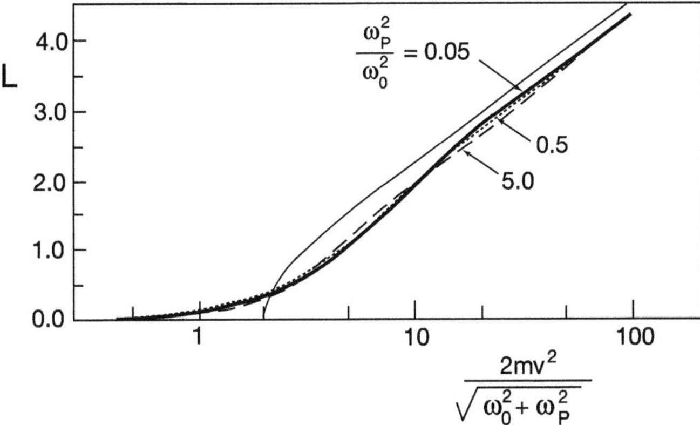

Fig. 5.3. Stopping number for static electron gas (thin line) and harmonic-oscillator gas at three densities corresponding to $\omega_{\mathrm{P}}^2/\omega_0^2 = 0.05, 0.5, 5$. From Belkacem and Sigmund (1990)

5.5.3 Stopping Force

Figure 5.3 shows stopping numbers, defined by (2.56), computed from this model for three densities, compared with the prediction for the static electron gas (5.77). Classical dispersion theory predicts a resonance frequency $\sqrt{\omega_0^2 + \omega_{\mathrm{P}}^2}$, cf. (5.56). Therefore this frequency has been inserted as the pertinent resonance frequency in the expression of the Bethe parameter defining the abscissa variable. The result shows satisfactory scaling for three finite values of $\omega_{\mathrm{P}}^2/\omega_0^2$. The case of a free static electron gas where $\omega_{\mathrm{P}}^2/\omega_0^2 = \infty$ does not obey this scaling property.

5.6 Relativistic Bethe Stopping Theory (⋆)

We are now well equipped to extend the Bethe stopping formula, derived first in Sect. 4.5.4 and rederived in Sect. 5.3.2, into the relativistic regime.

5.6.1 Regimes of Momentum Transfer

Two major features deserve attention,

- The transverse contribution to the stopping force (5.30) cannot be neglected, and
- Relativistic expressions are to be used for the dielectric function when the momentum transfer reaches the relativistic regime.

This suggests a division of the transferred-momentum space into three instead of only two intervals,

1. A low-Q regime, $0 < Q < Q_0$, in which the dipole approximation is expected to apply, just as in Sect. 4.5.4,
2. An intermediate-Q regime, $Q_0 < Q < Q_1$, within which nonrelativistic dynamics remains valid, and
3. A high-Q regime, $Q > Q_1$, where electrons are excited into the relativistic regime[6].

It is emphasized that the fundamental relation (5.30), which follows from Maxwell's equations and the definitions of charge and current density in the laboratory frame of reference, remains valid. This immediately implies that the contribution of the longitudinal field to the stopping force, presented in section 5.3.2, remains valid as far as regions 1 and 2 are concerned. The only modification here is the replacement of the upper limit Q_{max} by Q_1. What remains to be done, then, is

- To evaluate the contribution from the transverse field for $0 < Q < Q_1$ and
- To evaluate both the longitudinal and the transverse contribution for $Q > Q_1$.

5.6.2 Transverse Field: Low Momentum Transfers

To compute the contribution from transverse excitations to the stopping force (5.30), the procedure applied in Sect. 5.3.2 yields

$$
-\frac{\mathrm{d}E}{\mathrm{d}x}\bigg|_{\mathrm{trans}} = \frac{2\pi e_1^2 e^2 n}{mv^2} \sum_j \hbar\omega_{j0} \int_{Q>(\hbar\omega_{j0})^2/2mv^2} \frac{\mathrm{d}Q}{Q^2}
$$
$$
\times \left|F_{j0}(\mathbf{k})\right|^2 \frac{2mv^2 Q/(\hbar\omega_{j0})^2 - 1}{[2mc^2 Q/(\hbar\omega_{j0})^2 - 1]^2}. \quad (5.93)
$$

Figure 5.4 shows the weight factor

$$
w_{\mathrm{trans}} = \frac{2mv^2 Q/(\hbar\omega_{j0})^2 - 1}{(2mc^2 Q/(\hbar\omega_{j0})^2 - 1)^2} \quad (5.94)
$$

determining the contribution of the transverse field under the stopping integral with the corresponding factor $w_{\mathrm{long}} \equiv 1$ in the longitudinal contribution. It is seen that up to $\beta = v/c = 0.4$, $w_{\mathrm{trans}}/w_{\mathrm{long}} < 0.01$ for all values of Q. Even for $\beta = v/c = 0.9$, $w_{\mathrm{trans}} < w_{\mathrm{long}}$ everywhere, but for higher speeds, the transverse contribution becomes exceedingly important.

Now, substituting

$$
f_{j0}(Q) = \frac{1}{Z_2} \frac{\hbar\omega_{j0}}{Q} |F_{j0}(\mathbf{k})|^2 \quad (5.95)
$$

---

[6] We shall here keep to the definition (4.54) or (5.48) for Q. This notation differs from Fano (1963) who defines Q as the kinetic energy of a free electron.

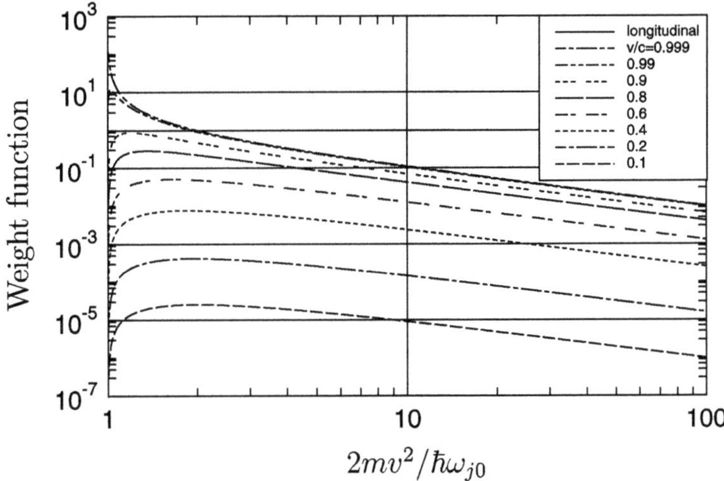

Fig. 5.4. Weight factor w_{trans}, (5.94) entering the transverse contribution to the stopping force, compared to the corresponding quantity $w_{\text{long}} \equiv 1$ in the longitudinal contribution. Curves for v/c approaching 1 are drawn for illustration only. Relativistic expressions need to be used there

in accordance with (4.107), we find

$$
-\left.\frac{\mathrm{d}E}{\mathrm{d}x}\right|_{\text{trans},Q<Q_1} = \frac{2\pi e_1^2 e^2 Z_2 N}{mv^2} \sum_j \int_{(\hbar\omega_{j0})^2/2mv^2}^{Q_1} \frac{\mathrm{d}Q}{Q} f_{j0}(Q)
$$
$$
\times \frac{2mv^2 Q/(\hbar\omega_{j0})^2 - 1}{[2mc^2 Q/(\hbar\omega_{j0})^2 - 1]^2} \quad (5.96)
$$

from (5.49) and (5.93).

Consider first the low-Q case,

$$
\frac{(\hbar\omega_{j0})^2}{2mv^2} < Q < Q_0, \quad (5.97)
$$

where the dipole approximation is taken to be valid. We may then take $f_{j0}(Q) \simeq f_{j0}(0) \equiv f_{j0}$ out of the integral. Introducing the integration variable

$$
\eta_j = \frac{2mv^2}{(\hbar\omega_{j0})^2} Q \quad (5.98)
$$

we find

$$
-\left.\frac{\mathrm{d}E}{\mathrm{d}x}\right|_{\text{trans},Q<Q_0} = \frac{2\pi e_1^2 e^2 Z_2 N}{mv^2} \sum_j f_{j0} \int_1^{\eta_{0j}} \frac{\mathrm{d}\eta_j}{\eta_j} \frac{\eta_j - 1}{[\eta_j c^2/v^2 - 1]^2}, \quad (5.99)
$$

where

$$\eta_{0j} = \frac{2mv^2}{(\hbar\omega_{j0})^2}Q_0.$$

(5.100)

The integral can be carried out and leads to

$$-\left.\frac{\mathrm{d}E}{\mathrm{d}x}\right|_{\mathrm{trans},Q<Q_0} = \frac{2\pi e_1^2 e^2 Z_2 N}{mv^2}\sum_j f_{j0}\left[-\ln(1-\beta^2)-\beta^2\right.$$
$$\left.+\ln\left(1-\frac{\beta^2}{\eta_{0j}}\right)+\frac{\beta^2(1-\beta^2)}{\eta_{0j}-\beta^2}\right].$$

(5.101)

Now, making use of the initial assumption that

$$Q_0 \gg \frac{(\hbar\omega_{j0})^2}{2mv^2},$$

(5.102)

i.e.,

$$\eta_{0j} \gg 1,$$

(5.103)

and applying the sum rule (4.114) reduces this to

$$-\left.\frac{\mathrm{d}E}{\mathrm{d}x}\right|_{\mathrm{trans},Q<Q_0} = \frac{2\pi e_1^2 e^2 Z_2 N}{mv^2}\left[-\ln(1-\frac{v^2}{c^2})-\frac{v^2}{c^2}\right].$$

(5.104)

Consider next the contribution from the intermediate-Q regime

$$Q_0 < Q < Q_1.$$

(5.105)

Going back to (5.93) we first note that the quantities $2mv^2Q/(\hbar\omega_{j0})^2$ and $2mc^2Q/(\hbar\omega_{j0})^2$ both are large compared to 1. The dominating part of the integral can, therefore, be written as

$$-\left.\frac{\mathrm{d}E}{\mathrm{d}x}\right|_{\mathrm{trans}} = \frac{2\pi e_1^2 e^2 n}{2(mc^2)^2}\sum_j (\hbar\omega_{j0})^3 \int_{Q_0}^{Q_1}\frac{\mathrm{d}Q}{Q^3}\left|F_{j0}(\boldsymbol{k})\right|^2.$$

(5.106)

The assumption that Q_1 lies in the nonrelativistic regime implies that integration over Q and summation over j leads to an energy $\ll mc^2$. This, together with one of the two factors $1/mc^2$ in front of the sum, implies that this integral is negligible compared to (5.104).

Now, from (5.49) we obtain the longitudinal contribution, carrying out the intagration only up to Q_1. Adding this to (5.104) we find

$$-\left.\frac{\mathrm{d}E}{\mathrm{d}x}\right|_{Q<Q_1} = \frac{2\pi e_1^2 e^2 Z_2 N}{mv^2}$$
$$\times\sum_j f_{j0}\left[\ln\frac{2mv^2Q_1}{(\hbar\omega_{j0})^2}-\ln(1-\frac{v^2}{c^2})-\frac{v^2}{c^2}\right].$$

(5.107)

5.6.3 High Momentum Transfers

For the high-Q regime, $Q > Q_1$ we go back to (5.30) and avoid separation into longitudinal and transverse components, because $\epsilon_l = \epsilon_t \equiv \epsilon(k,\omega)$ in this case. Setting

$$\epsilon(k,\omega) = 1 + \delta\epsilon(k,\omega), \tag{5.108}$$

and noting that the stopping force must be real, we may expand the integrand up to the first order in $\delta\epsilon(k,\omega)$ to obtain

$$\frac{1}{\epsilon(k,\omega)} - \frac{v^2}{c^2}\frac{k^2 - \omega^2/v^2}{k^2 - \epsilon_t(k,\omega)\omega^2/c^2}$$
$$\rightarrow -\delta\epsilon(k,\omega)\left[1 + \frac{\omega^2}{k^2c^2 - \omega^2} - \left(1 - \frac{v^2}{c^2}\right)\frac{k^2c^2\omega^2}{(k^2c^2 - \omega^2)^2}\right]. \tag{5.109}$$

Here, use has been made of the fact that for free independent electrons, $\epsilon_t(k,\omega) \equiv \epsilon(k,\omega)$ in the relativistic regime (Lindhard, 1954).

For $\delta\epsilon(k,\omega)$ we use the expression following from (5.71). Of the four poles in the denominator, only the two with $\omega = \pm\omega_k^+$ fall within the limits of integration for ω. Making use of (5.45) again and integrating over ω we then obtain

$$-\frac{\mathrm{d}E}{\mathrm{d}x}\bigg|_{Q>Q_1} = \frac{2\pi e_1^2 e^2 N Z_2}{mv^2}\int_{Q_1}^{Q_{\mathrm{max}}}\frac{\mathrm{d}Q}{\sqrt{1 + 2Q/mc^2}}$$
$$\times \frac{Q - \hbar^2\omega_k^{+2}/mc^2 + \hbar^2\omega_k^{+2}\beta^2/2mc^2}{(Q - \hbar^2\omega_k^{+2}/2mc^2)^2}. \tag{5.110}$$

Here, the maximum value

$$Q_{\mathrm{max}} = \frac{2mv^2}{(1 - v^2/c^2)^2} \tag{5.111}$$

follows directly from the limit of the ω-integration,

$$\omega_k^{+2} < k^2 v^2. \tag{5.112}$$

The integration becomes elementary by introduction of the variable

$$\zeta = \sqrt{1 + \frac{2Q}{mc^2}}, \tag{5.113}$$

$$-\frac{\mathrm{d}E}{\mathrm{d}x}\bigg|_{Q>Q_1} = \frac{2\pi e_1^2 e^2 N Z_2}{mv^2}\int_{\zeta_1}^{\zeta_{\mathrm{max}}}\mathrm{d}\zeta\left[\frac{1}{\zeta - 1} - \frac{1}{2}(1 - \beta^2)\right] \tag{5.114}$$

or

$$-\frac{\mathrm{d}E}{\mathrm{d}x}\bigg|_{Q>Q_1} = \frac{2\pi e_1^2 e^2 N Z_2}{mv^2}\left[\ln\frac{\zeta_{\max}-1}{\zeta_1-1} - \frac{1}{2}(1-\beta^2)(\zeta_{\max}-\zeta_1)\right],$$

(5.115)

where $\zeta_{\max}=\zeta(Q_{\max})$ and $\zeta_1=\zeta(Q_1)$.

Up to this point the calculation is exact. Now, remind that Q_1 was chosen to lie in the nonrelativistic regime, i.e., $Q_1/mc^2 \ll 1$ or

$$\zeta_1 \simeq 1 + \frac{Q_1}{mc^2},$$

(5.116)

higher-order terms being negligible. With this, and

$$\zeta_{\max} = \frac{1+\beta^2}{1-\beta^2}$$

(5.117)

you find

$$-\frac{\mathrm{d}E}{\mathrm{d}x}\bigg|_{Q>Q_1} = \frac{2\pi e_1^2 e^2 N Z_2}{mv^2}\left[\ln\frac{2mv^2}{Q_1} - \ln\left(1-\frac{v^2}{c^2}\right) - \frac{v^2}{c^2}\right]$$

(5.118)

and, after addition of the low-Q contribution (5.104),

$$-\frac{\mathrm{d}E}{\mathrm{d}x} = \frac{4\pi e_1^2 e^2 Z_2 N}{mv^2}\sum_j f_{j0}\left[\ln\frac{2mv^2}{\hbar\omega_{j0}} - \ln\left(1-\frac{v^2}{c^2}\right) - \frac{v^2}{c^2}\right],$$

(5.119)

which is the relativistic Bethe formula for a heavy projectile (Bethe, 1932, Fano, 1963).

Comparison with the relativistic Bohr formula (4.100) shows that with regard to the relativistic corrections, the two expressions differ only by a term $-v^2/2c^2$ in (5.119) which arises from close collisions, i.e., from the Dirac equation. Note in particular that the term $-\ln(1-v^2/c^2)$ in (5.118), which enters via $Q_{\max}$, already emerged from the Bohr theory.

5.6.4 Relativistic Density Effect

The existence of a substantial density effect on the stopping force at relativistic speeds was pointed out by Fermi (1940), who also provided the first quantitative estimate based on the flow of energy through a cylinder surrounding a trajectory at a suitably large distance. Clearly, such an effect must stem from the transverse interaction. Indeed, no sudden change can be expected from the term $1/\epsilon(k,\omega)$ in (5.30) once the projectile speed approaches the speed of light, in contrast to the transverse term with its denominator $k^2 - \epsilon\omega^2/c^2$.

Let us rewrite the transverse term in the form

$$\left(-\frac{\mathrm{d}E}{\mathrm{d}x}\right)_{\mathrm{trans}} = \frac{e_1^2}{\pi c^2}\mathrm{Im}\int_{-\infty}^{\infty}\omega\mathrm{d}\omega\int_{|\omega/v|}^{k_{\mathrm{max}}}\frac{\mathrm{d}k}{k}\frac{k^2-\omega^2/v^2}{k^2-\epsilon\omega^2/c^2}, \tag{5.120}$$

where k_{max} is large but real.

The relativistic density effect originates in interactions over a large distance due to the circumstance that the adiabatic radius becomes very large. Hence, only small values of k are of interest in the present context, and $\epsilon(k,\omega)$ can be replaced by $\epsilon(0,\omega)$, mostly abbreviated as ϵ in the rest of this section. The integration over k can then be carried out first (Fano, 1956) and yields

$$\left(-\frac{\mathrm{d}E}{\mathrm{d}x}\right)_{\mathrm{trans,dist}} = \frac{e_1^2}{2\pi v^2}\mathrm{Im}\int_{-\infty}^{\infty}\omega\mathrm{d}\omega\left\{\frac{1}{\epsilon}\left[\ln\frac{k_{\mathrm{max}}^2}{k_{\mathrm{max}}^2-\epsilon\omega^2/c^2}\right.\right.$$
$$\left.\left.+\ln\left(1-\epsilon\frac{v^2}{c^2}\right)\right]+\frac{v^2}{c^2}\ln\frac{v^2k_{\mathrm{max}}^2/\omega^2-\epsilon v^2/c^2}{1-\epsilon v^2/c^2}\right\}, \tag{5.121}$$

where the subscript 'trans,dist' indicates that we only deal with the distant part of the interaction.

For $k_{\mathrm{max}}^2\gg\epsilon\omega^2/c^2$, and noticing that only the imaginary part is of interest, this reduces to (Halpern and Hall, 1948, Fano, 1956)

$$\left(-\frac{\mathrm{d}E}{\mathrm{d}x}\right)_{\mathrm{trans,dist}} = \frac{e_1^2}{2\pi v^2}\mathrm{Im}\int_{-\infty}^{\infty}\omega\mathrm{d}\omega\left(\frac{1}{\epsilon}-\frac{v^2}{c^2}\right)\ln\left(1-\epsilon\frac{v^2}{c^2}\right). \tag{5.122}$$

You may insert (5.42) for the dielectric function, go to the low-k limit and expand up to the first order in the electron density in accordance with (5.44). Then you obtain

$$\left(-\frac{\mathrm{d}E}{\mathrm{d}x}\right)_{\mathrm{trans,dist,linear}} = -\frac{2\pi e_1^2 N Z_2 e^2}{mv^2}\left[\frac{v^2}{c^2}+\ln\left(1-\frac{v^2}{c^2}\right)\right], \tag{5.123}$$

as it must be according to (5.104). The difference between (5.122) and (5.123) represents the density correction. However, if you proceed to the next order in the electron density, you arrive at a vanishing contribution. Indeed, the density effect has a threshold velocity, as will be seen now.

Following Fano (1956), let us first assume the argument of the logarithm in (5.122) to be positive for all ω,

$$1-\beta^2\epsilon(0,\omega)>0; \quad \beta=\frac{v}{c}. \tag{5.124}$$

Then, singularities occur in the poles of ϵ which all lie in the lower half of the complex ω-plane, albeit close to the real axis. We may then deform the integration path to a large semi-circle,

$$\omega=\Omega e^{i\phi}; \quad \phi=\pi\dots 0. \tag{5.125}$$

With this we have

$$\epsilon \sim 1 - \frac{\omega_{\mathrm{P}}^2}{\omega^2}, \quad \frac{1}{\epsilon} \sim 1 + \frac{\omega_{\mathrm{P}}^2}{\omega^2}, \tag{5.126}$$

provided that $\Omega \gg \omega_\nu$ for all resonances. Then, (5.122) reduces to

$$\left(-\frac{\mathrm{d}E}{\mathrm{d}x} \right)_{\mathrm{trans,dist}} = \frac{e_1^2}{2\pi v^2} \mathrm{Im} \left\{ \int \omega \mathrm{d}\omega (1 - \beta^2) \ln(1 - \beta^2) \right.$$
$$\left. + \omega_{\mathrm{P}}^2 \int \frac{\mathrm{d}\omega}{\omega} \left[\ln(1 - \beta^2) + \beta^2 \right] \right\}, \tag{5.127}$$

while terms of higher order in $1/\omega$ vanish for $\Omega \to \infty$. Here the first integral is real and does not contribute. The second integral is easily carried out by making the substitution (5.125), leading to

$$\left(-\frac{\mathrm{d}E}{\mathrm{d}x} \right)_{\mathrm{trans,dist}} = -\frac{e_1^2 \omega_{\mathrm{P}}^2}{2v^2} \left[\ln(1 - \beta^2) + \beta^2 \right], \tag{5.128}$$

i.e., the low-density limit (5.123). This explains why it did not help to go beyond the leading term in a Taylor expansion in ϵ.

Next, consider the case where

$$1 - \beta^2 \epsilon(0, \omega) < 0 \tag{5.129}$$

for some range of ω. This is relevant when that range extends into the positive-imaginary half plane. As an example, take the harmonic oscillator

$$\epsilon = 1 + \frac{\omega_{\mathrm{P}}^2}{\omega_0^2 - \omega^2}, \tag{5.130}$$

for which $1 - \beta^2 \epsilon = 0$ at

$$\omega_1^2 = \omega_0^2 - \frac{\beta^2}{1 - \beta^2} \omega_{\mathrm{P}}^2. \tag{5.131}$$

Evidently, ω_1 is real for $\beta = 0$ but turns imaginary above a certain value of $\beta = v/c$. Let us consider the latter case, denote $\omega_1 = i\ell$ with $\ell > 0$ and go back again to a general dielectric function ϵ.

We can cope with the multi-valued character of the logarithm in (5.122) by introducing a cut for $\mathrm{Im}\,\omega < \ell$. Then, in order to compensate for having pulled the path of integration across the cut we need to add a term containing an integral around the cut,

$$\Delta \left(-\frac{\mathrm{d}E}{\mathrm{d}x} \right)_{\mathrm{trans}} = \frac{e_1^2}{2\pi v^2} \mathrm{Im} \left(\int_{0+}^{i\ell} - \int_{0-}^{i\ell} \right) (iy)\mathrm{d}(iy)$$
$$\left(\frac{1}{\epsilon(0, iy)} - \beta^2 \right) \ln \left[1 - \beta^2 \epsilon(0, iy) \right], \tag{5.132}$$

where we have set $\omega = iy$. Since $1/\epsilon(0, iy) - \beta^2$ is real, only the imaginary part of the logarithm contributes to the imaginary part of the integral. The two logarithms contribute a factor $-2\pi i$, and hence,

$$\Delta\left(-\frac{dE}{dx}\right)_{\mathrm{trans}} = -\frac{e_1^2}{v^2}\int_0^\ell y\,dy\left(\beta^2 - \frac{1}{\epsilon(0, iy)}\right) \tag{5.133}$$

in accordance with Fano (1956).

Now, consider an insulating medium so that $\epsilon(0, \omega)$ is given by (A.222),

$$\epsilon(\omega) = 1 + \omega_\mathrm{P}^2\sum_j\frac{f_j}{\omega_j^2 - \omega^2}, \tag{5.134}$$

a standard relation from electromagnetic theory[7]. This function has singularities at ω_j, $j = 1, 2, \ldots$ and zeros at $\tilde{\omega}_j$, where

$$\omega_j < \tilde{\omega}_j < \omega_{j+1}; \quad j = 1, 2, \ldots, \tag{5.135}$$

assuming $\omega_{j+1} > \omega_j$ for all j. With this, you can write the reciprocal in the form

$$\frac{1}{\epsilon(\omega)} = 1 + \omega_\mathrm{P}^2\sum_j\frac{\tilde{f}_j}{\omega^2 - \tilde{\omega}_j^2}, \tag{5.136}$$

where the constants $\tilde{\omega}_j, \tilde{f}_j$ are uniquely related to ω_j, f_j.

The integrations in (5.133) are elementary in that case and lead to

$$\Delta\left(-\frac{dE}{dx}\right)_{\mathrm{trans}} = -\frac{e_1^2\omega_\mathrm{P}^2}{2v^2}\left[(1-\beta^2)\frac{\omega_1^2}{\omega_\mathrm{P}^2} + \sum_j\tilde{f}_j\ln(\tilde{\omega}_j^2 - \omega_1^2)\right]. \tag{5.137}$$

Now, consider again the specific case of a harmonic oscillator with ϵ given by (5.130). Here,

$$\frac{1}{\epsilon} = 1 + \frac{\omega_\mathrm{P}^2}{\omega^2 - \tilde{\omega}_0^2} \tag{5.138}$$

with

$$\tilde{\omega}_0^2 = \omega_0^2 + \omega_\mathrm{P}^2 \tag{5.139}$$

and hence

$$\omega_1^2 = \omega_0^2 - \omega_\mathrm{P}^2\frac{\beta^2}{1 - \beta^2}, \tag{5.140}$$

---

[7] Damping constants have been dropped. This is justified here as long as they are small.

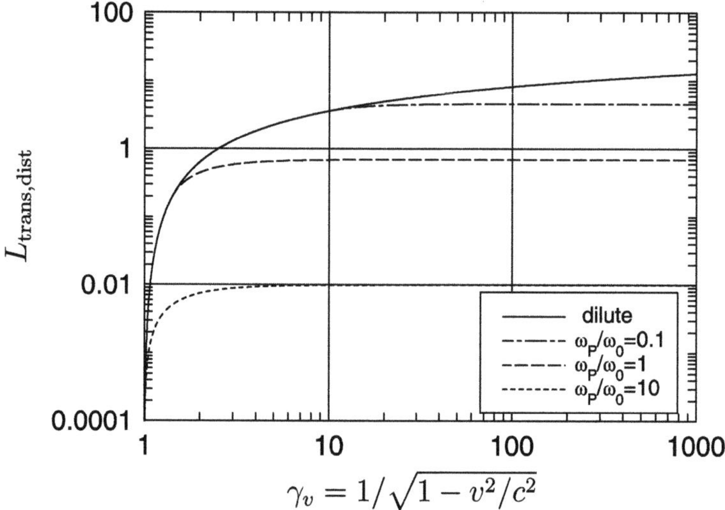

Fig. 5.5. Illustrating the density correction on a single-resonance spectrum (5.130). Contributions to stopping number from transverse excitations (distant part) for dilute medium (thin line) and including density correction for $\omega_\mathrm{P}/\omega_0 = 0.1$, 1.0 and 10 (top to bottom)

so that

$$\Delta\left(-\frac{\mathrm{d}E}{\mathrm{d}x}\right)_{\text{trans}} = -\frac{e_1^2 \omega_\mathrm{P}^2}{2v^2}\left[-\ln(1-\beta^2) - \beta^2 \right.$$
$$\left. +(1-\beta^2)\frac{\omega_0^2}{\omega_\mathrm{P}^2} - \ln\left(1 + \frac{\omega_0^2}{\omega_\mathrm{P}^2}\right)\right]. \quad (5.141)$$

Combined with (5.128) this yields

$$\left(-\frac{\mathrm{d}E}{\mathrm{d}x}\right)_{\text{trans,dist}} = -\frac{2\pi e_1^2 N Z_2 e^2}{mv^2}$$
$$\times\left[(1-\beta^2)\frac{\omega_0^2}{\omega_\mathrm{P}^2} - \ln\left(1 + \frac{\omega_0^2}{\omega_\mathrm{P}^2}\right)\right] \quad (5.142)$$

for a single resonance frequency.

Figure 5.5 demonstrates that the notion of a density effect is well justified: Indeed, the stopping number is sensitive to the density. However, the density correction is also sensitive to ω_0 and, hence, to the dispersive properties of the medium. Therefore, a single oscillator frequency cannot be expected to describe the phenomenon in a quantitative way.

The influence of the oscillator-strength spectrum on the density effect was analysed by several authors, in particular by Halpern and Hall (1948),

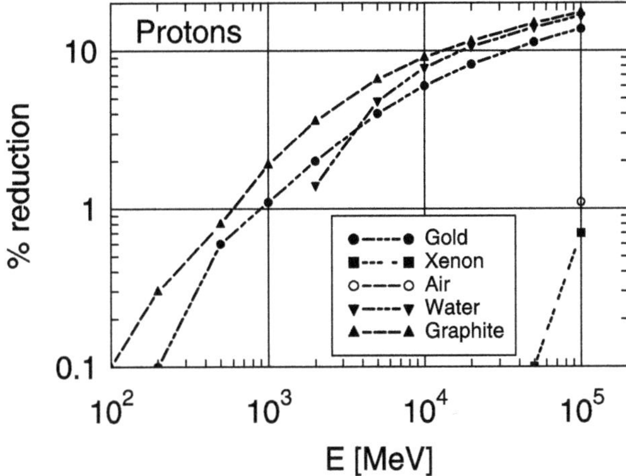

Fig. 5.6. Percentage correction to stopping force due to relativistic density effect according to Sternheimer et al. (1984)

A. Bohr (1948) and Sternheimer (1952). The step from (5.134) to (5.136) may be computationally intensive, dependent on how many frequencies are involved. More serious is the fact that knowledge of the spectrum was scarce in the 1950s and is still by no means complete. Therefore, a long series of papers appeared with tabulations for different materials as well as revised tables based on revised oscillator strengths, mainly by Sternheimer. A recent version of such data was published by Sternheimer et al. (1984)[8]. Figure 5.6 gives an impression of the significance of the effect for protons as a function of beam energy.

The relativistic density effect is intimately connected to the phenomenon of Cherenkov radiation, which will be discussed along with other radiation effects in Volume III. At this point it is emphasized that energy losses to Cherenkov radiation are fully accounted for in the general treatment presented in this section.

This entire complex has received much attention in the older literature because of its significance in particle physics. Comprehensive reviews may be found in articles by Uehling (1954) and Crispin and Fowler (1970).

---

[8] Sternheimer's notation differs from Fano's, which has been used in the above derivation. In particular, Sternheimer did not make use of (5.136). More importantly, the spectrum (5.134) is taken to be the real spectrum of the solid material, whereas Sternheimer (1952) makes reference to *atomic* excitation spectra to which a correction is applied. Therefore, the pertinent equations look different from the present ones. However, with the proper choice of parameters, Sternheimer's formulation can be shown to be equivalent with Fano's.

5.7 Fermi Gas

The models discussed up till now owe their relative simplicity to the neglect of
the Pauli exclusion principle. Ignoring this important requirement in a many-
electron system is by no means generally justified. The present section serves
to provide a description incorporating the Pauli principle for the special case
of a free-electron gas following Lindhard (1954), and various consequences
on particle stopping. In that case it is comparatively easy to generate a set
of orthogonal single-electron wave functions. A similar treatment for bound
particles must invoke assumptions on the structure of the medium.

5.7.1 Electronic States

Consider electronic states in a one-dimensional box of length L. Eigenstates
are characterized by wave functions $\sin kx$ with $k = \nu\pi/L$, $\nu = 1, 2, \ldots$ and
energies $\epsilon_k = \hbar^2 k^2/2m$. The quantization of k arises from the requirement
that the wave function vanish at $x = 0$ and L. Wave functions belonging to
different values of k are orthogonal. The Pauli principle requires that no more
than two electrons occupy one k-state, corresponding to spin up and down,
respectively.

For a system containing many electrons it is convenient to replace the
boundary condition on the wave function by a periodicity requirement such
that wave functions are forced to be periodic with a period L. This allows to
use wave functions of the form $\exp(ikx)$ with $k = \nu 2\pi/L$ and $\nu = \pm 1, \pm 2, \ldots$.
This means that the distance between adjacent k-levels doubles. Instead, there
are two independent states for each allowed value of $|k|$, and including spin
a maximum of four electrons may occupy a given energy level ϵ_k. This difference is significant for low-lying states but may be ignored in a system
containing many electrons (cf. problem 5.9).

This argument is readily extended into three dimensions. Here, individual
levels may be characterized by wave functions

$$\psi_{\boldsymbol{k}} = \frac{1}{\sqrt{\mathsf{V}}} e^{i\boldsymbol{k}\cdot\boldsymbol{r}} \tag{5.143}$$

with $\boldsymbol{k} = (\nu_x, \nu_y, \nu_z)2\pi/L$ and $\mathsf{V} = L^3$. These wave functions are orthonor-
malized,

$$\int_{\mathsf{V}} \mathrm{d}^3\mathbf{r}\, \psi_{\boldsymbol{k}}^{\star} \psi_{\boldsymbol{k}'} = \delta_{\boldsymbol{k}\boldsymbol{k}'}, \tag{5.144}$$

and the *density of states*, i.e., the maximum number of electrons per unit
volume in $\boldsymbol{k}$-space including spin, is $2\mathsf{V}/(2\pi)^3$.

In a degenerate Fermi gas all states with energies up to the Fermi energy
$\epsilon_{\mathrm{F}} = \hbar^2 k_{\mathrm{F}}^2/2m$ are occupied, and all states with higher energies are empty.

The wave vectors of these states all lie within a sphere of radius k_F, the Fermi wave number. For a total of N electrons per volume V this yields the relation

$$\mathsf{N} = 2 \sum_{|\boldsymbol{k}|<k_\mathrm{F}} = 2\frac{\mathsf{V}}{(2\pi)^3} \int_{k<k_\mathrm{F}} \mathrm{d}^3\boldsymbol{k} = \frac{\mathsf{V}k_\mathrm{F}^3}{3\pi^2} \tag{5.145}$$

or

$$k_\mathrm{F}^3 = 3\pi^2 n, \tag{5.146}$$

where n is the number of electrons per volume in real space. Thus, k_F^3 is a measure of the electron density. Equivalent parameters are the Fermi momentum $\hbar k_\mathrm{F}$, the Fermi speed $v_\mathrm{F} = \hbar k_\mathrm{F}/m$, or the Fermi energy ϵ_F.

A similar estimate yields the mean kinetic energy per electron,

$$\overline{\epsilon_k} = \frac{2}{\mathsf{N}} \sum_{|\boldsymbol{k}|<k_\mathrm{F}} \frac{\hbar^2 k^2}{2m} = \frac{3}{5}\epsilon_\mathrm{F} \tag{5.147}$$

for the degenerate Fermi gas.

5.7.2 Lindhard Function

Consider now a Fermi gas occupying a volume V at a uniform density of n electrons per volume and try to evaluate the response of this system to a perturbing field $\Phi(\boldsymbol{k},\omega)$. The total hamiltonian

$$\sum_{\nu=1}^{\mathsf{N}} \left[H_\nu - e\Phi(\boldsymbol{r}_\nu, t) \right] \tag{5.148}$$

is symmetric in the electron coordinates. The initial state of the system (at $t = -\infty$) is given by a Slater determinant of single-electron wave functions $\psi_{\boldsymbol{k}_0}(\boldsymbol{r}_\nu)$. This ensures satisfaction of the Pauli principle which requires the wave function to be antisymmetric in the electron coordinates. Moreover, since the hamiltonian is symmetric in these coordinates, the wave function remains antisymmetric at all times. Thus the Pauli principle is satisfied automatically.

This allows us to evaluate the response of a Fermi gas to a perturbing potential $\Phi(\boldsymbol{k},\omega)$ for one electron at a time. The sole difference between the present situation and the one discussed in Sect. 5.3 is the initial state which now is characterized by a nonvanishing wave vector $\boldsymbol{k}_0$. Thus, $|0\rangle$ is replaced by $\psi_{\boldsymbol{k}_0}$. Since the electrons are free, the origin $\boldsymbol{R}$ of the binding force loses its physical significance. In principle it enters into a phase factor $\exp(-\mathrm{i}\boldsymbol{k}_0 \cdot \boldsymbol{R})$. The assumption of $\boldsymbol{R}$ being distributed at random is replaced by the requirement that free-electron wave functions carry a random phase factor. This explains the notion of 'random phase approximation' (Pines, 1964) for this approximation scheme.

For the matrix element (5.39) entering (5.42) we find

$$F_{j0}(\boldsymbol{k}) = \delta_{\boldsymbol{k}_j,\boldsymbol{k}+\boldsymbol{k}_0}. \tag{5.149}$$

Moreover, the total polarization charge adds up linearly from the individual contributions of electrons characterized by different wave vectors $\boldsymbol{k}_0$. Thus, (5.42) is replaced by

$$\epsilon_l(k,\omega) = 1 + \frac{m\omega_{\mathrm{P}}^2}{\hbar k^2}$$
$$\times \sum_{\boldsymbol{k}_0} \frac{n_{\boldsymbol{k}_0}}{\mathsf{N}} \left(\frac{1}{\omega_{\boldsymbol{k}\boldsymbol{k}_0} - \omega - \mathrm{i}\Gamma} + \frac{1}{\omega_{\boldsymbol{k}\boldsymbol{k}_0} + \omega + \mathrm{i}\Gamma} \right), \tag{5.150}$$

where

$$\omega_{\boldsymbol{k}\boldsymbol{k}_0} = \omega_{|\boldsymbol{k}_0+\boldsymbol{k}|} - \omega_{\boldsymbol{k}_0} = \omega_k + \frac{\hbar}{m}\boldsymbol{k}\cdot\boldsymbol{k}_0 \tag{5.151}$$

and $n_{\boldsymbol{k}_0}$ is the occupation number (0, 1, or 2) for state $\boldsymbol{k}_0$. For a system containing many electrons we may replace the ratio $n_{\boldsymbol{k}_0}/\mathsf{N}$ by an occupation probability $f_{\boldsymbol{k}_0}$ obeying the normalization $\sum f_{\boldsymbol{k}_0} = 1$.

5.7.3 Degenerate Fermi Gas

For the specific case of the fully degenerate Fermi gas we may replace the summation in (5.150) by an integration over the Fermi sphere,

$$\sum_{\boldsymbol{k}_0} \frac{n_{\boldsymbol{k}_0}}{\mathsf{N}} \to \frac{3}{4\pi k_{\mathrm{F}}^3} \int_{k_0<k_{\mathrm{F}}} \mathrm{d}^3\boldsymbol{k_0}, \tag{5.152}$$

This integration can be carried out in closed form for infinitesimal Γ. The calculation, which is not trivial, has been written up in appendix A.5.4. With the variables

$$u = \frac{\omega}{kv_{\mathrm{F}}} \quad \text{and} \quad z = \frac{k}{2k_{\mathrm{F}}} \tag{5.153}$$

it takes on the form given by Lindhard (1954),

$$\epsilon_l(z,u) = 1 + \frac{\chi^2}{z^2}\left[f_1(z,u) + \mathrm{i}f_2(z,u) \right] \tag{5.154}$$

with

$$f_1(z,u) = \frac{1}{2} + \frac{1}{8z}\left[1-(z-u)^2\right]\ln\left|\frac{z-u+1}{z-u-1}\right|$$

$$+ \frac{1}{8z}\left[1-(z+u)^2\right]\ln\left|\frac{z+u+1}{z+u-1}\right| \tag{5.155}$$

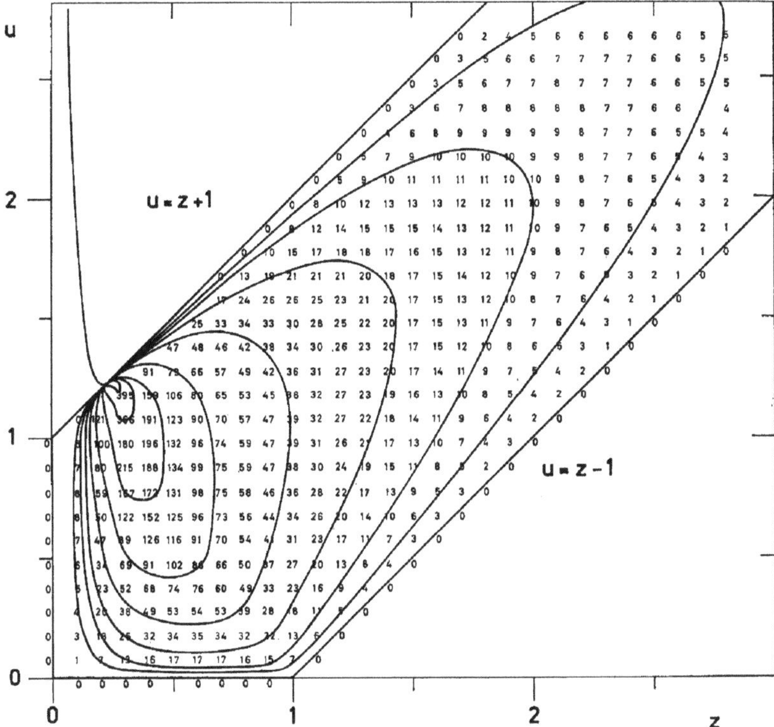

Fig. 5.7. Distribution of generalized oscillator strength in $\omega - k$ space, see text. From Lindhard and Winther (1964)

and

$$
f_2(z,u) = \begin{cases} \dfrac{\pi}{2}u & z + u < 1 \\[2mm] \dfrac{\pi}{8z}\left[1 - (z-u)^2\right] & \text{for } |z - u| < 1 < z + u\ , \\[2mm] 0 & |z - u| > 1 \end{cases} \tag{5.156}
$$

where

$$
\chi^2 = \frac{e^2}{\pi \hbar v_{\mathrm{F}}} \tag{5.157}
$$

is a measure of the electron density. An alternative measure of the electron density is the Wigner-Seitz radius r_s defined by

$$
\frac{4\pi}{3}(r_s a_0)^3 n = 1, \tag{5.158}
$$

The connection between the two measures reads.

$$
\chi^2 = (4/9\pi^4)^{1/3} r_s. \tag{5.159}
$$

Fig. 5.7 shows the analog of the generalized oscillator strength, cf. (4.106),

$$f(\hbar\omega)\mathrm{d}(\hbar\omega) = \frac{\mathrm{i}\omega\mathrm{d}\omega}{\pi\omega_{\mathrm{P}}^2}\left(\frac{1}{\epsilon(k,\omega)} - \frac{1}{\epsilon(k,-\omega)}\right) \tag{5.160}$$

in dimensionless units, ignoring the transverse contribution. This graph may be compared to Fig. 5.1. You may identify two distinct regions, a resonance line in the low-z end where there is a unique dispersion relation $\omega(k)$, similar to the left part of the curve shown in Fig. 5.1. This *plasma line* represents collective excitations, or *plasmon excitations* with an energy $\simeq \hbar\omega_{\mathrm{P}}$. The remainder of the oscillator strength is distributed over a diagonal area

$$\frac{\hbar^2 k^2}{2m} - \hbar k v_{\mathrm{F}} \leq \hbar\omega \leq \frac{\hbar^2 k^2}{2m} + \hbar k v_{\mathrm{F}}. \tag{5.161}$$

This reflects single-particle excitations of electrons with initial velocities distributed over the Fermi sphere.

We have seen in Sect. 5.4.4 that the two portions of the dispersion curve in Fig. 5.1 yield equal contributions to the stopping force of the medium. Lindhard and Winther (1964) showed that a related property holds for the two regimes that we just identified in Fig. 5.7. This is called *equipartition rule*. Here it does not involve the stopping cross section but the *stopping number*, and it does not apply to the absolute values of L_{close} versus L_{dist} but to the respective *increments* as the projectile speed increases. This is important since Fig. 5.7 shows that unlike in Fig. 5.1, where single-particle and plasma excitations set in at the same threshold velocity, single-particle excitations take place down to the lowest projectile speeds, while plasma resonances set in at a threshold speed. This implies that equipartition in the *stopping cross section* is only fulfilled asymptotically at high speed.

5.7.4 Stopping Force at High Projectile Speed

Eq. (5.79) on page 155 indicates that in the limit of high projectile speed the stopping number of a static electron gas is given by the Bethe logarithm augmented by an asymptotic series in inverse powers of v. The present section serves to demonstrate that such an expansion can also be found when the internal motion of the electrons in the medium is taken into account. We consider the limit of $v \gg v_{\mathrm{F}}$ and try to establish a series expansion in terms of the small parameter v_{F}/v.

Taylor expansion of one of the fractions in (5.150) yields

$$\sum_{\boldsymbol{k}_0} \frac{f_{\boldsymbol{k}_0}}{\omega_{\boldsymbol{k}\boldsymbol{k}_0} - \omega - \mathrm{i}\Gamma} = \frac{1}{\omega_k - \omega - \mathrm{i}\Gamma} - \frac{\sum_{\boldsymbol{k}_0} f_{\boldsymbol{k}_0}(\hbar\boldsymbol{k}\cdot\boldsymbol{k}_0/m)}{(\omega_k - \omega - \mathrm{i}\Gamma)^2}$$
$$+ \frac{\sum_{\boldsymbol{k}_0} f_{\boldsymbol{k}_0}(\hbar\boldsymbol{k}\cdot\boldsymbol{k}_0/m)^2}{(\omega_k - \omega - \mathrm{i}\Gamma)^3}\ldots. \tag{5.162}$$

For an isotropic distribution of electron velocities the linear average disappears so that

$$\epsilon_l(k,\omega) = \epsilon_{\text{stat}}(k,\omega)$$
$$+ \frac{m\overline{v_{\text{e}}^2}\omega_{\text{P}}^2}{3\hbar}\left(\frac{1}{(\omega_k - \omega - i\Gamma)^3} + \frac{1}{(\omega_k + \omega + i\Gamma)^3}\right)\ldots, \quad (5.163)$$

where $\epsilon_{\text{stat}}(\boldsymbol{k},\omega)$ is the dielectric function of the static electron gas, (5.60) and $\overline{v_{\text{e}}^2}$ the mean-square electron speed in the medium.

From (5.45) we may deduce that

$$\lim_{\Gamma=0} i\frac{1}{(\omega_k - \omega - i\Gamma)^3} = \frac{\pi}{2}\frac{\text{d}^2}{\text{d}\omega^2}\,\delta(\omega_k - \omega), \quad (5.164)$$

and with this, the integral (5.30) can be evaluated by partial integration, with the result

$$\frac{\text{d}E}{\text{d}x} = -\frac{4\pi e_1^2 e^2}{mv^2}N\left(\ln\frac{2mv^2}{\hbar\omega_{\text{P}}} - \frac{3}{5}\frac{v_{\text{F}}^2}{v^2}\ldots\right), \quad (5.165)$$

a result first derived by Lindhard and Winther (1964). The present derivation was found by Sigmund and Fu (1982). The next term in the expansion can be shown (Lindhard and Winther, 1964, Sigmund and Fu, 1982) to have the form $-3v_{\text{F}}^4/14v^4$.

Equation (5.165) shows a correction term to the Bethe logarithm which becomes significant when the projectile speed approaches the Fermi velocity. Indeed, for the degenerate electron gas, $3v_{\text{F}}^2/5 = \overline{v_e^2}$. This term represents Doppler broadening due to the zero-point motion of the target electrons (Sigmund, 1982). At a given projectile speed, the term is evidently greatest for inner target electrons and, therefore, is usually called a *shell correction*.

Figure 5.8 shows stopping numbers L calculated numerically for two densities compared with a straight Bethe logarithm. Dashed lines represent low- and high-speed approximations. The low-speed approximation – where $L \propto v^3$ – will be discussed in Volume II. The high-speed approximation takes into account the correction term shown in (5.165). Figure 5.9 shows a numerical evaluation of the stopping force for $r_2 = 2$, showing that there is a threshold velocity for resonance excitations.

Figure 5.10 shows velocity spectra of electrons excited by a penetrating particle, normalized to the Rutherford cross section according to Brice and Sigmund (1980). The region where $\text{d}\sigma$ goes to zero, $2v - v_{\text{F}} < v_1 < v + v_{\text{F}}$, has been ignored in this graph. Apart from that, differences from straight Rutherford scattering are seen in the velocity region up to $\sim v_{\text{F}}$, where a sharp peak is formed. While the Pauli principle forbids excitations within a fully occupied Fermi sphere, there is not even a singularity at small momentum transfers: Only electrons within a thin shell below the Fermi surface can be excited, and the number of such electrons goes to zero with decreasing momentum change.

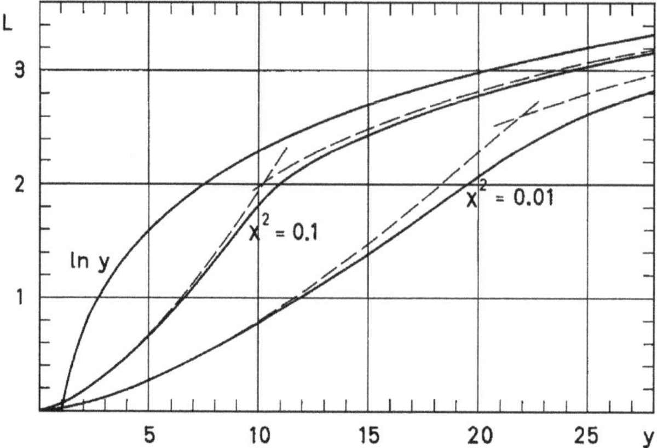

Fig. 5.8. Stopping in Fermi gas according to Lindhard and Winther (1964). Stopping number, defined by (2.56); $y = 2mv^2/\hbar\omega_{\mathrm{P}}$; $\chi^2 = v_0/\pi v_{\mathrm{F}}$. Dashed lines: Low- and high-speed asymptotic expansions. From Lindhard and Winther (1964).

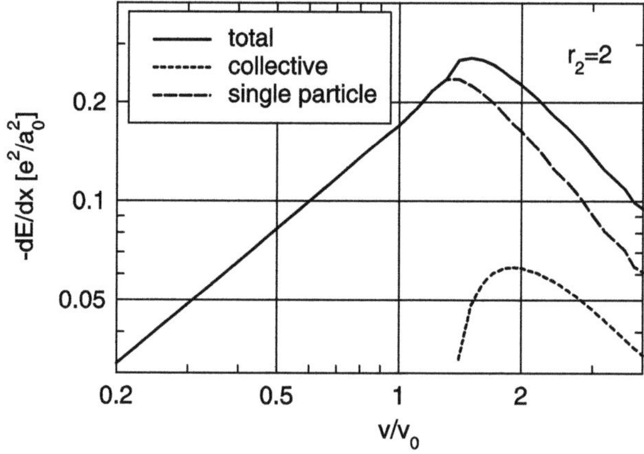

Fig. 5.9. Stopping in Fermi gas according to Lindhard and Winther (1964). Total stopping force for $r_s = 2$ or $\chi^2 = 0.332$, and contributions from single-particle and collective excitations

5.8 Discussion and Outlook

All results derived in this chapter are based on Maxwell's equations. Quantum theory, where appropriate, enters through the dielectric functions $\epsilon_l(\boldsymbol{k}, \omega)$ and $\epsilon_t(\boldsymbol{k}, \omega)$. It was Fermi (1940) – based on a suggestion by Swann (1938) – who first pointed at the usefulness of conventional electrodynamics in the context of stopping in condensed matter at relativistic velocities, and it was Lindhard (1954) who laid the ground for dielectric theory as a universal tool.

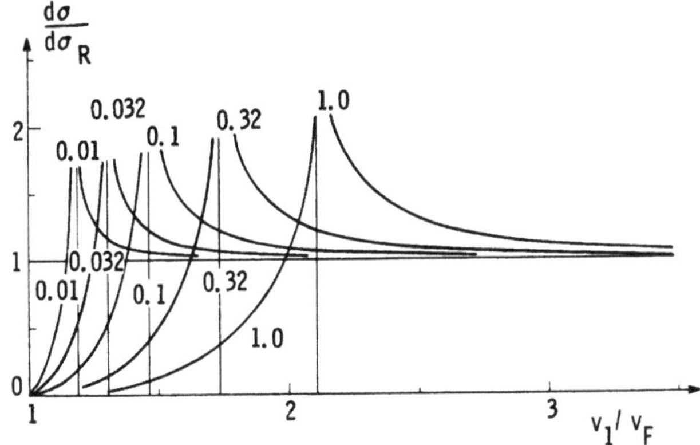

Fig. 5.10. Velocity spectrum of electrons excited in Fermi gas. v_1 denotes the velocity of an excited electron, v_F the Fermi speed, and $d\sigma_R$ the free-Coulomb spectrum. Numbers denote the density parameter $\chi^2 = v_0/\pi v_F$. From Brice and Sigmund (1980)

Although Lindhard's paper refers to the free electron gas in its title, the basic formalism is much more comprehensive. This should be clear from a large number of applications discussed in this chapter. In fact, in contrast to many existing surveys[9], the free electron gas has been given relatively little attention here.

At the time of writing of this monograph, the literature on stopping of charged particles increases at a pace of about 200 theoretical papers per year. Amongst those, calculations for the Fermi gas based on the Lindhard function or more simplified forms are by far dominating. More than anything else, this expresses a general experience in theoretical physics, that 'what can be calculated will be calculated'.

The dielectric function plays a central role in solid-state theory (Pines, 1964, Smith, 1983). Consequently, numerous modifications of the Lindhard function have been proposed over the years, many of which are improvements incorporating features that are important for electron dynamics. As an example, the plasmon-pole approximation has been mentioned above. You should keep in mind, however, that the majority of those modifications was not motivated by the needs of stopping theory.

Aspects not discussed in the present section are the use of the electron-gas model in practical stopping calculations in conjunction with the Thomas-

---

[9] The Fermi gas is one of the most popular playgrounds in theoretical physics, in particular in many-body theory. For a review of particle penetration through an electron gas from a rather different point of view, the reader is referred to Echenique et al. (1990). In addition, the author is preparing a monograph on stopping of light ions, together with N. R. Arista, where much attention will be given to aspects of particle penetration in the Fermi gas.

Fermi method, which will be discussed in Chapter 7, and the very central role it plays in low-velocity stopping, an aspect postponed to Volume II.

Problems

5.1. Derive the transverse dielectric function $\epsilon_t(k,\omega)$ for the system discussed in Sect. 5.3.1. Hint: Instead of $\rho_{e,ind}$, determine the induced current from the standard quantal expression for a particle current.

5.2. Discuss in detail the physical assumptions entering the derivations of the Bethe formula presented in sections 4.3 and 5.3.2 and convince yourself that the same result must emerge from either approach.

5.3. Verify (5.45) by writing the expression on the left-hand side as a Lorentzian with an infinitesimal width.

5.4. Derive the transverse dielectric function $\epsilon_t(k,\omega)$ for a static electron gas by making use of the result of problem 5.1.

5.5. Verify that $\epsilon_t(k,\omega)$ is given by eq. (5.73) under the assumptions made in Sect. 5.4.2.

5.6. Verify eq. (5.84).

5.7. Make a plot of (5.119) with and without the relativistic correction.

5.8. ($\star$) Try to evaluate eq. (5.133) for a material characterized by two resonance frequencies ω_1, ω_2 and dipole oscillator strengths f_1, f_2.

5.9. Show explicitly the equivalence of periodic boundary conditions to describe particles in a one-dimensional box and point out possible differences.

5.10. Assume two wave functions $\psi_1(\mathbf{r},t)$ and $\psi_2(\mathbf{r},t)$ to be governed by the same hamiltonian $\mathcal{H}$, and consider the integral

$$I_{12}(t) = \int d^3\mathbf{r}\psi_1^\star(\mathbf{r},t)\psi_2(\mathbf{r},t). \tag{5.166}$$

Show that $I_{12}(t) = 0$ for all $t > 0$ if $I_{12}(0) = 0$.

5.11. Go explicitly through all steps leading to the conclusion arrived at in section 5.7.2 that the Pauli principle is satisfied automatically in Lindhard's procedure. Utilize the result of problem 5.10.

5.12. Try to reproduce Fig. 5.8 by numerical integration using (5.154).

References

Abramowitz M. and Stegun I.A. (1964): *Handbook of mathematical functions.* Dover, New York

Alexandrov A.F., Bogdankevich L.S. and Rukhadze A.A. (1984): *Principles of plasma electrodynamics*, vol. 9 of *Springer Ser. Electrophys.* Springer, Berlin

Belkacem A. and Sigmund P. (1990): Stopping power and energy loss spectrum of a dense medium of bound electrons. Nucl Instrum Methods B **48**, 29–33

Bethe H. (1932): Bremsformel für Elektronen relativistischer Geschwindigkeit. Z Physik **76**, 293–299

Bohr A. (1948): Atomic interaction in penetration phenomena. Mat Fys Medd Dan Vid Selsk **24 no. 19**, 1–52

Bohr N. (1913): On the theory of the decrease of velocity of moving electrified particles on passing through matter. Philos Mag **25**, 10–31

Bohr N. (1915): On the decrease of velocity of swiftly moving electrified particles in passing through matter. Philos Mag **30**, 581–612

Brice D.K. and Sigmund P. (1980): Secondary electron spectra from dielectric theory. Mat Fys Medd Dan Vid Selsk **40 no. 8**, 1–34

Cherenkov P. (1934): Sichtbares Leuchten von reinen Flüssigkeiten unter der Einwirkung von Gammastrahlen. Doklady Akad NAUK **2**, 451–457

Crispin A. and Fowler G.N. (1970): Density effect in the ionization energy loss of fast charged particles in matter. Rev Mod Phys **42**, 290–316

Echenique P.M., Flores F. and Ritchie R.H. (1990): Dynamic screening of ions in condensed matter. Sol State Phys **43**, 229–318

Fano U. (1956): Atomic theory of electromagnetic interactions in dense materials. Phys Rev **103**, 1202–1218

Fano U. (1963): Penetration of protons, alpha particles, and mesons. Ann Rev Nucl Sci **13**, 1–66

Fermi E. (1939): The absorption of mesotrons in air and in condensed materials. Phys Rev **56**, 1242–1242

Fermi E. (1940): The ionization loss of energy in gases and in condensed materials. Phys Rev **57**, 485–493

Frank I. and Tamm I. (1937): Coherent visible radiation of fast electrons passing through matter. Compt Rend Acad Sci URSS **14**, 109–114

Halpern O. and Hall H. (1948): The ionization loss of energy of fast charged particles in gases and condensed bodies. Phys Rev **73**, 477–486

Jackson J.D. (1975): *Classical electrodynamics.* John Wiley & Sons, New York

Klimontovich Y.L. and Silin V.P. (1952): O spektrakh sistem vzaimodeistvuyushchikh chastits. Zh Eksp Teor Fiz **23**, 151–160

Kramers H.A. (1947): The stopping power of a metal for alpha-particles. Physica **13**, 401–412

Lifshitz E.M. and Pitaevskii L.P. (1981): *Physical kinetics*, vol. 10 of *Course of theoretical physics.* Pergamon Press, Oxford

Lindhard J. (1954): On the properties of a gas of charged particles. Mat Fys Medd Dan Vid Selsk **28 no. 8**, 1–57

Lindhard J. and Winther A. (1964): Stopping power of electron gas and equipartition rule. Mat Fys Medd Dan Vid Selsk **34 no. 4**, 1–22

Lundqvist B.I. (1967a): Single-particle spectrum of degenerate electron gas. 2. Numerical results for electrons coupled to plasmons. Phys Cond Matter **6**, 206–215

Lundqvist B.I. (1967b): Single-particle spectrum of the degenerate electron gas I. The structure of the spectral weight function. Phys Cond Matt **6**, 193–205

Pines D. (1964): *Elementary excitations in solids.* Benjamin, New York

Sigmund P. (1982): Kinetic theory of particle stopping in a medium with internal motion. Phys Rev A **26**, 2497–2517

Sigmund P. and Fu D.J. (1982): Energy loss straggling of a point charge penetrating a free-electron gas. Phys Rev A **25**, 1450–1455

Smith H. (1983): The Lindhard function and the teaching of solid state physics. Phys Scripta **28**, 287–293

Sternheimer R.M. (1952): The density effect for the ionization loss in various materials. Phys Rev **88**, 851–859

Sternheimer R.M., Berger M.J. and Seltzer S.M. (1984): Density Effect for the Ionization Loss of Charged Particles in Various Substances. Atomic Data Nucl Data Tables **30**, 261–271

Swann W.F.G. (1938): Theory of energy losses of high energy particles by L. W. Nordheim. J Franklin Inst **226**, 575

Tamm I. (1939): Radiation emitted by uniformly moving electrons. J Physics (Moscow) **1**, 439–454

Uehling E.A. (1954): Penetration of heavy charged particles in matter. Ann Rev Nucl Sci **4**, 315–350

6

Stopping of Swift Point Charge II: Extensions

6.1 Introductory Comments

The theory laid out in Chapters 4 and 5 provides a powerful basis for the theoretical treatment of the stopping of point charges, but the range of applicability of explicit expressions given so far turns out to be quite limited. This is particularly true for beam energies in the keV/u range, but important corrections are also necessary in the relativistic regime.

An illustrative example is given in Fig. 6.1 taken from a pioneering paper by Lindhard and Scharff (1953). Shown is the stopping number L defined by[1]

$$S = \frac{4\pi Z_1^2 Z_2 e^4}{mv^2} L \tag{6.1}$$

as a function of the parameter

$$x = \frac{v^2}{Z_2 v_0^2}. \tag{6.2}$$

Eq. (4.118) in conjunction with Fig. 4.7 suggests that in such a plot, experimental results for different materials and ions should lie on a single curve, and that this curve should be a straight line in a semilogarithmic plot.

The expected scaling property appears to be quite well fulfilled, but deviations from the predicted logarithmic shape are substantial. You may already have expected this from inspection of Fig. 4.6.

From a theoretical point of view we have to realize

- that the use of perturbation theory must restrict the range of projectile charges for which the theory can be valid. This must be true for both classical and quantal stopping theory, but most pronouncedly so for the quantal scheme, where perturbation theory has been applied to both close and distant interactions.

---

[1] With very few exceptions this chapter addresses heavy charged particles. To emphasize this feature, the projectile charge will be denoted as $Z_1 e$ instead of e_1, the notation used previously.

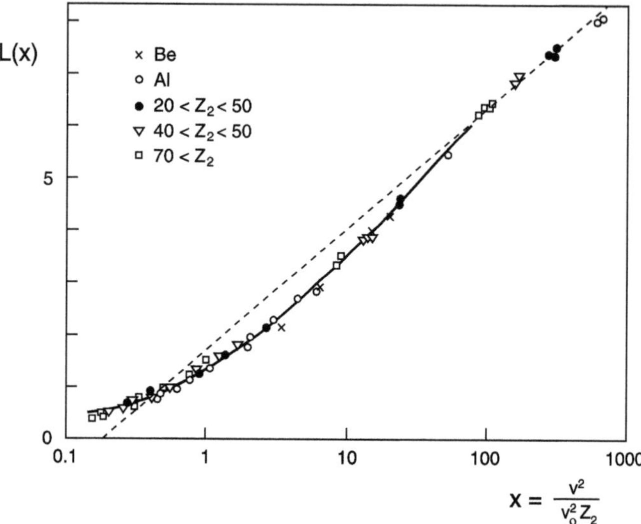

$$X = \frac{v^2}{v_0^2 Z_2}$$

Fig. 6.1. Stopping number versus beam velocity extracted from measured stopping forces on protons. From Lindhard and Scharff (1953)

- that neglecting the intrinsic motion of target electrons must set a lower limit on projectile velocity: Electron motion has been ignored from the beginning in the classical theory; although in quantal perturbation theory the effect has been incorporated initially, it was thrown away when sum rules were applied.
- Penetrating ions are not necessarily point charges but may carry electrons in bound states. This has a variety of implications which will be discussed in detail in Volume II.
- We need to be aware of two separate velocity regimes, a classical and a Born regime. This division depends on the atomic number of the projectile according to the Bohr criterion (2.80). Ambiguities must arise in the transition regime which need to be overcome.

An interesting practical aspect is the experimental determination of I-values which come mostly from stopping data (ICRU, 1984). Accurate stopping measurements have been performed mainly in the velocity range where the stopping number deviates from the logarithmic behavior predicted by the simple Bethe formula. Clearly, these deviations need to be under control in a reliable data analysis. Some aspects of this will be discussed in Chapter 7.

Extensions of the Bethe stopping theory have become a research field attracting considerable interest. As a result, the literature is very extensive[2]. The present survey is selective, yet an attempt has been made to provide

---

[2] According to the Web of Knowledge, the original paper of Bethe (1930) has been quoted 60 times just in 2004.

one explicit derivation for every main effect. While such derivations still are intended to be reasonably self-contained, references to the original literature will nevertheless be more frequent than in the preceding chapters. Consulting those references is not meant to be a prerequisite for going through the remainder of this book but should assist you in broadening and deepening your knowledge of the field.

Straight numerical simulation has become a standard tool in modern physics to overcome complexities in a theoretical description. In the present context, this implies numerical solution of the time-dependent Schrödinger or Dirac equation for the projectile-target interaction, where the target could be an atom, a molecule or cluster, an electron gas or a slab of solid matter. Such simulations have become possible for simple systems containing few electrons and may serve as reference standards for more approximate calculations. The necessary computational effort is substantial even for very small systems, especially since computations need to be made for a grid of impact parameters and projectile speeds. Considering the number of ion-target combinations as well as the number of electrons involved in the majority of collision systems of interest, it seems unlikely that stopping theory can become obsolete in the near future.

6.2 Bare and Dressed Projectiles

Before going any further, we need to specify conditions to be fulfilled in order that we may treat a projectile as a point charge. This is not a problem for energetic electrons, antiprotons or other negatively-charged elementary particles except in the exotic situation where such particles move through antimatter. It is a problem, on the other hand, for an ion which will be able to lose or capture electrons in collisions with the target atoms. In that way a dynamic equilibrium tends to be established which may be characterized by a *mean equilibrium charge* $\langle q_1 \rangle$ or, even better, by *equilibrium charge fractions* P_J which represent the probability for an ion to be found in a *charge state* q_J. These parameters are governed by cross sections for capture and loss of electrons and, hence, depend on speed and ion-target combination.

All explicit expressions for electronic stopping cross sections derived so far in this book assume a purely Coulombic interaction between the projectile and the target electrons. Electrons carried by the projectile give rise to *screening* of the Coulomb interaction. It is important, therefore, to identify an approximate velocity range where screening can be neglected.

6.2.1 Bohr Screening Criterion

The equilibrium charge state of an ion can be determined roughly from a famous argument by Bohr (1948). Consider an ion with an atomic number Z_1 moving through a stopping medium at a speed v. Seen from the reference frame of the ion there is a stream of target particles with speed v which may knock out electrons from the ion. This process is efficient as long as v exceeds the orbital speed v_e of a target electron. At lower beam velocities collisions tend to become adiabatic. Bohr actually expected ionization (and stopping) cross sections to rapidly drop to zero below this adiabaticity limit. We now know that the decrease is slow, but nevertheless, those cross sections have their maximum values around $v \simeq v_e$.

Disregarding complications like capture and the specific nature of the target, Bohr's argument implies that projectile electrons with orbital velocities below v tend to be stripped. Detailed implications of this as well as alternative arguments will be discussed in Volume II. According to this criterion, the projectile may be expected to behave like a point charge if its speed exceeds the orbital velocity of the K shell,

$$v > Z_1 v_0 \tag{6.3}$$

or $E > Z_1^2 \cdot 25 \text{ keV}/u$, where u is the atomic mass unit. This criterion is easily fulfilled for protons and alpha particles. However, already for neon ions, the lower limit is 50 MeV.

Even if an ion is not completely stripped, one may ask whether it acts like a point charge as far as penetration properties are concerned. Much confusion exists about this point in the literature. The topic has been discussed extensively in a recent monograph (Sigmund, 2004) and will be taken up again in Volume II. The brief answer is as follows: For weakly screened ions, when the majority of the projectile electrons are stripped, the error you make in replacing the nuclear charge by the ion charge, will be tolerable. In all other cases you may have to expect more or less pronounced errors.

6.3 Bloch Theory

6.3.1 Bloch Formula

From now on we shall frequently deal with stopping numbers instead of stopping cross sections. The former are related to the latter by (6.1).

Bloch (1933) derived a stopping formula on the basis of the semiclassical treatment discussed in Sect. 4.3 which may be written in the form

$$L_{\text{Bloch}} = L_{\text{Bethe}} + \psi(1) - \operatorname{Re}\psi\left(1 + i\,\frac{Z_1 e^2}{\hbar v}\right), \tag{6.4}$$

where L_{Bethe} is taken from (4.118),

$$\psi(x) = \frac{d}{dx}\ln\Gamma(x) \tag{6.5}$$

denotes the digamma function, and Re the real part. A derivation of this formula will be presented in the following section. Here let us have a look at its implications.

The following useful expansions for small and large arguments for the function appearing in (6.4) are listed in Abramowitz and Stegun (1964), 6.3.17 and 6.3.19,

$$\mathrm{Re}\,\psi(1+iy) = -\gamma + y^2 \sum_{n=1}^{\infty} \frac{1}{n(n^2+y^2)} \tag{6.6a}$$

$$\mathrm{Re}\,\psi(1+iy) = \ln y + \frac{1}{12y^2} + \frac{1}{120y^4}\cdots, \tag{6.6b}$$

where $\gamma = -\psi(1) = 0.5772$ is Euler's constant. With this we find

$$L_{\mathrm{Bloch}} = L_{\mathrm{Bethe}} - \left(\frac{Z_1 v_0}{v}\right)^2 \sum_{n=1}^{\infty} \frac{1}{n(n^2+(Z_1 v_0/v)^2)} \tag{6.7}$$

or

$$L_{\mathrm{Bloch}} = L_{\mathrm{Bethe}} + \ln\frac{v}{e^\gamma Z_1 v_0} - \frac{v^2}{12 Z_1^2 v_0^2}\cdots. \tag{6.8}$$

Eq. (6.7) reduces, for high velocity v and/or low charge $Z_1 e$, to

$$L_{\mathrm{Bloch}} \simeq L_{\mathrm{Bethe}} - 1.202\left(\frac{Z_1 v_0}{v}\right)^2, \tag{6.9}$$

while (6.8) reduces to

$$L_{\mathrm{Bloch}} \simeq \ln\frac{Cmv^3}{Z_1 e^2 \omega_0} - \frac{1}{12}\left(\frac{v}{Z_1 v_0}\right)^2\cdots \tag{6.10}$$

with $C = 2/e^\gamma = 1.1229$.

Eq. (6.9) shows a correction to the Bethe formula which goes to zero for $\kappa = 2Z_1 v_0/v$ going to zero. This is in agreement with (2.80), according to which κ is a measure of deviations from the Born approximation.

Conversely, (6.10) reflects the Bohr formula (4.93) and a correction which decreases as the velocity reaches the Bohr regime.

Figure 6.2 shows the prediction of the Bloch formula for the stopping number L, (6.4), compared with the Bethe and the Bohr logarithm as well as the expression (6.9). Since the Bloch correction only depends on the Bohr parameter κ, the velocity variable was chosen to $v/Z_1 v_0$. The remaining constants have been absorbed into the ordinate variable.

It is seen that the transition from the Bohr to the Bethe logarithm, although smooth, is characterized by quite a narrow transition region. The thin line, representing the Bethe logarithm plus a term $\propto Z_1^4$ in the stopping force, (6.9), is useful at high projectile speed where the Bloch correction is small. Extension of this approximation into the classical regime would evidently lead to absurd results.

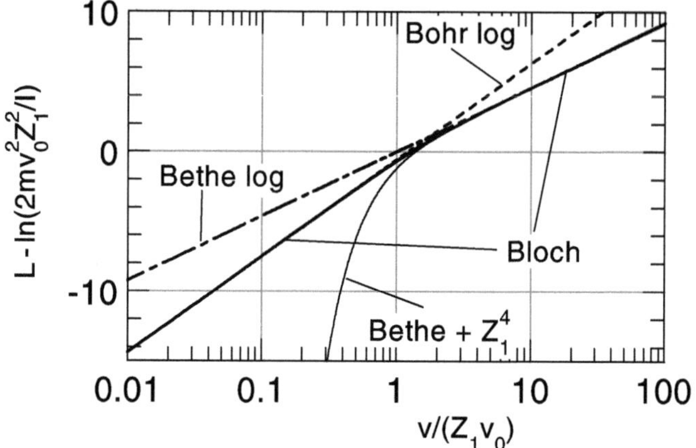

Fig. 6.2. Bloch formula (6.4), Bethe and Bohr logarithm, and high-velocity expansion (6.9). From Sigmund (1997)

6.3.2 Derivation

We have seen in Chapter 3 that the Born approximation delivers a correct prediction of Rutherford's law. On the other hand, the second Born approximation led to a divergence. It follows that this scheme cannot represent a rigorous description of Coulomb scattering. Problem 3.24 in Chapter 3 (page 103) shows that unlike the divergence in Rutherford's law, the divergence in the second Born approximation does not just affect small energy transfers or scattering angles.

In Chapter 4 the Born approximation was applied to the interaction with *bound electrons*. The effect of binding must be most pronounced in distant interactions. If there is a problem in close collisions between free collision partners, that problem is not going to evaporate merely by allowing for a binding force.

The Bloch theory represents a successful attempt to repair the shortcomings of the Born approximation in the description of close collisions. The basic idea is identical with the one underlying the Bohr theory, i.e., binding is unimportant for close collisions.

The following derivation of (6.4) is due to Lindhard and Sørensen (1996). The strategy of this calculation is to determine the *difference* in the stopping cross sections between free-Coulomb scattering and the Born approximation applied to free-Coulomb scattering,

$$\Delta S = S_{\text{free Coulomb}} - S_{\text{free Coulomb,Born}}. \tag{6.11}$$

Both quantities on the right-hand side have a divergence originating in small energy transfers, but since the Born approximation is accurate at small energy transfers, this divergence is expected to drop out in the difference. If the effect of binding, being pronounced in distant interactions, is assumed to be incorporated adequately in the Born approximation, one may obtain an improved stopping formula by setting

$$S = S_{\text{Bethe}} + \Delta S. \tag{6.12}$$

The starting point of the calculation is (3.118) which expresses the stopping cross section for free binary collisions by phase shifts. For a heavy projectile colliding with an electron we have

$$S = mv^2 \sigma^{(1)}, \tag{6.13}$$

with $\sigma^{(1)}$ being given by (3.119) and $\hbar k = mv$.

Phase shifts for free-Coulomb scattering are given by (3.137). The first term diverges, but since it is independent of ℓ it drops out in the difference,

$$\delta_\ell - \delta_{\ell+1} = \arg\left(\ell + 1 + i\kappa/2\right) = \arctan\frac{\kappa/2}{\ell+1}, \tag{6.14}$$

since the argument of a product is the sum of the arguments. From this we find

$$\sin^2\left(\delta_\ell - \delta_{\ell+1}\right) = \frac{(\kappa/2)^2}{(\ell+1)^2 + (\kappa/2)^2} \tag{6.15}$$

and hence

$$S_{\text{free Coulomb}} = \frac{4\pi Z_1^2 e^4}{mv^2} \sum_{\ell=0}^{\infty} \frac{\ell+1}{(\ell+1)^2 + \kappa^2/4} \tag{6.16}$$

for free-Coulomb interaction between a heavy projectile and an electron.

This expression diverges at large values of ℓ. Now, ℓ represents an angular momentum quantum number. In classical scattering, the angular momentum is given by $L = pmv$. Evidently, ℓ is a measure of the impact parameter p. Hence, the divergence of (6.16) at large ℓ is nothing but the wellknown logarithmic divergence of the stopping number at large impact parameters or small energy transfers. Note that unlike the simple estimate in Sect. 2.3.3 the present estimate does not show a divergence at small ℓ.

Now, the Born approximation to (6.16) requires expansion in powers of Z_1 up to second order. In view of the factor Z_1^2 in the front, this implies that the term $\kappa^2/4$ in the denominator needs to be dropped. This yields

$$S_{\text{free Coulomb,Born}} = \frac{4\pi Z_1^2 e^4}{mv^2} \sum_{\ell=0}^{\infty} \frac{1}{\ell+1}, \tag{6.17}$$

which again shows a logarithmic divergence at large ℓ. However, the difference

$$\Delta S = S_{\text{free Coulomb}} - S_{\text{free Coulomb,Born}}$$
$$= \frac{4\pi Z_1^2 e^4}{mv^2} \sum_{\ell=0}^{\infty} \frac{-\kappa^2/4}{(\ell+1)\left[(\ell+1)^2 + \kappa^2/4\right]} \tag{6.18}$$

is convergent and easily seen to reduce to (6.7).

de Ferrariis and Arista (1984) presented a simple approximation formula for (6.4),

$$L_{\mathrm{Bloch}} \simeq \ln \frac{2mv^2/\hbar\omega_0}{\sqrt{1+(\kappa/C)^2}}, \tag{6.19}$$

which is seen to reduce to the Bethe and Bohr logarithm for $\kappa \ll 1$ and $\kappa \gg 1$, respectively.

6.3.3 Inverse-Bloch Correction

According to (6.4) the *Bloch correction*

$$\Delta L_{\mathrm{Bloch}} = \psi(1) - \mathrm{Re}\,\psi\left(1 + \mathrm{i}\,\frac{Z_1 e^2}{\hbar v}\right) \tag{6.20}$$

extends the range of validity of the Bethe formula into the classical regime. Indeed, according to (6.8) it ensures that the stopping force approaches the Bohr expression at low speed.

One may also turn around the argument: According to (6.10) we may write L_{Bloch} as a sum of the Bohr stopping number plus a correction which decreases with decreasing velocity. Adding and subtracting a term $-\psi(1) + \ln(Z_1 v_0/v)$ to L_{Bloch} is easily seen to reduce (6.4) to

$$L_{\mathrm{Bloch}} = L_{\mathrm{Bohr}} + \ln \frac{Z_1 v_0}{v} - \mathrm{Re}\,\psi\left(1 + \frac{Z_1 v_0}{v}\right), \tag{6.21}$$

which defines the *inverse-Bloch correction*

$$\Delta L_{\mathrm{invBloch}} = \ln \frac{Z_1 v_0}{v} - \mathrm{Re}\,\psi\left(1 + \frac{Z_1 v_0}{v}\right), \tag{6.22}$$

and thereby allows to extend the range of validity of the classical into the Born regime (de Ferrariis and Arista, 1984, Sigmund, 1996).

6.3.4 Impact-Parameter Dependence

It is of interest to study the impact-parameter dependence underlying the Bloch formula. While Bloch (1933) derived his formula over the impact-parameter picture, a simpler procedure is to start at the Bohr theory, (4.22) and (4.23), and apply suitable modifications (Sigmund, 1997).

Note first that (4.22) diverges at $p = 0$. In accordance with free-Coulomb scattering, we may remove this divergence by replacing

$$p^2 \to p^2 + p_0^2 \tag{6.23}$$

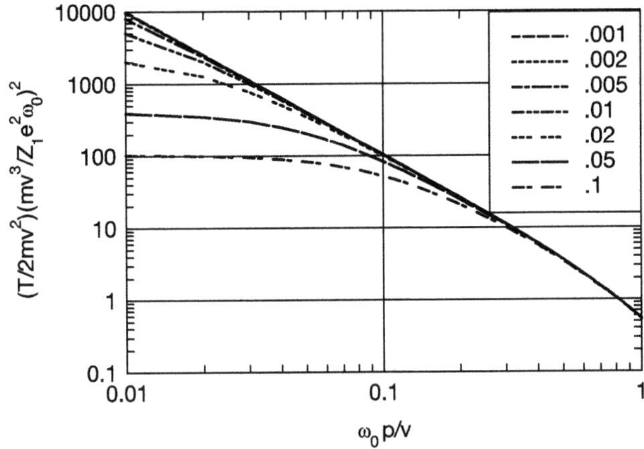

Fig. 6.3. Energy loss versus impact parameter in approximate Bloch model. Curves labelled by parameter ζ_0 defined in (6.27)

in the Coulomb factor $2Z_1^2 e^4/mv^2 p^2$. The parameter p_0 is left open for a moment, although we know from problem 3.6 that it must be of the order of the collision diameter b given in (3.43).

In Chapter 4, impact-parameter dependencies were characterized by functions $f(p)$ defined by (4.22). Since $f(p)$ differs from 1 only at impact parameters approaching the adiabatic radius, it must be justified to make the substitution (6.23) also in $f(p)$. With this we obtain the stopping cross section

$$S = \frac{4\pi Z_1^2 e^4}{mv^2} \int_0^\infty \zeta\, d\zeta \left\{ \left[K_0\left(\sqrt{\zeta^2 + \zeta_0^2}\right) \right]^2 + \left[K_1\left(\sqrt{\zeta^2 + \zeta_0^2}\right) \right]^2 \right\}, \quad (6.24)$$

where $\zeta_0 = \omega_0 p_0/v$. Introducing $\sqrt{\zeta^2 + \zeta_0^2}$ as a new variable you may carry out the integration, and the resulting stopping number reads

$$L_{\mathrm{approx}} = \zeta_0 K_0(\zeta_0) K_1(\zeta_0). \quad (6.25)$$

For $\zeta_0 \ll 1$ this reduces to

$$L_{\mathrm{approx}} \sim \ln \frac{1}{\zeta_0}. \quad (6.26)$$

Requiring this to be identical with (6.19) we then find

$$\zeta_0 = \frac{\sqrt{1 + (\kappa/C)^2}}{2mv^2/\hbar\omega_0} \quad (6.27)$$

or

$$p_0 = \frac{1}{2}\sqrt{\lambda^2 + (b/C)^2},\tag{6.28}$$

where $\lambda = \hbar/mv$ and $b = 2Z_1 e^2/mv^2$. Figure 6.3 shows the energy-loss function underlying (6.24) with ζ_0 given by (6.27) versus impact parameter in Bohr units as a function of $\omega_0 p/v$. It is seen that all curves coincide at large distances, while major differences are found for close collisions.

Note that the present treatment addresses a stationary, classical target electron initially at rest in the position of the nucleus. Grande and Schiwietz (1998) and Schiwietz and Grande (1999) have provided a similar scheme but allowing for the spatial distribution of the electron cloud. This is achieved by convoluting the Thomson equation (4.84) with the electron density. This is important when the relevant impact parameter is the one between the colliding nuclei such as in channeling.

In connection with integrated cross sections such as the stopping cross section and the straggling parameter, this effect averages out, while the related effect of *orbital motion* prevails, gives rise to shell corrections and needs to be considered whenever orbital speeds are not small compared to the projectile speed.

6.4 Barkas-Andersen Effect

6.4.1 Overview

According to the Bethe formula (4.118), the stopping force on a point charge is proportional to the square of the charge. Smith et al. (1953) found a minute difference between ranges in emulsion of positive and negative pions which they ascribed to either a difference in mass or in stopping between the particle and its antiparticle. Subsequent experiments by the same group (Barkas et al., 1963) ruled out the first option, and the difference in stopping was ascribed to higher-order contributions to the Born series underlying Bethe's treatment.

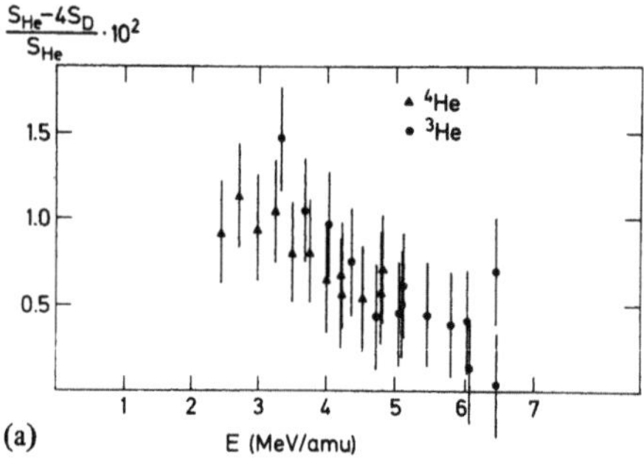

Fig. 6.4. Barkas-Andersen effect in aluminium. See text. From Andersen et al. (1969)

Related observations were made by Andersen et al. (1969) who found the stopping cross section for MeV alpha particles to exceed four times that for protons and deuterons. Their result for aluminium is shown in Fig. 6.4. It is seen that the deviation amounts to less than 1.5 pct, yet it is clearly outside experimental error and increases with decreasing projectile speed. These authors described their results in terms of a Z_1^3-correction to the Bethe stopping cross section.

A theoretical analysis by Ashley et al. (1972), carrying the classical Bohr model described in Chapter 4 to the next order in Z_1 and to be sketched below, led to a correction term governed by the dimensionless parameter

$$B' = \frac{Z_1 e^2 \omega_0}{mv^3}. \tag{6.29}$$

Lindhard (1976) derived the Barkas-Andersen factor B' from a dimensional argument and pointed out that a term proportional to Z_1^4 (shown in Fig. 6.2) cannot be neglected in the analysis of such experimental data. This was verified in subsequent measurements with hydrogen, helium and lithium ions by Andersen et al. (1977).

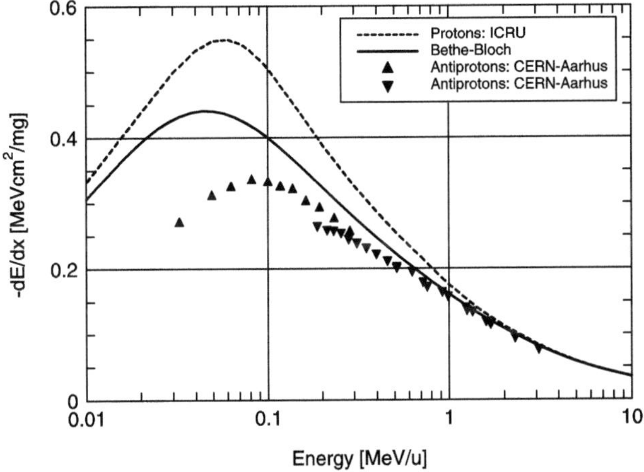

Fig. 6.5. Barkas effect in silicon. Proton data from ICRU (1993); antiproton data from Møller et al. (1997); curve labelled Bethe-Bloch calculated from binary theory to be described in Sect. 6.4.5 by taking the average between protons and antiprotons

More recent experiments at CERN (Møller, 1990, Møller et al., 1997) at lower velocities revealed very large differences between the stopping of protons and antiprotons (Fig. 6.5). These results were shown to be in rough agreement with theoretical predictions on the basis of classical and quantal perturbation theory (Mikkelsen et al., 1990).

Let us try to verify the sign of the correction. If the assertion of a higher-order term in the Born series is correct, we deal with a polarization effect: The penetrating projectile introduces a dipole moment in the target atom, i.e., the target electron moves towards a positively charged projectile. This decreases the effective interaction distance and hence increases the stopping force. The reverse is true for a negatively-charged projectile.

Figure 6.5 indicates that the relative magnitude of the Barkas effect might well keep on increasing with decreasing velocity. This is not only a complication at low projectile speed but more generally so for higher Z_1, a regime which we have barely touched upon so far.

6.4.2 Dimensional Arguments

An important hint toward a solution of the problem was given by Lindhard (1976) who observed that the Barkas parameter B' is the inverse of the parameter entering Bohr's stopping formula (4.93).

Moreover, Lindhard derived a useful scaling property from a dimensional argument. Consider Bohr's stopping model, i.e., a point charge Z_1e moving uniformly at a speed v and interacting with a classical electron bound harmonically with a resonance frequency ω_0. Coulomb's law tells us that Z_1 enters via the combination Z_1e^2. The only additional parameter that can enter the stopping cross section is the electron mass m. Other variables like time t, distance r from the force center and impact parameter p are integration variables that do not enter the final result.

Let us try to construct a general expression for the stopping cross section from these four quantities. By convention, the right dimension is ensured by pulling out the ratio $4\pi Z_1^2 e^4/mv^2$, which has the dimension of energy×area. This may be multiplied by an unknown function of all dimensionless parameters that can be constructed from the four quantities. It is easy to verify that there is only one such variable, the Bohr parameter

$$\xi_{\text{Bohr}} = \frac{mv^3}{Z_1e^2\omega_0}. \tag{6.30}$$

This implies that the stopping number in the Bohr model must obey the scaling relation

$$L = L(\xi_{\text{Bohr}}). \tag{6.31}$$

This is evidently obeyed by Bohr's original result, $L = \ln(C\xi_{\text{Bohr}})$. More important is the observation that within the range of validity of this model, the Barkas-Andersen effect is the same for all ions and targets when viewed in terms of the variable ξ_{Bohr}.

Going over from the classical to the quantum oscillator we dispose over $\hbar$ as an additional dimensional parameter. This introduces

$$\xi_{\text{Bethe}} = \frac{2mv^2}{\hbar\omega_0} \tag{6.32}$$

as a second, independent dimensionless variable. Note that the ratio between the two,

$$\frac{\xi_{\text{Bethe}}}{\xi_{\text{Bohr}}} = \frac{2Z_1 e^2}{\hbar v} \tag{6.33}$$

is Bohr's kappa parameter. Therefore, the interaction of a uniformly-moving point charge with a harmonic oscillator must deliver a stopping number that can be written in one of two general forms,

$$L = L^{\text{`Bohr'}}(\xi_{\text{Bohr}}, \kappa), \tag{6.34a}$$

$$L = L^{\text{`Bethe'}}(\xi_{\text{Bethe}}, \kappa). \tag{6.34b}$$

Here, L^{Bohr} and L^{Bethe} are unspecified functions of two variables which only asymptotically approach the logarithmic forms derived previously. We shall investigate in Sect. 6.4.5 how far these scaling rules survive the adoption of more realistic models for the stopping process.

There is a notable difference between the two forms in that an expansion in powers of Z_1 appears feasible in case of Bethe scaling, (6.34b), while this cannot be expected in case of Bohr scaling.

On the other hand, since the Barkas-Andersen effect increases with decreasing projectile speed, it must be most pronounced in the classical regime.

6.4.3 Binding and Screening

A simple but illuminating estimate of the Barkas-Andersen effect was presented by Lindhard (1976). The starting point is a reformulation of the Bohr theory. Indeed, Bohr's result for the stopping cross section can be derived from Rutherford's law for the scattering between a moving projectile and an initially stationary electron, if the range of interaction is limited to a maximum impact parameter close to the adiabatic radius v/ω_0, as was demonstrated in Sect. 2.3.4, (2.64). This limitation is caused by the binding of the target electron.

Lindhard suggested to reformulate this limitation by replacing the Coulomb potential by a screened potential with the screening radius being identified with the adiabatic radius so that

$$\mathcal{V}(r) = -\frac{Z_1 e^2}{r} \, \Phi\left(\frac{\omega_0 r}{v}\right). \tag{6.35}$$

One of several feasible choices for the screening function Φ is the Yukawa potential, i.e., a straight exponential.

Now, compare scattering on a screened and an unscreened Coulomb potential. Viewing the process from a reference frame moving with the projectile speed we find a target electron approaching the projectile, being accelerated and scattered and, after having passed the apsis, slowing down again and ultimately moving away at its initial speed. In the presence of screening, the electron will not accelerate until having reached the region limited by the adiabatic radius and hence move more slowly across the scattering potential than in the absence of screening. As a consequence, the collision time will increase and hence result in a greater scattering effect.

If an exponential screening function is chosen, the potential inside the screening radius may be approximated by

$$\mathcal{V}(r) \simeq -\frac{Z_1 e^2}{r}\left(1 - \frac{\omega_0 r}{v}\right) = -\frac{Z_1 e^2}{r} + \frac{Z_1 e^2 \omega_0}{v}, \tag{6.36}$$

i.e., a shifted Coulomb potential. Since the shift is independent of the impact parameter, it is equivalent with a shifted collision energy,

$$E' \simeq \frac{m}{2}v^2 - \frac{Z_1 e^2 \omega_0}{v}, \tag{6.37}$$

i.e.,

$$\frac{1}{E'} \simeq \frac{2}{mv^2}\left(1 + \frac{2Z_1 e^2 \omega_0}{mv^3}\right) \tag{6.38}$$

up to first order in Z_1. The collision energy enters the stopping cross section primarily via the quantity mv^2 in the denominator. Hence, a rough estimate of the stopping cross section including the Barkas-Andersen effect reads

$$S \simeq \frac{4\pi Z_1^2 e^4}{mv^2}\left(1 + \frac{2Z_1 e^2 \omega_0}{mv^3}\right)\ln\frac{2mv^2}{\hbar\omega_0} \tag{6.39}$$

applied to the Bethe formula. This confirms the dominating role of the Barkas-Andersen parameter, (6.29) which determines both the magnitude of the effect and its scaling properties. However, as it stands, the correction blows up at low velocities.

6.4.4 Higher-Order Perturbation Theory

It would be desirable to study some system that would allow a complete analytical treatment at least up to the third order in Z_1. Unfortunately, such a system has not yet been identified. Efforts by numerous authors have focused on the harmonic oscillator and the Fermi gas, but even for those idealized cases, numerical tools had to be invoked. More seriously, evaluation of stopping cross sections often required additional, unsupported physical assumptions. This has resulted in lively discussions in the literature for two decades starting 1972. The material presented in the following three subsections has been included with the main purpose to provide a reference standard for more comprehensive calculations.

Classical Oscillator ($\star$)

The theory of Ashley et al. (1972) represents a straight extension of the classical Bohr theory to the next order in Z_1 as far as the contribution from distant collisions is concerned. Close collisions were not treated theoretically but were assumed to be described adequately by Rutherford's law of free-Coulomb scattering. The calculation described here is equivalent with that of Ashley et al. (1972), but the detailed formulation follows Schinner and Sigmund (2000).

Starting at (4.11) in Sect. 4.2 we aim at an expression for the energy transfer $T(p)$ going up to the third power in Z_1. This implies that the second term in (4.17), which is $\propto Z_1$ to leading order, cannot be neglected.

With the notation

$$\boldsymbol{P}(\boldsymbol{p}) = \int_{-\infty}^{\infty} \mathrm{d}t\, \boldsymbol{F}(t)\mathrm{e}^{\mathrm{i}\omega_0 t} = \boldsymbol{P}^{(1)} + \boldsymbol{P}^{(2)} \dots \tag{6.40}$$

indicating expansion in powers of Z_1 we have, according to (4.11),

$$T = \frac{1}{2m}\left(|\boldsymbol{P}^{(1)}|^2 + 2\,\mathrm{Re}\,\boldsymbol{P}^{(1)\star}\boldsymbol{P}^{(2)} \dots\right) \tag{6.41}$$

up to third order, where $\boldsymbol{P}^{(1)}$ has been evaluated in Sect. 4.2.1 and reads

$$\boldsymbol{P}^{(1)} = \frac{2Z_1 e^2 \omega_0}{v^2}\left(\mathrm{i}K_0\left(\frac{\omega_0 p}{v}\right), \frac{\boldsymbol{p}}{p}K_1\left(\frac{\omega_0 p}{v}\right)\right) \tag{6.42}$$

according to (4.21). The notation has been changed slightly relative to Sect. (4.2.2). In particular, the impact parameter $\boldsymbol{p}$ has not been specified to point in the direction of the y-axis, although the beam direction still is adopted as the x-axis.

In order to evaluate $\boldsymbol{P}^{(2)}$ we need to write $\boldsymbol{r}(t)$ in a form that allows convenient integration. This is achieved with the Green function of the harmonic oscillator, (A.97) in Appendix A.2.5,

$$\boldsymbol{r}(t) = \int \mathrm{d}\omega\left(-e\boldsymbol{E}(\omega)\right)G(\omega)\mathrm{e}^{-\mathrm{i}\omega t}, \tag{6.43}$$

which is to be inserted into

$$\boldsymbol{P}^{(2)}(\boldsymbol{p}) = \frac{Z_1 e^2}{2\pi^2}\int \mathrm{d}^3 q \int \mathrm{d}t\, \mathrm{e}^{\mathrm{i}\omega_0 t}\, \frac{\mathrm{i}\boldsymbol{q}}{q^2}\, \mathrm{e}^{-\mathrm{i}\boldsymbol{q}\cdot(\boldsymbol{p}+\boldsymbol{v}t)}\, \mathrm{i}\boldsymbol{q}\cdot\boldsymbol{r}(t). \tag{6.44}$$

This leads to

$$\boldsymbol{P}^{(2)}(\boldsymbol{p}) = \frac{Z_1^2 e^4}{2\pi^3}\int_{-\infty}^{\infty}\mathrm{d}\omega \int \mathrm{d}^3 q \int \mathrm{d}^3 q'\,\delta(\omega - \boldsymbol{q}'\cdot\boldsymbol{v})\,\delta(\omega_0 - \boldsymbol{q}\cdot\boldsymbol{v} - \omega)$$

$$\times\, \mathrm{i}\boldsymbol{q}\,\mathrm{e}^{-\mathrm{i}(\boldsymbol{q}+\boldsymbol{q}')\cdot\boldsymbol{p}}\, G(\omega)\,\frac{\mathrm{i}\boldsymbol{q}}{q^2}\cdot\frac{\mathrm{i}\boldsymbol{q}'}{q'^2}. \tag{6.45}$$

We now split the q-vectors into longitudinal and transverse components,

$$q = (\kappa, q_\perp); \quad q' = (\kappa', q'_\perp) . \tag{6.46}$$

Before integrating over the transverse components we perform the replacements

$$q_\perp = \mathrm{i}\nabla_p; \quad q'_\perp = \mathrm{i}\nabla_{p'} . \tag{6.47}$$

At the same time we replace

$$\mathrm{e}^{-\mathrm{i}q'\cdot p} = \mathrm{e}^{-\mathrm{i}q'\cdot p'}\Big|_{p'=p} \tag{6.48}$$

with the intention that the substitution $p' = p$ be performed after the gradient operations.

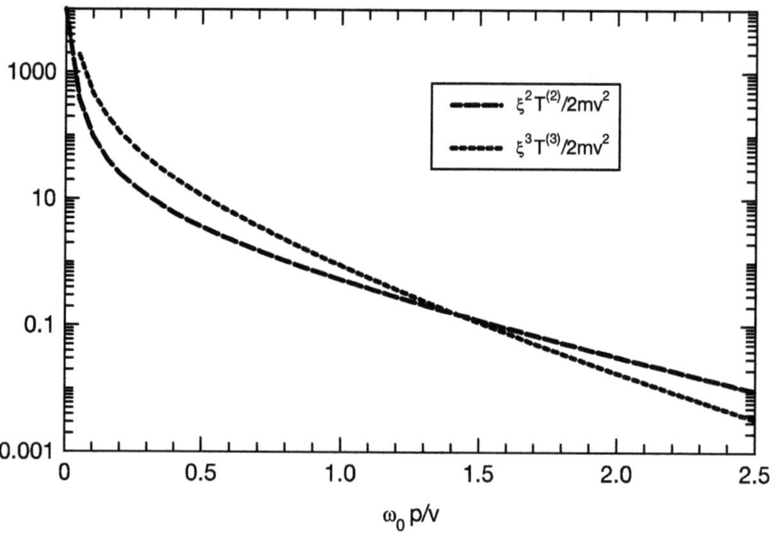

Fig. 6.6. Barkas-Andersen effect for the Bohr oscillator, distant collisions. Plotted is the energy transfer $T^{(2)}(p)$ resulting from the Bohr theory in units of $2mv^2/\xi^2$, whereas the result from second-order classical perturbation theory, $T^{(3)}(p)$, is plotted in units of $2mv^2/\xi^3$. From Sigmund and Schinner (2000)

We then obtain

$$P^{(2)}(p) = -\frac{2\mathrm{i}Z_1^2 e^4}{\pi} \int_{-\infty}^{\infty} \mathrm{d}\omega\, G(\omega) \int_{-\infty}^{\infty} \mathrm{d}\kappa \int_{-\infty}^{\infty} \mathrm{d}\kappa'\, (\kappa, \mathrm{i}\nabla_p)\, (\kappa\kappa'$$

$$- \nabla_p \cdot \nabla_{p'})\, \delta(\omega_0 - \kappa v - \kappa' v)\delta(\omega - \kappa' v) K_0(\kappa p)\, K_0(\kappa' p')\Big|_{p'=p} \tag{6.49}$$

or, after carrying out the integrations over κ and κ',

$$\boldsymbol{P}^{(2)}(\boldsymbol{p}) = -\frac{2\mathrm{i}Z_1^2 e^4}{\pi v^2} \int_{-\infty}^{\infty} \mathrm{d}\omega\, G(\omega) \left(\frac{\omega_0 - \omega}{v}, \mathrm{i}\nabla_p\right)$$
$$\left(\frac{(\omega_0 - \omega)\omega}{v^2} - \nabla_p \nabla_{p'}\right) K_0\left(\frac{(\omega_0 - \omega)p}{v}\right) K_0\left(\frac{\omega p'}{v}\right)\Bigg|_{\boldsymbol{p}'=\boldsymbol{p}}. \quad (6.50)$$

The integration over ω has not so far been carried out analytically.

Figure 6.6 shows the result from a numerical evaluation of (6.50), compared with the result from the Bohr theory. Both results are nominally independent of the projectile speed but apply only to distant collisions. The two results can be compared directly, since the Barkas-Andersen factor $1/\xi$ has been taken out from the third-order term. It is seen that the two functions have similar shape and magnitude, although $T^{(3)}(p)$ falls off slightly faster than $T^{(2)}(p)$.

This similarity has been known early on and led to the frequently-adopted assumption that the Barkas-Andersen term in the resulting stopping cross section should show a similar scaling property as the leading term (Ashley et al., 1972, Jackson and McCarthy, 1972). Quantitative details, however, hinge on the limitations of Fig. 6.6 toward small impact parameters and the behavior of the energy-loss function below that limit. This problem did not find a solution until much later.

Quantal Oscillator

The first complete calculation of a Z_1^3 correction to both $T(p)$ and the stopping cross section was presented by Mikkelsen and Sigmund (1989). This calculation is based on a harmonically-bound target electron, i.e., it carries the calculation described in Sect. 4.5.2 to the next order. The computation of $T(p)$ invokes two numerical integrations and a nontrivial summation over two quantum numbers. Stopping cross sections were determined by yet another numerical integration. In order to ensure rapid convergence of the infinite series, the Coulomb potential was represented by

$$V(r) = -\frac{Z_1 e^2}{\sqrt{\pi}} \int_0^{\infty} \frac{\mathrm{d}\eta}{\sqrt{\eta}} \mathrm{e}^{-\eta r^2} \quad (6.51)$$

instead of its Fourier transform (4.15).

Figure 6.7 shows $T(p)$ in a plot that can be compared with Fig. 4.4. As in the the second-order term $T^{(2)}$, the classical dipole approximation agrees with the quantal result at large impact parameters. The validity of this finding has been shown not to be limited to the harmonic-oscillator model by Hill and Merzbacher (1974). Pronounced discrepancies occur at impact parameters below the oscillator radius, but the qualitative behavior is very similar in the two graphs.

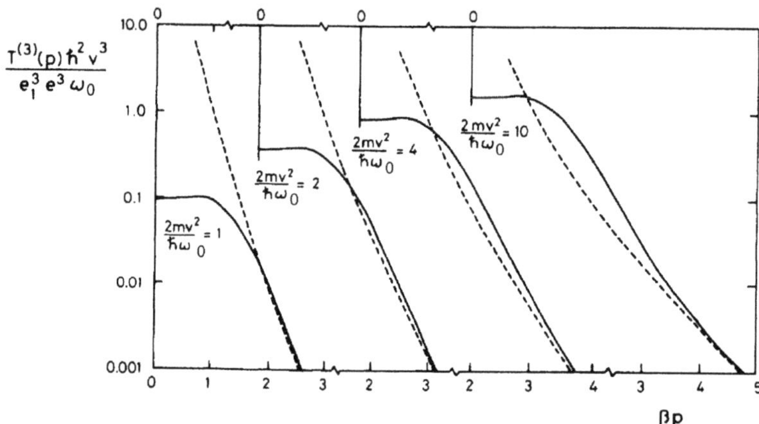

Fig. 6.7. Impact-parameter dependence of Z_1^3 correction to energy loss versus impact parameter for three-dimensional, spherical harmonic oscillator. The abscissa unit is the oscillator radius $\beta^{-1} = \sqrt{\hbar/m\omega_0}$. Solid lines: Quantum harmonic oscillator. Dashed lines: Classical oscillator, dipole limit (Ashley et al., 1972). From Mikkelsen and Sigmund (1989)

Figure 6.8 shows stopping numbers for the two leading orders defined by

$$S = \frac{4\pi Z_1^2 e^4}{mv^2} \left(L_0 + \frac{Z_1 e^2 \omega_0}{mv^3} L_1 \ldots \right) \tag{6.52}$$

It is seen that (6.39) agrees reasonably well in magnitude – although not in slope – with the computed result in the important range of $5 < 2mv^2/\hbar\omega_0 < 10$. However, it is also seen that L_1 becomes negative for $2mv^2/\hbar\omega_0 < 4$. This is easily seen to give rise to negative values of L at low velocities. While this indicates a failure of the perturbation expansion, we note that the domain of the Born expansion is the range $2mv^2/\hbar\omega_0 > 2mZ_1^2 v_0^2/(Z_2 \times 10\mathrm{eV}) \simeq 5Z_1^2/Z_2$, which does not include those low velocities.

Figure 6.9 shows three approximations to the stopping cross section of a harmonic oscillator for protons. While the result from the first Born approximation has a realistic behavior, the stopping cross section becomes negative below 2 keV when the Z_1^3 term is included, and negative below 3 keV when also the Z_1^4 term is allowed for. This phenomenon was already evident from Fig. 6.2. The Born series is a high-velocity expansion which breaks down at low speed.

Mikkelsen (1991) went one step further in the Born series and evaluated the term of fourth order in Z_1 for the spherical harmonic oscillator. For the energy loss versus impact parameter in the semiclassical model, the following scaling property was found for the Z_1^4-contribution,

$$T^{(4)}(p) = \left(\frac{Z_1 e^2}{\hbar v} \right)^4 \hbar\omega_0\, g\left(\frac{2mv^2}{\hbar\omega_0}, \beta p \right), \tag{6.53}$$

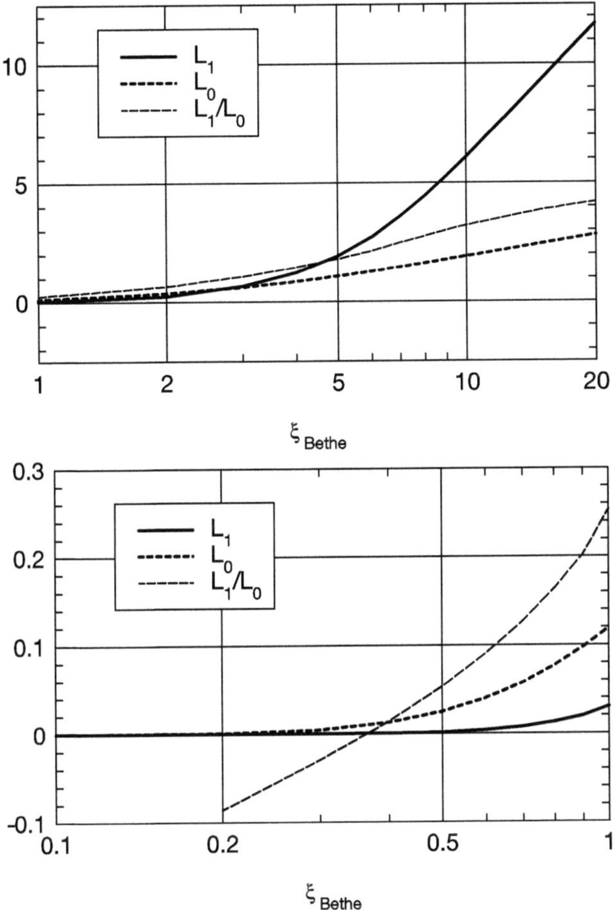

Fig. 6.8. Stopping numbers L_0 and L_1 for quantal harmonic oscillator according to Sigmund and Haagerup (1986) and Mikkelsen and Sigmund (1989), respectively. $\xi_{\mathrm{Bethe}} = 2mv^2/\hbar\omega_0$. See text. The two graphs represent different intervals oof the abscissa variable

where $\beta = \sqrt{m\omega_0/\hbar}$. You may derive the scaling law for the stopping number from this in problem 6.3.

Electron Gas ($\star$)

The Z_1^3 correction for a homogeneous electron gas has received much interest in the literature, both in the static approximation and taking into account the

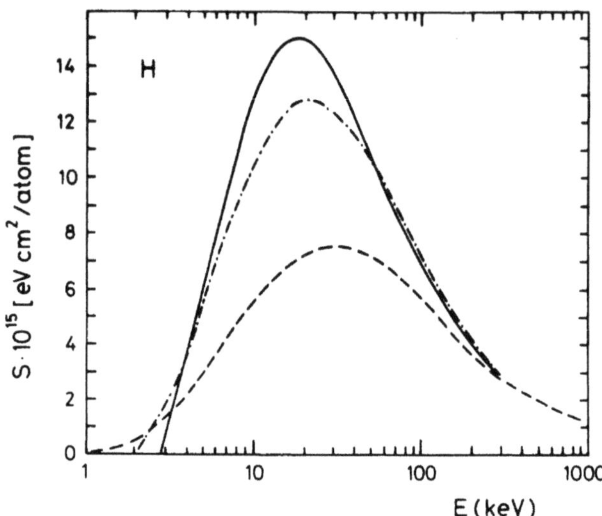

Fig. 6.9. Stopping cross section of atomic hydrogen for protons. The target atom is approximated by a spherical harmonic oscillator with $I = 15.0$ eV. Dashed line: Born approximation (Sigmund and Haagerup, 1986). Dot-dashed line: Including Z_1^3 term of Mikkelsen and Sigmund (1989). Solid line: Including Z_1^4 term. From Mikkelsen (1991)

Fermi motion. Pioneering work was done by Esbensen (1976) who extended the formalism of Lindhard (1954) to the next order in Z_1. Esbensen's evaluation of the Z_1^3 correction to the stopping force focused on the static electron gas.

This work reached the open literature only more than a decade later (Esbensen and Sigmund, 1990), together with several extensions, in particular an evaluation of the Z_1^3-correction to a dense gas of oscillators that was mentioned in Sect. 5.5. In the meantime, a study by Sung and Ritchie (1983) had appeared which seriously questioned the validity of Esbensen's results. This gave rise to further activity, in particular work by Hu and Zaremba (1988) and Pitarke et al. (1993a, 1995).

A key point in the discussion was the question of whether or not close collisions contribute substantially to the Barkas-Andersen correction. Lindhard's point had received strong support from Arista (1982), who argued on the basis of wellknown differences between electron-atom and positron-atom scattering. A more explicit result is seen in Fig. 6.7. Nevertheless, arriving at a conclusion in the electron-gas picture turned out to be more complex because of difficulties in the evaluation of the stopping force near the maximum energy loss.

6.4.5 Beyond Perturbation Theory

The two preceding sections have demonstrated that *higher-order Z_1 corrections* to the Bethe stopping theory fall into two categories, the Bloch correction characterized by the Bohr parameter $\kappa = 2Z_1 v_0/v$ and the Barkas-Andersen correction characterized by the parameter $Z_1 e^2 \omega_0/mv^3 = 1/\xi$. These two parameters become large as the speed becomes small. Hence, series expansions in powers of Z_1 must break down at small projectile speeds. For the Bloch correction this has already been seen in Fig. 6.2. The occurrence of the factor v^3 implies that an even more pronounced effect must be expected for the Barkas-Andersen correction. A nonperturbational approach appears necessary to estimate these corrections in the velocity regime where they cannot be expected to be small. The Bloch correction has been formulated nonperturbatively from the beginning. For the Barkas-Andersen effect, such approaches have become available during the past decade.

Much activity has been devoted to the regime of *slow ions*, i.e., ions in the velocity range $v \lesssim v_0$, where even protons can no longer be treated as point charges. Approaches for this regime will be discussed in Volume II.

Binary Stopping Theory

A comparatively simple theoretical scheme is the binary stopping theory (Sigmund and Schinner, 2000). The underlying physical model is identical with Bohr's, but series expansion in powers of Z_1 is avoided. This has been achieved by making use of the approximate equivalence between binding and screening discussed in Sect. 6.4.3. Let us tentatively adopt a screened potential (6.35) in Yukawa form,

$$\mathcal{V}(r) = -\frac{Z_1 e^2}{r}\, \mathrm{e}^{-\omega_0 r/v}, \tag{6.54}$$

following Lindhard (1976) and neglect binding. We could allow for an adjustable parameter in the exponent, but that will turn out not to be necessary in a moment.

Binary scattering governed by a spherically-symmetric potential has an exact solution which can be expressed by (3.34). Before making this step, let us try whether we can reproduce Bohr's results to the leading order in Z_1. This is achieved by the momentum approximation of classical scattering theory discussed in Appendix A.3.1. Specifically, for the potential (6.54), you arrive at the center-of-mass scattering angle

$$\Theta = -\frac{2Z_1 e^2 \omega_0}{mv^3}\, K_1\left(\frac{\omega_0 p}{v}\right). \tag{6.55}$$

by solving problem 3.8. From this we find the energy transfer

$$T = 2mv^2 \sin^2 \frac{\Theta}{2} = \frac{2Z_1^2 e^4 \omega_0^2}{mv^4}\left[K_1\left(\frac{\omega_0 p}{v}\right)\right]^2 \tag{6.56}$$

for small scattering angles. Comparison with (4.23) shows that the second contribution in Bohr's result is reproduced rigorously. With this pleasant surprise in mind, one may ask whether there is a way also to arrive at the first term in (4.23) from binary scattering theory.

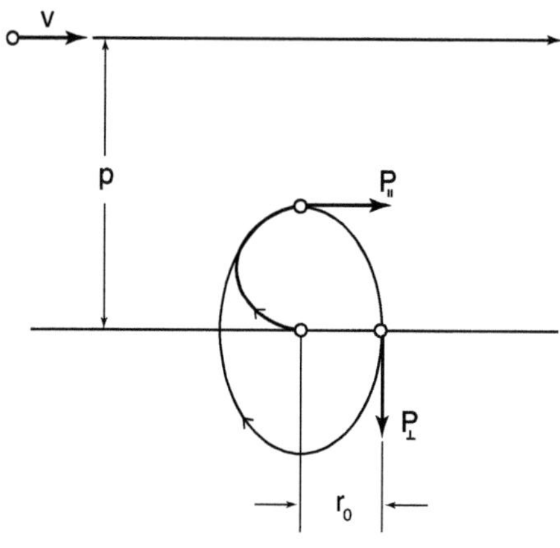

Fig. 6.10. Elliptic orbit of an excited target electron in the Bohr model. From Sigmund and Schinner (2000)

In binary scattering between free particles, at small scattering angles, the recoiling particle moves normally to the beam direction. In the Bohr model, where the recoiling particle (the electron) is bound, momentum is transferred both parallel and perpendicular to the beam direction. Equation (6.42) confirms that (6.56) indeed describes the momentum transfer normal to the beam direction.

Momentum transfer also in the direction parallel to the beam has the consequence that after the collision, the electron moves along an elliptic orbit around its binding site (Fig. 6.10). While this must be true for both close and distant collisions, (6.42) determines the axes in the limit of distant collisions,

$$\frac{2|Z_1|e^2}{mv^2}K_0\left(\frac{\omega p}{v}\right) \quad \text{and} \quad \frac{2|Z_1|e^2}{mv^2}K_1\left(\frac{\omega p}{v}\right), \tag{6.57}$$

as follows immediately from energy conservation. Now, at the marked point, where the electron moves perpendicular to the beam direction, its kinetic energy is determined entirely by (6.56), but in addition it also has a potential energy

$$E_{\text{pot}} = m\omega_0^2 r_0^2/2 \tag{6.58}$$

with

$$r_0 = \frac{2|Z_1|e^2}{mv^2}K_0\left(\frac{\omega p}{v}\right). \tag{6.59}$$

You can easily verify that (6.58) and (6.59) yield the first term in (4.21). Thus, in order to reproduce Bohr's result from binary scattering theory we need to find a way to determine r_0 without invoking a binding force.

Note first that the momentum transfer $P_\perp$ perpendicular to v determines the angular momentum transferred to the electron,

$$J_{\text{bound}} = r_0 P_\perp. \tag{6.60}$$

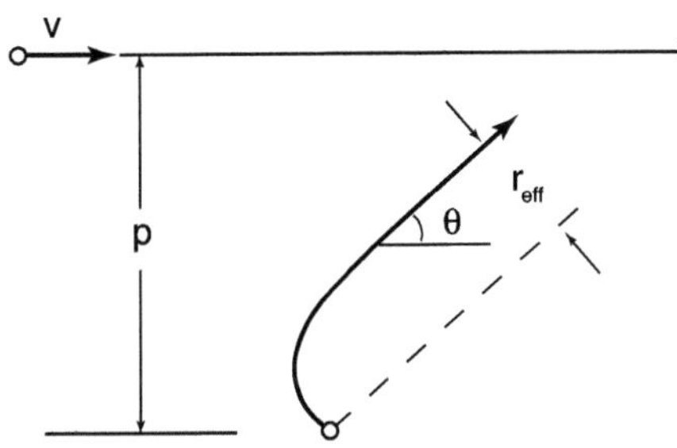

Fig. 6.11. Orbit of target electron in binary-scattering model. From Sigmund and Schinner (2000)

Now, in our binary-scattering problem, angular momentum is conserved in the center-of-mass frame, but in the laboratory frame we have

$$J_{\text{free}} = r_{\text{eff}} P, \tag{6.61}$$

where r_{eff} is indicated in Fig. 6.11 and P is the final momentum of the electron in the laboratory system. For a distant collision we have

$$P \to P_\perp \quad \text{and} \quad r_{\text{eff}} \to r_0. \tag{6.62}$$

While the relation for P has been utilized repeatedly from Chapter 1 on, the relation for r_{eff} is nontrivial:

The location of asymptotic trajectories in space involves the time integral discussed in Sect. 3.3.4. Comparison of Figs. 3.7 and 6.11 leads to

$$r_{\text{eff}} = 2\tau \cos \frac{\Theta}{2} - 2p \sin \frac{\Theta}{2} \to 2\tau - p\Theta, \tag{6.63}$$

where the limiting transition refers to large impact parameters p.

A perturbation expansion of the time integral in terms of Z_1 is given in Appendix A.3.1. Insertion of (A.133) and (6.62) into (6.63) yields

$$r_{\text{eff}} = \frac{2|Z_1|e^2}{mv^2} K_0 \left(\frac{p}{a_{\text{ad}}} \right). \tag{6.64}$$

This proves (6.62).

We may now conclude that the distant-collision part of Bohr's result can be reproduced by adopting binary scattering on the Yukawa potential (6.35).

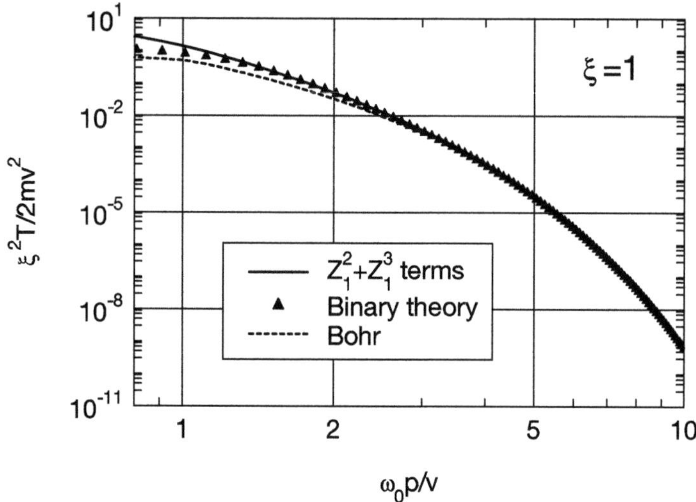

Fig. 6.12. Energy loss *versus* impact parameter: Comparison between Bohr, perturbed-Bohr and binary theory. $\xi = mv^3/Z_1 e^2 \omega_0 = 1$

Conversely, close collisions are described properly by the same model to the extent that screening becomes insignificant. However, there is a wide intermediate regime, where the predictions of binary scattering differ from those of both the Bohr model and free-Coulomb scattering. In order to test the validity of those predictions, Sigmund and Schinner (2000) made comparisons with estimates of the Z_1^3 effect discussed in Sect. 6.4.4.

Figure 6.12 shows a comparison between energy-loss functions $T(p)$ for protons predicted from the Bohr and the binary theory. Plotted is the dimensionless quantity $\xi^2 T/2mv^2$ *versus* $\omega_0 p/v$. Excellent agreement is found between the prediction of the binary theory and the Bohr model for $\omega_0 p/v \gtrsim 2.5$ and for the Z_1^3-corrected Bohr model for $\omega_0 p/v \gtrsim 1$.

Unlike Fig. 6.7, the binary theory predicts the Barkas-Andersen correction to approach zero toward $p = 0$. This is an inherent feature of the theory caused by the adoption of the screened-Coulomb potential (6.35): If the distance of closest approach is well below the adiabatic radius, free-Coulomb scattering is approached, and hence, the Barkas-Andersen correction must approach zero. Note, however, that at $\xi = 1$, it has its maximum slightly below $\omega_0 p/v = 1$ or $p \simeq b/2$, where b is the collision diameter, i.e., at a very small impact parameter.

Figure 6.13 shows predictions for the stopping number *versus* Bohr parameter $\xi = mv^3/Z_1 e^2 \omega_0$. Excellent agreement is found between the curve labelled 'average' and the Bohr theory. That curve was determined as the average between the predictions for protons and antiprotons, i.e., eliminating

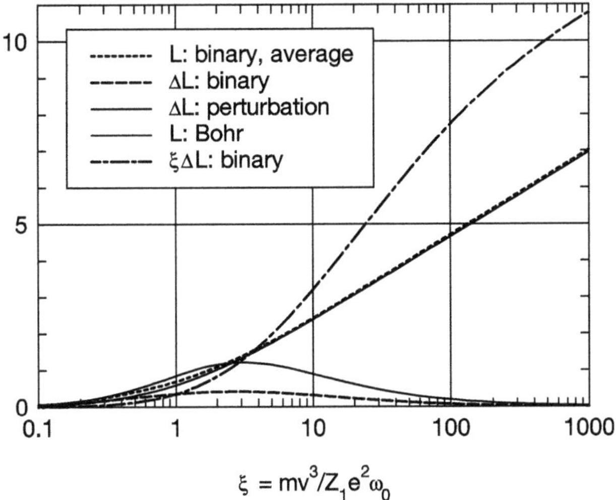

Fig. 6.13. Stopping number L *versus* velocity in terms of Bohr parameter $\xi = mv^3/Z_1 e^2 \omega_0$

the Barkas-Andersen correction. The curve labelled 'Bohr' does not make use of the high-speed expansion that leads to the Bohr logarithm.

Figure 6.13 also shows estimates of the Barkas-Andersen correction $\Delta L(\xi)$. The prediction of the binary theory (ΔL: binary) indicates a broad maximum around $\xi \simeq 3$. Also shown is a curve $\xi \Delta L(\xi)$ which indicates the remaining dependence after separating off the Barkas-Andersen factor ξ^{-1}. This function has frequently been asserted to have logarithmic dependence on v (Ashley et al., 1972, Jackson and McCarthy, 1972, Lindhard, 1976). Such behavior is observed only over very limited ranges of projectile speed.

Electron Gas

The Fermi gas is a favored playground in many-body theory. The stopping of a heavy particle is an example of a practical application of new or well-established theoretical tools to treat such many-body problems. In fact, it is presumably one of the best available examples, since the relation to experiment is close. The reader interested and experienced in many-body theory is kindly referred to papers by Hu and Zaremba (1988) and Pitarke et al. (1993b), Bergara et al. (1996), where further references may be found.

The key point in the application of Fermi-gas physics in stopping theory, in the author's opinion, is the basic simplicity of the model. Only few metals are genuinely free-electron metals for which the model provides an adequate description of the conduction electrons, and several of those, such as the alkalis, are not of particular relevance in particle stopping. We shall see in the following chapter that the free-electron model also may be applied to describe stopping in insulators, but it is not obvious, then, to what degree the accuracy of such a description can be improved by applying many-body descriptions instead of the mean-field theory described above.

6.5 Stopping Medium in Internal Motion

Shell corrections to the stopping force account for the orbital motion of target electrons. This effect has been ignored in Bohr theory. It is inherent in the general formulation of Bethe theory but has been eliminated in the final derivation of the stopping formula (4.118). These shortcomings will be repaired in Sect. 6.6. Here, as a preliminary exercise, we study stopping of a swift particle in a medium made up by *independently-moving electrons*. Early work in this area is due to Gryzinski (1957, 1965) and Gerjuoy (1966) who focused on classical Coulomb interaction from the beginning.

The treatment presented here makes use of conservation laws for elastic scattering and is equally valid in classical and quantum mechanics. Collisions are assumed elastic, and the interaction potential is taken to be spherically symmetric.

6.5.1 Nonrelativistic Regime

Consider an ion moving with a velocity v through a medium made up by electrons distributed uniformly in space at a density n, all moving with a velocity v_e. In the rest frame of the projectile – which, in practice, is the center-of-mass frame for $M_1 \gg m$ – the projectile is hit by electrons moving with a velocity

$$w_e = v_e - v \tag{6.65}$$

and a uniform current density $J' = nw_e$.

The mean number of collision events in a time interval $\mathrm{d}t$ leading to scattering of a target electron into an angular interval $\mathrm{d}^2\Omega$ is given by

$$J' \, \mathrm{d}t \, \mathrm{d}\sigma(w_e, \Theta), \tag{6.66}$$

where $\mathrm{d}\sigma(w_e, \Theta)$ is the differential cross section and Θ the center-of-mass scattering angle. Electrons deflected on the ion by a polar angle Θ transfer an average momentum

$$\Delta P = mw_e(1 - \cos\Theta) \tag{6.67}$$

to the ion. Only the component parallel to w_e matters, since the transverse components cancel for a uniform current density.

This determines the total force on the ion,

$$F = \frac{\mathrm{d}P}{\mathrm{d}t} = nmw_ew_e \int (1 - \cos\Theta) \, \mathrm{d}\sigma(w_e, \Theta). \tag{6.68}$$

Now, assume a distribution $g(\boldsymbol{v}_e)$ of electron velocities $\boldsymbol{v}_e$ in the laboratory frame of reference. We then find

$$\boldsymbol{F} = nm \left\langle w_e \boldsymbol{w}_e \int (1 - \cos \Theta) \, \mathrm{d}\sigma(w_e, \Theta) \right\rangle_{\boldsymbol{v}_e} , \tag{6.69}$$

where $\langle \ldots \rangle_{\boldsymbol{v}_e} = \int \mathrm{d}^3 v_e \, g(\boldsymbol{v}_e)(\ldots)$. If the velocity distribution is isotropic, $\boldsymbol{F}$ can only have a component along $\boldsymbol{v}$ which is the negative stopping force

$$-\frac{\mathrm{d}E}{\mathrm{d}x} = -\frac{\boldsymbol{v}}{v} \cdot \boldsymbol{F} = nS \tag{6.70}$$

with the stopping cross section per target electron

$$S = m \left\langle \frac{|\boldsymbol{v}_e - \boldsymbol{v}|}{v} \boldsymbol{v} \cdot (\boldsymbol{v} - \boldsymbol{v}_e) \sigma^{(1)}(|\boldsymbol{v}_e - \boldsymbol{v}|) \right\rangle_{\boldsymbol{v}_e} . \tag{6.71}$$

Here,

$$\sigma^{(1)}(v) = \int (1 - \cos \Theta) \, \mathrm{d}\sigma(v, \Theta) \tag{6.72}$$

is a transport cross section per target electron.

These arguments stem from kinetic gas theory. The derivations here are slight generalizations of the one presented in Lindhard's lectures. Eq. (6.71) has been common knowledge for many years. The earliest published reference that I am aware of is Trubnikov and Yavlinskii (1965).

6.5.2 Relativistic Extension ($\star$)

In order to extend (6.68) into the relativistic regime, assume the projectile velocity to be directed along the x-axis in the laboratory frame of reference $\mathcal{S}$. Then, in the rest frame $\mathcal{S}'$ of the projectile, the electron velocity has components

$$w_{e,x} = \frac{v_{e,x} - v}{1 - v v_{e,x}/c^2} \tag{6.73a}$$

$$w_{e,y} = \frac{v_{e,y}}{\gamma_v (1 - v v_{e,x}/c^2)} \tag{6.73b}$$

$$w_{e,z} = \frac{v_{e,z}}{\gamma_v (1 - v v_{e,x}/c^2)}, \tag{6.73c}$$

$$\gamma_v = \frac{1}{\sqrt{1 - v^2/c^2}} \tag{6.73d}$$

according to the addition theorem of velocities (cf. Appendix A.3.2). Then (6.66) reads

$$J' \mathrm{d}t' \mathrm{d}\sigma(w_e, \Theta), \tag{6.74}$$

where $\mathrm{d}t'$ is a time interval in $\mathcal{S}'$.

Since the current density is a four-vector we obtain

$$J'_x = n\gamma_v(v_{e,x} - v); \quad J'_y = nv_{e,y}; \quad J'_z = nv_{e,z}. \tag{6.75}$$

This yields

$$J' = \sqrt{J'^2_x + J'^2_y + J'^2_z} = n\gamma_v v_{\mathrm{M}}, \tag{6.76}$$

where $\gamma_v = 1/\sqrt{1 - v^2/c^2}$ and

$$v_{\mathrm{M}} = \sqrt{(\boldsymbol{v} - \boldsymbol{v}_e)^2 - \frac{v^2 v_e^2 - (\boldsymbol{v} \cdot \boldsymbol{v}_e)^2}{c^2}} \tag{6.77}$$

is called the *Møller speed* (Møller, 1972).

The relativistic extension of (6.67) now reads

$$\Delta\boldsymbol{P}' = m\gamma_{w_e}\boldsymbol{w}_e(1 - \cos\Theta), \tag{6.78}$$

where

$$\gamma_{w_e} = \frac{1}{\sqrt{1 - w_e^2/c^2}} = \gamma_v\gamma_{v_e}(1 - \boldsymbol{v} \cdot \boldsymbol{v}_e/c^2), \tag{6.79}$$

and

$$\gamma_{v_e} = \frac{1}{\sqrt{1 - v_e^2/c^2}} \tag{6.80}$$

(cf. problem 6.4), and the force equation becomes

$$\boldsymbol{F}' = \frac{\mathrm{d}\boldsymbol{P}'}{\mathrm{d}t'} = nm\gamma_v v_{\mathrm{M}}\gamma_{w_e}\boldsymbol{w}_e\sigma^{(1)}(w_e). \tag{6.81}$$

It has been assumed here that the relativistic projectile mass is large compared to the relativistic electron mass, i.e.,

$$\gamma_v M_1 \gg \gamma_{v_e} m. \tag{6.82}$$

For relativistic ions, violation of this assumption requires excessively hot targets. A derivation of a more general expression that does not imply this simplifying assumption was given by Tofterup (1983).

The relativistic extension of (6.71) reads

$$S = m\gamma_v \left\langle \gamma_{w_e} \frac{v_{\mathrm{M}}}{v} \frac{\boldsymbol{v} \cdot (\boldsymbol{v} - \boldsymbol{v}_e)}{1 - \boldsymbol{v} \cdot \boldsymbol{v}_e/c^2} \sigma^{(1)}(w_e) \right\rangle_{\boldsymbol{v}_e} \tag{6.83}$$

or, in the form given by Tofterup (1983),

$$S = m\gamma_v^2 \left\langle \gamma_{v_e} \frac{v_{\mathrm{M}}}{v} \boldsymbol{v} \cdot (\boldsymbol{v} - \boldsymbol{v}_e) \sigma^{(1)}(w_e) \right\rangle_{\boldsymbol{v}_e}, \tag{6.84}$$

where

$$w_e = \frac{v_{\mathrm{M}}}{1 - \boldsymbol{v} \cdot \boldsymbol{v}_e/c^2}. \tag{6.85}$$

6.5.3 A Useful Transformation

For $v_e \ll v$, (6.71) reduces to

$$S_0 = mv^2\sigma^{(1)}(v), \tag{6.86}$$

whereas the relativistic version (6.83) carries an additional factor γ_v^2.

We may use this relation to eliminate the transport cross sections (Sigmund, 1982). Solving (6.86) for $\sigma^{(1)}(v)$ and inserting it into (6.71) yields

$$S(v) = \left\langle \frac{\boldsymbol{v}\cdot(\boldsymbol{v}-\boldsymbol{v}_e)}{v|\boldsymbol{v}-\boldsymbol{v}_e|}S_0(|\boldsymbol{v}-\boldsymbol{v}_e|)\right\rangle_{v_e}. \tag{6.87}$$

This expresses the stopping cross section of a medium in motion by the equivalent quantity of a medium at rest. The relation is exact for binary elastic collisions at nonrelativistic velocities, but it also allows to estimate the effect of target motion under less restrictive conditions (Sigmund, 1982).

The extension into the relativistic regime reads (Tofterup, 1983)

$$S(v) = \left\langle \frac{\gamma_v}{\gamma_{w_e}}\frac{\boldsymbol{v}\cdot(\boldsymbol{v}-\boldsymbol{v}_e)}{v w_e}S_0(w_e)\right\rangle_{v_e} = \left\langle \frac{\boldsymbol{v}\cdot(\boldsymbol{v}-\boldsymbol{v}_e)}{vv_\mathrm{M}\gamma_{v_e}}S_0(w_e)\right\rangle_{v_e} \tag{6.88}$$

6.5.4 High-Speed Expansion: Nonrelativistic

Let us consider (6.87) in the limit where $v_e \ll v$ and apply Taylor expansion in powers of v_e up to second order. Then,

$$|\boldsymbol{v}-\boldsymbol{v}_e| = v\left(1 - \frac{\boldsymbol{v}\cdot\boldsymbol{v}_e}{v^2} + \frac{1}{2}\frac{v^2 v_e^2 - (\boldsymbol{v}\cdot\boldsymbol{v}_e)^2}{v^4}\cdots\right). \tag{6.89}$$

Insertion into (6.87), retaining all terms up to second order and assuming an isotropic velocity distribution,

$$g(\boldsymbol{v}_e) \equiv g(v_e), \tag{6.90}$$

you find (problem 6.5)

$$S(v) \simeq S_0(v) + \frac{\langle v_e^2\rangle}{v^2}\left(-\frac{1}{3}S_0(v) + \frac{1}{3}vS_0'(v) + \frac{1}{6}v^2 S_0''(v)\right)\cdots, \tag{6.91}$$

where $S_0'(v) = \mathrm{d}S_0(v)/\mathrm{d}v$. The corresponding term of fourth order in v_e was found by Sigmund (1982), where also the general case of arbitrary mass ratios M_1/m has been treated.

The conclusion emerges that motion of target electrons, as long as it is significant but not dominating, introduces correction terms of the order of $\langle v_e^2\rangle/v^2$. Such terms are categorized under the label *shell corrections*. We keep in mind, however, that all relations in this section have been derived under the assumption of binary elastic collisions and hence do not necessarily hold strictly for bound electrons.

6.5.5 Relativistic Orbital Speed ($\star$)

An expansion in powers of v_e/v has little meaning when v_e approaches the speed of light. However, the two cases $v_0 \ll v \ll v_e \simeq c$ and $v, v_e \simeq c$ are of interest.

In the first case we have

$$v_\mathrm{M} \simeq v_e - v\eta \tag{6.92a}$$

$$w_e \simeq v_e - v\eta/\gamma_{v_e}^2 \tag{6.92b}$$

$$\eta = \cos(\boldsymbol{v}, \boldsymbol{v}_e), \tag{6.92c}$$

up to first order in v/v_e and hence

$$S(v) \simeq \frac{v}{3\gamma_{v_e}} \left\langle \frac{2}{v} S_0(v_e) + \frac{1}{\gamma_{v_e}^2} S'(v_e) \right\rangle \quad \text{for } v_0 \ll v \ll v_e. \tag{6.93}$$

The nonrelativistic version of this formula was found long ago (Sigmund, 1982).

In the second case, when both v and v_e come close to c, you easily find that

$$v_\mathrm{M} \simeq c(1 - \eta) \tag{6.94a}$$

$$w_e \simeq c \tag{6.94b}$$

so that (6.88) reduces to

$$S(v) \simeq \left\langle \frac{1}{\gamma_{v_e}} S_0(w_e) \right\rangle_{\boldsymbol{v}_e}, \tag{6.95}$$

where w_e has been kept in $S_0(w_e)$ since the expression otherwise may diverge for $v \to c$ according to (5.119).

6.6 Shell Correction

6.6.1 Introduction

As mentioned previously, stopping formulae discussed so far, with the exception of the quantal-oscillator model in Sect. 4.5.3 and Fermi-gas theory in Sect. 5.7.4, ignore the orbital motion of the electrons in the material. While this must be justified at projectile speeds substantially above orbital speeds, that condition is barely fulfilled in measurements, in particular on heavy materials where inner electrons have velocities approaching the speed of light.

The explicit or implicit presence of a *shell correction* is a necessary ingredient in any stopping theory that is to describe stopping at projectile speeds that are not large compared to the velocities of all target electrons. Nevertheless, a precise definition of the shell correction is difficult because of the simultaneous presence of the Bloch and the Barkas correction[3] and because other approximations may have been made that break down at lower projectile speed.

According to (5.165), a high-velocity expansion of the stopping number receives additions $\propto v^{-2}$ and v^{-4} to the leading logarithmic term. This feature is common to all evaluations in the literature of the shell correction when a series expansion is made. You may note that the reference velocity is the Fermi velocity v_F in both terms. A term $\propto \hbar\omega_P/mv^2$ could have been expected but does not appear in the leading term. Such a term would imply a binding correction, while v_F^2/v^2 is a purely kinematic term. This is a strong indicator that the shell correction is primarily a kinematic correction.

6.6.2 Bohr Theory

Bohr stopping theory as outlined in Sect. 4.2 assumes the target electron initially at rest. This is unproblematic in a classical calculation. Hence, determining a shell correction implies inclusion of some initial motion. In a harmonic-oscillator potential, this is a Lissajou-type motion in three dimensions,

$$\boldsymbol{r}_0(t) = \boldsymbol{a}_0 \cos\omega_0 t + \boldsymbol{b}_0 \sin\omega_0 t, \tag{6.96}$$

with amplitude vectors $\boldsymbol{a}_0, \boldsymbol{b}_0$ directed at random such that

$$\langle \boldsymbol{a}_0 \rangle = \langle \boldsymbol{b}_0 \rangle = \langle \boldsymbol{a}_0 \cdot \boldsymbol{b}_0 \rangle = 0 \tag{6.97}$$

and distributed in magnitude such that

$$\langle \boldsymbol{a}_0^2 \rangle = \langle \boldsymbol{b}_0^2 \rangle = \frac{\langle v_e^2 \rangle}{\omega_0^2}, \tag{6.98}$$

where v_e^2 is the mean square orbital velocity of the target electron.

One may then define the shell correction as a correction to the stopping number,

$$\Delta L = L_{\mathrm{Bohr},v_e} - L_{\mathrm{Bohr},v_e=0}. \tag{6.99}$$

Such a calculation has become available recently (Sigmund, 2000), but only the first term in the expansion in terms of $\langle v_e^2 \rangle/v^2$ was evaluated. The calculation

---

[3] In addition to the substance there is also some ambiguity about the name. The frequently-used terminology 'inner-shell correction' dates from a time when projectile speeds in typical stopping experiments were far above those of valence and conduction electrons, so that the correction was only of interest for the innermost target shells. On the other hand, *on a relative scale*, the magnitude of the shell correction is very similar for all shells, and only their location on the velocity scale varies. Terms like 'kinematic correction' or 'Doppler broadening' are more descriptive but are only applicable as long as nothing else is corrected for.

is lengthy, and care is indicated with regard to the transition from distant to close interactions. Nevertheless, full documentation may be found in the original paper for the reader who wants to go through this exercise.

The main conclusion emerging from that work is that the dominating contribution to the shell correction stems from close collisions where the harmonic binding force can be neglected.

6.6.3 Bethe Theory

Unlike the Bohr theory, the Bethe scheme takes full account of the orbital motion from the beginning, and (4.106) represents an exact result within the leading term in the Born series. Therefore, within this scheme it is appropriate to define the shell correction via

$$\Delta L = L_{\text{Born}} - \ln \frac{2mv^2}{I},\tag{6.100}$$

where

$$L_{\text{Born}} = \frac{1}{2}\sum_j \int_{(\hbar\omega_{j0})^2/2mv^2}^{\infty} \frac{\mathrm{d}Q}{Q}f_{j0}(Q); \quad \hbar\omega_{j0} = \epsilon_j - \epsilon_0.\tag{6.101}$$

This *definition* does not explicitly make reference to orbital motion. After all, assuming negligible orbital motion would be in conflict with the quantum mechanics of a bound electron.

Splitting the integration in (6.101) at $Q = 2mv^2$ you may write ΔL in the form

$$\Delta L = \frac{1}{2}\sum_j \left(\int_{2mv^2}^{\infty} \frac{\mathrm{d}Q}{Q}f_{j0}(Q) \right.$$
$$\left. + \int_{(\hbar\omega_{j0})^2/2mv^2}^{2mv^2} \frac{\mathrm{d}Q}{Q}[f_{j0}(Q) - f_{j0}(0)] \right),\tag{6.102}$$

making use of the definition (4.119) for I. This may be rewritten as

$$\Delta L = \frac{1}{2}\sum_j \left(\int_{2mv^2}^{\infty} \frac{\mathrm{d}Q}{Q}f_{j0}(Q) \right.$$
$$\left. - \int_0^{(\hbar\omega_{j0})^2/2mv^2} \frac{\mathrm{d}Q}{Q}[f_{j0}(Q) - f_{j0}(0)] \right),\tag{6.103}$$

since $\sum_j \int_0^{2mv^2} (\mathrm{d}Q/Q)\,[f_{j0}(Q) - f_{j0}(0)] = 0$ because of the sum rule (4.114).

Fano-Turner Expansion

Eq. (6.103), which is due to Walske (1952), indicates that the shell correction vanishes as v becomes large. This suggests asymptotic expansion in powers of $1/v$ (Brown, 1950, Walske, 1952, 1956, Fano and Turner, 1964). The calculation makes use of sum rules, and care needs to be taken to include all terms up to the chosen order in $1/v$.

For a one-electron atom, Fano (1963) obtained

$$\Delta L = -\frac{\langle v_e^2 \rangle}{v^2} - \frac{\langle v_e^4 \rangle}{2v^4} - \frac{5\pi}{3} \left(\frac{v_0}{v} \right)^4 a_0^3 \rho_e(0), \tag{6.104}$$

where $\rho_e(\boldsymbol{r}) = |\psi(\boldsymbol{r})|^2$ is the electron density.

It is seen that the leading correction ($\propto 1/v^2$) is exclusively determined by the *motion* of the target electron and formally independent of its binding, irrespective of the fact that the two quantities are intimately related by quantum mechanics. Conversely, the second term ($\propto 1/v^4$) does contain an additional contribution which is not generally expressible in terms of the electron speed v_e. Furthermore, for a multi-electron atom, Fano (1963) also mentions terms involving pairwise correlation of target electrons.

We may conclude that in the Bethe theory, the *first shell correction* $-\langle v_e^2 \rangle / v^2$ reflects orbital motion, while the second term is more complex and depends on specific properties of the target atom.

Moreover, the fact that $\langle v_e^6 \rangle$ diverges for a hydrogen atom indicates that the asymptotic expansion cannot be extended in general to higher than fourth order terms in $1/v$.

Harmonic Oscillator

A rigorous evaluation of the Bethe theory has been performed for the spherical harmonic oscillator, cf. Fig. 4.6. Also an asymptotic expansion was found (Sigmund and Haagerup, 1986), which reads

$$L = \ln \frac{2mv^2}{\hbar\omega_0} - 3\frac{\hbar\omega_0}{2mv^2} - \frac{25}{2} \left(\frac{\hbar\omega_0}{2mv^2} \right)^2 \cdots \tag{6.105}$$

Unlike in the hydrogen atom, momentum spectra for a harmonic oscillator are gaussian-like and hence allow evaluation of $\langle v_e^\nu \rangle$ for arbitrarily high values of ν. Therefore, (6.105) is an infinite series, although its practical significance is limited to fairly large values of $2mv^2/\hbar\omega_0$.

In order to compare (6.105) with (6.104), note that for a three-dimensional harmonic oscillator,

$$\frac{1}{2}m\langle v_e^2 \rangle = \frac{1}{2}\epsilon_0 = \frac{3}{4}\hbar\omega_0, \tag{6.106}$$

where ϵ_0 is the ground-state energy, and hence

$$-3\frac{\hbar\omega_0}{2mv^2} = -\frac{\langle v_e^2\rangle}{v^2},$$

(6.107)

in agreement with the general result (6.104).

For the fourth moment you find in problem 6.7 that

$$\langle v_e^4\rangle = \frac{15}{4}\left(\frac{\hbar\omega_0}{m}\right)^2,$$

(6.108)

and hence

$$-\frac{25}{2}\left(\frac{\hbar\omega_0}{2mv^2}\right)^2 = -\frac{5}{6}\frac{\langle v_e^4\rangle}{v^4}.$$

(6.109)

This result appears to be specific for the harmonic oscillator.

Fermi Gas

The stopping force of a Fermi gas was evaluated in Sect. 5.7.4. Those results can be compared with the present findings.

Note first that here,

$$\langle v_e^2\rangle = \frac{3}{5}v_F^2$$

(6.110)

according to (5.147). With this, (5.165) becomes identical with the leading term in (6.104), as it should. Moreover,

$$\langle v_e^4\rangle = \frac{3}{7}v_F^4$$

(6.111)

and hence,

$$-\frac{3}{14}\frac{v_F^4}{v^4} = -\frac{\langle v_e^4\rangle}{2v^4},$$

(6.112)

which is in agreement with the kinetic term in (6.104).

Atomic Wave Functions

Equation (6.103) has been evaluated numerically with hydrogenic wave functions for K and L electrons by Walske (1952, 1956) and for M electrons by Khandelwal and Merzbacher (1966). Early literature has been reviewed by Fano (1963). Numerous computations have been performed more recently on excitation cross sections for point projectiles within the first Born approximation. Amongst those addressing stopping forces, the work of Inokuti et al. (1981), McGuire (1983) and Bichsel (2002) may be mentioned.

6.6.4 Kinetic Theory

According to the previous section, a common feature of the Bethe theory is the first shell correction, $-\langle v_e^2 \rangle / v^2$ which confirms the dominance of the orbital motion of the target electron. Now, orbital motion was ignored not only in the Bohr theory and in the approximation underlying the Bethe logarithm, but also in the Bloch theory and in several estimates discussed above of the Barkas-Andersen correction. It is desirable, therefore, to have available a general scheme, albeit approximate, to incorporate shell corrections into a theory that neglects orbital motion. A scheme that is suitable for this purpose is the transformation (6.87) discussed in Sect. 6.5. This transformation is rigorous for free target particles. Errors must be expected in the presence of binding, but their magnitude can be estimated by means of comparisons with reliable direct estimates of shell corrections in specific cases.

High-Speed Expansion

Let us explore the high-speed expansion (6.91), and introduce the stopping number $L_0(v)$ representing the system ignoring target motion via

$$S_0(v) = \frac{4\pi Z_1^2 Z_2 e^4}{mv^2} L_0(v). \tag{6.113}$$

Then, (6.91) reduces to

$$\Delta L = \frac{\langle v_e^2 \rangle}{v^2} \left(-\frac{1}{3} v \frac{dL_0}{dv} + \frac{1}{6} v^2 \frac{d^2 L_0}{dv^2} \right) \dots \tag{6.114}$$

It is easily seen that for $L_0 = \ln(2mv^2/I)$ you obtain $\Delta L = -\langle v_e^2 \rangle / v^2$, i.e., the correct first shell correction in the Born approximation.

This result is actually more general. In Sect. 6.3.4 a stopping formula was derived which approximates the Bloch formula. Setting

$$L_0(v) = \zeta_0 K_0(\zeta_0) K_1(\zeta_0) \tag{6.115}$$

with

$$\zeta_0 = \frac{\hbar \omega_0}{2mv^2} \sqrt{1 + \left(\frac{2Z_1 v_0}{Cv} \right)^2}, \tag{6.116}$$

you may evaluate (6.91) (problem 6.9) and obtain

$$\Delta L = \frac{\langle v_e^2 \rangle}{v^2} \zeta_0^2 \left\{ -\left[\frac{7}{3} + \frac{2}{1 + C^2/\kappa^2} + \frac{1}{6} \frac{1}{(1 + C^2/\kappa^2)^2} \right] [K_0(\zeta_0)]^2 \right.$$

$$- \left[1 + \frac{2}{3} \frac{1}{1 + C^2/\kappa^2} - \frac{1}{6} \frac{1}{(1 + C^2/\kappa^2)^2} \right] [K_1(\zeta_0)]^2$$

$$\left. + \frac{2}{3} \left[2 + \frac{1}{1 + C^2/\kappa^2} \right]^2 \zeta_0 K_0(\zeta_0) K_1(\zeta_0) \right\} \tag{6.117}$$

For large v, i.e., small ζ_0 and small κ, the leading term is $\propto [K_1(\zeta_0)]^2$. With $\zeta_0 K_1(\zeta_0) \to 1$ and $1/\kappa \to \infty$ you obtain

$$\Delta L = -\frac{\langle v_e^2 \rangle}{v^2} \ldots \tag{6.118}$$

to leading order for the approximate Bloch formula, which agrees with (6.104).

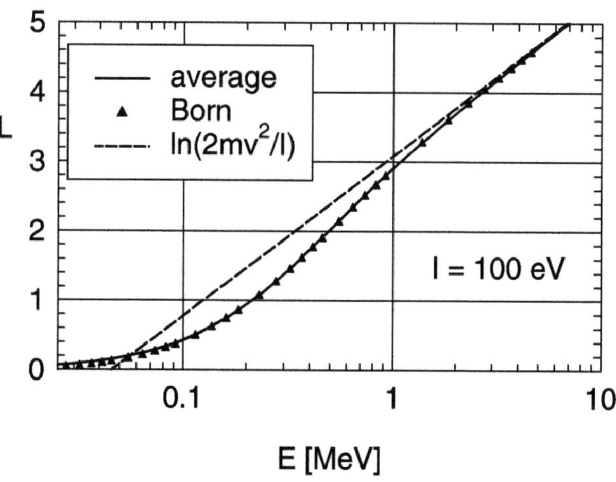

Fig. 6.14. Stopping number of a quantum oscillator with $I = 100$ eV for proton. Solid line: First Born approximation (exact) according to Sigmund and Haagerup (1986). Points: Calculated from binary theory, average between result for proton and antiproton. Also included is the Bethe logarithm. From Sigmund and Schinner (2006)

In view of this, we may tentatively redefine the shell correction ΔL as

$$\Delta L = L - L_{\text{static}}, \tag{6.119}$$

where L_{static} ignores the orbital motion of target electrons. It is clear that there is no way to *measure* a shell correction defined in this manner. However, an advantage of this definition, compared to (6.11), is the inclusion of *all competing effects* such as Bloch and Barkas correction.

6.6.5 Is the Shell Correction Purely Kinematic?

Figure 6.14 shows the stopping number of a quantum oscillator with $I = \hbar\omega_0 = 100$ eV for protons. In addition to the Bethe logarithm you see the exact result within the Born approximation (Sigmund and Haagerup, 1986) and a prediction from binary theory averaged between the predictions for protons and antiprotons. Excellent agreement is found within the Born regime ($E \gtrsim 100$ keV). In other words, the Born approximation can be well simulated by classical theory plus shell and inverse-Bloch correction. We note that the applied shell correction is purely kinematic. Thus, from a physical point of view we may safely accept the notion of the shell correction being of a purely kinematic nature.

The traditional definition of the shell correction makes reference to the Bethe logarithm as the reference (Fano, 1963). Within that notion, the shell correction is composed of a kinematic correction as well as a mathematical correction for an asymptotic expansion at high speed (Sigmund and Schinner, 2006).

6.7 Relativistic Projectile Speed

We have seen in Sect. 3.5.3 that the interaction between a relativistic electron and a point charge deviates from Rutherford's law. In fact, the cross section depends on the sign and magnitude of the charge, and on spin. These features were first studied by Mott (1929), but in view of the success of the Bethe theory, they were considered insignificant for the stopping of charged particles. Indeed, deviations from Rutherford's formula increase with increasing charge and, hence, become important mainly for high-Z_1 projectiles.

6.7.1 General Observations

Relativistic corrections were considered in connection with the Barkas-Andersen effect by Jackson and McCarthy (1972) who referred to unpublished work by Fermi. They were, however, discarded as a main contributor in view of clear experimental evidence in favor of a low-velocity effect.

A clear need for higher-order relativistic corrections became evident in experimental work on ranges of relativistic heavy ions in matter by Tarlé and Solarz (1978) with 600 MeV/u Fe and Ahlen and Tarlé (1983) with 955 MeV/u U ions. Figure 6.15 shows a 6 % deviation of the energy loss for U ions from the calculated value. It is also evident that the deviation is systematic for the four ions involved. Nevertheless, these measurements were indirect, involving ranges in more than one target material, but direct stopping measurements by Scheidenberger et al. (1994) and Datz et al. (1996) confirmed and extended this evidence.

Incorporation of the Mott cross section into Bethe theory by Ahlen (1978) substantially improved the agreement between calculated and measured ranges (cf. Fig. 6.15). Similar conclusions emerge from the measurements by Scheidenberger et al. (1994).

However, although the Bloch correction discussed in Sect. 6.3.1 becomes small at high velocities, this does not imply that this trend survives into the

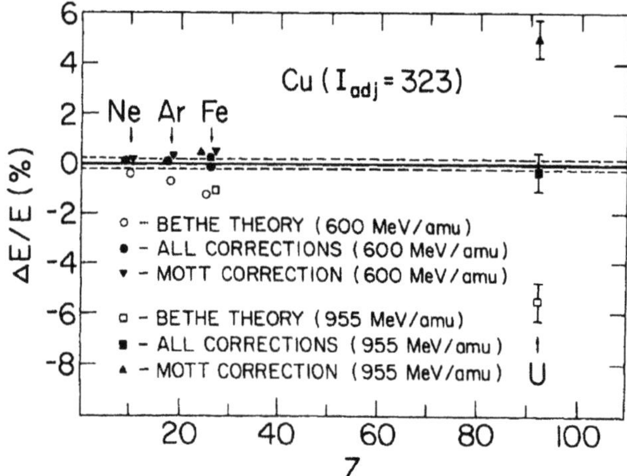

Fig. 6.15. Relative deviation (%) in energy loss of heavy ions (Ne, Ar, Fe, U) in Cu from predictions of the relativivistic Bethe formula. From Ahlen and Tarlé (1983)

relativistic regime. This aspect was studied by Ahlen (1982). The range of validity of Ahlen's calculation is restricted to moderately heavy projectiles such as Ne and Ar, while noticeable errors were expected already for Fe ions.

6.7.2 Lindhard-Sørensen Theory (⋆)

A more comprehensive theory incorporating the above effects as well as deviations from straight Coulomb interaction due to the structure of the projectile nucleus has been presented by Lindhard and Sørensen (1996). The basic strategy of this theory has been discussed already in Sect. 6.3.2, where the correction to the Bethe formula was identified with the difference in free-Coulomb scattering between an exact evaluation and the first Born approximation. That calculation, based on nonrelativistic quantum mechanics, was then generalized to Dirac theory.

As a first step, the transport cross section for scattering of Dirac particles on an arbitrary spherically symmetric potential is evaluated. In the notation of Sect. 3.5.3 it may be written in the form

$$\sigma^{(1)} = 4\pi\lambda^2 \sum_{\varkappa} |\varkappa| \left[\frac{\varkappa - 1}{2\varkappa - 1} \sin^2\left(\delta_{\varkappa} - \delta_{\varkappa-1}\right) \right.$$

$$\left. + \frac{1/2}{4\varkappa^2 - 1} \sin^2\left(\delta_{\varkappa} - \delta_{-\varkappa}\right) \right] \quad (6.120)$$

instead of (6.16), where $\lambda = \hbar/\gamma m v$ and the sum over $\varkappa$ includes all nonzero integer values of $\varkappa$.

Next, phase shifts for Coulomb scattering quoted in Sect. 3.5.3 are inserted into (6.120), and the perturbation limit is determined and subtracted. This yields a correction to the relativistic Bethe stopping number of the form

$$\Delta L = \sum_{\varkappa \neq 0} \left[\frac{|\varkappa|}{\eta^2} \frac{\varkappa - 1}{2\varkappa - 1} \sin^2 (\delta_\varkappa - \delta_{\varkappa-1}) - \frac{1}{2|\varkappa|} \right]$$

$$+ \frac{1}{\gamma_v^2} \sum_{\varkappa=1}^{\infty} \frac{\varkappa}{4\varkappa^2 - 1} \frac{1}{\varkappa^2 + (\eta/\gamma_v)^2} + \frac{v^2}{2c^2} \quad (6.121)$$

instead of (6.18). This expression was evaluated numerically.

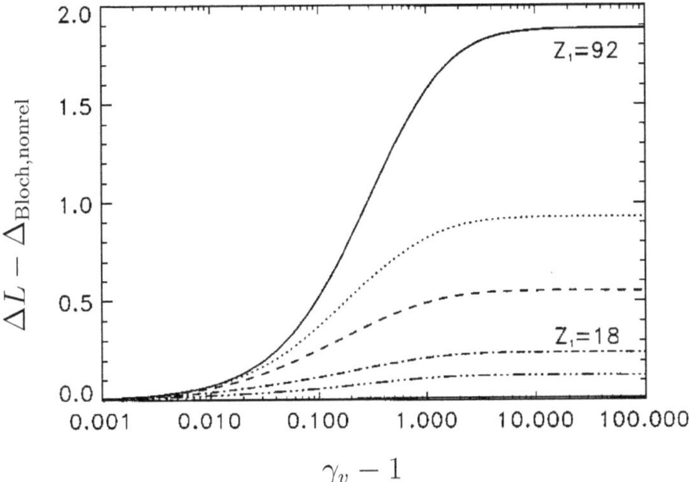

Fig. 6.16. Deviation of the Lindhard-Sørensen correction ΔL to the relativistic Bethe formula from the nonrelativistic Bloch correction (6.20) for $Z_1 = 1, 10, 18, 36, 54, 92$ (bottom to top) versus $\gamma_v - 1 = 1/\sqrt{1 - v^2/c^2} - 1$. From Lindhard and Sørensen (1996)

Figure 6.16 shows the quantity

$$\Delta L - \Delta L_{\text{NR}} = L_{\text{rel}} - L_{\text{Bethe,rel}} - \Delta L_{\text{Bloch,nonrel}} \quad (6.122)$$

for several values of Z_1. It is seen that the correction is small for light ions ($\simeq 0.1$ for Ne). The correction saturates at $\gamma_v = 1/\sqrt{1 - v^2/c^2} \gtrsim 2$, and the magnitude of the saturation value is approximately proportional to Z_1.

As a final step, phase shifts are obtained for the potential of a uniformly charged sphere with a standard-size nuclear radius $R = 1.18A_1^{1/3}$, where A_1 is the mass number of the projectile, following a procedure outlined by Bhalla and Rose (1962).

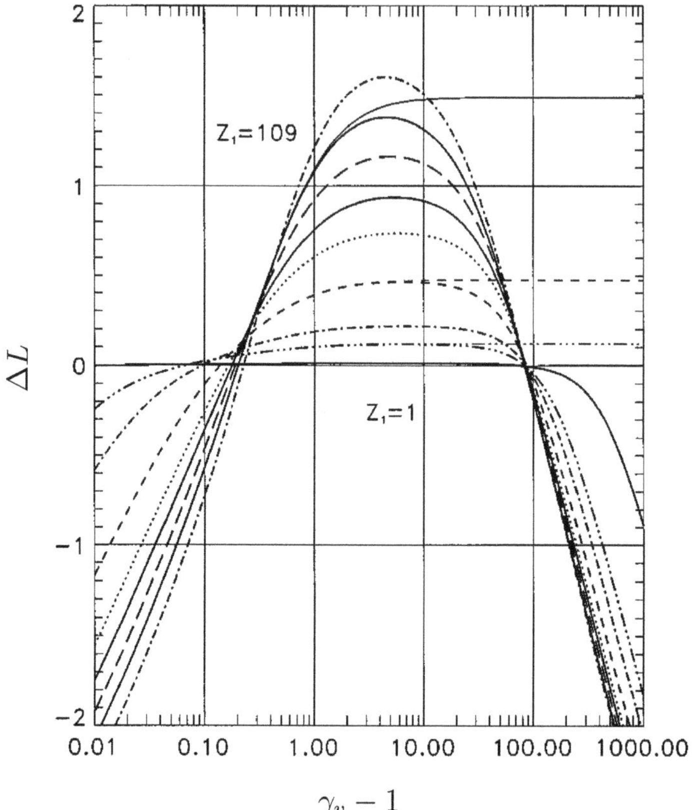

Fig. 6.17. Lindhard-Sørensen correction for projectile nucleus modelled as a uniformly charged sphere for $Z_1 = 1, 10, 18, 36, 54, 66, 79, 92, 109$. Horizontal lines for $Z_1 = 10, 36, 92$ refer to point nuclei. From Lindhard and Sørensen (1996)

Figure 6.17 shows the quantity ΔL for this case. Unlike in Fig. 6.16 the nonrelativistic Bloch correction has not been subtracted. It is seen that instead of saturating at high values of γ_v, the correction starts decreasing at $\gamma_v \simeq 10$ and eventually becomes negative.

A particularly appealing result was dervied for the stopping number at ultrarelativistic velocities,

$$L \to \ln \frac{1.64c}{R\omega_{\mathrm{P}}}, \tag{6.123}$$

where R is the radius of the projectile nucleus and ω_{P} the plasma frequency for all target electrons. This expression includes a density correction.

6.7.3 Additional Effects

Other effects contributing to energy loss need to be considered in the relativistic regime, in particular for heavy projectiles.

While energy loss going into Cherenkov radiation has not been considered explicitly, it has been fully acounted for in the evaluation of the density effect in Sect. 5.6.4. Another radiative contribution, energy going into *Bremsstrahlung*, needs to be accounted for separately. Bremsstrahlung is the leading stopping mechanism for high-energy electrons. However, for heavier projectiles, the larger mass reduces dramatically the significance of angular deflections of the projectile. In fact, Sørensen (2005) demonstrated that the leading additional contribution to target excitation is electron-positron pair production at high projectile speed.

The heaviest ions may capture target electrons also at velocities in the relativistic regime. Hence, screening by projectile electrons needs to be considered in the analysis of precision measurements (Weick et al., 2002).

6.8 Discussion and Outlook

The exposition of the Bohr-Bethe theory in Chapter 4 left open several problems. Collective effects in the stopping medium were discussed in Chapter 5 which, as a by-product, also delivered a relativistic extension of the Bethe formula for an individual target atom, yet still within the first Born approximation. The present chapter may roughly be categorized into three parts,

- The Bloch theory providing the link between the classical Bohr theory and Bethe's theory based on the Born approximation,
- Shell and Barkas-Andersen corrections which become significant at low projectile speeds, and
- Relativistic corrections.

The most prominent omissions concern

1. Stopping of ions below the Bohr velocity,
2. Stopping of ions carrying electrons,
3. Stopping of light particles, in particular electrons.

While the third topic is outside the scope of this monograph as far as detailed quantitative estimates are concerned, the first two topics will play key roles in Volume II.

In view of the neglect of projectile screening, stopping theory as outlined so far is applicable to light ions, in particular protons and antiprotons, and to heavier ions at increasingly high projectile speeds. For those particles, the relativistic Bethe formula amended by the density correction is the universal reference standard for comparisons with experiment. Examples will be seen in the following chapter. At projectile speeds approaching orbital speeds of target electrons, the shell correction causes a gradual decrease in the stopping force below the value following from the straight Bethe formula. On the other hand, with a properly incorporated shell correction, the drop-off to zero at $2mv^2 = I$ changes into a smooth decrease of the stopping force into friction-like behavior.

The Barkas-Andersen correction acts, by and large, in the same velocity regime as the shell correction. For protons and other positively charged projectiles it opposes the shell correction, while it enhances the effect of the latter for antiprotons. An estimate of the relative significance of Barkas-Andersen effect and shell correction will be given in the following chapter.

Finally, I like to mention a development which has been going on for some time but, in the author's opinion, has not yet led to a final satisfactory result. You may remind that in Sect. 5.6, the stopping cross section was evaluated for three regimes of momentum transfer delimited by Q_0 and Q_1. While the evaluation for $Q_0 < Q < Q_1$ was based on sum rules, explicit matrix elements calculated from the Dirac equation for *plane waves* were employed for $Q > Q_1$. This was necessitated by the breakdown of the Bethe sum rule in the relativistic regime.

Problems with relativistic sum rules have been discussed extensively in the literature, most recently so by Cohen (2004). As it turns out, meaningful results can be obtained only by eliminating the effect of negative-energy states. Various schemes have been designed to do this in a justifiable manner which would not contradict basic principles in quantum mechanics. Results for practical use are, however, on the way (Cohen, 2003).

Problems

6.1. Make a plot of Bohr's kappa parameter and Bohr's screening criterion in suitable variables and identify the regimes of Bohr and Bethe stopping theory as well as the regime where projectile screening may not be neglected.

6.2. Make a plot of the Bloch correction as a function of Bohr's κ parameter and compare it with successive approximations based on partial sums of the expansions (6.6a).

6.3. Determine a scaling law for the Z_1^4-contribution to the stopping cross section of a harmonic oscillator by the pertinent integral over eq. (6.53), and show that the stopping number can be written as

$$L^{(4)} = \left(\frac{Z_1 e^2}{\hbar v} \right)^2 g\left(\frac{2mv^2}{\hbar \omega} \right). \tag{6.124}$$

6.4. Give a proof of (6.79).

6.5. Derive (6.91) by the indicated procedure.

6.6. Derive (6.103).

6.7. Derive (6.108). Hint: Use the relation

$$\left(\frac{p^2}{2m}\right)^2 = \left(H - \frac{1}{2}m\omega^2 r^2\right)^2. \tag{6.125}$$

6.8. Compare (6.105) with Fig. 6.1 and explain why an expansion like (6.105) can only provide a realistic estimate at moderately high speed. At low speed, the stopping force is often assumed to be proportional to v, i.e., , $L \propto v^3$. A simple procedure to produce a realistic stopping curve is based on the ansatz (Varelas and Biersack, 1970, Andersen and Ziegler, 1977)

$$\frac{1}{L} = \frac{1}{L_{\text{low}}} + \frac{1}{L_{\text{high}}}, \tag{6.126}$$

where L_{low} and L_{high} denote the approximate L-values at low and high speed, respectively. Make a plot of (6.126) by selecting a plausible factor in $L_{\text{low}} = \text{const} \times v^3$ and include L_{high} and L_{low}.

6.9. Derive eq. (6.117). Make use of the relations

$$\frac{dK_0(\zeta_0)}{d\zeta_0} = -K_1(\zeta_0); \quad \frac{d}{d\zeta_0}(\zeta_0 K_1(\zeta_0)) = -\zeta_0 K_0(\zeta_0), \tag{6.127}$$

which you may find in Abramowitz and Stegun (1964).

References

Abramowitz M. and Stegun I.A. (1964): *Handbook of mathematical functions.* Dover, New York

Ahlen S.P. (1978): Z_1^7 stopping-power formula for fast heavy ions. Phys Rev A **17**, 1236–1239

Ahlen S.P. (1982): Calculation of the relativistic Bloch correction to stopping power. Phys Rev A **25**, 1856–1867

Ahlen S.P. and Tarlé G. (1983): Observation of large deviations from the Bethe-Bloch formula for relativistic uranium ions. Phys Rev Lett **50**, 1110–1113

Andersen H.H., Bak J.F., Knudsen H. and Nielsen B.R. (1977): Stopping power of Al, Cu, Ag and Au for MeV hydrogen, helium, and lithium ions. Z_1^3 and Z_1^4 proportional deviations from the Bethe formula. Phys Rev A **16**, 1929

Andersen H.H., Simonsen H. and Sørensen H. (1969): An experimental investigation of charge-dependent deviations from the Bethe stopping power formula. Nucl Phys **A125**, 171–175

Andersen H.H. and Ziegler J.F. (1977): *Hydrogen stopping powers and ranges in all elements*, vol. 2 of *The Stopping and Ranges of Ions in Matter*. Pergamon, New York

Arista N.R. (1982): Z^3 corrections to the scattering of electrons and positrons in atoms and to the energy loss of fast particles in solids. Phys Rev A **26**, 209–216

Ashley J.C., Ritchie R.H. and Brandt W. (1972): Z_1^3 Effect in the stopping power of matter for charged particles. Phys Rev B **5**, 2393–2397

Barkas W.H., Dyer J.N. and Heckman H.H. (1963): Resolution of the Σ^--mass anomaly. Phys Rev Lett **11**, 26–28

Bergara A., Pitarke J.M. and Ritchie R.H. (1996): Nonlinear quantum hydrodynamical model of the electron gas. Nucl Instrum Methods B **115**, 70–74

Bethe H. (1930): Zur Theorie des Durchgangs schneller Korpuskularstrahlen durch Materie. Ann Physik **5**, 324–400

Bhalla C.P. and Rose M.E. (1962): Finite nuclear size effects in β decay. Phys Rev **128**, 774–778

Bichsel H. (2002): Shell corrections in stopping powers. Phys Rev A **65**, 052709-1–11

Bloch F. (1933): Zur Bremsung rasch bewegter Teilchen beim Durchgang durch Materie. Ann Physik **16**, 285–320

Bohr N. (1948): The penetration of atomic particles through matter. Mat Fys Medd Dan Vid Selsk **18 no. 8**, 1–144

Brown L.M. (1950): Asymptotic expression for the stopping power of K-electrons. Phys Rev **79**, 297–303

Cohen S.M. (2003): Bethe stopping power theory for heavy-element targets and relativistic projectiles. Phys Rev A **68**, 012720

Cohen S.M. (2004): Aspects of relativistic sum rules. Adv Quantum Chem **46**, 241–265

Datz S., Krause H.F., Vane C.R., Knudsen H., Grafström P. and Schuch R.H. (1996): Effect of nuclear size on the stopping power of ultrarelativistic heavy ions. Phys Rev Lett **77**, 2925–2928

Esbensen H. (1976): Contributions to detailed perturbation theory for slowing-down of charged particles. Ph.D. thesis, Aarhus University

Esbensen H. and Sigmund P. (1990): Barkas effect in a dense medium: Stopping power and wake field. Annals of Physics **201**, 152–192

Fano U. (1963): Penetration of protons, alpha particles, and mesons. Ann Rev Nucl Sci **13**, 1–66

Fano U. and Turner J.E. (1964): Contributions to the theory of shell corrections. In *Studies in penetration of charged particles in matter*, vol. 1133, 49–67. NAS-NRC, Washington

de Ferrariis L. and Arista N.R. (1984): Classical and quantum-mechanical treatments of the energy loss of charged particles in dilute plasmas. Phys Rev A **29**, 2145–2159

Gerjuoy E. (1966): Cross section for energy transfer between two moving particles. Phys Rev **148**, 54–59

Grande P.L. and Schiwietz G. (1998): Impact-parameter dependence of the electronic energy loss of fast ions. Phys Rev A **58**, 3796–3801

Gryzinski M. (1957): Stopping power of a medium for heavy, charged particles. Phys Rev **107**, 1471–1475

Gryzinski M. (1965): Classical theory of atomic collisions. 1. Theory of inelastic collisions. Phys Rev **138**, A336–358

Hill K.W. and Merzbacher E. (1974): Polarization in distant Coulomb collisions of charged particles with atoms. Phys Rev A **9**, 156–165

Hu C.D. and Zaremba E. (1988): Z_1^3 correction to the stopping power of ions in an electron gas. Phys Rev B **37**, 9268–9277

ICRU (1984): *Stopping powers for electrons and positrons*, vol. 37 of *ICRU Report*. International Commission of Radiation Units and Measurements, Bethesda, Maryland

ICRU (1993): *Stopping powers and ranges for protons and alpha particles*, vol. 49 of *ICRU Report*. International Commission of Radiation Units and Measurements, Bethesda, Maryland

Inokuti M., Dehmer J.L., T. B. and Hanson J.D. (1981): Oscillator-strength moments, stopping powers, and total inelastic- scattering cross sections of all atoms through strontium. Phys Rev A **23**, 95–109

Jackson J.D. and McCarthy R.L. (1972): Z^3 corrections to energy loss and range. Phys Rev B **6**, 4131–4141

Khandelwal G.S. and Merzbacher E. (1966): Stopping power of M electrons. Phys Rev **144**, 349–352

Lindhard J. (1954): On the properties of a gas of charged particles. Mat Fys Medd Dan Vid Selsk **28 no. 8**, 1–57

Lindhard J. (1976): The Barkas effect – or Z_1^3, Z_1^4-corrections to stopping of swift charged particles. Nucl Instrum Methods **132**, 1–5

Lindhard J. and Scharff M. (1953): Energy loss in matter by fast particles of low charge. Mat Fys Medd Dan Vid Selsk **27 no. 15**, 1–31

Lindhard J. and Sørensen A.H. (1996): On the relativistic theory of stopping of heavy ions. Phys Rev A **53**, 2443–2456

McGuire E.J. (1983): Extraction of shell corrections from Born-approximation stopping-power calculations in Al. Phys Rev A **28**, 49–52

Mikkelsen H.H. (1991): The Z_1^4-Term in electronic stopping - impact parameter dependence and stopping cross section for a quantal harmonic oscillator. Nucl Instrum Methods B **58**, 136–148

Mikkelsen H.H., Esbensen H. and Sigmund P. (1990): The Z_1^3 contribution to the stopping power of protons and antiprotons in silicon - 2 theoretical predictions. Nucl Instrum Methods B **48**, 8–9

Mikkelsen H.H. and Sigmund P. (1989): Barkas effect in electronic stopping power: Rigorous evaluation for the harmonic oscillator. Phys Rev A **40**, 101–116

Møller C. (1972): *The theory of relativity.* Oxford University Press, Oxford, 2 edn.

Møller S.P. (1990): Measurement of the Barkas effect using MeV antiprotons and protons and an active silicon target. Nucl Instrum Methods B **48**, 1–7

Møller S.P., Uggerhøj E., Bluhme H., Knudsen H., Mikkelsen U., Paludan K. and Morenzoni E. (1997): Direct measurement of the stopping power for antiprotons of light and heavy targets. Phys Rev A **56**, 2930–2939

Mott N.F. (1929): The scattering of fast electrons by atomic nuclei. Proc Roy Soc London A **124**, 425–442

Pitarke J.M., Ritchie R.H. and Echenique P.M. (1993a): Z_1^3 correction to the stopping power of an electron gas for ions. Nucl Instrum Methods B **79**, 209–212

Pitarke J.M., Ritchie R.H. and Echenique P.M. (1995): Quadratic response theory of the energy-loss of charged-particles in an electron-gas. Phys Rev B **52**, 13883–13902

Pitarke J.M., Ritchie R.H., Echenique P.M. and Zaremba E. (1993b): The Z_1^3 correction to the Bethe-Bloch energy loss formula. Europhys Lett **24**, 613–619

Scheidenberger C., Geissel H., Mikkelsen H.H., Nickel F., Brohm T., Folger H., Irnich H., Magel A., Mohar M.F., Pfützner G.M.M. et al. (1994): Direct observation of systematic deviations from the Bethe stopping theory for relativistic heavy ions. Phys Rev Lett **73**, 50–53

Schinner A. and Sigmund P. (2000): Polarization effect in stopping of swift partially screened heavy ions: perturbative theory. Nucl Instrum Methods B **164-165**, 220–229

Schiwietz G. and Grande P.L. (1999): A unitary convolution approximation for the impact-parameter dependent electronic energy loss. Nucl Instrum Methods B **153**, 1–9

Sigmund P. (1982): Kinetic theory of particle stopping in a medium with internal motion. Phys Rev A **26**, 2497–2517

Sigmund P. (1996): Low-velocity limit of Bohr's stopping-power formula. Phys Rev A **54**, 3113–3117

Sigmund P. (1997): Charge-dependent electronic stopping of swift nonrelativistic heavy ions. Phys Rev A **56**, 3781–3793

Sigmund P. (2000): Shell correction in Bohr stopping theory. Europ Phys J D **12**, 111–116

Sigmund P. (2004): *Stopping of heavy ions*, vol. 204 of *Springer Tracts of Modern Physics.* Springer, Berlin

Sigmund P. and Haagerup U. (1986): Bethe stopping theory for a harmonic oscillator and Bohr's oscillator model of atomic stopping. Phys Rev A **34**, 892–910

Sigmund P. and Schinner A. (2000): Binary stopping theory for swift heavy ions. Europ Phys J D **12**, 425–434

Sigmund P. and Schinner A. (2006): Shell correction in stopping theory. Nucl Instrum Methods B **243**, 457–460

Smith F.M., Birnbaum W. and Barkas W.H. (1953): Measurements of meson masses and related quantities. Phys Rev **91**, 765–766

Sørensen A.H. (2005): Energy loss of relativistic heavy ions due to pair production and emission of bremsstrahlung. Nucl Instrum Methods B **230**, 12–16

Sung C.C. and Ritchie R.H. (1983): The Z_1^3 dependence of the energy loss of an ion passing through an electron gas. Phys Rev A **28**, 674–681

Tarlé G. and Solarz M. (1978): Evidence for higher-order contributions to the stopping power of relativistic iron nuclei. Phys Rev Lett **41**, 483–486

Tofterup A.L. (1983): Relativistic binary-encounter and stopping theory: general expressions. J Phys B **16**, 2997–3003

Trubnikov B.A. and Yavlinskii Y.N. (1965): Energy loss of slow protons in metals. Sov Phys JETP **48**, 253–260. Engl. transl. Sov. Phys. JETP **21**, 167-171 (1965)

Varelas C. and Biersack J. (1970): Reflection of energetic particles from atomic or ionic chains in single crystals. Nucl Instrum Methods **79**, 213–218

Walske M.C. (1952): The stopping power of K-electrons. Phys Rev **88**, 1283–1289

Walske M.C. (1956): Stopping power of L-electrons. Phys Rev **101**, 940–944

Weick H., Sørensen A.H., Geissel H., Maier M., Münzenberg G., Nankov N. and Scheidenberger C. (2002): Stopping power of partially ionized relativistic heavy ions. Nucl Instrum Methods B **193**, 1–7

Arriving at Numbers

7.1 Introductory Comments

Common to all theoretical schemes presented in Chapters 4–6 is the need for numerical input characterizing the physical properties of the stopping material. Only one single parameter, the I-value, enters Bohr's and Bethe's asymptotic stopping formulae. Following a wide-spread tradition in physics and chemistry, this single parameter has most often been determined by comparison with measurements. At Bohr's and Bethe's time only range data were available, while direct measurements of stopping forces on thin targets became common in the second half of the past century.

Deviations from the velocity dependence predicted by the Bethe formula were found long ago. Frequently, such deviations were described in terms of a dependence of the I-value on projectile speed. This is obviously an unsatisfactory interpretation of a material property.

The significance of the shell correction was realized in the early 1950s, and expressions ready to use in the analysis of experiments became available in the 1960s. The importance of the Barkas-Andersen correction was recognized only gradually in the 1970s. Again, numerical coefficients entering theoretical expressions were frequently determined by fitting to measured stopping forces.

This chapter provides a survey of methods to determine input parameters by

- Fitting theoretical expressions to measured stopping forces,
- Utilizing pertinent parameters extracted from optical measurements or other data unrelated to particle stopping, or
- Utilizing theoretical input,

and to provide comparisons with experimental stopping data.

Scaling properties play an important role in all three procedures. The emphasis will be on the second and third method. Fitting to stopping data is still popular and may well lead to more accurate fits for practical applications than the other two methods. Those, however, are more attractive from the point of view of understanding the physics and whenever extrapolation is necessary. Moreover, empirical stopping data are available only for a comparatively small fraction of ion-target combinations. The coverage with experimental data varies drastically over the periodic table. Interpolation becomes questionable in areas of poor coverage.

The chapter concludes with a survey of existing data compilations of stopping forces as well as pertinent computer codes.

7.2 Stopping Models I: Statistical Method

Stopping data are needed for a large number of elemental and compound target materials. It is highly desirable, therefore, to have scaling relations allowing to predict stopping parameters by interpolation between elements or compounds where reliable experimental or theoretical data are available. The statistical model of Thomas (1926) and Fermi (1927, 1928) provides such an opportunity on the basis of the atomic number(s) of the atoms making up the target material.

7.2.1 Thomas-Fermi Model of the Atom

The Thomas-Fermi model of the atom considers the electron cloud as a Fermi gas with a position-dependent density. This is called a *local-density approximation*. The density profile is determined by minimizing the total energy of the atom which is made up of a kinetic and a potential contribution. While potential energy is assumed identical with the classical Coulomb energy of a continuous charge distribution in the field of the nucleus, quantum theory enters into the kinetic energy via the relation between electron density n and Fermi wave number k_{F}, cf. (5.146),

$$n = \frac{k_{\mathrm{F}}^3}{3\pi^2}. \tag{7.1}$$

This model is suitable to describe the global behavior of atomic parameters that depend primarily on the atomic number. Variations over a smaller scale, such as over a row of the periodic table, are only accounted for if special precautions are taken.

7.2.2 Scaling Properties

To illustrate the general principle, approximate a neutral atom with Z electrons by a point nucleus surrounded by a homogeneously charged sphere of radius a with a potential energy, composed of the self-energy of the electron cloud and the potential energy of the electron cloud in the field of the nucleus,

$$E_{\mathrm{pot}} = -\frac{9}{10}\frac{Z^2 e^2}{a}, \tag{7.2}$$

and a kinetic energy

$$E_{\mathrm{kin}} = Z \,\frac{3}{5}\,\frac{\hbar^2 k_{\mathrm{F}}^2}{2m}. \tag{7.3}$$

With (7.1) and $n = 3Z/4\pi a^3$ we find $E_{\mathrm{kin}} \propto \hbar^2 Z^{5/3}/ma^2$. The total energy $E_{\mathrm{kin}} + E_{\mathrm{pot}}$ of this model atom has a minimum at

$$a_{\mathrm{TF}} = \mathrm{const}\,\frac{a_0}{Z^{1/3}}, \tag{7.4}$$

where $a_0 = \hbar^2/me^2$ is the Bohr radius, with a numerical constant of the order of 1 which, in general, is related to the adopted charge distribution.

With this, both the potential, the kinetic and the total energy of the atom scale as

$$E \sim \frac{Z^{7/3}e^2}{a_0}. \tag{7.5}$$

From the energy per electron, $\sim Z^{4/3}e^2/a_0$, we then extract the electron velocity

$$v_{\mathrm{TF}} = Z^{2/3}v_0, \tag{7.6}$$

where $v_0 = e^2/\hbar$ is the Bohr velocity. Finally, for the angular frequency we obtain

$$\omega \sim \frac{v_{\mathrm{TF}}}{a_{\mathrm{TF}}} \sim Z\,\frac{v_0}{a_0}. \tag{7.7}$$

From (7.7) we find the scaling rule predicted by Bloch (1933) for the I-value,

$$I = \hbar\omega = Z_2 I_0 \tag{7.8}$$

with a constant I_0 that was determined empirically to lie around 10 eV (Lindhard and Scharff, 1953). It is important to note that even though I has the dimension of an energy it does not scale as either $Z_2^{4/3}$ or $Z_2^{7/3}$. Indeed, I has the character of a frequency both in classical and quantum theory.

Similarly, we may find a scaling property of the first shell correction following (6.104),

$$-\frac{\langle v_e^2 \rangle}{v^2} \sim -\frac{Z_2^{4/3}v_0^2}{v^2}, \tag{7.9}$$

which indicates a fairly rapid increase with the atomic number of the stopping material.

Finally, for the Barkas-Andersen parameter (6.29) we find

$$B = \frac{Z_1 e^2 \omega}{mv^3} \sim Z_1 Z_2 \frac{I_0}{mv_0^2}\left(\frac{v_0}{v}\right)^3. \tag{7.10}$$

Such relations are very helpful in assessments of the relative importance of various effects, as well as for interpolation.

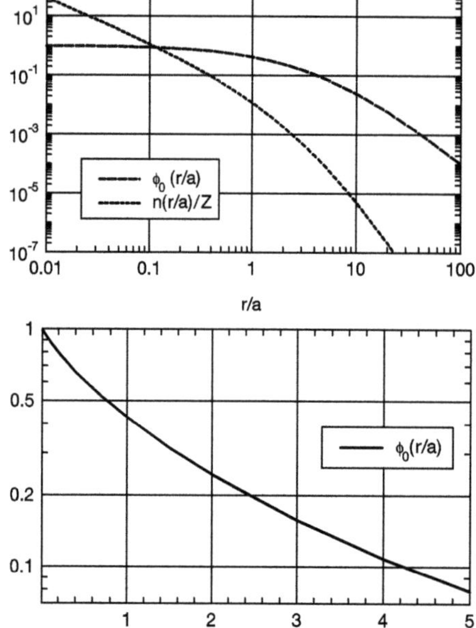

Fig. 7.1. Screening function $\phi_0(r/a)$ and electron density $n(r/a)$ of a neutral atom according to the Thomas-Fermi model. Numerical evaluation by Bush and Caldwell (1931). The graphs differ in scale

7.2.3 Charge and Velocity Distributions

A more quantitative picture is found by minimizing the total energy on the basis of an arbitrary electron distribution $n(r)$ normalized to the number of electrons in the atom. This yields the fundamental Thomas-Fermi equation (Gombas, 1956)

$$\nabla^2(\Phi - \Phi_0) = 4\pi\sigma_0 e(\Phi - \Phi_0)^{3/2}, \tag{7.11}$$

where

$$\sigma_0 = \frac{2^{3/2}}{3\pi^2} \frac{1}{(ea_0)^{3/2}}. \tag{7.12}$$

Here, Φ_0 is a Lagrange multiplier ensuring proper normalization of the electron density.

Equation (7.11) has been solved numerically for both neutral atoms and ions. Tabulations of $\Phi(r)$ and its derivative as well as analytical approximation formulas were compiled by Gombas (1956). The electron distribution is related to the potential by Poisson's law, i.e.,

$$n(r) = \sigma_0 \left(\Phi - \Phi_0\right)^{3/2} \tag{7.13}$$

in view of (7.11).

The potential of a neutral atom may be written in the form

$$\Phi(r) = \frac{Ze}{r}\phi_0\left(\frac{r}{a_{\mathrm{TF}}}\right),$$ (7.14)

where

$$a_{\mathrm{TF}} = \frac{1}{(4\pi\sigma_0)^{2/3}eZ^{1/3}} = \frac{0.8853a_0}{Z^{1/3}},$$ (7.15)

$\Phi_0 = 0$, and ϕ_0 a screening function shown in Fig. 7.1. Power-like behavior is found at large distances, $\phi_0(x) \simeq 144/x^3$. At small distances, ϕ_0 reminds of an exponential.

The density reads

$$n(r) = \sigma_0\left[\frac{Ze}{r}\phi_0\left(\frac{r}{a_{\mathrm{TF}}}\right)\right]^{3/2} = \frac{32Z^2}{9\pi^3a_0^3}\left[\frac{a_{\mathrm{TF}}}{r}\phi_0\left(\frac{r}{a_{\mathrm{TF}}}\right)\right]^{3/2}$$ (7.16)

according to (7.13). It is also shown in Fig. 7.1.

The velocity distribution in an atom may be found as follows. For an inhomogeneous electron gas, the electron density in real and velocity space is given by

$$F(r,v) = \begin{cases} \dfrac{3n(r)}{4\pi v_{\mathrm{F}}^3(r)} & \text{for } v < v_{\mathrm{F}}(r), \\[2ex] 0 & \text{for } v > v_{\mathrm{F}}(r), \end{cases}$$ (7.17)

where $3n(r)/4\pi v_{\mathrm{F}}(r)^3 = m^3/4\pi^3\hbar^3$ is constant. Integration over the spatial variable yields the velocity spectrum normalized to 1,

$$f(v)\mathrm{d}^3\boldsymbol{v} = \frac{m^3}{4\pi^3\hbar^3}\frac{1}{Z}\mathrm{d}^3\boldsymbol{v}\int_{v_{\mathrm{F}}(r)>v}\mathrm{d}^3\boldsymbol{r}$$ (7.18)

or

$$f(v) = \frac{1}{3\pi^2 Z}\left(\frac{mr_1}{\hbar}\right)^3,$$ (7.19)

where r_1 is defined by

$$n(r_1) = \frac{1}{3\pi^2}\left(\frac{mv}{\hbar}\right)^3.$$ (7.20)

In dimensionless units

$$x = \frac{r_1}{a_{\mathrm{TF}}}; \quad y = \frac{ma_{\mathrm{TF}}v}{\hbar Z^{1/3}},$$ (7.21)

this reads

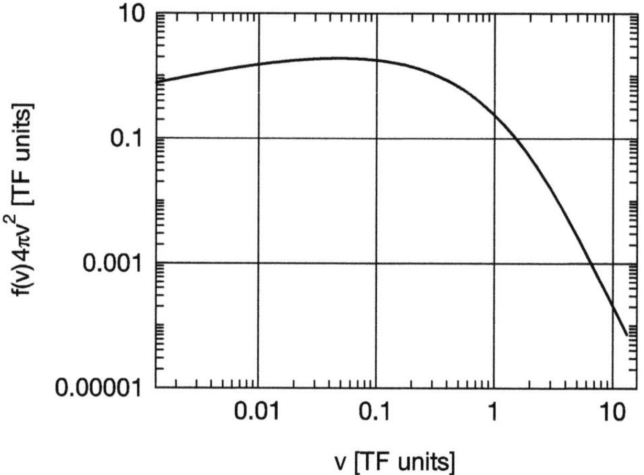

Fig. 7.2. Velocity distribution calculated from Thomas-Fermi electron distribution. v in Thomas-Fermi units according to (7.21)

$$f(v)\mathrm{d}^3 v = \frac{x^3}{3\pi^2}\mathrm{d}^3 y \rightarrow \frac{4}{3\pi}x^3 y^2 dy \tag{7.22}$$

with

$$y^2 = 1.7706\,\frac{\phi_0(x)}{x}. \tag{7.23}$$

The function $f(v)4\pi v^2$ is shown in Fig. 7.2; the plot has been constructed using the approximation of Sommerfeld (1932).

$$\phi_0(x) \simeq \left[1 + \left(\frac{x^3}{144}\right)^{\lambda/3}\right]^{-3/\lambda} \;\; ; \quad \lambda = 0.8034. \tag{7.24}$$

7.2.4 The Lindhard-Scharff Model and its Implementation

Following the spirit of the Thomas-Fermi model of the atom, Lindhard and Scharff (1953) proposed to evaluate the stopping cross section of an atom from the energy-loss function of a free-electron gas,

$$S = \int \mathrm{d}^3 r\,\frac{4\pi Z_1^2 e^4 n(r)}{mv^2}\,L\left(\frac{2mv^2}{\chi\hbar\omega_P(r)}\right), \tag{7.25}$$

where $n(r)$ is the electron density of a target atom, and $\omega_P(r)$ the 'local plasma frequency'

$$\omega_P(r) = \sqrt{\frac{4\pi n(r)e^2}{m}}. \tag{7.26}$$

A factor χ has been inserted in (7.25) that will be discussed below.

Since the electron density is high for inner shells, (7.25) implies high I-values for those shells, and vice versa. The procedure determines an effective I-value given by

$$\ln I = \int \mathrm{d}^3 r \, \frac{n(\boldsymbol{r})}{Z_2} \ln\left[\chi \hbar \omega_P(\boldsymbol{r})\right].\tag{7.27}$$

According to Lindhard and Scharff (1953), replacing $\omega_P(\boldsymbol{r})$ in (7.27) by $\chi \omega_P(\boldsymbol{r})$ with a numerical coefficient $\chi > 1$, should account for electron binding. The value $\chi = \sqrt{2}$ was adopted for not too light atoms.

In order to rationalize this adjustment, consider a Thomas-Fermi atom with a characteristic electron density

$$n \simeq \frac{Z/2}{4\pi a_{\mathrm{TF}}^3/3} \simeq \frac{3Z^2}{8\pi a_0^3},\tag{7.28}$$

assuming that about half of the electrons are located inside the Thomas-Fermi radius. This implies a characteristic plasma frequency

$$\omega_P = \sqrt{\frac{4\pi n e^2}{m}} \simeq \sqrt{\frac{3}{2}} Z \frac{v_0}{a_0}.\tag{7.29}$$

The characteristic revolution frequency of an electron in a Thomas-Fermi atom, on the other hand, is given by

$$\omega_{\mathrm{rev}} \simeq \frac{v_{\mathrm{TF}}}{a_{\mathrm{TF}}} \simeq Z \frac{v_0}{a_0},\tag{7.30}$$

i.e., very close to ω_P.

Employing a simplified atomic model, Lindhard and Scharff (1953) arrived at the result that $\omega_P \simeq \omega_{\mathrm{rev}}$ even from shell to shell. We have seen already in (5.57) and Fig. 5.3 that the effective resonance frequency is determined by the square sum, as is known from classical electrodynamics. Hence,

$$\omega_{\mathrm{eff}} \simeq \sqrt{2}\, \omega_P.\tag{7.31}$$

One may be tempted to evaluate the mean excitation potential from (7.27) by insertion of the Thomas-Fermi density of the atom. This route was taken by Bonderup (1967) who, however, concluded that uncertainties in the procedure would show up primarily in the calculated I-value, while shell corrections calculated in a similar manner would be more reliable.

On the other hand, Chu and Powers (1972a), employing Hartree-Fock electronic charge densities rather than Thomas-Fermi densities, found quite reasonable agreement with I-values deduced from measured stopping forces, as is seen in Fig. 7.3.

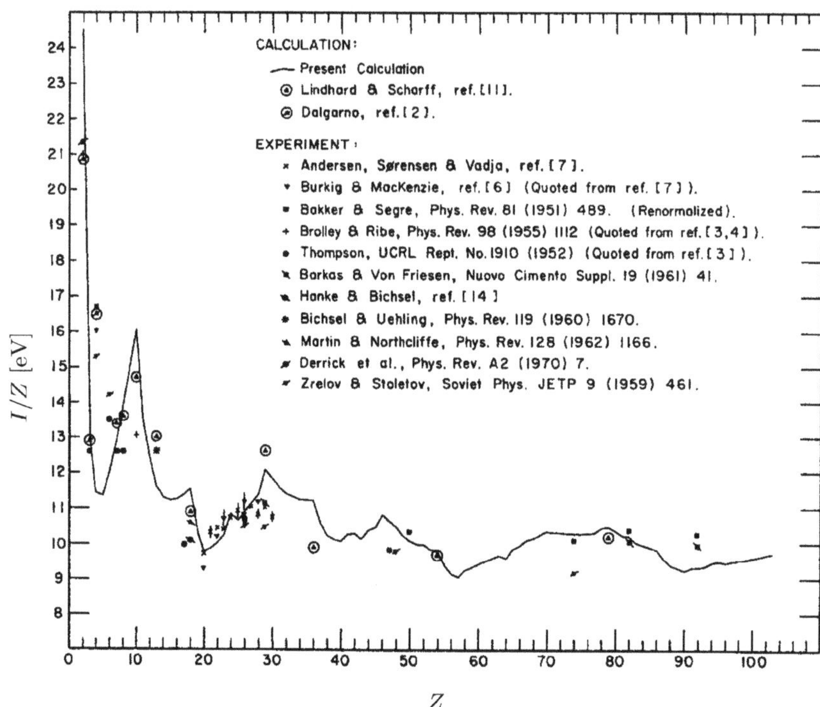

Fig. 7.3. I-values calculated from (7.27) utilizing atomic Hartree-Fock electron densities according to Herman and Skillman (1963). From Chu and Powers (1972a)

A similar adjustment is not necessary for the first shell correction,

$$\langle v_e^2 \rangle = \frac{1}{Z} \int d^3r \int d^3v \, v^2 F(r,v) = 1.3554 \, v_{\text{TF}}^2 \int_0^\infty \frac{dx}{\sqrt{x}} \left[\phi_0(x) \right]^{5/2} \qquad (7.32)$$

(cf. problem 7.2), since this correction is purely kinematic, cf. (6.104) and similar relationships in Sect. 6.6.

Figure 7.4 shows calculated stopping cross sections for alpha particles. Agreement with a large number of experimental data – mostly at 1.0 MeV/u – is quite good.

Figures 7.3 and 7.4 show a nonmonotonic dependence of the I-value as well as the stopping cross section on Z_2. This important phenomenon, called Z_2 structure or Z_2 oscillations, will be discussed in some detail in Sect. 7.5.3.

A derivation of Eq. (7.25) from first principles has never been presented, even though it has been looked for (Johnson and Inokuti, 1983). Evidently, the model seeks to replace an average in oscillator-strength space by an average in real space, i.e., it determines an oscillator-strength spectrum $f(\omega)$ from the relation

$$f(\omega)d\omega = n(r)4\pi r^2 dr \qquad (7.33)$$

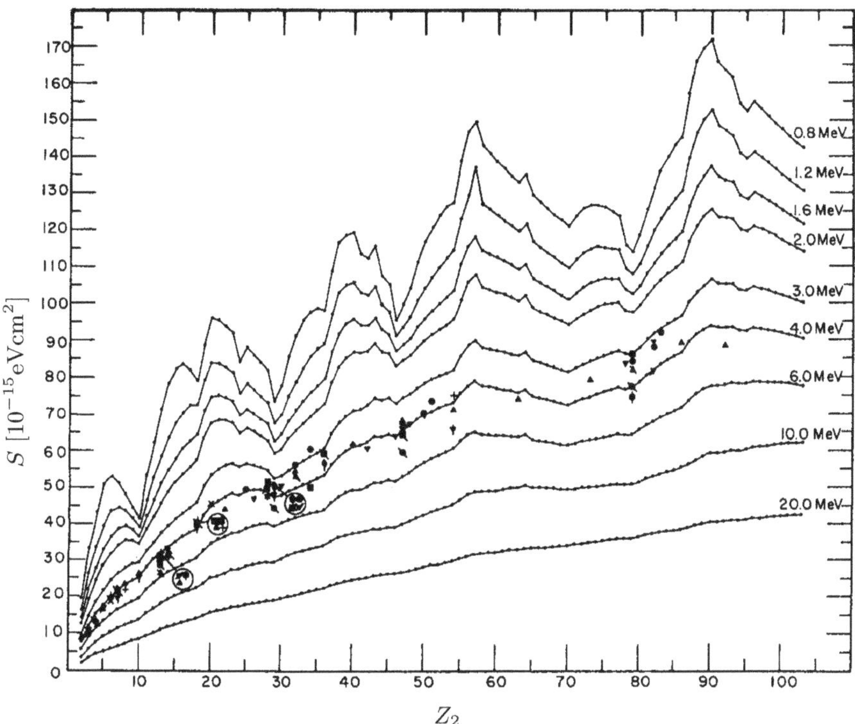

Fig. 7.4. Stopping cross section for alpha particles in a large number of stopping media. Beam energies ranging from 0.2 to 5.0 MeV/u. Calculations based on (7.25) allowing for shell correction expansion according to Lindhard and Winther (1964) but excluding Barkas-Andersen correction. Experimental data from numerous sources. From Chu and Powers (1972b)

with

$$\omega = \sqrt{\frac{4\pi n(r)e^2}{m}}. \tag{7.34}$$

Figure 7.5 compares this function, assuming the Thomas-Fermi density of a neutral atom, with currently available data for aluminium. The statistical approximation is found to reflect the average behavior above ~ 70 eV, while major discrepancies are found at lower energies.

A hint on its range of validity may be found in the underlying dependence on impact parameter. Assume, for a moment, that all energy loss in an electron gas be local, i.e., energy transfer only to electrons in the immediate vicinity of the trajectory of the penetrating ion. Then, the mean energy transfer $T(p)$ to an atom at an impact parameter p to the target nucleus is given by

$$T(\boldsymbol{p}) = \int_{-\infty}^{\infty} dx' \, \frac{4\pi Z_1^2 e^4 n(x', \boldsymbol{p})}{mv^2} L\left(\frac{2mv^2}{\chi \hbar \omega_P \left(n(x', \boldsymbol{p})\right)}\right), \tag{7.35}$$

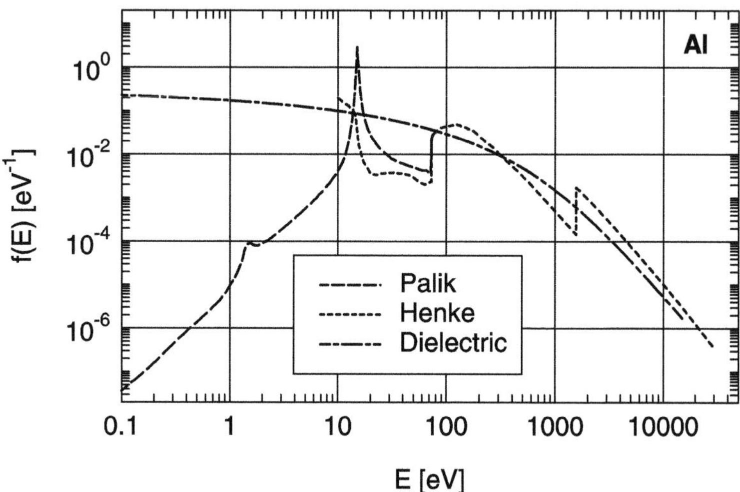

Fig. 7.5. Optical excitation spectrum of metallic aluminium according to Palik (2000) (data taken from Smith et al. (1985)) and Henke et al. (1993). The curve labelled 'dielectric' reflects the Lindhard-Scharff model assuming Thomas-Fermi electron density of a neutral atom using (7.34)

where $\boldsymbol{p}$ is the vectorial impact parameter and x' a coordinate along the trajectory. For a random medium one then obtains the stopping force

$$-\frac{\mathrm{d}E}{\mathrm{d}x} = N \int \mathrm{d}^2\boldsymbol{p}\, T(\boldsymbol{p}), \qquad (7.36)$$

N being the number of atoms per volume. With $\boldsymbol{r} = (x', \boldsymbol{p})$, (7.36) readily reduces to (7.25).

Now, we know that energy transfer is not local, and the higher the projectile speed, the larger the difference between the true interaction range – given by the adiabatic radius which increases with projectile speed – and the spatial extension of the electron cloud of the target atom. Conversely, while the local-density approach should be more reliable at low projectile speed, the validity of Bethe stopping theory is limited downward, as discussed in the previous section.

A direct comparison between an 'exact' result and the local-density approximation is shown in Fig. 7.6 on the example of the quantal harmonic oscillator. The local-density approximation overestimates the energy loss in close collisions by about a factor of two but drops to zero much faster than the exact result at large impact parameters. In the case in question, with $2mv^2/\hbar\omega = 100$, the adiabatic radius is equal to 10 Bohr radii for $\hbar\omega = 13.6$ eV.

Even though the local-density approximation appears highly questionable as far as the prediction of impact-parameter-dependent energy loss is con-

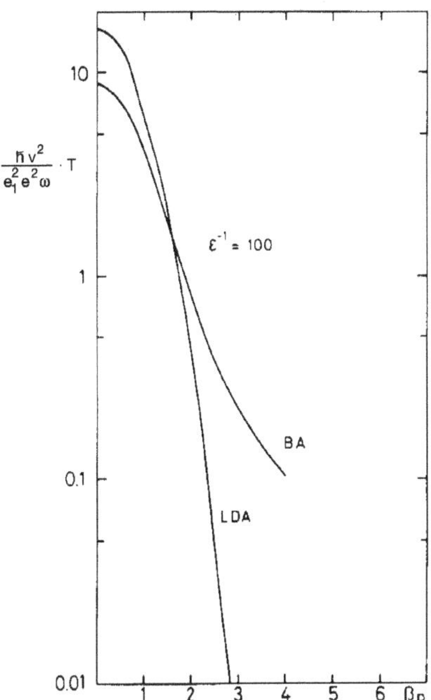

Fig. 7.6. Mean energy loss versus impact parameter $T(p)$ for a spherical harmonic oscillator: Comparison between semiclassical calculation in Born approximation and local-density approximation based on the stopping function of Lindhard and Winther (1964). $\beta = \sqrt{m\omega/\hbar}$. $\epsilon^{-1} = 2mv^2/\hbar\omega$. From Mikkelsen and Sigmund (1987)

cerned, stopping cross sections found from (7.25) tend to show reasonable agreement with experiment. Indeed, in view of the simplicity of the model, numerous calculations have been performed and still are being performed for stopping materials where reliable electron densities are available.

While the proposal by Lindhard and Scharff (1953) provided a way to arrive at absolute predictions of stopping parameters at a time when feasible alternatives were unavailable, use of this approximation in the 21st century would appear to warrant more of a theoretical foundation than what is available here and now.

7.2.5 Generalizations

A large number of modifications of (7.25) has been explored in the literature. Most frequently, the Lindhard function $\epsilon(k, \omega)$ governing stopping in a Fermi gas has been replaced by an alternative expression. This does usually not give rise to major changes (Pathak and Yussouff, 1972) but may facilitate analytical or numerical evaluations. The plasmon pole approximation discussed in Sect. 5.4.5 is a frequently applied example. An expression proposed by Mermin (1970) is thought to provide a more realistic account of plasmon damping than the Lindhard function with infinitesimal damping.

Substantial progress has been made by the application of density functional theory (Echenique et al., 1981) which takes into account the influence of the electron gas on the electronic structure of a target atom. This has a substantial affect on outer target electrons and hence is most significant in low-velocity stopping, to be discussed in Volume II.

The local density picture can be avoided altogether by considering an inhomogeneous electron gas from the beginning. A simple example was discussed in Sect. 5.5. The reader interested in the recent development in this area is referred to Pitarke and Campillo (2000).

7.3 Stopping Models II

7.3.1 Shell and Subshell Splitting

It is tempting to determine the stopping number of a multi-electron atom as

$$Z_2 L(v) = \sum_\nu Z_\nu L_\nu(v), \tag{7.37}$$

where $L_\nu(v)$ is the stopping number of the νth electron, characterized by a single resonance frequency ω_ν and the number Z_ν of such electrons in the atom. This has been common practice ever since deviations from the universal Bethe logarithm in the stopping number became apparent (Brown, 1950, Walske, 1952, 1956, Khandelwal and Merzbacher, 1966). In practice, the ω_ν denote a characteristic resonance frequency of a principal shell or, perhaps, a subshell of the target atom.

A scheme of this type has been proposed by Sternheimer et al. (1982) to compute the Fermi density effect discussed in Sect. 5.6.4. The scheme also offers a unique procedure to relate the frequency ω_ν to the ionization energy of the respective shell.

This type of shell summation is well defined and transparent, even though it is approximate in several respects: A single resonance frequency per shell may not be adequate; excitations of individual electrons are assumed to be uncorrelated; collective excitations are ignored.

Equation (4.117) suggests a generalization of (7.37) so that

$$L(v) = \sum_j f_j L(v, \omega_j), \tag{7.38}$$

where $L(v, \omega_j)$ is the contribution to the stopping number from an electron characterized by a resonance frequency ω_j, taking into account pertinent effects such as shell and Barkas-Andersen corrections. Equation (7.38) is exact within the first Born approximation disregarding shell corrections. Inclusion of shell and Barkas-Andersen corrections appears tempting and perhaps reasonable, but this is not rigorously justified. In particular, shell corrections draw heavily on large energy transfers which are not weighted accurately by dipole oscillator strengths.

Figure 7.5 shows the optical excitation spectrum of metallic aluminium. The absorption edges of the K shell at $\hbar\omega_\mathrm{K} = 1559$ eV and the L shell at $\hbar\omega_\mathrm{L} = 72.59$ eV are clearly visible. Integration from the K absorption edge upward yields[1]

$$\sum_{\omega_{j0} \geq \omega_\mathrm{K}} Z f_j = 1.70, \tag{7.39}$$

i.e., less than 2.0, the number of electrons in the K shell. This is a general feature: The dipole oscillator strength has a slight preference for outer-shell electrons.

Spectra of the type shown in figure 7.5 are available for numerous elements and compounds, although few have been studied in as much detail as that of aluminium. Pertinent tabulations, based on a variety of experiments and calculations were published by Henke et al. (1993), Palik (1985, 1991, 1996), followed up by an electronic version (Palik, 2000), and Berkowitz (1979, 2002).

Such data can be used to evaluate the Bethe stopping formula (4.117). This is equivalent with determining the I-value.

In general, prior inspection of such data is necessary. More or less pronounced deviations from the sum rule (4.114) are common, and uncritical evaluation of I-values may lead to erroneous results.

Although (7.38) can be evaluated for a spectrum like the one in Fig. 7.5 after an appropriate expression for $L(v, \omega_j)$ has been chosen, splitting up and bunching the spectrum into contributions from principal shells or subshells is advisable for computational reasons and, most often, not a cause of major error compared with more basic uncertainties. Evidently, bunching must be performed in a way to preserve sum rule and I-value. This can be achieved by the following definitions of f_ν and ω_ν,

$$f_\nu = \sum_{\omega_{0\nu} \leq \omega_j \leq \omega_{1\nu}} f_j \tag{7.40a}$$

$$\ln \omega_\nu = \frac{\displaystyle\sum_{\omega_{0\nu} \leq \omega_j \leq \omega_{1\nu}} f_j \ln \omega_j}{\displaystyle\sum_{\omega_{0\nu} \leq \omega_j \leq \omega_{1\nu}} f_j}, \tag{7.40b}$$

where $\omega_{0\nu}$ and $\omega_{1\nu}$ limit the frequency interval assigned to the νth shell or subshell. In practice, the sums turn into integrals in energy space.

---

[1] In the literature, dipole oscillator strengths are found normalized either to 1, as is done in this monograph, or to Z. The latter notation allows the interpretation of f_j as the number of j-electrons in an atom, even though it is typically not an integer. In the present notation, this quantity reads $Z f_j$, as e.g., in (7.39).

The use of experimentally determined or accurately calculated oscillator strengths ensures a considerable degree of reliability of the material dependence of calculated stopping cross sections. Errors may be caused by inaccuracies of the shell stopping numbers employed in the evaluation of $L(v) = \sum_\nu f_\nu L(v, \omega_\nu)$.

If the stopping material contains free electrons, one of the shells is typically treated as a homogeneous Fermi gas, and its contribution to the total stopping force is represented by a pertinent expression such as the one by Lindhard and Winther (1964).

As mentioned already, weighting contributions from target resonances by the spectrum of dipole oscillator strengths is justified within the range of validity of the first Born approximation. The fact that the Bohr theory delivers the same dependence of the energy loss at high impact parameters as the Bethe theory suggests to also apply this weighting to calculations based on classical theory (Sigmund and Schinner, 2002a). However, dipole oscillator strengths have no place in close, classical Coulomb collisions. Applying (7.38) in the Bohr regime can at most be approximate.

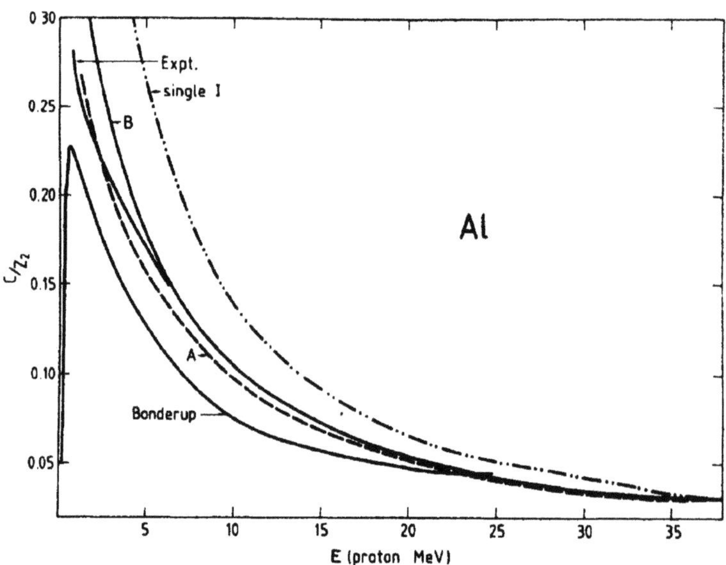

Fig. 7.7. Shell correction for aluminium. Curves A and B: evaluated from kinetic theory (see text) for two different sets of input parameters ω_j, f_j. Single I: Same without shell splitting. Bonderup: Calculated from Lindhard-Scharff dielectric theory (Bonderup, 1967). Expt: Extracted from measurements by Andersen et al. (1977). From Sabin and Oddershede (1982)

7.3.2 Kinetic Theory

Kinetic theory (Sigmund, 1982) is primarily a procedure to determine shell corrections. A high-speed expansion was discussed in Sect. 6.6.4, but a more rigorous description may be found from the transformation (6.87) or, if necessary, its relativistic extension (6.88).

Equation (6.87) has been explored in connection with the asymptotic Bethe formula by Sabin and Oddershede (1982), Oddershede and Sabin (1984) and in later work by the same group. Here, a simple Bethe-type stopping cross section $S_{j0}(v)$ was assigned to every target subshell with the stopping number

$$
L_{j0}(v) = \begin{cases} \ln\left(\dfrac{2mv^2}{\hbar\omega_j}\right) & \text{for } 2mv^2 > \hbar\omega_j, \\[2ex] 0 & \text{for } 2mv^2 < \hbar\omega_j. \end{cases}
\tag{7.41}
$$

For an isotropic velocity distribution of the target atoms, the average over $\boldsymbol{v}_e$ reduces to a double integral which was evaluated numerically, using Hartree-Fock velocity spectra according to Herman and Skillman (1963). An example is shown in Fig. 7.7.

More recently, (6.87) has been adopted by Sigmund and Schinner (2002a) as the standard procedure to determine shell corrections in binary stopping theory introduced in Sect. 6.4.5. Results will be presented below.

7.3.3 Harmonic-Oscillator Model

In the harmonic-oscillator model (Sigmund and Haagerup, 1986), the mean energy loss to an atom versus impact parameter is represented as

$$
T(p, v) = \sum_j Z_2 f_j T_{\text{osc}}\left(p, v, \omega_j\right),
\tag{7.42}
$$

and the atomic stopping cross section as

$$
S(v) = \sum_j Z_2 f_j S_{\text{osc}}\left(v, \omega_j\right),
\tag{7.43}
$$

where $T(v, \omega)$ and $S(v, \omega)$ refer to a spherical quantal harmonic oscillator with a resonance frequency ω.

Equation (7.42) is asymptotically exact at large impact parameters as follows from the Bloch theory, cf. (4.49). Also (7.43) has a considerable amount of rigor. Indeed, according to (6.105), this relation represents the correct Bethe logarithm as well as the first shell correction, as follows from (6.107).

The stopping cross section for atomic hydrogen predicted by (7.42) has been compared with alternative predictions by Sigmund and Haagerup (1986)

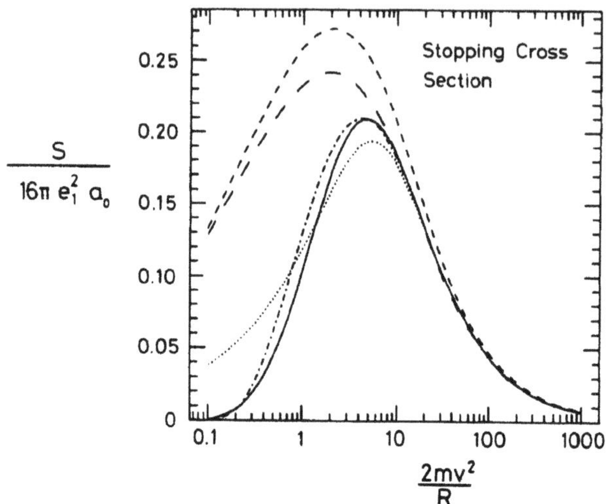

Fig. 7.8. Stopping cross section of atomic hydrogen calculated from harmonic-oscillator model, compared to alternative approaches. All calculations within the first Born approximation. Solid line: Direct integration; short-dashed line: Lindhard-Scharff model, uncorrected ($\chi = 1$); long-dashed line: Lindhard-Scharff model, corrected ($\chi = \sqrt{2}$); dotted line: kinetic theory; dash-dotted line: harmonic-oscillator model. $R = 13.6$ eV. From Mikkelsen et al. (1992)

and Mikkelsen et al. (1992). Results are shown in Fig. 7.8. It is seen that within the first Born approximation, rather good agreement is obtained over the velocity range covered between a rigorous numerical evaluation and the oscillator model. All approaches represent the high-velocity regime very well, while both dielectric and kinetic theory produce discrepancies around and below the stopping maximum.

Figure 7.9 shows similar results for straggling. Serious discrepancies are found mostly for predictions of the kinetic theory at rather low beam energies, ($\lesssim 10$ keV/u). Similar conclusions emerge from higher moments up to fifth order.

Note that higher moments are determined mostly by close collisions. Therefore, we may conclude that the harmonic-oscillator model, even though derived and justified initially for distant collisions, incorporates essential features also for close interactions. This appears reasonable in view of the fact that binding forces have little significance in close interactions, as argued already by Bohr (1913).

On the background of these findings, the oscillator model has been applied also to situations where theoretical justification is less comprehensive, first of all target atoms containing more than one electron, and, moreover, calculations going beyond the first Born approximation.

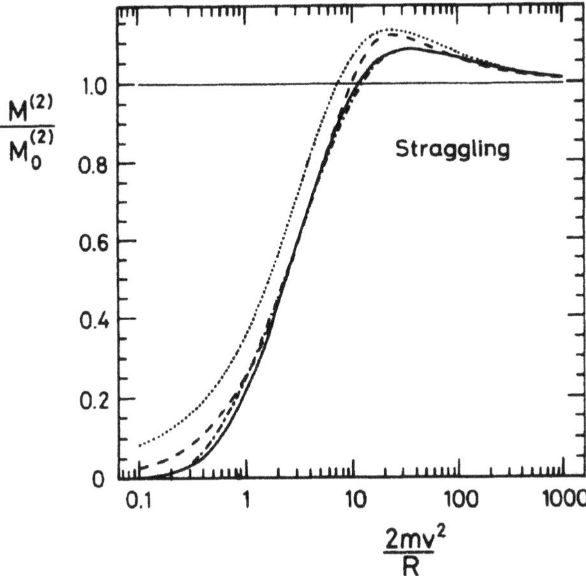

Fig. 7.9. Same as Fig. 7.8 for relative straggling parameter. From Mikkelsen et al. (1992)

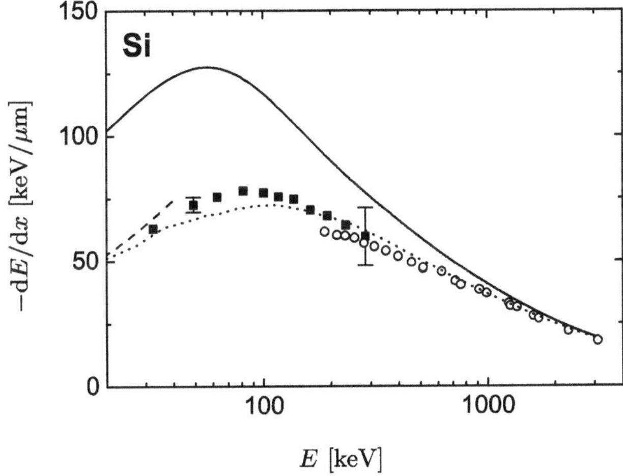

Fig. 7.10. Measured stopping force on antiprotons in Si (points) and comparison to calculations by dielectric theory (Sørensen, 1990) (dashed line) and harmonic-oscillator model (Mikkelsen and Sigmund, 1989). The solid line represents proton stopping according to ICRU (1993). Extracted from Møller et al. (1997)

Figure 7.10 shows stopping forces on antiprotons in Si measured by Møller et al. (1997) compared with calculations from dielectric theory by Sørensen (1990) and the oscillator model of Mikkelsen and Sigmund (1989). Good agreement is found over a wide range of beam energies.

Finally, the harmonic-oscillator model, as expressed in particular by (7.42), is a very useful tool for calculating impact-parameter-dependent energy losses. Insertion of results like those depicted in Fig. 4.4 yields asymptotically exact results at large impact parameters. Practical results were reported by Mortensen et al. (1991). Addition of third-order contributions like those shown in Fig. 6.7 extends the range of validity toward lower impact parameters.

7.3.4 Binary-Collision Models

We have seen in Chapter 4 that close collisions obey the laws of Coulomb binary scattering, in particular in the classical regime where Bohr's kappa parameter $\kappa = 2Z_1 v_0/v > 1$. It is tempting, therefore, to employ this feature in stopping theory. This, however, requires careful consideration of distant collisions which, at not loo low projectile speeds, contribute substantially to the stopping force.

Following the example of Darwin (1912), it has been assumed frequently that energy transfers lower than the ionization energy – or, after the development of atomic theory, the first excitation potential – are forbidden and hence do not contribute to stopping. This leads to a stopping cross section per target electron of the form

$$S = \frac{2\pi Z_1^2 e^4}{mv^2} \ln \frac{2mv^2}{U}, \tag{7.44}$$

where U is the pertinent binding energy. Here, the Coulomb factor in the front has half the magnitude of the one appearing in Bohr's and Bethe's stopping formulae.

You may find Darwin's argument appealing and wonder what could be wrong. Well, first of all we have seen that for distant collisions, i.e., soft interactions, classical and quantum mechanics yield results that agree asymptotically at large impact parameters. Clearly, if you go to large enough impact parameters, $T(p)$ will be less than the lowest excitation level. The point is that we talk about a *mean energy transfer*, which is governed by excitation probabilities which may become much smaller than unity.

Thus, energy transfers below the ionization energy do contribute to stopping: In classical theory, an electron can evidently start an oscillatory motion without being kicked off. In quantum theory, a mean energy transfer below the lowest excitation level just signals an excitation probability less than unity.

With this in mind, the range of validity of conventional binary-collision models for stopping (Gryzinski, 1965, Harberger et al., 1974b,a, Kührt and

Wedell, 1983, Kührt et al., 1985) must be limited to the velocity range around and below the stopping maximum, where the adiabatic radius is of the order of atomic dimensions.

This limitation has been overcome in the binary stopping theory of Sigmund and Schinner (2000), where electron binding has been taken into account via screening of the Coulomb interaction, as described in Sect. 6.4.5. As mentioned already, this scheme incorporates a Barkas-Andersen correction, and shell corrections are allowed for via kinetic theory. Although the theory is intrinsically classical, its range of validity can be extended into the Born regime via the inverse-Bloch correction (6.22) discussed in Sect. 6.3.3.

Other options concern static screening of dressed projectiles, i.e., projectiles carrying electrons as well as excitation and ionization of the projectile. Such projectile processes will be discussed in detail in Volume II.

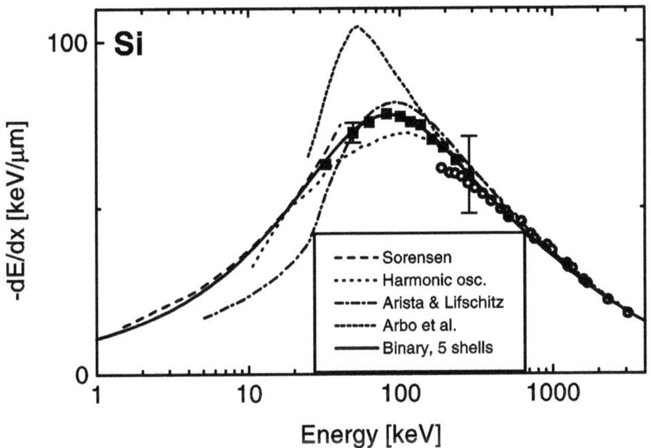

Fig. 7.11. Stopping of antiprotons in Si. Measurements from Møller et al. (1997). Calculations by Sørensen (1990), Arista and Lifschitz (1999), Arbó et al. (2000), Sigmund and Schinner (2001). From Sigmund and Schinner (2001)

Figure 7.11 shows stopping forces on antiprotons in Si compared with results from several calculational schemes. Agreement with the binary theory is nearly perfect. Antiproton stopping is a most appropriate standard of reference, since unlike protons, antiprotons do not carry electrons during passage through matter at any speed.

Binary stopping theory has been developed primarily for applications involving ions heavier than helium. However, allowing routinely for the inverse-Bloch correction, it is a powerful scheme to treat stopping of light ions at projectile speeds around and above the stopping maximum, as discussed by Sigmund and Schinner (2002b).

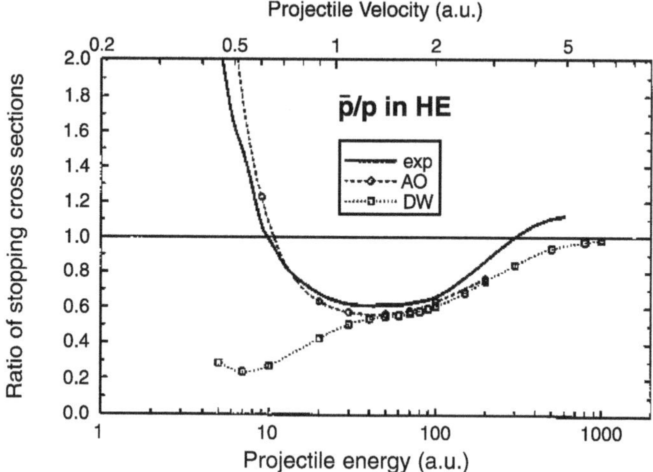

Fig. 7.12. Ratio of antiproton/proton stopping cross section in helium: Experimental data (solid line) extracted from measurements on antiprotons from Agnello et al. (1995) and protons from Golser and Semrad (1991). Theoretical ratios extracted from calculations for antiprotons from Schiwietz et al. (1996) and for protons from Grande and Schiwietz (1993) (AO) and Olivera et al. (1994) (DW). From Schiwietz et al. (1996)

7.3.5 Numerical Simulations

Unlike in other areas of physics research, numerical simulation is a comparatively slowly developing tool in the theory of charged-particle stopping. In contrast to atomic-collision physics, where a specific scattering geometry tends to limit the parameter space for which calculations are required, determining stopping forces tends to be computationally intensive. Considering the amount of brain power invested in genuine theory, it is not a trivial task to make simulation competitive.

Simulations have been carried out by numerically solving Newton's second law or the Schrödinger equation.

In view of the Bohr criterion, classical-trajectory simulations (Olson, 1989, 1996, Grüner et al., 2004) have been geared toward heavy ions. The crucial task is establishing an atomic model allowing for meaningful calculations on the basis of classical mechanics (Olson, 1996). A given scattering event is initiated by assigning initial positions and velocities to all target electrons in accordance with tabulated orbital densities and velocities. To these stochastic parameters adds the starting time of the projectile, whereas the impact parameter can be varied systematically.

Numerical solution of collision problems on the basis of the Schrödinger equation has been possible for a long time. An extensive effort directed at stopping problems is due to Schiwietz (1990), Grande and Schiwietz (1991) and

Schiwietz et al. (1996). More recently, the electron-nuclear-dynamics scheme developed by Deumens et al. (1994) has been applied to a number of stopping problems (Cabrera-Trujillo et al., 2004).

Figure 7.12 shows experimental and theoretical data on the Barkas-Andersen effect in helium gas. The results are seen to be sensitive to the method employed for calculating proton stopping cross sections. As documented by Schiwietz (1990), Grande and Schiwietz (1991), this quantity hinges on a proper treatment of electron capture at moderately low projectile speed. You may note that the stopping force on antiprotons may exceed that on protons at projectile speeds less $\sim 0.5v_0$ and below. This reflects the fact that in contrast to the antiproton, the proton is essentially neutral in this velocity range.

As a rule of thumb, the strength of numerical simulation in this area lies in the low-speed regime, where only a limited number of excitation levels contributes to the stopping cross section. Therefore, more attention will be given to this topic in connection with low-velocity stopping in Volume II.

7.4 Remarks on Stopping Measurements

It is a well-established tradition for theoreticians in the field of particle penetration to keep informed about experimental developments and to communicate with experimentalists. This, however, does not imply an ability for a theoretician to give a comprehensive account of experimental techniques and their strengths and limitations. In fact, excellent accounts are available which are warmly recommended to the interested reader (Andersen, 1991, ICRU, 1993, Geissel et al., 2002, ICRU, 2005). The present account will be very brief and focus on some problems and pitfalls which need to be kept in mind in the analysis of stopping measurements.

7.4.1 Energy-Loss Spectra in Transmission

Most common are measurements on thin foils, in which an incoming (almost) monoenergetic beam loses a small fraction of its energy. The energy distribution of the emerging beam is then measured by an energy-sensitive detector, or by some other detecting device combined with electrostatic or magnetic deflection, or by time of flight.

As far as energy-sensitive detectors are employed, calibration problems have frequently given rise to serious discussions about the analysis of such measurements.

Foil thickness is a crucial aspect of this technique. First of all, it needs to be known with a resonable precision, and it should not vary too much across the beam area. Moreover, for small beam energies it may be difficult to produce thin enough foils to prevent excessive energy loss.

Foil thicknesses can be measured by a variety of techniques. It is an advantage in the present context to use accelerator-based techniques, such as energy loss by a beam of a species and energy for which reliable stopping data exist, or by Rutherford scattering of a transmitted beam. In this way, it is the product Nx, i.e., the mass per area that is measured. This makes it possible to determine the main *atomic* parameter, the stopping cross section, irrespective of possible uncertainties about the density of the target.

The quantity to be compared with theory described so far is the mean energy loss, i.e., the mean value over the energy-loss spectrum. Experimentally it may in fact be easier to determine the peak of the energy-loss spectrum. The two quantities are identical for a gaussian spectrum, but this is not the case when the spectrum becomes skew. This aspect will be discussed in Chapter 9.

7.4.2 Other Measurements on Thin Foils

Instead of measuring the energy lost by the projectile one may determine the energy deposited in the target. This is the principle of the calorimetric method pioneered by Andersen et al. (1966). In the original setup, the target was kept at liquid-helium temperature in order to maximize sensitivity of the measurement of temperature increase, but alternative solutions have been developed in particular in connection with very-high-energy beams (Geissel et al., 2002).

This technique avoids ambiguities related to the difference between peak and mean energy loss. However, not all energy lost by the projectile goes into heating the target, so that corrections for nuclear and chemical reactions, electron, photon and atom or molecule emission as well as structural changes by radiation damage need to be estimated.

7.4.3 Reflection Geometry

It may be difficult or impossible to provide self-supporting films of adequate quality at sufficiently low thickness to make meaningful stopping measurements, in particular at low beam energy. In that case, films of a light target material evaporated on a heavy substrate may be the solution. The edge of the Rutherford spectrum (cf. Fig. 1.7) of projectiles reflected at a given, well-defined scattering angle then defines the minimum energy loss. Careful data analysis is still required, and homogeneity of the evaporated film is crucial. Nevertheless, numerous measurements especially with 1–2 MeV alpha particles have been performed with this geometry.

Several alternatives are available such as replacing the substrate by an implanted layer of heavy material. Even the spectrum of ions reflected from a homogeneous sample contains information about the stopping force of the material.

7.4.4 Doppler-Shift Attenuation

Consider an atom recoiling with a definite energy from some nuclear reaction. That atom may undergo gamma decay at a certain time, following an exponential decay law governed by the lifetime of the nucleus. If the recoil is still in motion during the decay, the emitted photon will be Doppler-shifted when observed in the laboratory frame of reference. If the slowing-down behavior of the nucleus is known, one may determine the lifetime from the observed gamma spectrum. If the lifetime is known, one may determine the slowing-down behavior in the stopping medium.

Unlike other techniques, this method determines the *energy loss per unit time*. Hence, the target density needs to be known for a reliable determination of the stopping cross section

7.4.5 Pitfalls

Apart from a possible difference between mean and peak energy loss, to be discussed in Chapter 9, a crucial point in the analysis of a stopping measurement is the role of elastic scattering and nuclear stopping. It was shown in Chapter 2 that in the Bethe regime, nuclear stopping makes up less than 0.1 % of the energy loss. This is normally without interest in the determination of stopping forces. However, at lower energies, when shell corrections close several excitation channels, the effect may become significant and needs to be considered. Moreover, elastic scattering may give rise to significant angular deflection and hence to increased pathlength through at a given foil thickness.

7.4.6 Range Measurements

Initially, stopping forces were extracted from range measurements. Clearly, if

$$R(E) = \int_0^E \frac{\mathrm{d}E'}{NS(E')} \tag{7.45}$$

is known for several values of the initial beam energy E, one may extract the stopping force $NS(E)$ for a certain energy interval. This technique is only in use for very special situations now.

7.5 Extraction of Input Parameters from Stopping Measurements

Extracting input parameters from experimental data requires high-precision measurements with an accuracy of ~ 1 % or better. Typically, experimental data taken at a particular setup will cover no more than one order of magnitude in beam energy. It is advisable to analyze the stopping number, i.e., to divide measured stopping cross sections by the Coulomb factor $4\pi Z_1^2 Z_2 e^4/mv^2$. This leaves a slowly varying function $L(v)$ of the form

$$L(v) \simeq \ln \frac{2mv^2}{I} + \text{relativistic correction} + \text{Fermi density correction}$$
$$+ \text{Bloch correction} + \text{shell correction}$$
$$+ \text{Barkas-Andersen correction.} \quad (7.46)$$

Remind that the projectile is assumed to be a *point charge* throughout this book, i.e., the Bohr screening criterion (6.3) has to be fulfilled. This implies measurements in the upper keV/u and MeV/u energy range, i.e., at cyclotrons and van de Graaf accelerators, where the second and third term represent small and well-known relativistic corrections, while the remaining three get more and more significant the lower the beam energy.

7.5.1 *I*-Values and Shell Correction

I-values may be found in principle by performing measurements at high proton velocities, applying relativistic corrections and determining the slope of the Bethe logarithm in a semi-logarithmic plot. In practice, shell corrections are negligible only for the lightest materials, although they may be well approximated by the asymptotic expression $-\langle v_e^2 \rangle/v^2$. *I*-values determined in this manner long ago by Andersen et al. (1969) are still competitive and very close to present-day standard (ICRU, 1993).

On the other hand, for heavy target atoms, the stopping number – corrected for relativistic effects – cannot be expected ever to reach the asymptotic Bethe logarithm. Indeed, for the K shell of a heavy atom, the quantity $Z^2 v_0^2/v^2$ will not be $\ll 1$ at any speed.

One may, in fact, argue that the total *I*-value is of little interest for heavy elements: If the stopping number is split into contributions from different target shells, those from inner shells cannot be characterized by asymptotic expansions. Hence, the Bethe-type logarithm containing the *I*-value of the respective shell does not occur, and with this, the total *I*-value defined by (4.119) does not enter explicitly into the stopping cross section.

Note that the *absolute value* of the shell correction needs to be known in order that I can be determined: Variation of I just introduces an additive constant into the stopping number. It is impossible to empirically determine *both I*-values *and* shell correction from the same set of experimental data.

7.5.2 Barkas-Andersen and Bloch Correction

According to Thomas-Fermi scaling laws, the shell correction becomes significant when the projectile speed is not large compared to $Z_2^{2/3} v_0$. The

Barkas-Andersen correction becomes significant when the parameter $B = Z_1 e^2 \omega / m v^3$ is not $\ll 1$, i.e., when the projectile speed is not large compared to $(Z_1 Z_2)^{1/3} v_0$. The Bloch correction becomes significant when the parameter $\kappa = 2 Z_1 v_0 / v$ approaches 1. Finally, a projectile may carry electrons if its speed is not large compared to the orbital speed $Z_1^{2/3} v_0$. Thus, for protons, the above sequence happens to be the order in which various corrections to the asymptotic Bethe logarithm become significant for the stopping number as we lower the projectile speed.

Andersen et al. (1977) carried out an analysis of stopping data for H, He and Li ions on aluminium and gold, allowing for shell, Barkas-Andersen and Bloch corrections but disregarding projectile screening. Adopting shell corrections from Bonderup (1967), Barkas-Andersen corrections were extracted under different assumptions regarding the Bloch correction. A main goal of that study was an experimental test of the prediction of Lindhard (1976) that the Barkas-Andersen correction was about twice as large as predicted by Ashley et al. (1972) due to the contribution from close collisions. A definite answer was not reached because of doubts on the accuracy of the Bloch correction.

In an official report on stopping of protons and alpha particles (ICRU, 1993), I-values were taken over from earlier work, and adopting theoretical shell corrections as well as the Bloch correction, the authors determined a constant in the Barkas-Andersen correction by fitting to experimental data. This is not fully consistent: At least part of the I-values entering the description were determined by analysis of experimental data either neglecting the Barkas-Andersen correction or adopting a definite value.

In a long series of papers (e.g., Porter and Jeppesen (1983)), stopping data mainly for alpha particles were fitted to what is called the modified Bethe-Bloch theory. This scheme suffers from inconsistencies:

1. The Barkas-Andersen correction of Ashley et al. (1972) is adopted, deliberately ignoring the generally accepted fact that it is too small by about a factor of two, cf. especially Porter (2004),
2. Projectile screening is incorporated via an 'effective charge', i.e., the projectile charge $Z_1 e$ is replaced by a velocity-dependent quantity $q_1(v)e$ in order to account for the presence of electrons on the projectile. However, the Coulomb factor preceding the stopping number has been derived for unscreened Coulomb interaction, and the same is true for the quantity κ entering the Bloch term.

Since screening by projectile electrons is indeed a problem in the extraction of input parameters – albeit more so for alpha particles than for protons – the best way to escape the problem is to use an antiproton beam. With reliable Barkas-Andersen and shell corrections available, and reassured of the validity of the Bloch correction by the work of Lindhard and Sørensen (1996), a new attack on determining I-values should promise considerable success.

7.5.3 Z_2 Structure

According to the Thomas-Fermi estimate of the frequency mentioned in Sect. 7.2.2, the mean excitation energy should vary as $I \simeq Z_2 I_0$ with some universal constant I_0. Figure 7.3 demonstrates that this behavior, predicted by Bloch (1933), is well confirmed as a first approximation. However, superimposed on this monotonic increase with Z_2 is an oscillatory behavior, 'Z_2 structure', which was found long ago by Burkig and MacKenzie (1957) and studied in considerable detail by Andersen et al. (1969) and others (ICRU, 1984, 1993).

Figure 7.4 shows the resulting structure in the stopping cross section versus atomic number at constant velocity. As one would expect from (4.118), the amplitude of observable oscillations in the stopping cross section is seen to increase with decreasing velocity. As a matter of fact,

- The effect of a variation of I with Z_2 becomes the more pronounced the smaller the numerator $2mv^2$ in the Bethe logarithm,
- Variations must be most pronounced in outer target shells. Since inner-shell excitation channels close one by one with decreasing projectile speed, only those shells contribute at low velocity which produce the most pronounced oscillations.
- Shell corrections tend to enhance Z_2-structure caused by the variation of ω_j with Z_2: A low value of ω_j is accompanied by a low orbital velocity and hence by a low shell (negative) correction, and vice versa (Oddershede et al., 1983).

7.6 Input Parameters from Other Sources

7.6.1 Theory

Dipole oscillator-strength spectra for a large number of elements have been estimated theoretically on the basis of Hartee-Slater orbitals by (Dehmer et al., 1975), Inokuti et al. (1978) and Inokuti et al. (1981). Results are reported in the form of moments over the distribution of the form

$$\int \mathrm{d}\omega\, \omega^\mu f(\omega) \quad \text{or} \quad \int \mathrm{d}\omega\, \omega^\mu f(\omega) \ln \omega; \quad \mu = -2, -1 \ldots \tag{7.47}$$

in suitable atomic units. I-values obtained by this procedure are shown in Fig. 7.13.

On the basis of the full spectra – which are not publicly available – Oddershede and Sabin (1984) extracted I-values for individual subshells and associated oscillator strengths by bunching according to (7.40a) for $1 \leq Z_2 \leq 36$. These tables are unreliable, in particular with regard to inner shells where

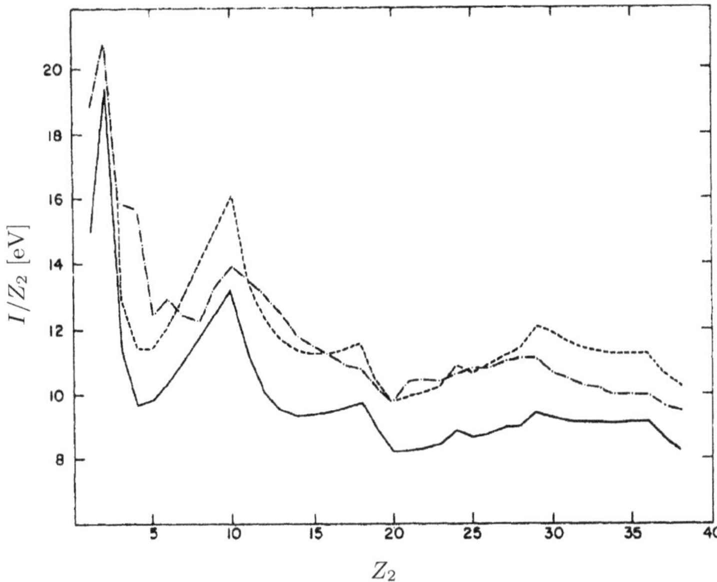

Fig. 7.13. I-values for elemental targets versus atomic number. Solid line: Calculated from Hartree orbitals (Inokuti et al., 1981). Dashed line: Dielectric theory involving Hartree-Fock densities (Chu and Powers, 1972a). Dot-dashed line: Empirical values (Andersen and Ziegler, 1977). From Inokuti et al. (1981)

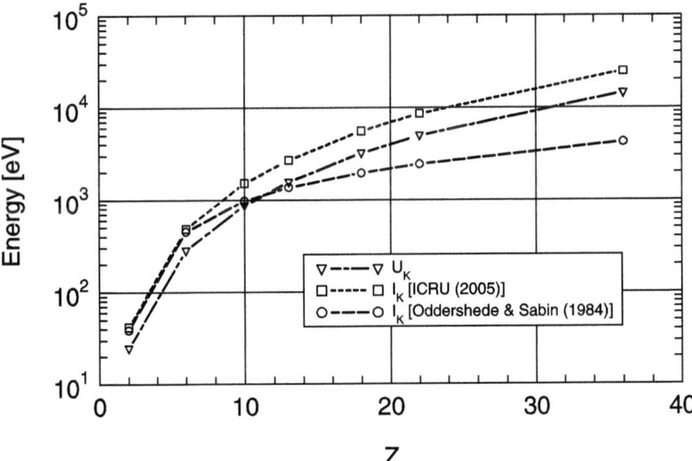

Fig. 7.14. I-value and ionization energy U for K-shell of selected elements according to Oddershede and Sabin (1984) and ICRU (2005)

tabulated resonance frequencies tend to be too low. Figure 7.14 shows a comparison of I-values for the K shell, given by Oddershede and Sabin (1984) and in a recent ICRU report (ICRU, 2005). Increasing discrepancies, going to more than a factor of two, are seen with increasing atomic number. For argon and heavier elements, the I_K as given by Oddershede and Sabin (1984) happens to lie even below the ionization energy U_K.

7.6.2 Optical and X-Ray Data

A large amount of information is available on oscillator-strength spectra extracted from empirical data such as photoabsorption cross sections, optical dispersion and x-ray absorption.

The spectrum of dipole oscillator strengths $f(\omega)$ is related to the long-wavelength dielectric function $\varepsilon(\omega)$ through

$$f(\omega) = -\frac{m}{2\pi^2 n_e e^2}\, \omega \, \mathrm{Im}\, \frac{1}{\varepsilon(\omega)} \tag{7.48}$$

where Im denotes the imaginary part. and $n_e = nZ_2$ the number of electrons per volume. Since $\varepsilon(\omega)$ can be expressed by the complex refractive index $\mathrm{n}(\omega) + \mathrm{i}\mathrm{k}(\omega)$, the oscillator strength spectrum may also be written in the form

$$f(\hbar\omega) = 1.5331 \times 10^{-3}\, \frac{A_2}{Z_2\rho}\, \frac{\hbar\omega\,\mathrm{nk}}{\left(\mathrm{n}^2 + \mathrm{k}^2\right)^2} \tag{7.49}$$

$\hbar\omega$ in eV; $f(\hbar\omega)$ in eV^{-1}.

Optical constants for numerous solids including covalent and ionic compounds have been tabulated over a wide frequency range (Palik, 1985, 1991, 1996, 2000). Equivalent information may be extracted from a compilation of x-ray scattering and absorption data (Henke et al., 1993) and from photoabsorption and photoelectron spectroscopy (Berkowitz, 1979, 2002).

Table 7.1 shows input data for 25 elements determined from oscillator-strength spectra by bundling according to (7.40a) and (7.40b). Notation is explained in the caption[2].

7.7 Compound Materials and Bragg Additivity

Before discussing stopping in chemical compounds, consider first a mixture of noble gas atoms at low pressure. Then, from the fundamental statistical considerations discussed in Sect. 2.2.3 you will realize that (2.19) is generalized

---

[2] A special remark is indicated regarding I-values: I-values listed in table 7.1 are those listed in ICRU (1993) and are not consistent with values listed for different shells. The difference is seen in table 7.2. The reason for this seeming inconsistency is that I-values from ICRU (1993) are a widely accepted standard and are used in the PASS code as a unit to compute the Bohr parameter $\xi = mv^3/Z_1 e^2\omega$. Actual calculations hinge on I-values for individual shells, while the total I-value drops out again when results are plotted as a function of velocity instead of ξ.

Table 7.1. The first row in a block shows atomic number Z, atomic weight A, I-value [eV] according to ICRU Report 49 and the mass density [g/cm^3]. Subsequent rows in a block show principal quantum number n, azimuthal quantum number ℓ, subshell occupation Zf_j, subshell I-value $\hbar\omega_j$ and subshell binding energy U_j. From ICRU (2005)

Z / n	ℓ	A / Zf_j	I / $\hbar\omega_j$	density / U_j
1		1.00794	19.2	0.00008988
1	0	1.000	19.2	15.42
2		4.0026	41.8	0.0001785
1	0	2.000	41.8	24.588
4		9.0122	63.7	1.848
1	0	1.930	209.11	114.3
0	0	2.070	21.68	9.32
6		12.0111	81.0	1.90
1	0	1.992	486.2	288.2
2	0	1.841	60.95	16.59
2	1	2.167	23.43	11.26
7		14.0067	82.0	.001250
1	0	1.741	732.61	403.8
2	0	1.680	100.646	20.33
2	1	3.579	23.550	14.534
8		15.9994	95.0	.001429
1	0	1.802	965.1	538.2
2	0	1.849	129.85	28.7
2	1	4.349	31.60	13.618
10		20.180	137.0	0.0008999
1	0	1.788	1525.9	869.5
2	0	2.028	234.9	47.7
2	1	6.184	56.18	21.564
13		26.9815	166.	2.699
1	0	1.623	2701.	1564.1
2	0	2.147	476.5	121.5
2	1	6.259	150.42	76.75
0	1	2.971	16.89	9.08
14		28.086	173.	2.329
1	0	1.631	3206.1	1844.1
2	0	2.094	586.4	154.04
2	1	6.588	186.8	103.71
3	0	2.041	23.52	13.46
3	1	1.646	14.91	8.1517
18		39.948	188.	.0017837
1	0	1.535	5551.6	3206.2
2	1	8.655	472.43	266.85
3	0	1.706	124.85	29.24
3	1	6.104	22.332	15.937
22		47.88	233.	4.508
1	0	1.581	8554.6	4969.9
2	1	8.358	850.58	487.5
3	1	8.183	93.47	44.37
3	2	2.000	39.19	8.1
0	0	1.878	19.46	6.8282
26		55.847	286.	7.873
1	0	1.516	12254.7	7112.
2	1	8.325	1279.29	750.8
3	1	8.461	200.35	68.85
3	2	6.579	49.19	9.34
0	0	1.119	17.66	7.90
28		58.7	311.	8.907
1	0	1.422	14346.9	8337.8
2	1	7.81	1532.28	903.01
3	1	8.385	262.71	84.88
3	2	8.216	74.37	10.213
0	0	2.167	23.03	7.6398
29		63.546	322.	8.933
1	0	1.458	15438.5	8983.9
2	1	8.049	1667.96	984.3
3	1	8.79	294.1	92.0
3	2	9.695	70.69	10.62
0	0	1.008	16.447	7.7264
32		72.59	350.	5.323
1	0	1.442	19022.1	11105.
2	1	7.791	2150.79	1276.7
3	1	7.837	455.79	140.37
3	2	10.122	179.87	31.82
4	0	2.463	57.89	14.3
4	1	2.345	20.95	7.9
36		83.80	352.	.003743
1	0	1.645	24643.	14328.
2	1	7.765	2906.4	1753.2
3	1	19.192	366.85	157.25
4	1	7.398	22.24	17.38

Z / n	ℓ	A / Zf_j	I / $\hbar\omega_j$	density / U_j
42		95.94	424.	10.22
1	0	1.313	34394.	20003.
2	1	6.409	4365.3	2639.0
3	1	19.229	589.36	321.46
4	1	8.633	129.42	47.98
4	2	5.036	35.59	8.56
0	0	1.380	18.42	7.09
47		107.868	470.	10.500
1	0	1.295	43664.3	25518.
2	1	6.219	5824.91	3513.9
3	1	18.751	909.79	484.5
4	1	8.748	175.47	73.71
4	2	10.184	54.89	10.4
0	0	1.803	19.63	7.576
50		118.69	488.	7.285
1	0	1.277	49948.	29203.
2	1	6.099	6818.2	4123.9
3	1	20.386	1036.1	616.19
4	1	8.011	172.65	104.44
4	2	10.007	70.89	28.12
5	0	2.272	33.87	12.
0	1	1.948	14.54	7.34
54		131.3	482.	.005895
1	0	1.563	58987.	34563.
2	1	6.312	8159.	5032.2
3	1	21.868	1296.6	831.38
4	1	5.762	356.75	168.77
4	2	11.245	101.03	68.15
5	1	7.250	16.52	15.27
74		183.85	727.	19.254
1	0	1.202	115025.9	69525.
2	1	5.582	17827.44	11181.625
3	2	19.527	3214.36	2125.98
4	2	18.741	750.41	354.33
5	1	8.411	305.21	49.6875
4	3	14.387	105.50	32.5
5	2	4.042	38.09	9.0
6	0	2.108	21.25	7.89
78		195.09	790.	21.45
1	0	1.159	128342.	78395.
2	1	5.467	20254.	12570.
3	1	18.802	3601.8	2488.
4	2	33.905	608.1	280.9
5	1	8.300	115.0	69.0
5	2	9.342	42.75	9.6
0	0	1.025	17.04	9.
79		196.9665	790.	19.291
1	0	1.124	131872.	80725.
2	1	5.331	20903.	12981.
3	1	18.078	3757.4	2585.7
4	2	34.604	682.1	300.8
5	1	8.127	105.2	74.30
5	2	10.414	44.89	11.66
0	0	1.322	17.575	9.226
82		207.2	823.	11.343
1	0	2.000	154449.	88005.
2	1	8.000	25067.	14283.
3	2	18.000	5105.0	2908.78
4	2	18.000	987.44	562.49
4	3	14.000	247.59	140.59
5	1	8.000	188.1	106.8098
5	2	10.000	40.61	20.904
6	0	2.000	19.2	10.
0	1	2.000	15.17	7.4167
92		238.0289	890.	19.05
1	0	2.000	167282.	115606.
2	1	8.000	27868.	19259.5
3	2	18.000	6022.7	4162.13
4	3	32.000	1020.4	704.71
5	2	18.000	244.81	164.15
6	1	8.000	51.33	33.
7	0	2.000	13.	6.1941
6	2	1.000	11.06	6.1
0	3	3.000	14.43	6.

Table 7.2. Selected I-values calculated from oscillator-strength spectra (Palik, 1985, 1991, Henke et al., 1993) compared with recommended values from ICRU (1993). Brackets indicate estimated values

Element	ICRU (2005)	ICRU (1993)	Element	ICRU (2005)	ICRU (1993)
Li	50.0	40.0	Cl	189.8	188.
Be	64.7	63.7	Ar	189.8	188.
B	75.4	76.0	K	179.6	(190)
C	86.0	81.0	Ca	209.4	191.
N	79.5	82.0	Sc	217.7	191.
O	94.6	95.0	Ti	238.6	233.
F	114.3	115.0	V	236.6	245
Ne	135.5	137.0	Cr	242.6	257.
Na	141.6	(149)	Mn	255.2	272.
Mg	144.6	(156)	Fe	291.1	286.
Al	158.3	166.	Co	290.2	297.
Si	169.5	173.	Ni	301.3	311.
P	165.0	173.	Cu	326.3	322.
S	168.9	180.			

into

$$\langle \Delta E \rangle = \Delta x \sum N_\nu \sum_j T_j \sigma_{\nu j}, \tag{7.50}$$

where N_ν is the number of ν-atoms per volume and $\sigma_{\nu j}$ the cross section for energy transfer T_j of a ν-atom.

We may express this relation by an effective stopping cross section per atom,

$$S_{\text{eff}} = \sum_\nu \frac{N_\nu}{N} S_\nu = \sum_\nu c_\nu S_\nu, \tag{7.51}$$

where $S_\nu = \sum_j T_j \sigma_{\nu j}$ and

$$c_\nu = \frac{N_\nu}{N} \tag{7.52}$$

is the relative abundance of species ν.

Now, (7.51) is not only applied to gas mixtures but also to metallic alloys and chemical compounds, both gaseous and solid. It is then called 'Bragg's additivity rule'. This can only be an approximate relationship, since the electronic excitation spectrum of a molecule differs from that of a dilute mixture of the constituent atoms. However, deviations from this relationship are frequently quite small.

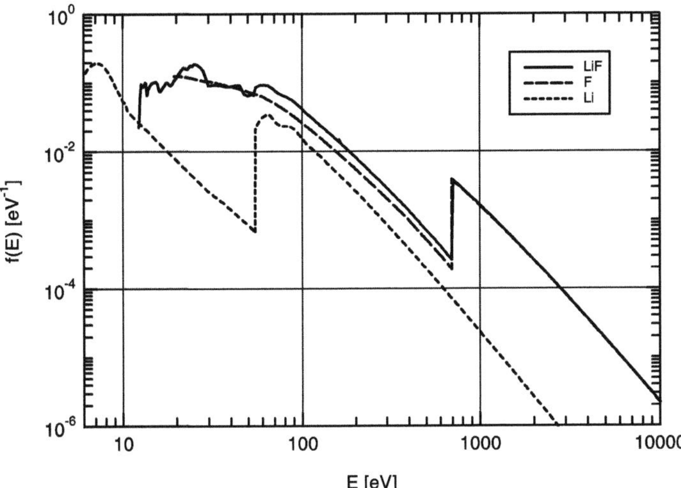

Fig. 7.15. Excitation spectrum of LiF compared with spectra for Li metal and fluorine. Data extracted from Palik (2000)

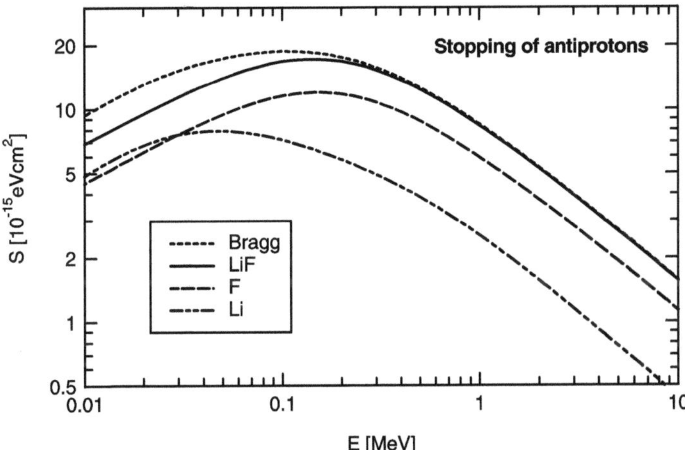

Fig. 7.16. Predicted stopping forces for LiF, metallic lithium and fluorine, including the prediction of Bragg's rule. From Sharma et al. (2004)

Let us start with a few qualitative considerations (Sigmund et al., 2005):

— At high velocity, where the stopping number is governed by the Bethe logarithm, valence structure effects enter logarithmically through changes in the total I-value. For heavy target atoms, the contribution of the valence shell to the I-value is small. Significant valence effects can mainly be expected for atoms of the first and second row of the periodic table.

— The valence contribution makes up a fixed fraction of the I-value. With increasing speed, its relative contribution to the Bethe logarithm $\ln(2mv^2/I)$

decreases. Hence, significant valence effects should be looked for in the region where $L \lesssim 1$.

- Shell corrections tend to close inner-shell excitation channels one by one with decreasing projectile speed. Clearly, valence effects must be most important in the velocity range where the stopping force is dominated by valence electrons.
- Also the shell and the Barkas-Andersen correction must be influenced by valence effects. As a general rule, increased binding implies increased electron velocity and increased I-value. Therefore, both corrections tend to increase in magnitude with decreasing velocity. However, for positive ions they act in opposite directions. Therefore, the sign of the joint effect depends on their relative significance.
- For negative particles, such as antiprotons, all three effects act in the direction of *decreasing* stopping cross section from the isolated atom to the compound[3].
- The magnitude of valence effects should be greatest for strongly bound compounds such as alkali halides.

Figure 7.15 shows oscillator-strength spectra for LiF as well as for Li and F. Pronounced deviations from Bragg additivity are seen around and below the bandgap of LiF.

Figure 7.16 shows stopping forces on antiprotons calculated for LiF by means of the binary theory (Sharma et al., 2004). Pronounced deviations from Bragg additivity are found from around 100 keV downward. However, there is hardly any ion-target combination for which greater valence effects can be expected: LiF has light constituents, the electronic structure deviates drastically from those of the constituent atoms, the velocity range covered extends far down, and for antiproton bombardment the effects of shell and Barkas-Andersen correction add up.

7.8 Data Compilations and Codes

The experimental literature on proton and alpha-particle stopping is enormous. On the basis of experimental results available until 1977, Andersen and Ziegler (1977) established a set of proton stopping data for all elements by cautious inter- and extrapolation which is still useful today. Extensive tables covering a wide velocity range and numerous elements and compounds were published by Janni (1982a,b). More recently, experimental data have been compiled by Ziegler and Paul. Ziegler's data are underlying the SRIM code (Ziegler, 2005), which is freely available on the internet, while Paul's data can be downloaded freely in the form of graphs or Origin files (Paul, 2005).

---

[3] Empirical corrections for deviations from Bragg additivity are not always negative: Positive corrections can occur when the standard of comparison is the stopping cross section of a *solid* monoatomic medium.

Certified data on proton and alpha-particle stopping were published by the International Commission on Radiation Units and Measurements (ICRU, 1993), following up on a similar tabulation for electrons (ICRU, 1984). These data were produced by the codes Pstar, Astar and Estar (for protons, alphas and electrons, respectively), which implement the Bethe theory, employing empirical input for I-values and Barkas-Andersen corrections. These codes are freely available on the internet (Berger et al., 2005).

Figures 7.17–7.20 show comparisons between measured and calculated stopping forces on protons and alpha particles for solid and gaseous materials, respectively. Also included are theoretical curves generated by the PASS code which implements binary stopping theory (Sigmund and Schinner, 2002a) and which is planned to become freely available on the internet. Only the energy range from 25 keV upward is covered, i.e., ions with $v > v_0$, the Bohr velocity, because other stopping mechanisms in addition to Coulomb excitation become significant at lower velocities, as to be discussed in Volume II.

Binary theory is essentially classical, quantum theory entering via the inverse-Bloch correction, and hence geared toward heavier ions. Nevertheless, differences from Pstar/Astar are minor and, in most cases, less than the scatter between different experimental data. On the other hand, these calculations do not involve fitting to stopping measurements, unlike SRIM and Pstar/Astar.

A useful code that is also freely available on the internet is the CasP code implementing the unitary convolution approximation by Grande and Schiwietz (1998, 2002, 2004). A code by Arista and Lifschitz (1999), Arista (2002) has likewise proved useful but is not yet publicly available.

7.9 Discussion and Outlook

The reader who needs reliable values of stopping cross sections for a given system over a given velocity range has a number of options that range between purely empirical and purely theoretical solutions. As a rule of thumb, the domain of theory is at the high-speed end, while empirical solutions tend to be increasingly useful the lower the projectile speed. At the high-velocity end, measurements are rare and conceptionally difficult, and theory is at least conceptionally simple because of well-established physical principles, even though there may be technical difficulties, as has been seen in the discussion in Sect. 6.7.2.

The regime of low-velocity stopping, for projectile speeds below the Bohr velocity v_0, has been reserved for Volume II, but it may suffice to mention at this point that the mechanism of Coulomb excitation of the target, underlying most of the theoretical treatment in this monograph, is generally believed to lose significance with decreasing projectile speed, while other processes, obeying quite different scaling relations, get dominant. But even at intermediate projectile speeds, around the stopping maximum, deviations from straight Bethe or Bohr behavior get pronounced. Even though theoretical treatments for these effects are available, a variety of material parameters enter that may or may not be well known.

Full ab initio treatments of particle stopping are possible at low projectile speeds, where only the lowest excitation levels of target and projectile are relevant. However, accurate treatments of this type are feasible mostly for the lightest ion-target combinations and, hence, serve mostly as benchmarks for comparison with more approximate schemes.

Amongst approximate theoretical schemes, the one by Lindhard and Scharff (1953) has been by far dominating for several decades and is still very popular, although it has not been utilized as a basis for current tabulations such as Andersen and Ziegler (1977), ICRU (1993), Ziegler et al. (1985) or codes (Ziegler, 2005, Grande and Schiwietz, 2004). While Andersen and Ziegler (1977) and Ziegler (2005) essentially rely on measurements as far as protons are concerned, ICRU (1993) is based on a shellwise implementation of the Bethe formula, making use of modified hydrogenic wave functions in the computation of shell corrections and a somewhat questionable Barkas-Andersen correction, as explained in Sect. 7.5.2.

Evidently, the reliability of empirical interpolations must vary substantially over the periodic table, dependent on the amount and quality of available experimental data. Extrapolation into velocity regimes where data are scarce or nonexistent evidently requires theoretical support.

Simple and theoretically well-supported approaches such as the harmonic-oscillator model have been useful for gaining qualitative insight but have only rarely been utilized for tabulation of stopping data for a wide class of materials. Binary theory, developed primarily for stopping of heavy ions, has been explored also for antiprotons and protons with considerable success. A Windows version of the PASS code implementing this theory is expected to be publicly available shortly.

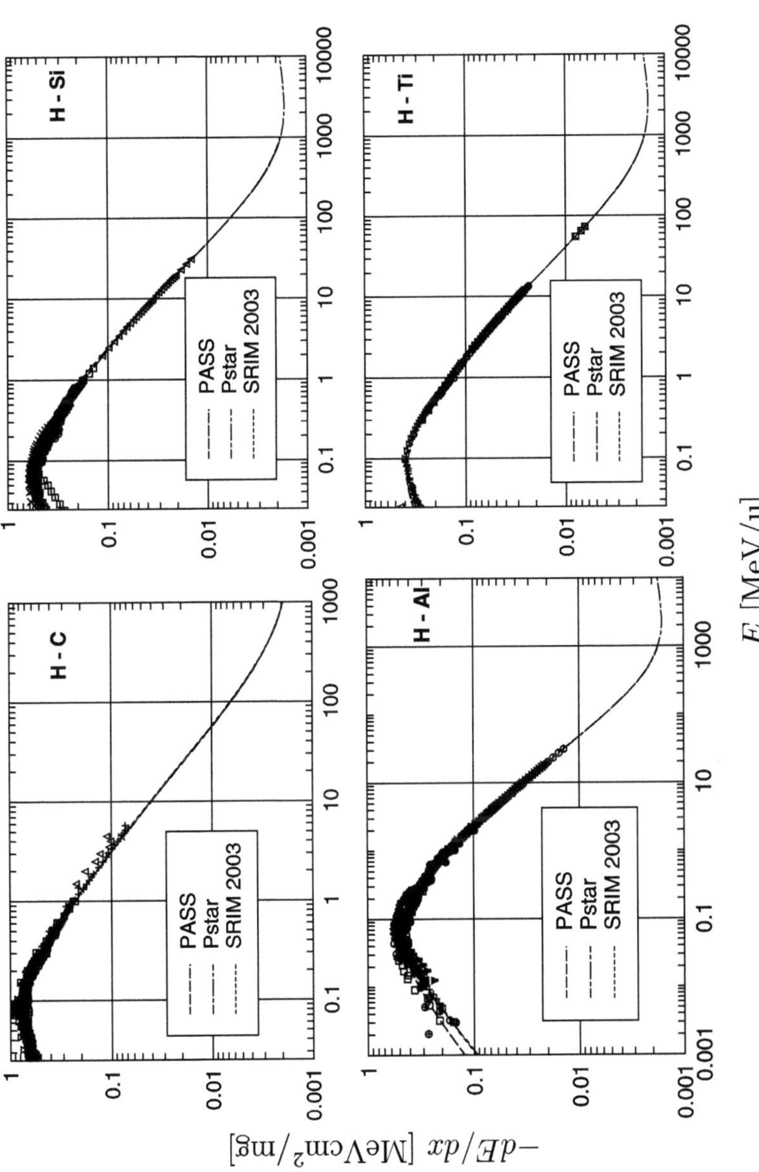

Fig. 7.17. Stopping forces on hydrogen ions in solids calculated by PASS, Pstar and SRIM 2003 compared with experimental data compiled by Paul (2005)

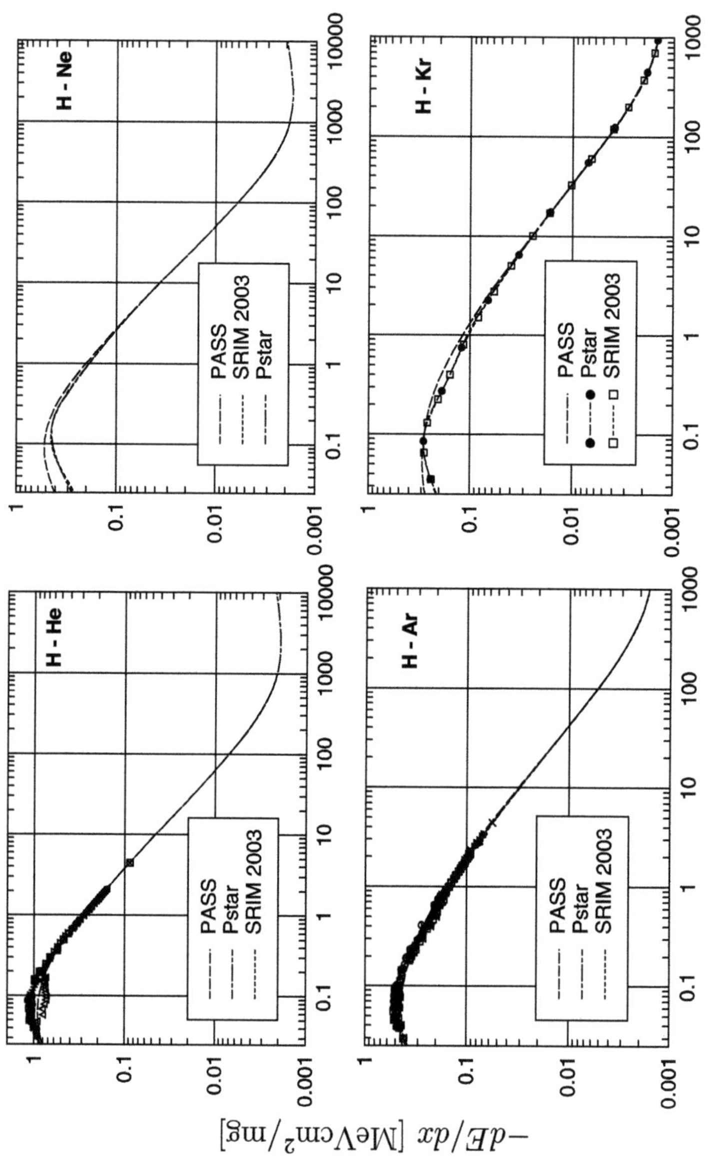

Fig. 7.18. Stopping forces on hydrogen ions in gases calculated by PASS, Pstar and SRIM 2003 compared with experimental data compiled by Paul (2005)

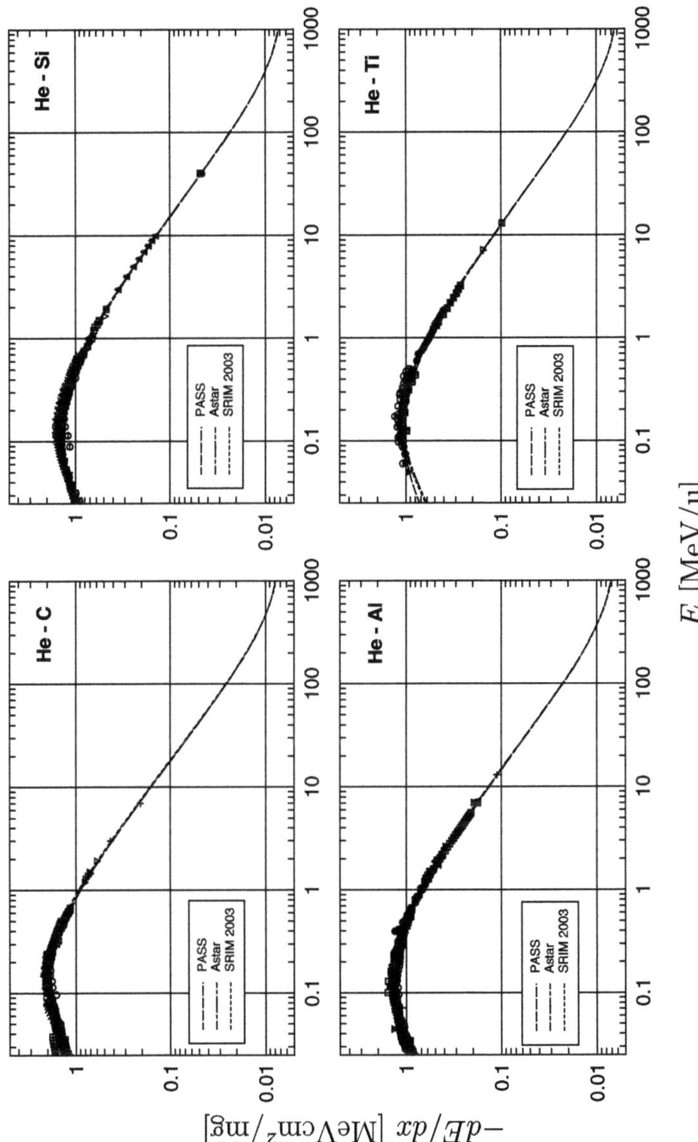

Fig. 7.19. Stopping forces on helium ions in solids calculated by PASS, Pstar and SRIM 2003 compared with experimental data compiled by Paul (2005)

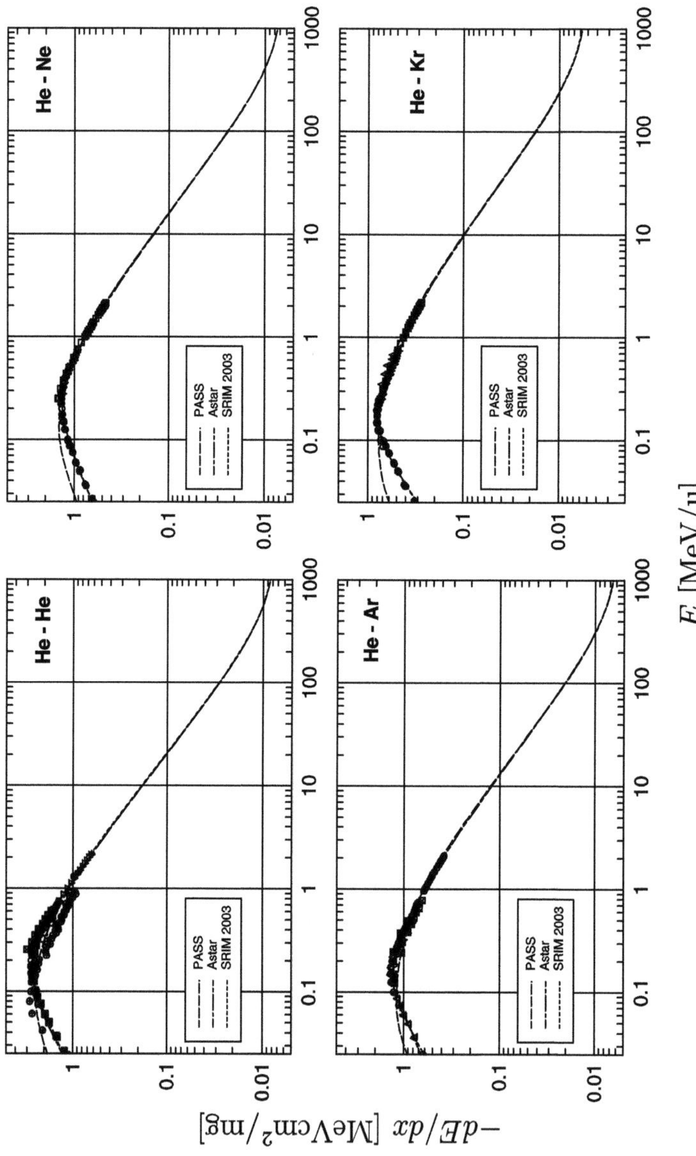

Fig. 7.20. Stopping forces on helium ions in gases calculated by PASS, Pstar and SRIM 2003 compared with experimental data compiled by Paul (2005)

Problems

7.1. Calculate an approximate I-value for atomic hydrogen from (7.27) making use of the electron density $|\psi(\boldsymbol{r})|^2$ of the ground state of the hydrogen atom.

7.2. Derive (7.32) making use of (7.17).

7.3. Why does the curve labelled BA in Fig. 7.6 not have the value $2mv^2$ at $p = 0$?

7.4. Try to identify at least one of the techniques described in this chapter for measuring stopping cross sections which does not require precise knowledge of the density N of the stopping medium (atoms per volume). Also point out at least one technique that does require such knowledge.

7.5. Determine the I-value of an aluminium atom by the procedure of Sternheimer et al. (1982) described on page 240 and compare the result with the standard value of 166 eV for metallic aluminium. [Hint: Find a tabulation of subshell ionization energies.]

7.6. Derive the expression

$$S(v) = \frac{\pi}{v^2} \int_0^\infty f(v_e) v_e \, \mathrm{d}v_e \int_{|v-v_e|}^{v+v_e} \mathrm{d}v' \, S_0(v')(v^2 - v_e^2 + v'^2) \qquad (7.53)$$

from (6.87), assuming an isotropic velocity distribution $f(v_e)$.

7.7. Derive (7.51) for a gas mixture.

7.8. Find the numerical table of stopping cross sections for a harmonic oscillator, computed in the first Born approximation by Sigmund and Haagerup (1986) and use data from Table 7.1 to compute stopping cross sections for protons in aluminium and argon by means of (7.42) for 25 keV$< E <$10 MeV.

7.9. Download the CasP code from www.hmi.de/people/schiwietz/casp.html, calculate stopping forces for some of the systems listed in Figs. 7.17–7.20 and compare them with experimental results and the other codes.

7.10. Develop a strategy for determining the I-value of a material from measurements of the stopping cross section S for protons over a certain energy interval:

1. Get tabulated values of S from ICRU (1993) or an equivalent source and treat them as 'experimental' data.
2. Extract the stopping number L from S and subtract the relativistic correction (which is independent of the material).
3. In a semilogarithmic plot you will find a straight line at high velocities, from which you may determine the I-value.

7.11. In continuation of problem 7.10,

1. try to extract a shell correction from the difference between the 'experimental' $L(v)$ curve and the Bethe logarithm.
2. Show that its onset is consistent with its presumed origin in the K shell.
3. Estimate the magnitude of the Barkas correction around the onset of the K-shell correction.

References

Agnello M., Belli G., Bendiscioli G., Bertin A., Botta E., Bressani T., Bruschi M., Bussa M.P., Busso L., Calvo D. et al. (1995): Antiproton slowing down in H_2 and He and evidence of nuclear stopping power. Phys Rev Lett **74**, 371–374

Andersen H.H. (1991): Accelerators and stopping power experiments. In A. Gras-Marti, H.M. Urbassek, N.R. Arista and F. Flores, editors, *Interaction of charged particles with solids and surfaces*, vol. B 271 of *NATO ASI Series*, 145–192. Plenum, New York

Andersen H.H., Bak J.F., Knudsen H. and Nielsen B.R. (1977): Stopping power of Al, Cu, Ag and Au for MeV hydrogen, helium, and lithium ions. Z_1^3 and Z_1^4 proportional deviations from the Bethe formula. Phys Rev A **16**, 1929

Andersen H.H., Garfinkel A.F., Hanke C.C. and Sørensen H. (1966): Stopping power of aluminium for 5-12 MeV protons and deuterons. Mat Fys Medd Dan Vid Selsk **35 no. 4**, 1–24

Andersen H.H., Sørensen H. and Vajda P. (1969): Excitation potentials and shell corrections for elements. Phys Rev **180**, 373–380

Andersen H.H. and Ziegler J.F. (1977): *Hydrogen stopping powers and ranges in all elements*, vol. 2 of *The Stopping and Ranges of Ions in Matter*. Pergamon, New York

Arbó D.G., Gravielle M.S. and Miraglia J.E. (2000): Second-order Born collisional stopping of ions in a free-electron gas. Phys Rev A **62**, 032901-1–7

Arista N.R. (2002): Energy loss of heavy ions in solids: non-linear calculations for slow and swift ions. Nucl Instrum Methods B **195**, 91–105

Arista N.R. and Lifschitz A.F. (1999): Nonlinear calculation of stopping powers for protons and antiprotons in solids: the Barkas effect. Phys Rev A **59**, 2719–2722

Ashley J.C., Ritchie R.H. and Brandt W. (1972): Z_1^3 Effect in the stopping power of matter for charged particles. Phys Rev B **5**, 2393–2397

Berger M.J., Coursey J.S. and Zucker M.A. (2005): Stopping-power and range tables for electrons, protons, and helium ions. URL http://physics.nist.gov/PhysRefData/Star/Text/

Berkowitz J. (1979): *Photoabsorption, photoionization and photoelectron spectroscopy*. Academic Press, New York

Berkowitz J. (2002): *Atomic and molecular photoabsorption. Absolute total cross sections.* Academic Press, San Diego

Bloch F. (1933): Bremsvermögen von Atomen mit mehreren Elektronen. Z Physik **81**, 363–376

Bohr N. (1913): On the theory of the decrease of velocity of moving electrified particles on passing through matter. Philos Mag **25**, 10–31

Bonderup E. (1967): Stopping of swift protons evaluated from statistical atomic model. Mat Fys Medd Dan Vid Selsk **35 no. 17**, 1–20

Brown L.M. (1950): Asymptotic expression for the stopping power of K-electrons. Phys Rev **79**, 297–303

Burkig V.C. and MacKenzie K.R. (1957): Stopping power of some metallic elements for 19.8-MeV protons. Phys Rev **106**, 848–851

Bush V. and Caldwell S.H. (1931): Thomas-Fermi equation solution by the differential analyzer. Phys Rev **38**, 1898–1902

Cabrera-Trujillo R., Sabin J.R., Deumens E. and Öhrn Y. (2004): Calculation of cross sections in electron-nuclear dynamics. Adv Quantum Chem **47**, 253–274

Chu W.K. and Powers D. (1972a): Calculations of mean excitation energy for all elements. Phys Lett **40A**, 23–24

Chu W.K. and Powers D. (1972b): On the Z_2 dependence of stopping cross sections for low energy alpha particles. Phys Lett A **38**, 267–268

Darwin C.G. (1912): A theory of the absorption and scattering of the α rays. Phil Mag (6) **23**, 901–921

Dehmer J.L., Inokuti M. and Saxon R.P. (1975): Systematics of dipole oscillator-strength distributions for atoms of the first and second row. Phys Rev A **12**, 102–121

Deumens E., Diz A., Longo R. and Öhrn Y. (1994): Time-dependent theoretical treatments of the dynamics of electrons and nuclei in molecular systems. Rev Mod Phys **66**, 917–983

Echenique P.M., Nieminen R.M. and Ritchie R.H. (1981): Density functional calculation of stopping power of an electron gas for slow ions. Sol St Comm **37**, 779–781

Fermi E. (1927): Un metodo statistico per la determinazione di alcune proprietá dell' atomo. Rend Acad Lincei **6**, 602

Fermi E. (1928): Eine statistische Methode zur Bestimmung einiger Eigenschaften des Atoms und ihre Anwendungen auf die Theorie des periodischen Systems der Elemente. Z Physik **48**, 73

Geissel H., Weick H., Scheidenberger C., Bimbot R. and Gardès D. (2002): Experimental studies of heavy-ion slowing down in matter. Nucl Instrum Methods B **195**, 3–54

Golser R. and Semrad D. (1991): Observation of a striking departure from velocity proportionality in low-energy electronic stopping. Phys Rev Lett **66**, 1831–1833

Gombas P. (1956): Statistische Behandlung des Atoms. In S. Flügge, editor, *Handbuch der Physik*, vol. 36, 109–231. Springer, Berlin

Grande P.L. and Schiwietz G. (1991): Impact-parameter dependence of electronic energy loss and straggling of incident bare ions on H and He atoms by using the coupled-channel method. Phys Rev A **44**, 2984–2992

Grande P.L. and Schiwietz G. (1993): Nonperturbative stopping-power calculation for bare and neutral hydrogen incident on He. Phys Rev A **47**, 1119–1122

Grande P.L. and Schiwietz G. (1998): Impact-parameter dependence of the electronic energy loss of fast ions. Phys Rev A **58**, 3796–3801

Grande P.L. and Schiwietz G. (2002): The unitary convolution approximation for heavy ions. Nucl Instrum Methods B **195**, 55–63

Grande P.L. and Schiwietz G. (2004): CasP version 3.1. URL `www.hmi.de/people/schiwietz/casp.html`

Grüner F., Bell F., Assmann W. and Schubert M. (2004): Integrated approach to the electronic interaction of swift heavy ions with solids and gases. Physical Review Letters **93**, 213201

Gryzinski M. (1965): Classical theory of atomic collisions. 1. Theory of inelastic collisions. Phys Rev **138**, A336–358

Harberger J., Johnson R.E. and Boring J.W. (1974a): Binary-encounter stopping cross-sections. 2. Calculations for helium in multielectron atoms, target z dependence. Phys Rev A **9**, 1172–1181

Harberger J.H., Johnson R.E. and Boring J.W. (1974b): Binary-encounter stopping cross-sections. 1. Basic theory and calculations for helium in hydrogen. Phys Rev A **9**, 1161–1171

Henke B.L., Gullikson E.M. and Davies J.C. (1993): X-ray interactions: photoabsorption, scattering, transmission, and reflection at $E = 50$–30,000 eV, $Z = 1$–92. At Data & Nucl Data Tab **54**, 181–342

Herman F. and Skillman S. (1963): *Atomic structure calculations*. Prentice Hall, New Jersey

ICRU (1984): *Stopping powers for electrons and positrons*, vol. 37 of *ICRU Report*. International Commission of Radiation Units and Measurements, Bethesda, Maryland

ICRU (1993): *Stopping powers and ranges for protons and alpha particles*, vol. 49 of *ICRU Report*. International Commission of Radiation Units and Measurements, Bethesda, Maryland

ICRU (2005): *Stopping of ions heavier than helium*, vol. 73 of *ICRU Report*. Oxford University Press, Oxford

Inokuti M., Baer T. and Dehmer J.L. (1978): Addendum: Systematics of moments of dipole oscillator-strength distributions for atoms in the first and second row. Phys Rev A **17**, 1229–1231

Inokuti M., Dehmer J.L., T. B. and Hanson J.D. (1981): Oscillator-strength moments, stopping powers, and total inelastic- scattering cross sections of all atoms through strontium. Phys Rev A **23**, 95–109

Janni J.F. (1982a): Proton range-energy tables, 1 keV – 10 GeV, part 1, for 63 compounds. At Data Nucl Data Tab **27**, 147–339

Janni J.F. (1982b): Proton range-energy tables, 1 keV – 10 GeV, part 2. elements. At Data Nucl Data Tab **27**, 341–529

Johnson R.E. and Inokuti M. (1983): The local-plasma approximation to the oscillator-strength spectrum: How good is it and why? Comm At Mol Phys **14**, 19–31

Khandelwal G.S. and Merzbacher E. (1966): Stopping power of M electrons. Phys Rev **144**, 349–352

Kührt E. and Wedell R. (1983): Energy-loss straggling and higher-order moments of energy-loss distributions for protons. Phys Lett A **96**, 347–349

Kührt E., Wedell R., Semrad D. and Bauer P. (1985): Theoretical description of the stopping power of light-ions in the intermediate energy-range. phys stat sol B **127**, 633–639

Lindhard J. (1976): The Barkas effect – or Z_1^3, Z_1^4-corrections to stopping of swift charged particles. Nucl Instrum Methods **132**, 1–5

Lindhard J. and Scharff M. (1953): Energy loss in matter by fast particles of low charge. Mat Fys Medd Dan Vid Selsk **27 no. 15**, 1–31

Lindhard J. and Sørensen A.H. (1996): On the relativistic theory of stopping of heavy ions. Phys Rev A **53**, 2443–2456

Lindhard J. and Winther A. (1964): Stopping power of electron gas and equipartition rule. Mat Fys Medd Dan Vid Selsk **34 no. 4**, 1–22

Mermin N.D. (1970): Lindhard dielectric function in the relaxation-time approximation. Phys Rev B **1**, 2362–2363

Mikkelsen H.H., Meibom A. and Sigmund P. (1992): Intercomparison of atomic models for computing stopping parameters from the Bethe theory - atomic hydrogen. Phys Rev A **46**, 7012–7018

Mikkelsen H.H. and Sigmund P. (1987): Impact parameter dependence of electronic energy loss: Oscillator model. Nucl Instrum Methods B **27**, 266–275

Mikkelsen H.H. and Sigmund P. (1989): Barkas effect in electronic stopping power: Rigorous evaluation for the harmonic oscillator. Phys Rev A **40**, 101–116

Møller S.P., Uggerhøj E., Bluhme H., Knudsen H., Mikkelsen U., Paludan K. and Morenzoni E. (1997): Direct measurement of the stopping power for antiprotons of light and heavy targets. Phys Rev A **56**, 2930–2939

Mortensen E.H., Mikkelsen H.H. and Sigmund P. (1991): Impact parameter dependence of light-ion electronic energy loss. Nucl Instrum Methods B **61**, 139–148

Oddershede J. and Sabin J.R. (1984): Orbital and whole-atom proton stopping power and shell corrections for atoms with $Z < 36$. At Data Nucl Data Tab **31**, 275–297

Oddershede J., Sabin J.R. and Sigmund P. (1983): Predicted Z_2-structure and gas-solid difference in low-velocity stopping power of light ions. Phys Rev Lett **51**, 1332–1335

Olivera G.H., Martínez A.E., Rivarola R.D. and Fainstein P.D. (1994): Electron-capture contribution to the stopping power of low-energy hydrogen beams passing through helium. Phys Rev A **49**, 603–606

Olson R. (1989): Energy deposition by energetic heavy-ions in matter. Radiat Eff **110**, 1–5

Olson R.E. (1996): Classical trajectory and Monte Carlo techniques. In *Atomic, Molecular & Optical Physics Handbook*, 664–668. American Institute of Physics

Palik E.D. (1985): *Handbook of optical constants*, vol. 1 of *Academic Press Handbook Series*. Academic Press, Orlando

Palik E.D. (1991): *Handbook of optical constants of solids*, vol. 2. Academic Press, Boston

Palik E.D. (1996): *Handbook of optical constants of solids*, vol. 3. Academic Press, Boston

Palik E.D. (2000): *Electronic handbook of optical constants of solids – version 1.0*. SciVision – Academic Press

Pathak A.P. and Yussouff M. (1972): Charged particle energy loss to electron gas. phys stat sol B **49**, 431–441

Paul H. (2005): Stopping power graphs. URL `www.exphys.uni-linz.ac.at/stopping/`

Pitarke J.M. and Campillo I. (2000): Band structure effects on the interaction of charged particles with solids. Nucl Instrum Methods B **164**, 147–160

Porter L.E. (2004): *The Barkas-effect correction to Bethe-Bloch stopping power*, vol. 46 of *Adv. Quantum Chem.*, 91–119. Elsevier, New York

Porter L.E. and Jeppesen R.G. (1983): Mean excitation energies and Barkas-effect parameters for Ne, Ar, Kr, and Xe. Nucl Instrum Methods B **204**, 605–613

Sabin J.R. and Oddershede J. (1982): Shell corrections to electronic stopping powers from orbital mean excitation energies. Phys Rev A **26**, 3209–3219

Schiwietz G. (1990): Coupled-channel calculation of stopping powers for intermediate-energy light ions penetrating atomic H and He targets. Phys Rev A **42**, 296–306

Schiwietz G., Wille U., Diez Muino R., Fainstein P.D. and Grande P.L. (1996): Comprehensive analysis of the stopping power of antiprotons and negative muons in He and H2 gas targets. J Physics B **29**, 307–321

Sharma A., Fettouhi A., Schinner A. and Sigmund P. (2004): Electronic stopping of swift ions in compounds. Nucl Instrum Methods B **218**, 19–28

Sigmund P. (1982): Kinetic theory of particle stopping in a medium with internal motion. Phys Rev A **26**, 2497–2517

Sigmund P. and Haagerup U. (1986): Bethe stopping theory for a harmonic oscillator and Bohr's oscillator model of atomic stopping. Phys Rev A **34**, 892–910

Sigmund P. and Schinner A. (2000): Binary stopping theory for swift heavy ions. Europ Phys J D **12**, 425–434

Sigmund P. and Schinner A. (2001): Binary theory of antiproton stopping. Europ Phys J D **15**, 165–172

Sigmund P. and Schinner A. (2002a): Binary theory of electronic stopping. Nucl Instrum Methods B **195**, 64–90

Sigmund P. and Schinner A. (2002b): Binary theory of light-ion stopping. Nucl Instrum Methods B **193**, 49–55

Sigmund P., Sharma A., Schinner A. and Fettouhi A. (2005): Valence structure effects in the stopping of swift ions. Nucl Instrum Methods B **230**, 1–6

Smith D.Y., Shiles E. and Inokuti M. (1985): The optical constants of metallic aluminium. In E.D. Palik, editor, *Handbook of optical constants of solids*, vol. 1, 369. Academic Press, Orlando

Sommerfeld A. (1932): Asymptotische Integration der Differentialgleichung des Thomas-Fermischen Atoms. Z Physik **78**, 283–308

Sørensen A.H. (1990): Barkas effect at low velocities. Nucl Instrum Methods B **48**, 10–13

Sternheimer R.M., Seltzer S.M. and Berger M.J. (1982): Density effect for the ionization loss of charged particles in various substances. Phys Rev B **26**, 6067–6076

Thomas L.H. (1926): The calculation of atomic fields. Proc Cambr Philos Soc **23**, 542–547

Walske M.C. (1952): The stopping power of K-electrons. Phys Rev **88**, 1283–1289

Walske M.C. (1956): Stopping power of L-electrons. Phys Rev **101**, 940–944

Ziegler J.F. (2005): Particle interactions with matter. URL `www.srim.org`

Ziegler J.F., Biersack J.P. and Littmark U. (1985): The stopping and range of ions in solids. 1–319. Pergamon, New York

Part III

Straggling

Energy-Loss Straggling: Variance and Higher Cumulants

8.1 Introductory Comments

Energy-loss straggling denotes the development of the width and shape of the energy spectrum of an initially monochromatic beam as a function of time or pathlength.

Straggling is an inherent feature of stopping measurements which cannot be reduced indefinitely by making more measurements.

In many applications, information on the scatter of data is just as important as mean values.

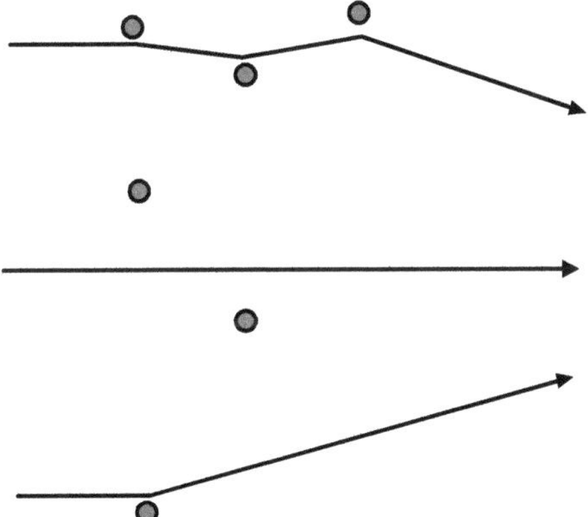

Fig. 8.1. The number of collisions and the distribution of impact parameters fluctuates from trajectory to trajectory

Fluctuations in energy loss were introduced in Sect. 2.2.4. There they were ascribed to the fact that penetrating particles have different trajectories (cf. Fig. 8.1. This implies that the number of atoms met during passage as well as the strength of individual encounters – governed by the impact parameter – differs from projectile to projectile. While such fluctuations do not vanish in the limit of an infinite number of beam particles, the *relative* fluctuation must decrease with an increasing number of interactions per projectile, i.e., increasing pathlength. That decrease, however, cannot go on indefinitely since the projectile slows down and eventually comes to rest.

There are other potential origins of straggling. Quantum mechanics causes the energy loss to fluctuate from one collision event to another even at one and the same impact parameter. For composite projectiles such as heavy ions carrying electrons, the charge and excitation state may vary over a trajectory, and since the energy loss typically depends on the projectile state, additional fluctuations arise.

In analysing measured energy spectra one also needs to consider initial beam spread, angular deflection, and target nonuniformities such as surface roughness, pinholes and inhomogeneous chemical composition. These effects are not categorized as straggling but may be difficult to separate experimentally from straggling effects.

Energy-loss straggling has atomistic and statistical aspects which will be discussed in this and the following chapter. On the atomistic side, a careful study of (2.30) is needed along the same line as the discussion of (2.29) in Chapters 4–6. After all, the fundamental relation (2.57) is based on Rutherford's law, disregarding binding of target particles which results in shell and Barkas-Andersen corrections. These effects need to be considered along with quantum mechanics and relativity.

On the statistical side, information is needed on the *shape* of the energy-loss profile. The profile is not necessarily gaussian, and hence the relation between mean and peak energy loss needs to be studied. It is often the latter one that is actually measured. A similar statement is true for the relation between standard deviation and half-width. Moreover, attention needs to be paid to limitations of Poisson statistics that entered as an essential ingredient into the treatment in Sect. 2.2.4.

Several results presented in this chapter can be derived by procedures discussed in Chapters 4–6. Therefore, some of the calculations are less explicitly presented here than in previous chapters. However, problems in the end of the chapter will encourage you to go through important derivations on your own, supported by references to the pertinent literature.

8.2 Classical versus Quantum Theory

The width of an energy-loss profile is determined primarily by the straggling parameter

$$W = \sum_j T_j^2 \sigma_j = \int T^2 \, \mathrm{d}\sigma(T), \tag{8.1}$$

which is related to the variance of the energy-loss profile by

$$\left\langle (\Delta E - \langle \Delta E \rangle)^2 \right\rangle = N x W \tag{8.2}$$

according to (2.26) and (2.30). This may be written in the form

$$W = \int \langle T^2(\boldsymbol{p}) \rangle \, \mathrm{d}^2 \boldsymbol{p}, \tag{8.3}$$

where $\boldsymbol{p}$ is an impact parameter either to a target electron (in classical theory) or a target nucleus (in semiclassical theory).

In classical theory, the energy loss at a given impact parameter is uniquely defined if the electron is at rest initially. In semiclassical theory, the energy loss is a fluctuating quantity even at a given impact parameter. However, for a uniform distribution of impact parameters that fluctuation is immaterial after integration according to (8.3).

In case of the stopping force we have seen in chapter 4 that Bohr's kappa criterion (2.80) provided a clear separation between the classical and the Born regime. This distinction must be expected to also affect straggling. The link between the two regimes is provided by Bloch theory as presented in Sect. 6.3.1.

Lindhard and Sørensen (1996) evaluated the Bloch correction to the straggling parameter using the formalism described in Sect. 6.3.2 and found that it vanishes in the nonrelativistic regime. This implies that the regime of applicability of classical scattering theory in straggling is not limited by the Bohr criterion. This is a central result. Therefore, the calculation will be reproduced explicitly here.

Notation and general procedure follow Sect. 6.3.2 closely. Starting at (3.118) and (3.120) we need to evaluate

$$W = (mv^2)^2 \frac{4\pi}{k^2} \sum_\ell (\ell + 1)$$

$$\times \left[2 \sin^2(\delta_\ell - \delta_{\ell+1}) - \frac{\ell + 2}{2\ell + 3} \sin^2(\delta_\ell - \delta_{\ell+2}) \right]. \tag{8.4}$$

Now, according to (3.137),

$$\delta_\ell - \delta_{\ell+2} = \arg \left([\ell + 1 + i\kappa/2][\ell + 2 + i\kappa/2] \right)$$

$$= \arctan \frac{\kappa/2}{\ell + 1} + \arctan \frac{\kappa/2}{\ell + 2}, \tag{8.5}$$

and hence,

$$\sin^2(\delta_\ell - \delta_{\ell+2}) = \frac{(2\ell + 3)^2 \kappa^2/4}{\left[(\ell + 1)^2 + \kappa^2/4 \right] \left[(\ell + 2)^2 + \kappa^2/4 \right]}. \tag{8.6}$$

Insertion of this as well as (6.15) into (8.4) leads to

$$W = \left(mv^2 \right)^2 \frac{4\pi}{k^2} \frac{\kappa^2}{4} \sum_\ell \frac{(\ell + 1)(\ell + 2 + \kappa^2/2)}{\left[(\ell + 1)^2 + \kappa^2/4 \right] \left[(\ell + 2)^2 + \kappa^2/4 \right]}. \tag{8.7}$$

Unlike (6.16), this series converges. It is, therefore, unnecessary to subtract the Born limit.

You may easily verify – by separating the above fraction into two – that (8.7) is equivalent with

$$W = 4\pi Z_1^2 e^4 \sum_\ell \left(\frac{\ell+1+\kappa^2/4}{(\ell+1)^2+\kappa^2/4} - \frac{\ell+2+\kappa^2/4}{(\ell+2)^2+\kappa^2/4} \right). \tag{8.8}$$

Here, terms cancel pairwise, and only the first term in the parentheses for $\ell = 0$ remains. This yields

$$W = 4\pi Z_1^2 e^4 \tag{8.9}$$

for the straggling parameter per target electron. Since there is evidently no Bloch correction to W, Bohr and Bethe theory must be expected to lead to equivalent or identical results within the degree of accuracy of the Bloch theory. In other words, the choice of theoretical scheme to treat straggling is much less influenced – if at all – by Bohr's kappa criterion than the description of the mean energy loss.

The above argument does not invoke shell corrections and other significant effects. It will, therefore, still be necessary to evaluate straggling separately in the Bohr and the Bethe scheme with various extensions. However, the absence of a Bloch correction strongly encourages an emphasis on classical theory.

8.3 Bohr Theory

The evaluation of the straggling parameter in the Bohr model[1] follows that of the stopping cross section presented in Sect. 4.5.1. Recall that in the Bohr stopping model, a target electron is initially at rest, bound harmonically to a fixed position in space, and hit at an impact parameter p by a heavy point charge moving uniformly with a velocity v. The range of impact parameters is divided up into a close- and a distant-collision regime. The limiting impact parameter p_0 separating the two regimes introduced in Sect. 4.5.1 remains unchanged. However, the factor T^2 under the integral indicates a dominance of large energy transfers and, hence, close collisions.

In the notation of Sect. 4.5.1 you find (problem 8.1)

$$W = \int_0^\infty 2\pi p \, \mathrm{d}p \, T^2(p) = \frac{W_{\mathrm{B}}}{\xi^2} \int_0^\infty \frac{2\mathrm{d}\zeta}{\zeta^3} \left[f(\zeta) \right]^2, \tag{8.10}$$

---

[1] Although the calculation presented in this section is thought to strictly follow the model outlined by Bohr (1913), Bohr himself in his study of straggling (Bohr, 1915) did not go so far and only evaluated straggling for binary Coulomb scattering. This was perfectly adequate for the range of applications that Bohr had in mind. To avoid confusion, a distinction will be made in the following between 'Bohr straggling' and 'straggling in the Bohr model'.

where $f(\zeta)$ is a function of $\zeta = \omega_0 p/v$ introduced in (4.22) on page 114, $\xi = mv^3/Z_1 e^2 \omega_0$ and W_{B} is given by

$$W_{\mathrm{B}} = 4\pi Z_1^2 e^4 \tag{8.11}$$

for a projectile ion with atomic number Z_1 hitting a one-electron atom.

From (4.84) and (8.10) you find the following contribution from close collisions,

$$W_{\mathrm{close}} = \frac{W_{\mathrm{B}}}{1 + (b/2p_0)^2} \tag{8.12}$$

by straight integration from 0 to $\zeta_0 = \omega_0 p_0/v$ (problem 8.1), while the contribution from distant interactions,

$$W_{\mathrm{dist}} = \frac{W_{\mathrm{B}}}{\xi^2} 2 \int_{\zeta_0}^{\infty} \zeta \, \mathrm{d}\zeta \left([K_1(\zeta)]^2 + [K_0(\zeta)]^2 \right)^2 , \tag{8.13}$$

is conveniently evaluated numerically.

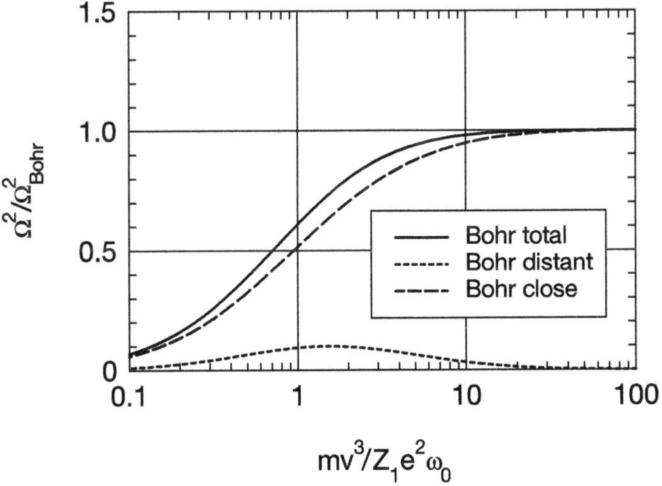

Fig. 8.2. Straggling in Bohr model. Ratio W/W_{B} for close and distant interactions and the sum

Figure 8.2 shows the ratio $\Omega^2/\Omega^2_{\mathrm{Bohr}} = W/W_{\mathrm{B}}$ as a function of the dimensionless variable $\xi = mv^3/Z_1 e^2 \omega_0$. Inclusion of binding in the determination of straggling is seen to result in a decrease of the straggling parameter toward

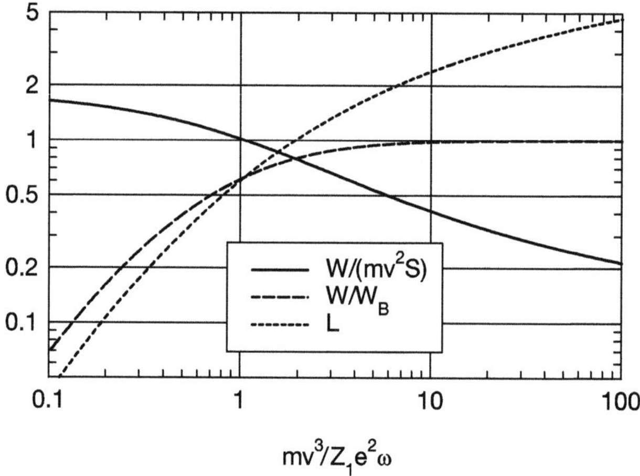

Fig. 8.3. Straggling and stopping in Bohr model. Dashed line: Relative straggling W/W_{B}; dotted line: stopping number L; solid line: ratio $W/(mv^2S) = W/(LW_{\mathrm{B}})$

low energies. The underlying reason is the fact that the single-collision spectrum gets narrower. In contrast to the corresponding results for the stopping number shown in Fig. 4.3, the contribution from distant interactions is at least an order of magnitude smaller than that from close interactions, even at low projectile speed. This is a direct manifestation of the significance of the factor T^2 in the integrand.

The simple expression W_{B} appears to accurately represent W from $\xi \simeq 10$ upward, except for corrections that are not allowed for in the original Bohr model and that will be discussed in the following sections. In that velocity regime, the *straggling parameter of an atom* with Z_2 electrons becomes

$$W_{\mathrm{B}} = 4\pi Z_1^2 Z_2 e^4, \tag{8.14}$$

leading to the variance

$$\Omega_{\mathrm{Bohr}}^2 = NxW_{\mathrm{B}} = 4\pi Z_1^2 Z_2 e^4 Nx. \tag{8.15}$$

This quantity is conventionally called *Bohr straggling* after Bohr (1915). However, Fig. 8.2 demonstrates that the Bohr model of stopping, when applied to straggling, delivers a result that only at high projectile speeds approaches Bohr straggling.

Figure 8.3 shows that the ratio between relative straggling and stopping number according to the curves labelled 'total' in Figs. 4.3 and 8.2 varies fairly slowly with projectile speed.

You will see in Sect. 8.10 that Bohr straggling (8.15), despite the simplicity of the underlying model, describes straggling experiments reasonably well. Nevertheless, a number of effects needs to be considered which cannot generally be ignored in the stopping cross section, in particular orbital motion (shell correction), Barkas-Andersen effect, relativity, and density effect.

Moreover, for atoms containing more than a single electron, correlations between target electrons need to be taken into account.

Finally, effects of electron promotion and charge exchange, which often can be ignored in the stopping cross section, may become more pronounced in straggling because of the higher significance of large energy transfers. A discussion of this last group of effects, as well as others associated with electrons accompanying the projectile, will be postponed to Volume II.

8.4 Born Approximation

Just as the calculation of straggling on the basis of the Bohr model, also the corresponding step in the Born approximation follows the same procedure as the evaluation of the stopping cross section. Recall that two different procedures led to the excitation cross section σ_j (4.82) which also forms the starting point for a straggling calculation.

8.4.1 Harmonic oscillator

Consider first the particularly simple case of a spherical harmonic oscillator. The calculation leading to (4.110) may be generalized to straggling (problem 8.2). The main change is the addition of a factor $\epsilon_j - \epsilon_0 = j\hbar\omega_0$ under the sum. This leads to

$$\frac{W}{W_{\rm B}} = \frac{1}{B} \sum_{j=1}^{\infty} \frac{j}{(j-1)!} \int_{j^2/B}^{\infty} dt\, t^{j-2} e^{-t}, \tag{8.16}$$

where

$$B = \frac{2mv^2}{\hbar\omega_0} \tag{8.17}$$

is the Bethe parameter.

Figure 8.4 shows this function, evaluated numerically by Sigmund and Haagerup (1986). Apart from the different abscissa variables, the qualitative behavior is similar to that in Fig. 8.2, but unlike there, we now observe a slight overshoot above the Bohr value between $B \simeq 10$ and 100. This effect, which will be shown to arise from the shell correction, was first found by Livingston and Bethe (1937).

8.4.2 Bethe Approximation

Next, consider the case of a general target atom but apply the Bethe approximation presented in Sect. 4.5.4.

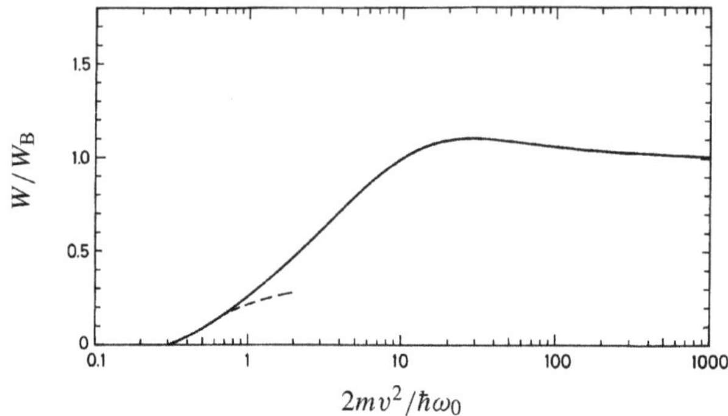

Fig. 8.4. Relative straggling W/W_B for quantal harmonic oscillator. Dotted line: Only the first excited level taken into account. From Sigmund and Haagerup (1986)

Addition of a factor $\epsilon_j - \epsilon_0$ to (4.106) leads to

$$\frac{W}{W_B} = \frac{1}{2mv^2} \sum_j (\epsilon_j - \epsilon_0) \int_{(\epsilon_j - \epsilon_0)^2/2mv^2}^{\infty} \frac{dQ}{Q} f_{j0}(Q), \tag{8.18}$$

which is exact within the Born approximation. Now, splitting the integral into a close- and a distant-interaction portion at some boundary Q_0, the approximations made in Sect. 4.5.4 lead to

$$\left(\frac{W}{W_\mathrm{B}}\right)_{\mathrm{close}} = \frac{1}{2mv^2} \int_{Q_0}^{2mv^2} \frac{dQ}{Q} \sum_j (\epsilon_j - \epsilon_0) f_{j0}(Q) \tag{8.19}$$

and

$$\left(\frac{W}{W_\mathrm{B}}\right)_{\mathrm{dist}} = \frac{1}{2mv^2} \sum_j (\epsilon_j - \epsilon_0) f_{j0} \ln \frac{2mv^2 Q_0}{(\epsilon_j - \epsilon_0)^2}$$

$$= \frac{1}{2mv^2} \left(\sum_j (\epsilon_j - \epsilon_0) f_{j0} \right) \ln \frac{2mv^2 Q_0}{I_1^2}, \tag{8.20}$$

where the quantity I_1 is defined via

$$\sum_j (\epsilon_j - \epsilon_0) f_j \ln I_1 = \sum_j (\epsilon_j - \epsilon_0) f_j \ln (\epsilon_j - \epsilon_0). \tag{8.21}$$

For a one-electron atom, the following sum rule holds rigorously (problem 8.3),

$$\sum_j (\epsilon_j - \epsilon_0) f_{j0}(Q) = Q + \frac{2}{3m} \langle 0 | p^2 | 0 \rangle \equiv Q + \frac{2}{3} m \overline{v_e^2}, \tag{8.22}$$

where

$$\overline{v_e^2} = \left\langle 0 \left| \frac{p^2}{m^2} \right| 0 \right\rangle \tag{8.23}$$

is the mean square velocity of an electron in the ground state.

Equation (8.22) enters in both (8.19) and (8.20), and after integration you find

$$\left(\frac{W}{W_{\mathrm{B}}} \right) = 1 + \frac{2}{3} \frac{\overline{v_e^2}}{v^2} \ln \frac{2mv^2}{I_1} - \frac{Q_0}{2mv^2}. \tag{8.24}$$

It is seen that Q_0 again drops out under the logarithm. In view of the intial assumption that $Q_0 \ll Q_{\max} = 2mv^2$, the last term needs to be dropped.

As expected from the absence of a Bloch correction, (8.24) approaches Bohr straggling at high speed.

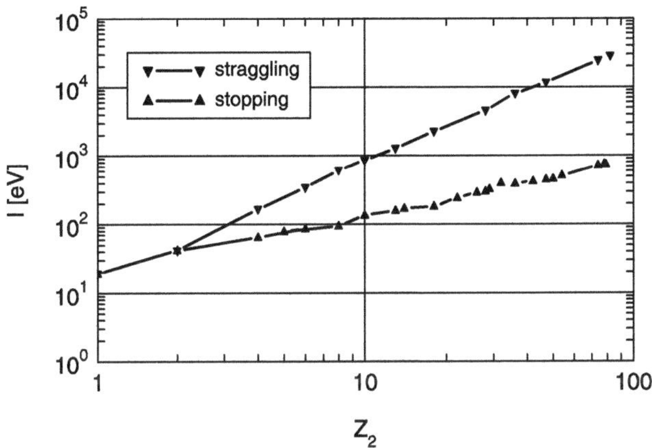

Fig. 8.5. Values of I_1 for selected elements calculated from data listed in Table 7.1, compared with I-values from ICRU (2005) which were calculated with equivalent input

Note that the quantity I_1 entering into (8.24) and defined by (8.21) differs from the I-value entering the Bethe logarithm in the stopping cross section, cf. (4.119). Figure 8.5 shows values for specific elements, based on realistic oscillator-strength spectra. Also included are I-values evaluated from the same source, i.e., Table 7.1. It is seen that I_1 increases approximately like $Z_2^{5/3}$ and that, for large Z_2 it may exceed I by more than an order of magnitude.

The leading correction term in (8.24) is governed by the orbital speed of the target electrons and, hence, has to be interpreted as a shell correction. It has a maximum at $2mv^2/I_1 = 2.718$, i.e., it produces a shoulder, often called the *Bethe-Livingston correction*.

It is of interest to identify the reason why straggling can exceed the Bohr value. When orbital motion of target electrons is taken into account, the energy-loss spectrum for individual collisions extends over a broader interval. In case of the stopping cross section, this results in a decrease at high speed but an increase at low speed. In case of straggling, with the dominance of high energy transfers, an increase is found at intermediate energies due to the fact that energy transfers exceeding the kinematic limit $2mv^2$ become possible. This becomes insignificant at velocities much higher than orbital velocities. The result is the Bethe-Livingston shoulder.

Although the energy loss fluctuates in a quantum system even at a fixed impact parameter[2], that fluctuation is immaterial for a homogeneous beam after integration over all impact parameters, as mentioned in Sect. 8.2. Equation (8.24) shows that the main quantal effect giving rise to fluctuations in the Bethe scheme is the orbital motion of the target electrons. This phenomenon is enforced by the uncertainty principle.

8.4.3 Relativistic Extension ($\star$)

The relativistic extension can be achieved in straight analogy to Sect. 5.6. Consider again three regimes in the Q-integration, separated by values $Q_0 \gg (\hbar\omega_{j0})^2/2mv^2$ and $Q_1 \ll mc^2$.

For $Q < Q_0$, the longitudinal interaction yields

$$\frac{W}{W_{\mathrm{Bohr}}}\bigg|_{\mathrm{long},Q<Q_0} = \frac{\overline{v_e^2}}{3v^2}\ln\frac{2mv^2 Q_0}{I_1^2} \tag{8.25}$$

as a slight modification of the expression derived in the previous section. While this is a noticeable correction at nonrelativistic projectile speeds $v \ll c$, it becomes small in the relativistic regime, except for inner shells in heavy target atoms.

Similarly, for $Q_0 < Q < Q_1$ the longitudinal interaction yields

$$\frac{W}{W_{\mathrm{Bohr}}}\bigg|_{\mathrm{long},Q_0<Q<Q_1} = \frac{Q_1 - Q_0}{2mv^2} + \frac{\overline{v_e^2}}{3v^2}\ln\frac{Q_1}{Q_0}. \tag{8.26}$$

The evaluation of the contribution from the transverse interaction follows closely that for the stopping force in Sect. 5.6. All that is needed is the addition of a factor $\hbar\omega_{j0}$ in the integrand of the stopping integral.

---

[2] When talking about an impact parameter in quantum collision theory we imply the semiclassical scheme so that the pertinent quantity is the impact parameter to the target *nucleus*. Bohr theory deals with the impact parameter to a target *electron*, but as long as orbital motion is not allowed for, the two quantities are identical.

For the low-Q regime you find

$$\frac{W}{W_{\text{Bohr}}}\bigg|_{\text{trans},Q<Q_0} = \frac{1}{2mv^2}\sum_j \hbar\omega_{j0}f_{j0}\bigg[-\ln(1-\beta^2)-\beta^2$$

$$+\ln\left(1-\frac{\beta^2}{\eta_{0j}}\right)+\frac{\beta^2(1-\beta^2)}{\eta_{0j}-\beta^2}\bigg], \quad (8.27)$$

where $\eta_{0j} = 2mv^2 Q_0/(\hbar\omega_{j0})^2$ was introduced by (5.100) in Sect. 5.6.2. Making use of the fact that $\eta_{0j} \gg 1$, the two last terms become small of order Q_0/mc^2 in comparison with the two first terms and will thus be neglected. This results in

$$\frac{W}{W_{\text{Bohr}}}\bigg|_{\text{trans},Q<Q_0} = \frac{\overline{v_e^2}}{3v^2}\left[-\ln(1-\beta^2)-\beta^2\right] \quad (8.28)$$

by means of the sum rule (8.22). From the intermediate regime, already the contribution to the stopping force was negligible. In the present case it is small of order $(Q_1/mc^2)^2$ and will thus be neglected.

For high momentum transfers, contributions from longitudinal and transverse excitations are not treated separately. Multiply the integrand in (5.114) by a factor[3] $\hbar\omega_k^+$. Integration leads to

$$\frac{W}{W_{\text{Bohr}}}\bigg|_{Q>Q_1} = \frac{1}{2mv^2}mc^2\left[\zeta-\frac{1}{4}(1-\beta^2)(\zeta-1)^2\right]_{\zeta_1}^{\zeta_{\max}}. \quad (8.29)$$

Making use again of the fact that $Q_1 \ll mc^2$ you find

$$\frac{W}{W_{\text{Bohr}}}\bigg|_{Q>Q_1} = \frac{1-\beta^2/2}{1-\beta^2}-\frac{Q_1}{2mv^2}. \quad (8.30)$$

Here, the term $\propto Q_1$ cancels against the corresponding term from the intermediate-Q regime. Summing all contributions we arrive at

$$\frac{W}{W_{\text{Bohr}}} = \frac{1-\beta^2/2}{1-\beta^2}-\frac{Q_0}{2mv^2}$$

$$+\frac{\overline{v_e^2}}{3v^2}\left[\ln\frac{2mv^2 Q_1}{I_1^2}-\ln(1-\beta^2)-\beta^2\right]. \quad (8.31)$$

While the uncompensated term $Q_0/2mv^2$ can be dropped because it is small ($Q_0 \ll Q_1 \ll 2mv^2$), the absence of a high-Q term compensating for $\ln Q_1$, corresponding to what was found for the stopping cross section, warrants attention.

The origin of this deficiency is the use of free-particle Dirac functions which evidently fail to reproduce the term containing $\overline{v_e^2}$ in the sum rule (8.22).

---

[3] Recall that the expressions $\hbar\omega_{j0}$ and $\epsilon_j - \epsilon_0$ denote the same quantity and will be used synonymously.

Straight extensions of Bethe-type sum rules do not exist, although numerous authors have shown interest in the problem. For a recent survey the reader is referred to Cohen (2004). Apart from the single-particle nature of the Dirac equation, a major obstacle is the existence of negative-energy states which prevents meaningful application of the closure property of a set of complete of single-particle states.

The range of validity of sum rules like (8.22) can be extended into the relativistic regime by series expansion in powers of $1/c^2$. The first relativistic correction term to (8.22) has been evaluated by Cohen and Leung (1998), although no numerical results were given. More elaborate calculations for the *Bethe sum rule* indicate corrections up to 1 pct. (Cohen, 2005).

We note that at relativistic speeds, a term $\propto \overline{v_e^2}/v^2$ is a small correction except for inner shells of heavy target atoms. Allowing for a minor inaccuracy we may, therefore, adopt the sum rule (8.22) also for the relativistic regime. This is equivalent with setting

$$Q_1 \to Q_{\max} = \frac{2mv^2}{(1-\beta^2)^2} \tag{8.32}$$

according to (5.111) and leads to

$$\frac{W}{W_{\text{Bohr}}} = \frac{1 - \beta^2/2}{1-\beta^2} + \frac{\overline{v_e^2}}{v^2}\left[\frac{2}{3}\ln\frac{2mv^2}{I_1} - \ln(1-\beta^2) - \frac{\beta^2}{3}\right]. \tag{8.33}$$

This result is identical with the one derived by Fano (1963) as far as the main term is concerned. Fano explicitly neglected relativistic corrections to the second term.

8.4.4 Density Effect

A relativistic density correction in the stopping force was determined in Sect. 5.6.4. That correction originated exclusively in the zero-k limit of the dielectric function, i.e., in very distant interactions. Such interactions do not contribute significantly to straggling. The density correction in straggling has, therefore, always been neglected (Sternheimer, 1960, Fano, 1963).

Since the relativistic density correction in the stopping force is a threshold effect, one might even wonder whether such a threshold also exists in straggling, or whether there is no corresponding correction in straggling at all.

8.5 Fermi Gas ($\star$)

8.5.1 Expression for Straggling

As an analog to (5.30) one may, according to Lindhard (1954), write the following expression for straggling for a medium characterized by a dielectric function $\epsilon(k,\omega)$,

$$\frac{\mathrm{d}\Omega^2}{\mathrm{d}x} = -\frac{e_1^2}{\pi v^2}\mathrm{Im}\int_0^\infty \frac{\mathrm{d}k}{k}\int_0^{kv}\mathrm{d}\omega\,2\hbar\omega^2\left(\frac{1}{\epsilon(k,\omega)}-1\right). \tag{8.34}$$

Superficially, the two expressions differ by a factor $2\hbar\omega$, where $\hbar\omega$ represents an individual energy transfer and the factor 2 accounts for a change in integration limit.

Just as the transition from (2.29) to (2.30), the physics behind this innocently-looking generalization is by no means trivial. In Sect. 2.2.4, the central assumption was the mutual independence of individual collision events which allowed the application of Poisson statistics. Similarly, (8.34) requires individual energy transfers $\hbar\omega$, whether being collective or not, to be mutually independent. This assumption also enters into the argument of Lindhard (1954) which rests on the quantization of the electromagnetic field. A simpler, although less rigorous procedure is to show that (8.34) leads to the standard expression in the limit of a dilute medium (problem 8.5).

The limitations of (8.34) are not clear. We shall see in Sect. 8.9 that correlations between individual atomic collisions produce corrections to (2.30). Similar effects must exist in other systems, in particular in the Fermi gas.

8.5.2 Static Electron Gas

Following Sect. 5.4.3, let us consider the static electron gas as an example. With the dielectric function (5.60) one finds

$$\mathrm{Im}\left(\frac{1}{\epsilon(\boldsymbol{k},\omega)}-1\right) = -\frac{\pi\omega_\mathrm{P}^2}{2\alpha_k}\left[\delta(\omega-\alpha_k)-\delta(\omega+\alpha_k)\right] \tag{8.35}$$

according to (5.74), with

$$\alpha_k^2 = \omega_k^2 + \omega_\mathrm{P}^2. \tag{8.36}$$

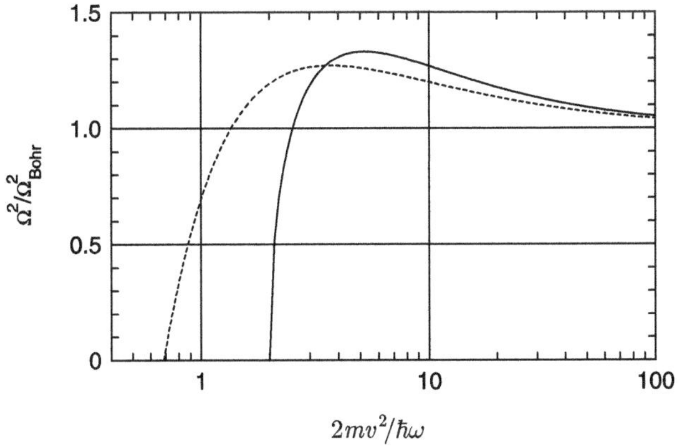

Fig. 8.6. Relative straggling in static electron gas. Solid line: Exact evaluation (cf. problem 8.8. Dashed line: equation (8.38)

This leads to the straggling parameter

$$\frac{W}{W_{\mathrm{B}}} = \frac{\hbar}{mv^2} \int_{kv > \alpha_k} \frac{\alpha_k}{k} \mathrm{d}k.$$
(8.37)

This integral is elementary (problem 8.8), and the result has been plotted in Fig. 8.6. However, the static electron gas can only be a reasonable approximation to a Fermi gas as long as the projectile speed is high compared with the Fermi velocity. Hence, following Sigmund and Fu (1982), we also make a Taylor expansion for small values of the dimensionless parameter $\hbar\omega_P/mv^2$, and go only up to the linear term. This yields (problem 8.6)

$$\frac{W}{W_{\mathrm{B}}} = 1 + \frac{\hbar\omega_P}{2mv^2} \left(\ln \frac{4mv^2}{\hbar\omega_P} - 1 \right),$$
(8.38)

which also has been included in Fig. 8.6. It is seen that the approach to asymptotic behavior is fairly slow.

Equation (8.38) may be compared to (8.24). You may notice the term $-\hbar\omega_P/2mv^2$ which, unlike the term $-Q_0/2mv^2$ in (8.24), is well-defined here. The logarithmic correction term, on the other hand, being proportional to ω_P, is density-dependent, while the logarithmic term in (8.24) hinges on the orbital velocity which is ignored in the present estimate.

8.5.3 Degenerate Fermi Gas: High Projectile Speed

In Sect. 5.7.4 the dielectric function was expanded in powers of v_{F}^2, thus enabling a shell-correction expansion of the stopping force. The same method may also be applied to straggling. Series expansion of (8.34) in terms of $\hbar\omega/mv^2$ and v_{F}^2/v^2, both up to first order, yields

$$\frac{W}{W_{\mathrm{B}}} = 1 + \left(\frac{\hbar\omega_P}{2mv^2} + \frac{v_{\mathrm{F}}^2}{5v^2} \right) \ln \frac{4mv^2}{\hbar\omega_P} - \left(\frac{\hbar\omega_P}{2mv^2} + \frac{21}{40} \frac{v_{\mathrm{F}}^2}{v^2} \right) + \dots$$
(8.39)

according to Sigmund and Fu (1982) (problem 8.6)[4]. This function exhibits a maximum (problem 8.7) which lies at

$$\left. \frac{4mv^2}{\hbar\omega_P} \right|_{\mathrm{max}} = 7.39$$
(8.40)

for a dilute target. The number on the right-hand side increases only slowly with increasing density of the medium.

---

[4] A somewhat different formula was found earlier by Bonderup and Hvelplund (1971) by expansion of $\varepsilon(k, \omega)$ for large k. That approximation was not meant to be an expansion in powers of $1/v^2$ but nevertheless led to the same coefficients of logarithmic terms as those in (8.39).

It is of interest to identify the origin of the two logarithmic terms in (8.39). The first of the two is present already in (8.38) for the static electron gas and is evidently associated with the plasma resonance. Let us, for a moment, consider a projectile that loses energy exclusively to plasma resonances, so that

$$d\sigma'(T) = \frac{S'}{\hbar\omega_P}\delta(T - \hbar\omega_P)dT,\tag{8.41}$$

where S' is the stopping cross section. This leads to a straggling parameter

$$W' = \hbar\omega_P S'.\tag{8.42}$$

If we disregard a numerical factor $\ln 2 - 1$, this is equivalent with the term $\propto \hbar\omega_P$ in (8.38), if S' is identified with *half* the stopping cross section S. Thus, (8.38) accounts for the fact that in comparison to binary Coulomb scattering, small energy losses have been rearranged and bunched into a sharp plasma resonance.

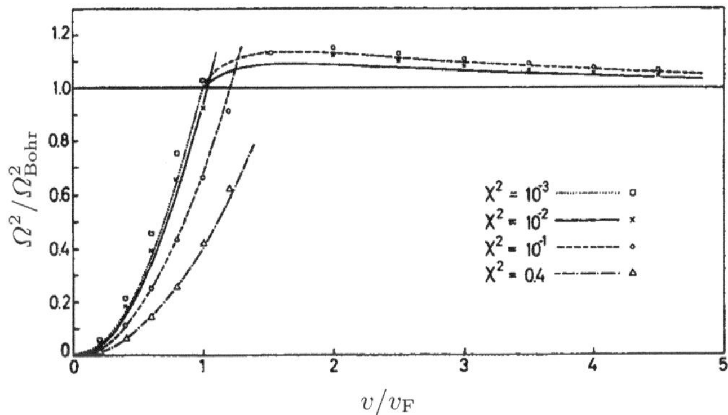

Fig. 8.7. Relative straggling in Fermi gas for different electron densities. $\chi = \sqrt{v_0/\pi v_F}$. From Bonderup and Hvelplund (1971)

The second logarithmic term in (8.39) is evidently a shell correction. Its significance becomes evident by writing

$$\frac{v_F^2}{5v^2} = \frac{1}{3}\frac{\langle v_e^2\rangle}{v^2}.\tag{8.43}$$

The main difference from the second term in (8.24) is again a factor $1/2$. This is consistent with equipartition discussed on page 174, i.e., for a static electron gas at high speed, half of the energy transfers goes into excitation of individual electrons.

Figure 8.7 shows relative straggling as a function of projectile speed according to Bonderup and Hvelplund (1971), determined by numerical evaluation.

8.6 Shell Correction: Kinetic Theory $(\star)$

The treatment of shell corrections by kinetic theory, based on the assumption that shell corrections are due to the orbital motion of target electrons, can be readily extended to straggling on the basis of the discussion in Sect. 6.5.

Note first that by dividing (6.66) by $nvdt$ you get an effective cross section in the laboratory frame of reference,

$$d\sigma_{\text{eff}} = \frac{w_e}{v} d\sigma(w_e, \theta). \tag{8.44}$$

The energy transfer from the projectile to a target electron is given by

$$T = m\boldsymbol{v} \cdot (\boldsymbol{w}'_e - \boldsymbol{w}_e), \tag{8.45}$$

according to (3.30) in Sect. 3.2.2, where $\boldsymbol{w}_e$ and $\boldsymbol{w}'_e$ are the electron velocities in the center-of-mass frame before and after interaction.

Now, choose a coordinate frame such that

$$\boldsymbol{w}_e = w_e(0, 0, 1) \tag{8.46a}$$

$$\boldsymbol{w}'_e = w_e(\sin\theta\cos\psi, \sin\theta\sin\psi, \cos\theta) \tag{8.46b}$$

$$\boldsymbol{v} = v(\sin\phi, 0, \cos\phi). \tag{8.46c}$$

This yields the straggling parameter per target electron (problem 8.9),

$$W = \int T^2 d\sigma_{\text{eff}}(v, \theta)$$

$$= \left\langle m^2 \frac{|\boldsymbol{v} - \boldsymbol{v}_e|}{v} \left\{ \left[(v^2 - \boldsymbol{v} \cdot \boldsymbol{v}_e)^2 - \frac{1}{2} \left(v^2 v_e^2 - (\boldsymbol{v} \cdot \boldsymbol{v}_e)^2 \right) \right] \sigma^{(2)}(|\boldsymbol{v} - \boldsymbol{v}_e|) \right. \right.$$

$$\left. \left. + \left[v^2 v_e^2 - (\boldsymbol{v} \cdot \boldsymbol{v}_e)^2 \right] \sigma^{(1)}(|\boldsymbol{v} - \boldsymbol{v}_e|) \right\} \right\rangle_{\boldsymbol{v}_e}, \tag{8.47}$$

where

$$\sigma^{(2)}(v) = \int (1 - \cos\theta)^2 d\sigma(v, \theta) \tag{8.48}$$

is a higher-order transport cross section. A more general relation was derived by Sigmund (1982) for arbitrary mass ratio M_1/m. For $v_e \ll v$, (8.47) reduces to a well-known result (3.118),

$$W_0 = (mv^2)^2 \sigma^{(2)}(v). \tag{8.49}$$

The transport cross sections can be eliminated by means of (6.86) and (8.49),

$$W(v) = \left\langle \frac{(v^2 - \boldsymbol{v} \cdot \boldsymbol{v}_e)^2 - \frac{1}{2}(v^2 v_e^2 - (\boldsymbol{v} \cdot \boldsymbol{v}_e)^2)}{v|\boldsymbol{v} - \boldsymbol{v}_e|^3} W_0(|\boldsymbol{v} - \boldsymbol{v}_e|) \right.$$

$$\left. + m\frac{v^2 v_e^2 - (\boldsymbol{v} \cdot \boldsymbol{v}_e)^2}{v|\boldsymbol{v} - \boldsymbol{v}_e|} S_0(|\boldsymbol{v} - \boldsymbol{v}_e|) \right\rangle_{\boldsymbol{v}_e}. \tag{8.50}$$

8.6.1 High-Speed Expansion

You may then perform a series expansion in powers of $\langle v_e^2 \rangle$ in analogy to (6.91), leading to

$$W(v) \simeq W_0(v) + \frac{\langle v_e^2 \rangle}{v^2}\left(-W_0(v) + \frac{1}{6}v^2 W_0''(v) + \frac{2}{3}mv^2 S_0(v)\right)\dots \quad (8.51)$$

This relation will be used shortly.

8.6.2 Relativistic Extension

A relativistic version of (8.50) has been derived by Tofterup (1983),

$$W(v) = \left\langle \frac{\gamma_v^2(v^2 - \boldsymbol{v}\cdot\boldsymbol{v}_e)^2 - \frac{1}{2}(v^2 v_e^2 - (\boldsymbol{v}\cdot\boldsymbol{v}_e)^2)}{\gamma_v^2 \gamma_{v_e}^2 v v_{\mathrm{M}}^3} W_0(w_e) \right.$$
$$\left. + m\frac{v^2 v_e^2 - (\boldsymbol{v}\cdot\boldsymbol{v}_e)^2}{v v_{\mathrm{M}}} S_0(w_e) \right\rangle_{\boldsymbol{v}_e} , \quad (8.52)$$

in the notation of Sect. 6.5.2.

As in Sect. 6.5.5, we may try an expansion for $v_0 \ll v \ll v_e$, which yields

$$W(v) \simeq \frac{2}{3}mv \langle v_e S_0(v_e)\rangle_{\boldsymbol{v}_e} \quad (8.53)$$

and is identical with the nonrelativistic relation derived by Sigmund (1982). For $v, v_e \simeq c$, the occurrence of the Møller speed in the denominators precludes the simple approximation made in case of the stopping cross section.

8.6.3 Bohr Theory

For the specific case of the Bohr model, where W_0 and $mv^2 S_0$ depend on $\xi = mv^3/Z_1 e^2\omega$, (8.51) reads

$$\frac{W}{W} = \frac{W_0}{W_{\mathrm{B}}} + \frac{\langle v_e^2 \rangle}{v^2}\left[\left(-1 + \xi\frac{\mathrm{d}}{\mathrm{d}\xi} + \frac{3}{2}\xi^2\frac{\mathrm{d}^2}{\mathrm{d}\xi^2}\right)\frac{W_0}{W_{\mathrm{B}}} + \frac{2}{3}L_0\right] \quad (8.54)$$

Straight insertion of the Bohr straggling parameter W_{B} (8.11) into (8.51) yields

$$\frac{W}{W_{\mathrm{B}}} = 1 + \frac{\langle v_e^2 \rangle}{v^2}\left(\frac{2}{3}\ln(C\xi) - 1\right)\dots \quad (8.55)$$

A more accurate estimate could in principle be found by inserting W/W_B and L_0 from Fig. 8.3. However, a Thomas-Fermi estimate of the range of validity of this expansion (problem 8.11) leads to

$$\xi \gg 2\frac{Z_2}{Z_1}. \quad (8.56)$$

As long as we deal with light ions, where $Z_1 \leq Z_2$, this implies that (8.55) is accurate enough within the range of validity of (8.54).

8.6.4 Bethe Theory

The corresponding relation to (8.55) for the Bethe theory is found by inserting $L_0 = \ln(2mv^2/I)$ into (8.50), yielding

$$\frac{W}{W_{\mathrm{B}}} = 1 + \frac{\langle v_e^2 \rangle}{v^2} \left(\frac{2}{3} \ln \frac{2mv^2}{I} - 1 \right) \dots, \tag{8.57}$$

which differs from (8.24) only in the constant under the logarithm.

8.6.5 Quantum Oscillator

For the spherical quantal oscillator, Sigmund and Haagerup (1986) derived the shell correction expansion directly without going over kinetic theory and found

$$\frac{W}{W_{\mathrm{B}}} = 1 + \frac{\hbar\omega_0}{mv^2} \left(\ln \frac{2mv^2}{\hbar\omega_0} - \frac{3}{4} \right) \dots. \tag{8.58}$$

With

$$\frac{1}{2} m \langle v_e^2 \rangle = \frac{3}{2} \hbar\omega_0, \tag{8.59}$$

this goes over into

$$\frac{W}{W_{\mathrm{B}}} = 1 + \frac{\langle v_e^2 \rangle}{v^2} \left(\frac{2}{3} \ln \frac{2mv^2}{\hbar\omega_0} - 1 \right) \dots \tag{8.60}$$

in exact agreement with (8.57). This provides strong support for the use of kinetic theory in straggling calculations.

8.6.6 Bloch Theory

We have seen in Sect. 8.2 that there is no Bloch correction to the straggling parameter. However, that calculation did not include a shell correction. Equation (8.51), on the other hand, contains the stopping cross section which does have a Bloch correction. Hence, the shell correction receives a Bloch term

$$\Delta \left(\frac{W}{W_{\mathrm{B}}} \right) = \frac{2}{3} \frac{\langle v_e^2 \rangle}{v^2} \left[\psi(1) - \mathrm{Re}\psi \left(1 + \mathrm{i} \frac{Z_1 e^2}{\hbar v} \right) \right]. \tag{8.61}$$

This result, mentioned by Sigmund (1982), recovers the result of a direct evaluation by Titeica (1937) of the Bloch correction to straggling.

8.6.7 Fermi Gas

Application of (8.50) to the Fermi gas is easily seen to lead to an incorrect result. Indeed, since L_0 has the form of a Bethe logarithm according to (5.79), the first shell correction would lead to a term

$$\frac{2}{3}\frac{\overline{v_e^2}}{v^2}\ln\frac{4mv^2}{\hbar\omega_{\mathrm{P}}} = \frac{2}{5}\frac{v_F^2}{v^2}\ln\frac{4mv^2}{\hbar\omega_{\mathrm{P}}} \tag{8.62}$$

according to (8.24), i.e., twice the result from a direct expansion, (8.39). The reason for this discrepancy is the fact that no account is taken in this estimate of the rearrangement of the Coulomb cross section at low energy transfers, as discussed in Sect. 8.5. This discrepancy can be removed by proper consideration of the Pauli principle (Sigmund, 1982).

8.6.8 Full Integration

Equation (8.51) can be evaluated without recourse to series expansion for any system where adequate expressions for W_0 and S_0 covering a velocity range from zero up to $v + v_e$ are available. However, no systematic effort has been reported in this area.

Let us see what happens within the Bohr theory. Although the results emerging from Sect. 8.3 could be utilized, an operationally simpler procedure is based on binary stopping theory described in Sect. 6.4.5. Unlike Bohr theory, that scheme incorporates the Barkas-Andersen effect which we want to disregard for the moment. In order to eliminate that effect, an average is taken between the predictions for a positive point charge and the equivalent negative charge. This will be denoted as *modified Bohr theory* in the following.

Figure 8.8 shows a comparison between straggling parameters calculated from the genuine Bohr model and modified Bohr theory. Agreement is not perfect but reasonable for the purpose in mind. After all, binary theory contains corrections to all orders in Z_1, and taking the particle-antiparticle average eliminates only contributions from uneven orders, while even terms of order Z_1^4 and higher remain.

Figure 8.9 shows the effect of shell correction. Compared are the prediction of modified Bohr theory with and without shell correction and the expansion (8.24). Excellent agreement is found between the high-speed expansion and modified-Bohr theory well above the Bethe-Livingston shoulder, but a significant discrepancy is found around the maximum and below. However, in the absence of a Barkas-Andersen correction, the question of what is right or wrong is not necessarily meaningful.

8.7 Barkas-Andersen Correction ($\star$)

We have seen in Chapter 6 that the velocity domains of significant shell correction and significant Barkas-Andersen correction overlap. While it is of interest

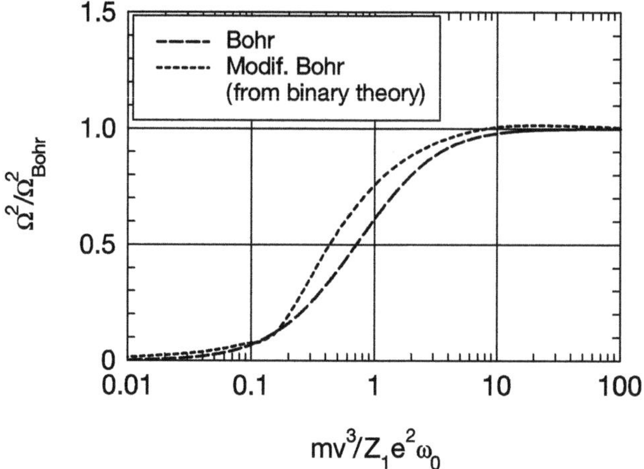

Fig. 8.8. Comparison of relative straggling calculated from Bohr model, Fig. 8.2, and modified Bohr theory (see text). Shell correction ignored

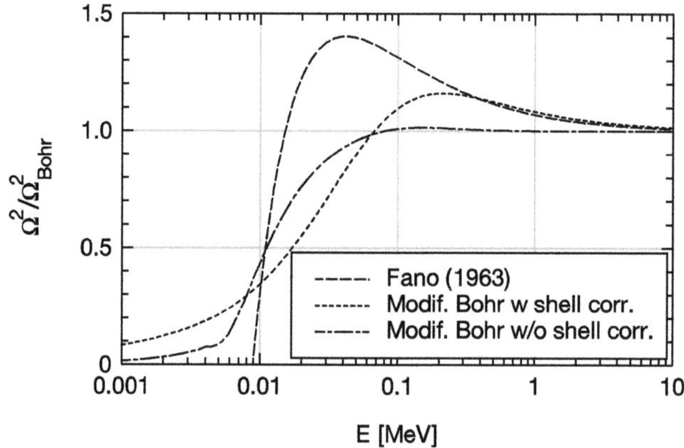

Fig. 8.9. Predicted straggling for H^+ in H. Dashed line: Equation (8.24) excluding the last term (Fano, 1963). Dot-dashed line: Modified Bohr theory without shell correction; dotted line: Modified Bohr theory with shell correction

to separately study the two phenomena, valid predictions allowing comparison with experiment need to incorporate the two effects combined. The problem has not received much attention in the literature. The present survey is based upon studies by Wang and Pitarke (1998) on the free-electron gas and by Sigmund and Schinner (2002) who employed binary stopping theory.

The Barkas-Andersen effect has its origin in the binding of a target electron. Bohr straggling as given by (2.57) is an exact result for straight Coulomb scattering, and the dominating contribution stems from close collisions which are insensitive to binding. Hence, one might expect the Barkas-Andersen effect in straggling to be less significant than in the stopping cross section.

This expectation is strengthened by applying Lindhard's procedure described in Sect. 6.4.3 – where the Barkas effect is simulated as a change in effective collision energy – to the Bohr straggling parameter W_B. Indeed, as W_B does not depend on the projectile speed, the operation (6.37) has no effect. However, Fig. 8.2 indicates that this simple result can only be true for $\xi \gtrsim 10$, whereas a Barkas-Andersen correction should be expected in the regime where the straggling parameter starts declining below the Bohr value.

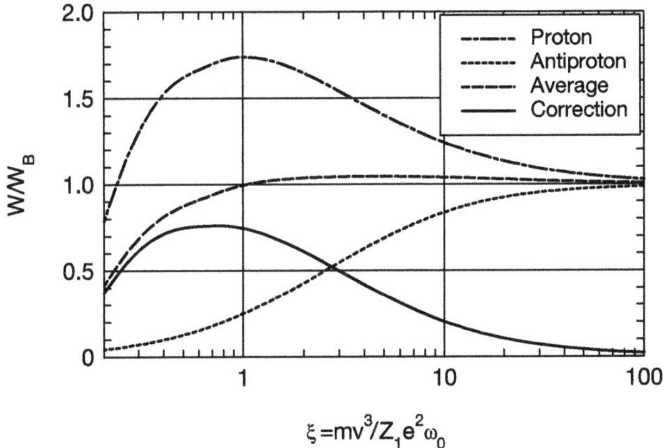

Fig. 8.10. Straggling for protons and antiprotons interacting with a classical Bohr oscillator, evaluated by binary theory. Shell correction neglected

Figure 8.10 shows predictions for the Bohr oscillator evaluated by the binary theory described in Sect. 6.4.5, disregarding shell correction. The line labelled 'average' denotes modified Bohr theory, and the line labelled 'correction' denotes the difference between the result for protons and the average. As expected, the Barkas-Andersen correction decreases rapidly for $\xi \gtrsim 10$. Conversely, it is seen to reach almost 100 % for $\xi \lesssim 1$.

Figure 8.11 shows predictions of binary theory for protons in atomic hydrogen. The curve labelled 'plain' denotes modified Bohr theory, and the remaining three curves incorporate shell correction and Barkas-Andersen correction, respectively, as well as the combination. It is seen that the shoulders produced by the two corrections do not strictly lie in the same energy interval (cf. problem 8.12). There is a high-speed regime where the Barkas-Andersen correction is insignificant while the shell correction is still essential. Conversely, at velocities below the shoulder, the shell correction reduces the enhancement due to the Barkas-Andersen correction.

Wang and Pitarke (1998) applied a second-order response function established by Pitarke et al. (1995) to determine the straggling parameter of a Fermi

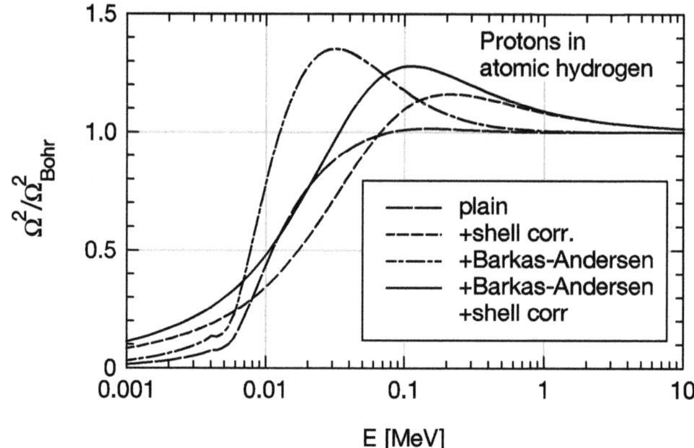

Fig. 8.11. Straggling for protons in atomic hydrogen, calculated from binary theory

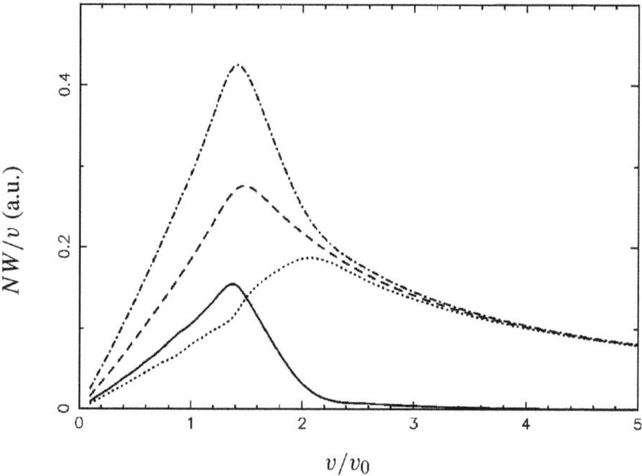

Fig. 8.12. Straggling parameter of a Fermi gas up to third order in Z_1 according to Wang and Pitarke (1998). Dashed line: Z_1^2 term; solid line: Z_1^3 term; dash-dot line: Total for protons; dotted line: Total for antiprotons. r_s=2. From Wang and Pitarke (1998)

gas up to the third order in Z_1. A typical result is shown in Fig. 8.12[5]. It is seen that the Z_1^3 contribution is negligible for $v \gtrsim 2v_0$ but increases rapidly at lower speed, reaching a maximum of $\simeq 60$ % at $v = 1.4v_0$.

---

[5] The different shape is due to the factor $1/v$ in the plotted quantity. It is easily verified that Bohr stopping is approached at the upper end of the velocity range.

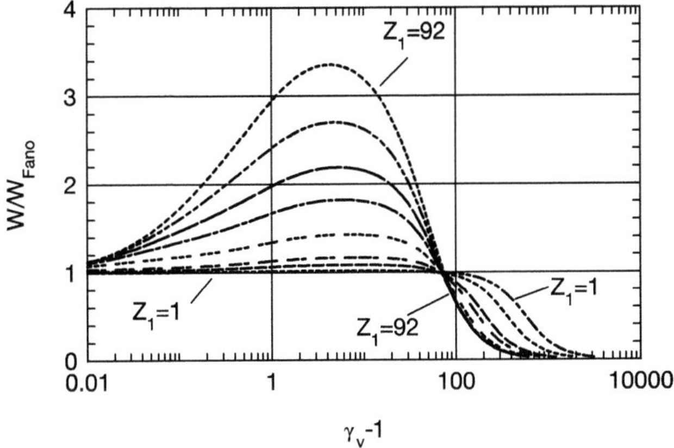

Fig. 8.13. Straggling in units of $W_{\mathrm{Bohr}}(1 - \beta^2/2)/(1 - \beta^2)$ in the relativistic regime according to Lindhard and Sørensen (1996) for $Z_1 = 92, 79, 66, 54, 36, 18, 10, 2$ and 1. All corrections, including finite nuclear size incorporated. $\gamma_v = 1/\sqrt{1 - v^2/c^2}$

8.8 Relativity: Lindhard-Sørensen Theory ($\star$)

For $Z_1 \gg 1$, a correction of the kind considered in Sect. 6.7.2 must be expected to play a role, even though we have seen in Sect. 8.2 that unlike in the stopping cross section it decreases toward zero with decreasing projectile speed. The treatment of Lindhard and Sørensen (1996) includes straggling and has been carried out in straight analogy to the one of the stopping cross section. The result of this treatment is expressed in terms of the transport cross section $\sigma^{(2)}$ defined in (3.120),

$$\sigma^{(2)} = 2\sigma^{(1)} - 4\pi\lambda^2 \sum_{\varkappa \neq 0} |\varkappa| \left[\frac{(\varkappa - 1)(\varkappa - 2)}{(2\varkappa - 1)(2\varkappa - 3)} \sin^2\left(\delta_\varkappa - \delta_{\varkappa - 2}\right) \right.$$

$$+ \frac{\varkappa - 1}{(2\varkappa - 3)(4\varkappa^2 - 1)} \sin^2\left(\delta_\varkappa - \delta_{-\varkappa + 1}\right)$$

$$\left. + \frac{1}{2} \frac{\varkappa + 1}{2\varkappa + 1} \left(\frac{1}{4\varkappa^2 - 1} + \frac{1}{4(\varkappa + 1)^2 - 1} \right) \sin^2\left(\delta_\varkappa - \delta_{-\varkappa - 1}\right) \right] \quad (8.63)$$

for a spherically symmetric potential, in the notation of Sect. 6.3.2.

Figure 8.13 shows results for $Z_1 = 1$–92 taking into account the finite-nuclear-size effect. The quantity plotted is the relative straggling parameter $(W/W_{\mathrm{Bohr}})(1 - \beta^2)/(1 - \beta^2/2)$, i.e., relative to the result from the high-speed limit of the Born approximation (8.31). It is seen that for $Z_1 \lesssim 10$, the Born approximation is very accurate except for $\gamma_v > 200$, where the finite-nuclear-size effect becomes important. However, for $\gamma_v < 100$, straggling becomes significantly larger than the Born result for heavy nuclei over a wide range of beam energies.

8.9 Bunching

Estimates of straggling presented above ignore all spatial correlation between target electrons. This implies that the electron density only enters as a proportionality factor. This is obvious in case of W_B which assumes Coulomb scattering on independent target electrons, but estimates based on the Bohr model and its extensions still operate with one target electron per atom and hence ignore the fact that electrons are bunched in atoms. Although the expression (8.18) based on the Born approximation fully takes into account all correlations within an atom that give contributions to straggling $\propto Z_1^2$, such correlations are ignored in (8.24), as mentioned explicitly by Fano (1963). Even in case of the Fermi gas, straggling comes out proportional to the electron density at high projectile speed, and low-speed corrections going as v_F^2 or ω_P^2 account for zero-point motion and effective binding but not for proximity in space.

A simple model may illustrate the potential importance of correlation effects. Consider first an ideal gas of N atoms per volume, each containing 2 electrons which can be raised to an excitation level ϵ. Let the cross section for exciting both electrons be σ, while excitation of just one electron is supposed to have vanishing probability. Then the stopping force and straggling are given by

$$-\frac{dE}{dx} = Nx\, 2\epsilon\, \sigma; \quad \Omega^2 = Nx\, (2\epsilon)^2\, \sigma. \tag{8.64}$$

Compare this system to an ideal gas of $2N$ atoms per volume, each with one electron and unchanged excitation energy and cross section. For this system we have

$$-\frac{dE}{dx} = 2N\, x\, \epsilon\, \sigma; \quad \Omega^2 = 2N\, x\, \epsilon^2\, \sigma, \tag{8.65}$$

i.e., the same stopping force but only half the straggling. Clearly, more collisions and correspondingly less energy loss per collision reduce straggling.

This kind of effect was first studied as a pairing effect in diatomic molecules (Sigmund, 1976a), but it must occur wherever the interelectronic distance is comparable to or smaller than the effective maximum impact parameter contributing to energy loss.

8.9.1 Classical Estimate

The bunching effect is most easily illustrated in a classical model where collisions may be described in terms of an impact parameter to an individual electron. In such a model the straggling parameter is given by

$$W = \int d^2\boldsymbol{p} \left\langle \left(\sum_{i=1}^{Z_2} T_i(\boldsymbol{p}_i) \right)^2 \right\rangle = \int d^2\boldsymbol{p} \sum_{i,j}^{Z_2} \left\langle T_i(\boldsymbol{p}_i) T_j(\boldsymbol{p}_j) \right\rangle \tag{8.66}$$

according to Sigmund and Schinner (2002), where $T_i(\boldsymbol{p}_i)$ is the energy loss to the ith target electron at an impact parameter $\boldsymbol{p}_i$, and $\boldsymbol{p}$ the impact parameter with respect to the nucleus. The brackets indicate an average over the spatial distribution of target electrons around the nucleus.

We may split the sum into two parts with $i = j$ and $i \neq j$, respectively,

$$W = \int \mathrm{d}^2\boldsymbol{p} \sum_{i=1}^{Z_2} \left\langle T_i^2(\boldsymbol{p}_i) \right\rangle + \Delta W, \tag{8.67}$$

where

$$\Delta W = \int \mathrm{d}^2\boldsymbol{p} \sum_{i \neq j}^{Z_2} \left\langle T_i(\boldsymbol{p}_i) T_j(\boldsymbol{p}_j) \right\rangle. \tag{8.68}$$

The first term on the right-hand side of (8.67) reflects straggling due to spatially uncorrelated electrons, while the correlation term ΔW is nonvanishing whenever there is a range of impact parameters where both $T_i(\boldsymbol{p}_i)$ and $T_j(\boldsymbol{p}_j)$ differ significantly from zero. Note that this term cannot become negative, even in case of anticorrelation between target electrons as e.g. in the ground state of a helium atom.

To further simplify the situation, take an independent-electron model of the atom so that

$$\Delta W = \int \mathrm{d}^2\boldsymbol{p} \sum_{i \neq j}^{Z_2} \langle T_i(\boldsymbol{p}_i) \rangle \langle T_j(\boldsymbol{p}_j) \rangle \tag{8.69}$$

or

$$\Delta W = \int \mathrm{d}^2\boldsymbol{p} \left(\sum_{i=1}^{Z_2} \langle T_i(\boldsymbol{p}_i) \rangle \right)^2 - \int \mathrm{d}^2\boldsymbol{p} \sum_{i=1}^{Z_2} \langle T_i(\boldsymbol{p}_i) \rangle^2. \tag{8.70}$$

The key operation here is the spatial averaging

$$T_i'(\boldsymbol{p}) = \langle T_i(\boldsymbol{p}_i) \rangle = \int \mathrm{d}^3\boldsymbol{r}_i \, n_i(\boldsymbol{r}_i) T_i(\boldsymbol{p} - \boldsymbol{\rho}_i), \tag{8.71}$$

where $n_i(\boldsymbol{r}_i)$ is the electron density and $\boldsymbol{\rho}_i$ the component of the electron coordinate $\boldsymbol{r}_i$ perpendicular to the projectile velocity. The difference between $T_i(\boldsymbol{p}_i)$ and $T_i'(\boldsymbol{p})$ is substantial: T_i depends on the impact parameter to a target *electron* and hence approaches $2mv^2$ – or a shell-corrected maximum energy transfer – at $\boldsymbol{p} = 0$. Conversely, $T_i'(\boldsymbol{p})$ may well have a minimum at $\boldsymbol{p} = 0$ because the chance of hitting a target electron head-on may be greater at a nonvanishing distance from the nucleus.

With this we may write

$$\Delta W = \int \mathrm{d}^2\boldsymbol{p} \, T'(\boldsymbol{p})^2 - \int \mathrm{d}^2\boldsymbol{p} \sum_i T_i'(\boldsymbol{p})^2. \tag{8.72}$$

Here, $T'(p) = \sum_i T_i'(p)$ is the average energy transfer to the atom at an impact parameter p to the nucleus.

Eq. (8.72) without the second term was stated by Besenbacher et al. (1980) and applied to an electron-gas model. However, the second term is essential at least for few-electron atoms since in its absence also a one-electron atom would show correlation.

8.9.2 Bunching in Helium

As an example, consider straggling in helium in the modified-Bohr model, where (8.67) and (8.72) reduce to

$$W = 2 \int \mathrm{d}^2 p \langle T_1^2(\boldsymbol{p}) \rangle = 2W_1. \tag{8.73}$$

Here W_1 represents the straggling parameter per electron ignoring correlation, and

$$\Delta W = 2 \int \mathrm{d}^2 p \, \langle T_1(\boldsymbol{p}_1) \rangle \, \langle T_1(\boldsymbol{p}_2) \rangle \tag{8.74}$$

$$= 2 \int \mathrm{d}^2 p \int \mathrm{d}^3 r_1 \, n_1(\boldsymbol{r}_1) \int \mathrm{d}^3 r_2 \, n_1(\boldsymbol{r}_2) T_1(\boldsymbol{p} - \boldsymbol{\rho}_1) T_1(\boldsymbol{p} - \boldsymbol{\rho}_2). \tag{8.75}$$

In the unmodified Bohr model, where the target electron is at rest in the origin, we have $\boldsymbol{p} = \boldsymbol{p}_1 = \boldsymbol{p}_2$, and hence ΔW becomes equal to $2W_1$, i.e., straggling would be doubled compared to the Bohr value. The correlation is more or less reduced by the fact that the two electrons are likely to be apart. It is even further reduced when anticorrelation caused by the Pauli principle is taken into consideration.

Introducing two additional variables $\boldsymbol{p}_1, \boldsymbol{p}_2$,

$$\Delta W = 2 \int \mathrm{d}^2 p \int \mathrm{d}^3 r_1 \, n_1(\boldsymbol{r}_1) \int \mathrm{d}^3 r_2 \, n_1(\boldsymbol{r}_2)$$

$$\times \int \mathrm{d}^2 p_1 \, T_1(\boldsymbol{p}_1) \delta(\boldsymbol{p}_1 - \boldsymbol{p} + \boldsymbol{\rho}_1) \int \mathrm{d}^2 p_2 \, T_1(\boldsymbol{p}_2) \delta(\boldsymbol{p}_2 - \boldsymbol{p} + \boldsymbol{\rho}_2), \tag{8.76}$$

and inserting Fourier transforms of the Dirac functions, you identify Fourier transforms of $T_1(\boldsymbol{p})$ and of $n_1(\boldsymbol{r})$. The integration over $\boldsymbol{p}$ can be carried out directly, and as a result you arrive at

$$\Delta W = \frac{1}{2\pi^2} \int \mathrm{d}^2 k \, |P_1(\boldsymbol{k})|^2 \, |T_1(\boldsymbol{k})|^2 \,, \tag{8.77}$$

where

$$P_1(\boldsymbol{k}) = \int \mathrm{d}^3 r \, n_1(r) \mathrm{e}^{-\mathrm{i}\boldsymbol{k}\cdot\boldsymbol{\rho}} \tag{8.78a}$$

$$T_1(\boldsymbol{k}) = \int \mathrm{d}^2 p \, T_1(p) \mathrm{e}^{-\mathrm{i}\boldsymbol{k}\cdot\boldsymbol{p}}. \tag{8.78b}$$

Since $n_1(r)$ is independent of orientation, further simplification is possible,

$$P_1(k) = \int_0^\infty 4\pi r^2 \, dr \, n_1(r) \, \frac{\sin kr}{kr}. \tag{8.79}$$

For a $1s$ state with $\psi_1(r) = \exp(-r/a')/\sqrt{\pi(a')^3}$ we may set $n_1(r) = |\psi_1(r)|^2$ and find

$$P_1(k) = \frac{1}{(1 + a'^2 k^2/4)^2}. \tag{8.80}$$

As an illustration let us assume an exponential dependence of the energy-loss function $T_1(p)$ (Grande and Schiwietz, 1991),

$$T_1(p) = \frac{S_1}{2\pi a^2} e^{-p/a}, \tag{8.81}$$

where S_1 is the atomic stopping cross section and $1/a$ a suitably chosen slope. This yields

$$T(k) = \frac{S_1}{(1 + k^2 a^2)^{3/2}} \tag{8.82}$$

and, hence,

$$\Delta W = \frac{S^2}{2\pi^2} \int_0^\infty \frac{2\pi k \, dk}{(1 + a^2 k^2)^3 \, (1 + a'^2 k^2/4)^4} \tag{8.83}$$

or

$$\frac{\Delta W}{2W_1} = 2 \int_0^\infty \frac{d\eta}{(1 + \eta)^3 (1 + (a'/2a)^2 \eta)^4}. \tag{8.84}$$

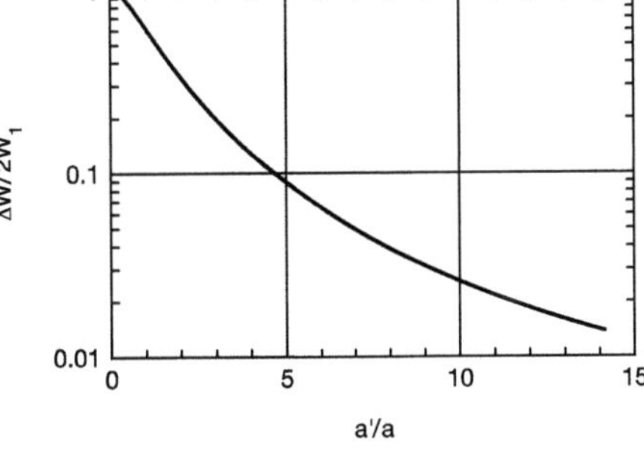

Fig. 8.14. Relative correlation, (8.84), for a simple model of a helium atom

This ratio is shown in figure 8.14. As expected, correlation would cause a 100 % increase in straggling for $a' \ll a$, i.e., if the interaction range of the beam – as defined by (8.81) – were much greater than the radius of the atom.

In order to appreciate the significance of this result, let us try to identify the physical significance of the radius a introduced in (8.81). It would be a mistake to identify it with the straight adiabatic radius, because what is important is the behavior around impact parameter zero, where the slope is governed rather by $b/2$, the classical collision radius according to (3.46), or perhaps the de Broglie wavelength $\hbar/mv$, whichever is larger. Unlike the adiabatic radius, these quantities become large at low speed.

8.9.3 Molecular Gas

An illuminating case of bunching – allowing a fairly direct experimental test – is that of a diatomic target molecule (Sigmund, 1976a). Assume energy loss functions $T_1(p)$ and $T_2(p)$ for the two constituent atoms and ignore all fluctuations at a given impact parameter. The energy loss may be electronic or nuclear or a combination of the two. In an individual collision,

$$T(\boldsymbol{p}) = T_1(\boldsymbol{p}_1) + T_2(\boldsymbol{p}_2), \tag{8.85}$$

leading to a stopping cross section

$$S = S_1 + S_2 \tag{8.86}$$

and a straggling parameter

$$W = W_1 + W_2 + \Delta W \tag{8.87}$$

with

$$\Delta W = 2 \int \mathrm{d}^2\boldsymbol{p}\, T_1(\boldsymbol{p}_1)T_2(\boldsymbol{p}_2). \tag{8.88}$$

While this is very similar to (8.76), the evaluation of this integral is simplified if we remember that the internuclear distance in a molecule is a well-defined quantity. Disregarding molecular vibrations we may set

$$\boldsymbol{p}_1 - \boldsymbol{p}_2 = \boldsymbol{b}, \tag{8.89}$$

where $\boldsymbol{b}$ is the projection of the internuclear axis $\boldsymbol{d}$ on the impact plane, i.e., a plane perpendicular to the beam velocity $\boldsymbol{v}$ (Fig. 8.15). With this, (8.88) may be rewritten as

$$\Delta W = 2 \int \mathrm{d}^2\boldsymbol{p}_1 \int \mathrm{d}^2\boldsymbol{p}_2\, T_1(\boldsymbol{p}_1)T_2(\boldsymbol{p}_2)\delta(\boldsymbol{p}_1 - \boldsymbol{p}_2 - \boldsymbol{b}). \tag{8.90}$$

Except for specially prepared systems, we may assume target molecules to be randomly oriented, i.e., ΔW has to be averaged over all orientations of $\boldsymbol{d}$.

Moreover, we may assume the atomic energy-loss functions to be isotropic[6].
Then, the only directionally-dependent quantity is the vector $\boldsymbol{b}$, the projection
of $\boldsymbol{d}$ on the impact plane. Hence, rotational averaging only affects the Dirac
function in (8.90).

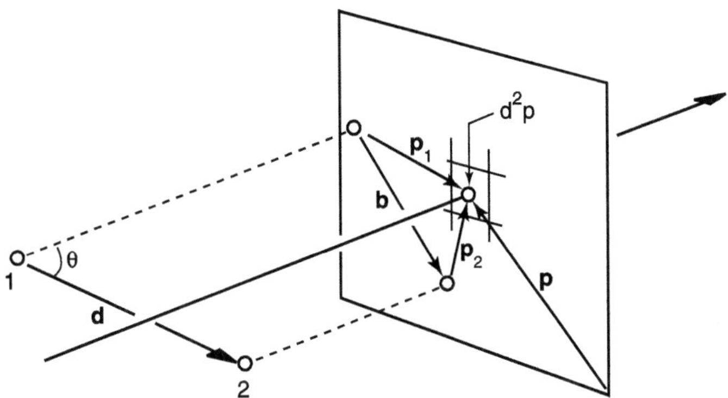

Fig. 8.15. Geometry of ion-molecule interaction. From Sigmund (1976b)

Writing

$$\boldsymbol{b} = (d \sin \theta \cos \psi, d \sin \theta \sin \psi, 0),\tag{8.91}$$

you find (cf. problem 8.13) that

$$\langle \delta(\boldsymbol{p}_1 - \boldsymbol{p}_2 - \boldsymbol{b}) \rangle_{\text{orientation}} = \frac{1}{4\pi} \int_0^{2\pi} \mathrm{d}\psi \int_0^{\pi} \sin \theta \, \mathrm{d}\theta \, \delta(\boldsymbol{p}_1 - \boldsymbol{p}_2 - \boldsymbol{b})$$

$$= \frac{1}{2\pi d^2 \sqrt{1 - (\boldsymbol{p}_1 - \boldsymbol{p}_2)^2/d^2}}\tag{8.92}$$

and hence

$$\langle \Delta W \rangle_{\text{orientation}} = \frac{1}{\pi d^2} \int \mathrm{d}^2 \boldsymbol{p}_1 \int \mathrm{d}^2 \boldsymbol{p}_2 \, \frac{T_1(\boldsymbol{p}_1)T_2(\boldsymbol{p}_2)}{\sqrt{1 - (\boldsymbol{p}_1 - \boldsymbol{p}_2)^2/d^2}}.\tag{8.93}$$

As a rough approximation, valid if the effective range of interaction is less
than the internuclear distance, one may approximate the denominator in the
integrand by $\simeq 1$ and obtain

$$\langle \Delta W \rangle_{\text{orientation}} \simeq \frac{S_1 S_2}{\pi d^2}.\tag{8.94}$$

---

[6] Evidently, the only difference between a molecule and two independent atoms in
this model is the correlation in internuclear distance, while valence effects on the
electronic structure, called deviations from Bragg additivity and discussed briefly
in Sect. 7.7, are ignored.

Even though this result may not be quantitatively accurate, it indicates that the strength of correlation has its maximum in the energy regime around the stopping maximum.

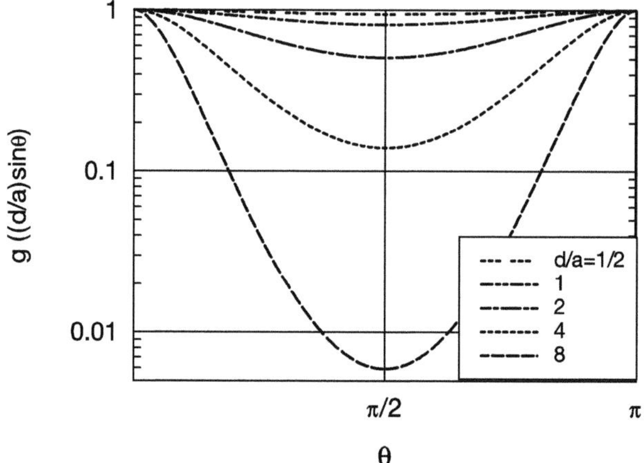

Fig. 8.16. Molecular-correlation effect according to (8.101) for exponential dependence of the energy loss on impact parameter: Dependence on angle θ between the molecular axis and the beam. $g(\xi) = 9\xi^2/2)K_2(\xi)$

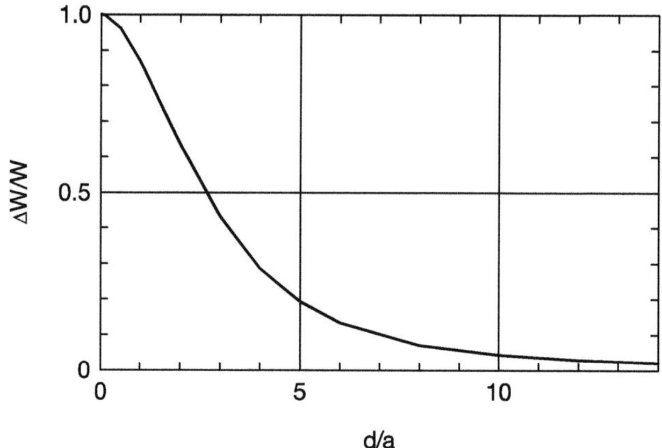

Fig. 8.17. Molecular-correlation effect for exponential dependence of the energy loss on impact parameter. Averaged over orientation

You can come further with a quantitative estimate for (8.90) by an alternative procedure. Replacing the Dirac function by its Fourier transform you find

$$\Delta W = \frac{1}{2\pi^2} \int \mathrm{d}^2 \boldsymbol{k}\, \mathrm{e}^{-\mathrm{i}\boldsymbol{k}\cdot\boldsymbol{b}}\, |T(\boldsymbol{k})|^2 \tag{8.95}$$

with

$$T(\boldsymbol{k}) = \int \mathrm{d}^2 \boldsymbol{p}\, T(\boldsymbol{p}) \mathrm{e}^{-\mathrm{i}\boldsymbol{k}\cdot\boldsymbol{p}}. \tag{8.96}$$

The orientational average can then be carried out (cf. problem 8.14) and yields

$$\left\langle \mathrm{e}^{-\mathrm{i}\boldsymbol{k}\cdot\boldsymbol{p}} \right\rangle_{\text{orientation}} = \frac{\sin kd}{kd}. \tag{8.97}$$

For an estimate, assume a model energy-loss function (Grande and Schiwietz, 1991)

$$T(\boldsymbol{p}) \equiv T(p) = \frac{S}{2\pi a^2} \mathrm{e}^{-p/a}, \tag{8.98}$$

where S is the stopping cross section and a characterizes the slope of some given energy-loss function. Then,

$$T(\boldsymbol{k}) = \frac{S}{(1 + k^2 a^2)^{3/2}} \tag{8.99}$$

and

$$\Delta W = \frac{S^2 b^2}{8\pi a^4} K_2\left(\frac{b}{a}\right) \tag{8.100}$$

or

$$\frac{\Delta W}{2W} = \frac{d^2}{2a^2} \sin^2 \theta\, K_2\left(\frac{d}{a}\sin\theta\right), \tag{8.101}$$

where K_2 is a modified Bessel function of the second kind. Figure 8.16 shows the angular dependence for a number of values of d/a. It is seen that the united-atom limit, where $\Delta W = 2W_1$, is reached essentially for $d \lesssim a$, while for $d \gg a$, correlation is observed only when the molecule is nearly aligned to the beam. Figure 8.17 shows the average over all orientations as a function of the ratio d/a.

8.9.4 Dense Matter

In condensed matter, atoms are arranged in some densely packed structure. Even in the presence of lattice vibrations, a penetrating particle moves most often so fast that it will see a static structure of target nuclei. In crystalline solids the regular structure may affect the trajectory, as mentioned in Chapter 1, cf. Fig. 1.3. In amorphous and liquid matter, scattering processes may

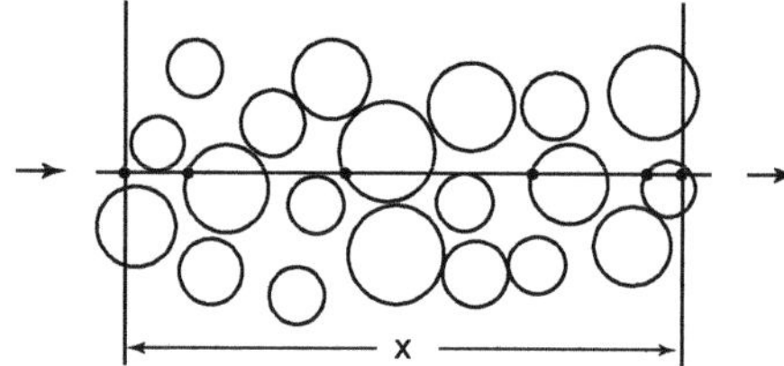

Fig. 8.18. Trajectory of a fast particle in a solid. Angular deflection neglected. Redrawn after Dixmier et al. (1982).

be expected to be irregular, so that a straight trajectory will be a most feasible first approximation (Fig. 8.18). Then the target atoms hit by a penetrating particle are those close enough to that trajectory, and their positions relative to the trajectory are well defined once the impact point of the projectile has been determined (Fig. 8.19). Statistical elements in such a model are

- The impact point on the entrance surface, denoted p in Fig. 8.19,
- Variations in static target configuration due to lattice vibrations, and
- Quantal fluctuations of individual energy losses at a given impact parameter.

There is no reason to a priori expect statistical independence of successive collision events. Indeed, in a random medium there would be a possibility of two atoms lying on top of each other. If the average interatomic distance is large compared to the atomic radius, this is unlikely and hence of little

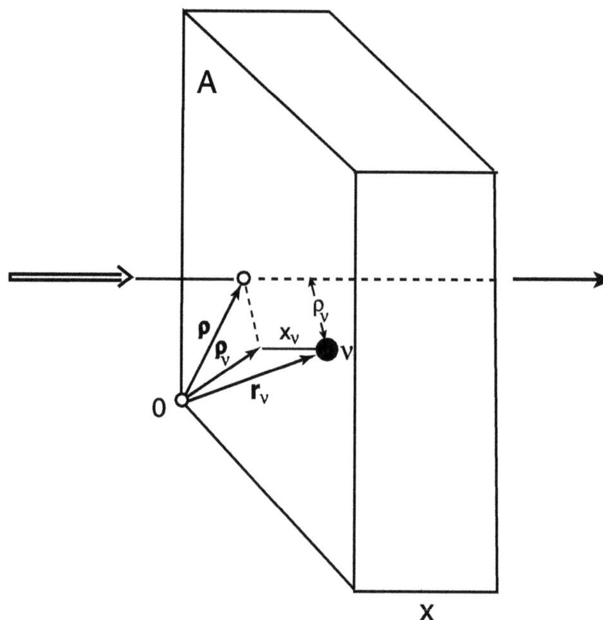

Fig. 8.19. Definition of geometric quantities determining energy loss in solids

concern. In the opposite case of a dense medium a noticeable error may be expected when such a possibility is allowed for.

Once the fundamental equation (9.4) is no longer applicable, we need another starting point for the discussion of energy loss and straggling where the primary element of randomness is the impact point of the beam particle. The scheme reported here was developed by the author (Sigmund, 1978).

Consider a solid target with z atoms in a slab of thickness x penetrated by the beam. These atoms, $\nu = 1, 2, \ldots z$ are located in positions $\mathbf{r}_\nu$, and an individual trajectory is assumed to be a straight line defined by a lateral position $\mathbf{p}$ (Fig. 8.19). The total energy loss in a given trajectory is given by

$$\Delta E = \sum_\nu T(\mathbf{p} - \mathbf{p}_\nu), \tag{8.102}$$

where $\mathbf{p}_\nu$ is the lateral component of the position vector $\mathbf{r}_\nu$.

The average energy loss is found by randomizing the point of impact,

$$\langle \Delta E \rangle = \frac{1}{A} \int d^2\mathbf{p}\, \Delta E, \tag{8.103}$$

where A is the cross sectional area of the beam. If A is a macroscopic area, it is much wider than the region where $T(\mathbf{p})$ is nonvanishing. Therefore, it is justified to extend the integration in (8.103) over the infinite plane, so that

$$\langle \Delta E \rangle = \frac{1}{A} \sum_\nu S = \frac{z}{A} S = NxS \tag{8.104}$$

for a monoatomic medium, where S is the atomic stopping cross section and $z = NxA$. Evidently, the mean energy loss is unaffected by the structure of the solid.

Consider now the mean-square energy loss,

$$\langle (\Delta E)^2 \rangle = \frac{1}{A} \int d^2\mathbf{p} \sum_{\mu,\nu} T(\mathbf{p} - \mathbf{p}_\mu) T(\mathbf{p} - \mathbf{p}_\nu). \tag{8.105}$$

You may replace the integration variable $\mathbf{p}$ by $\mathbf{p} + \mathbf{p}_\mu$ in each individual term in (8.105) and split the double sum into terms $\mu = \nu$ and $\mu \neq \nu$, respectively. This yields

$$\langle (\Delta E)^2 \rangle = \frac{z}{A} W + \frac{z}{A} \sum_\nu{}' \int d^2\mathbf{p}\, T(\mathbf{p}) T(\mathbf{p} + \mathbf{b}_\nu), \tag{8.106}$$

where $\mathbf{b}_\nu$ indicates the lateral component of an internuclear distance vector, and the apostrophe indicates omission of the vector pointing from one target nucleus onto itself. $W = \int d^2\mathbf{p}\, T^2(\mathbf{p})$ is the atomic straggling parameter.

In the limiting case of a random medium, the sum over ν can be replaced by

$$\sum_{\nu} \rightarrow \frac{z}{A} \int \mathrm{d}^2 \boldsymbol{b}. \tag{8.107}$$

Then, the second term on the right-hand side of (8.106) factorizes into

$$(zS/A)^2 = (NxS)^2 = \langle \Delta E \rangle^2 , \tag{8.108}$$

as it must be.

In a real solid, internuclear distances are distributed in accordance with a *pair distribution function* $g_2(\boldsymbol{r})$, where $N g_2(\boldsymbol{r})\, \mathrm{d}^3 \boldsymbol{r}$ is the probability to find a target atom in a volume element $\mathrm{d}^3 \boldsymbol{r}$ at a vector distance $\boldsymbol{r}$ from a given target atom, averaged over all target atoms. We may then make the following replacement,

$$\sum_{\nu}' \rightarrow N \int \mathrm{d}^3 \boldsymbol{r}\, g_2(\boldsymbol{r}) = N \int \mathrm{d}^3 \boldsymbol{r}\, [g_2(\boldsymbol{r}) - 1] + N \int \mathrm{d}^3 \boldsymbol{r}, \tag{8.109}$$

so that

$$\langle (\Delta E)^2 \rangle = NxW + N^2 x \int \mathrm{d}^3 \boldsymbol{r} \int \mathrm{d}^2 \boldsymbol{p}\, T(\boldsymbol{p})\, T(\boldsymbol{p} + \boldsymbol{b})\, [g_2(\boldsymbol{r}) - 1]$$
$$+ N^2 x \int \mathrm{d}^3 \boldsymbol{r} \int \mathrm{d}^2 \boldsymbol{p}\, T(\boldsymbol{p})\, T(\boldsymbol{p} + \boldsymbol{b}). \tag{8.110}$$

In the last term on the right-hand side we may integrate successively over $\boldsymbol{b}$ and $\boldsymbol{p}$, noticing that $\boldsymbol{b}$ is the lateral component of $\boldsymbol{r}$, whereafter that term reduces to $(NxS)^2$ as above. In the second term we may introduce an additional variable $\boldsymbol{p}' = \boldsymbol{p} + \boldsymbol{b}$ so that, altogether,

$$\Omega^2 = \left\langle (\Delta E - \langle \Delta E \rangle)^2 \right\rangle = NxW + N^2 x \int \mathrm{d}^2 \boldsymbol{p}\, T(\boldsymbol{p}) \int \mathrm{d}^2 \boldsymbol{p}'\, T(\boldsymbol{p}')$$
$$\int \mathrm{d}^3 \boldsymbol{r}\, [g_2(\boldsymbol{r}) - 1]\, \delta(\boldsymbol{p}' - \boldsymbol{p} - \boldsymbol{b}). \tag{8.111}$$

This expression reminds of (8.90). Indeed, assuming $T(p)$ and $g_2(r)$ to be directionally independent we may perform an angular average just as in (8.92) and obtain, finally,

$$\Omega^2 = Nx(W + \Delta W) \tag{8.112}$$

with

$$\Delta W = 2N \int \mathrm{d}^2 \boldsymbol{p}\, T(p) \int \mathrm{d}^2 \boldsymbol{p}'\, T(p') \int \mathrm{d}r\, \frac{g_2(r) - 1}{\sqrt{1 - (\boldsymbol{p} - \boldsymbol{p}')^2/r^2}} \tag{8.113}$$

for the variance in a monoatomic, isotropic, but nonrandom stopping medium.

Before discussing implications of (8.113) for dense stopping materials, let us briefly go back to the case of a dilute molecular gas. Here, the pair distribution function has the form

$$g_2(r) = 1 + \frac{\delta(r-d)}{4\pi N d^2},\tag{8.114}$$

as can be verified from the normalization. d is the internuclear distance in the molecule and N the density of *atoms*. Then, insertion of (8.114) into (8.113) yields (8.93), as it should.

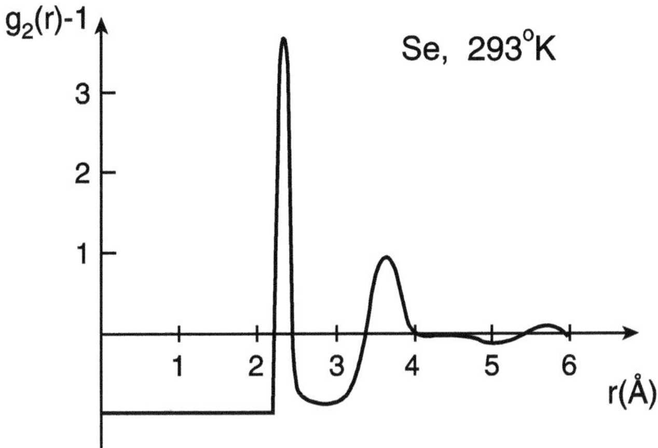

Fig. 8.20. Measured pair correlation function $g_2(r) - 1$ for selenium according to Hansen and Carneiro (1977)

For an amorphous solid, the *pair correlation function $g_2(r) - 1$* looks like the one shown in Fig. 8.20. It can be considered as being composed mainly of a region where $g_2(r) - 1 = -1$ at small distances, a rather sharp peak at the nearest-neighbor distance and a somewhat broader peak at the second-nearest-neighbor distance, whereafter it remains close to zero. Evidently, the most pronounced feature is a sphere around an atom where there are no other atoms. Thus, we may expect that generally, ΔW will be negative in dense matter. Indeed, while in a molecular gas, correlation forces atoms to pair at a distance less than average, the opposite situation appears to prevail in dense matter.

As in the case of the molecular gas, following (8.98) on page 307, a more specific analytical estimate can be made by assuming an exponential dependence of the energy-loss function on impact parameter. In that case you find

$$\frac{\Delta(\Omega^2)}{\Omega^2} = N \int d^3r \, [g_2(r) - 1] \, \frac{r^2}{2a^2} \sin^2\theta \, K_2\left(\frac{r}{a}\sin\theta\right)\tag{8.115}$$

in the notation of (8.101). Since the correlation function does not depend on orientation, we may perform the angular average and hence arrive at the function shown in Fig. 8.17, which may be approximated by $\exp(-0.15(r/a)^{3/2})$, and hence,

$$\frac{\Delta(\Omega^2)}{\Omega^2} \simeq N \int 4\pi r^2 \, \mathrm{d}r \Big[g_2(r) - 1\Big] \mathrm{e}^{-0.15(r/a)^{3/2}} . \tag{8.116}$$

The factor r^2 in this integral is essential. It implies that the nearest-neighbor peak in the correlation function may play a significant role.

8.10 Straggling Measurements

In comparisons between straggling measurements, it is vital to note what actually has been measured. Calculations presented in this chapter address the *variance* of an energy-loss spectrum, just as calculations presented in Chapters 4–6 address the mean energy loss. Experimentalists, when having to characterize a spectrum by two parameters, are tempted to quote peak position and halfwidth. This is not only easier: Determining a genuine mean value and variance may be impossible if there is a significant background in the detecting system.

The connection between halfwidth and variance is elementary for a gaussian profile but quite delicate for other spectral shapes. This aspect will be discussed in the following chapter. At this point, we just recall that spectra tend toward gaussian shape with increasing pathlength but tend to skew again when particles have lost a sizable fraction of their energy.

The width of a measured energy spectrum is not only determined by straggling: If a target foil has a nonuniform thickness, the mean energy loss may still be measured reliably, while the scatter will appear increased in comparison with what would be measured on a uniform foil of equal thickness in terms of the number of atoms per foil area. This effect would be extreme (and readily detectable) in the presence of pinholes, but separating it from true straggling may in general require considerable care. Here, it may be wise to recall the golden rule of straggling, quoted to the author many years ago by John A. Davies,

> If measured straggling is less than predicted theoretically, at least one of the two is wrong.

8.10.1 Gas Targets

Extensive measurements of energy-loss straggling for protons and He ions penetrating gases have been performed by Besenbacher et al. (1977, 1980). These measurements provide supporting evidence for the existence of the

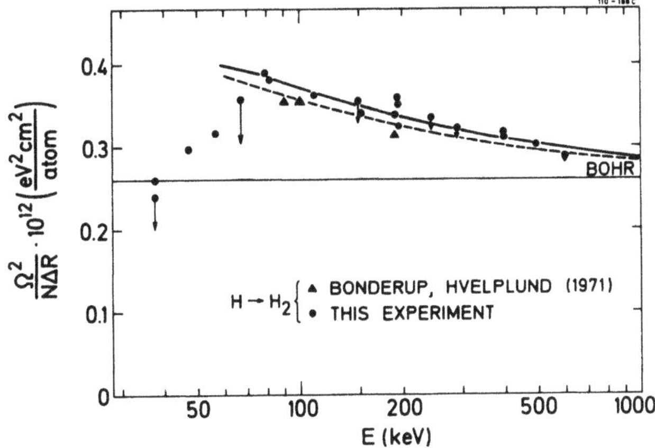

Fig. 8.21. Straggling of H^+ in H_2. The horizontal line represents Bohr straggling NxW_{Bohr}. Theoretical curves according to Bonderup and Hvelplund (1971) from electron-gas model averaged over Thomas-Fermi atomic-charge distribution allowing for two values of the Lindhard-Scharff parameter $\chi = \sqrt{2}$ and 1, cf. (7.25). From Besenbacher et al. (1981)

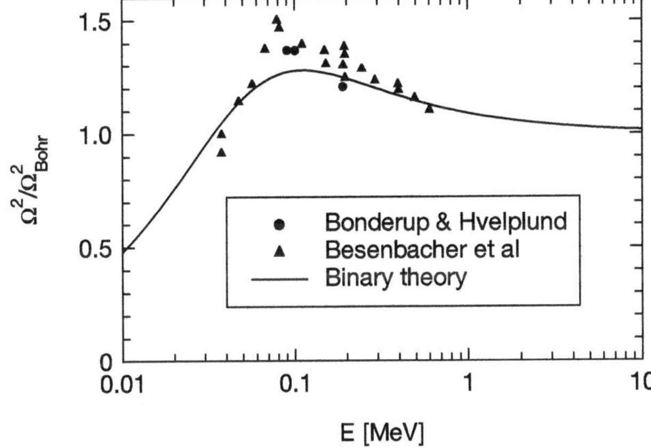

Fig. 8.22. Same experimental data as in Fig. 8.21 compared to calculations from binary theory. From Sigmund and Schinner (2002)

Bethe-Livingston shoulder (Fig. 8.21). The theoretical curve is a high-speed expansion by Bonderup and Hvelplund (1971) similar to (8.39), the validity of which is evidently limited to reasonably high projectile speeds.

Figure 8.22 shows the same experimental data compared to calculations by Sigmund and Schinner (2002). While the general behavior is described well by the calculation, the measured shoulder appears more pronounced. This is

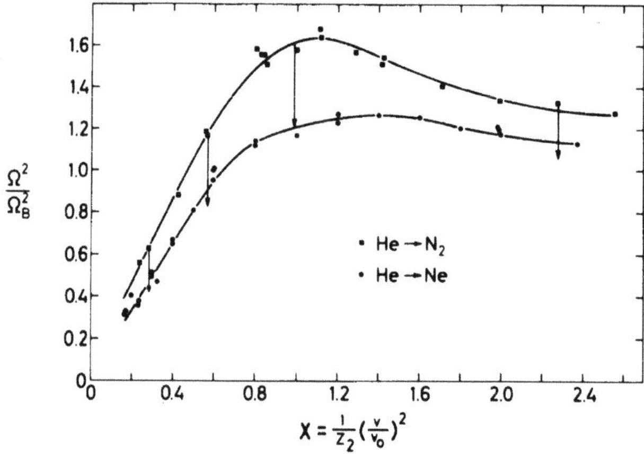

Fig. 8.23. Molecular correlation (bunching) in straggling. Comparison between straggling of He ions in N_2 and Ne gas. From Besenbacher et al. (1981)

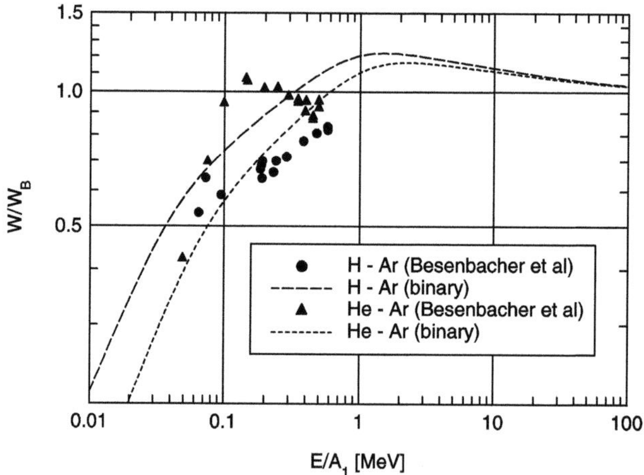

Fig. 8.24. Straggling of H and He ions in argon (Besenbacher et al., 1981) compared to predictions from binary theory

likely to be caused by molecular correlation (bunching), which has not been included in the theoretical estimate[7].

Independent experimental evidence of molecular correlation is shown in Fig. 8.23. Electronic properties of atomic nitrogen and neon are sufficiently similar to ascribe the pronounced enhancement near the maximum to molecular correlation (Besenbacher et al., 1981).

---

[7] For $Z_1 > 1$ there is usually a contribution to straggling from charge exchange which, like the bunching effect, has its peak around the stopping maximum and hence may be difficult to isolate (Sigmund, 2004).

Finally, Fig. 8.24 shows straggling of H and He ions in argon compared with calculations by binary theory. Enhanced straggling below the Bethe-Livingston maximum is thought to be caused by the atomic bunching effect which has been ignored in the calculation.

This set of measurements, the major component of a m.sc. thesis (Besenbacher, 1977), represents a unique source of information on straggling. Analysis of these data was performed on the basis of available knowledge by Besenbacher et al. (1980) and taken up again recently (Sigmund and Schinner, 2002), but much more is to be done here.

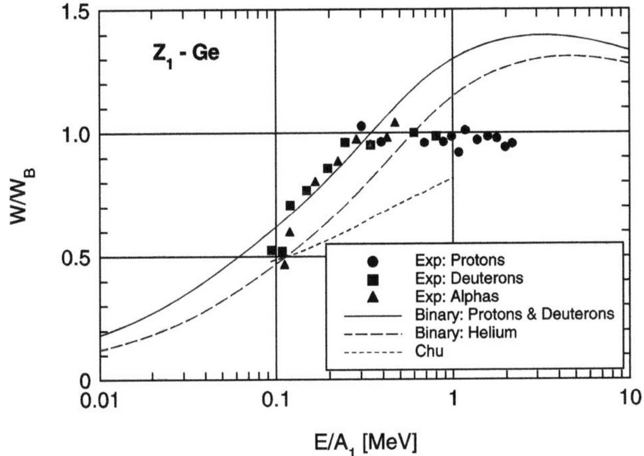

Fig. 8.25. Straggling of protons and helium ions in germanium. Experimental data from Malherbe and Albertz (1982a). Theoretical curves from Chu (1976) based on Bonderup and Hvelplund (1971) and binary theory. From Sigmund and Schinner (2002)

8.10.2 Solid Targets

The experimental literature on measurements of straggling in solids is extensive. A large amount of data has been compiled and, to some extent, systematized by Yang et al. (1991). A general trend emerges from Fig. 8.25: A shoulder is normally not observed, and Bohr straggling appears to describe the data surprisingly well in the region where a shoulder would be expected. This despite the fact that foil nonuniformities would generate seemingly enhanced straggling.

It has been mentioned by Sigmund and Schinner (2002) that experimental geometry may affect the outcome of a straggling measurement more than a simultaneous measurement of the stopping force. Indeed, straggling is influenced by close collisions, and close collisions also give rise to angular deflection.

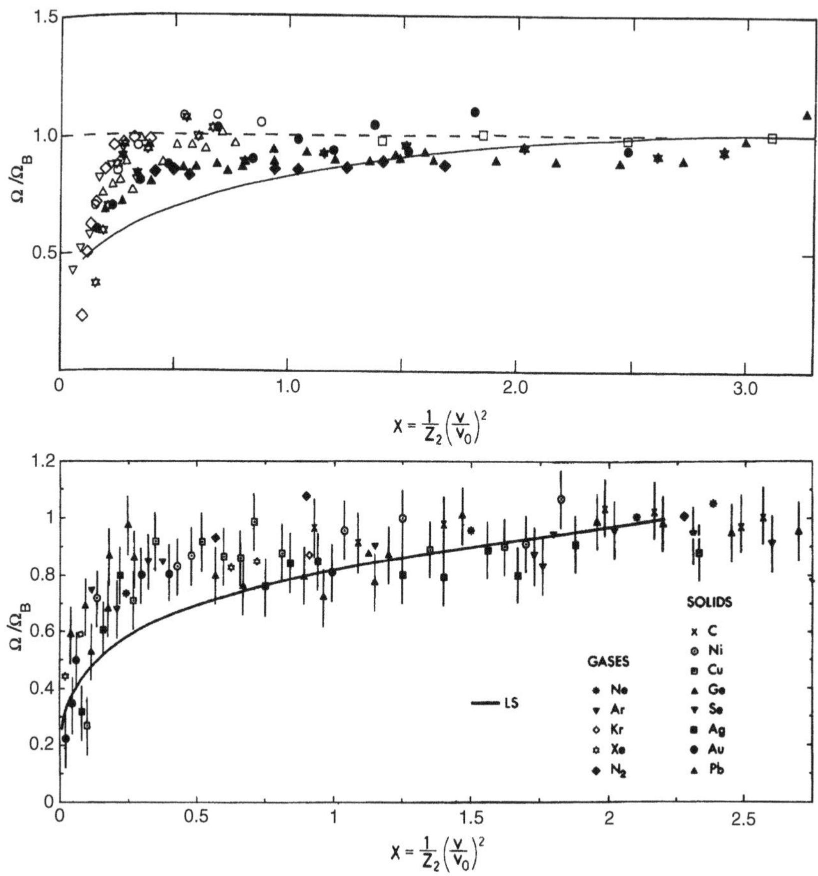

Fig. 8.26. Measured rms relative straggling $\sqrt{W/W_{\mathrm{B}}}$ versus $v^2/Z_2 v_0^2$. Upper graph: Hydrogen (filled symbols) and helium ions (empty symbols). Targets are C (squares), Ni (circles), Cu (up-triangles), Se (stars), Ag (rhombs) and Au (down-triangles. The line is a theoretical prediction of Lindhard and Scharff (1953). From Lombaard et al. (1983). Lower graph: Data for H and He ions in five gas and eight solid materials. From Malherbe and Albertz (1982b)

Another point of consideration is the correlation effect which, for solid matter is predicted to give rise to *decreased* straggling, cf. Sect. 8.9.4. However, exact compensation of the Bethe-Livingston shoulder against the correlation effect, as indicated in Fig. 8.25, would seem to be an unlikely coincidence.

Figure 8.26 shows a large number of experimental straggling data for both H and He ions in a number of solid materials (upper graph) and both gaseous and solid materials (lower graph). The only systematical trend emerging from the upper graph is a steeper increase of the He compared to H data. This appears indicative of charge exchange straggling, to be discussed in Volume II. The lower graph does not suggest any systematics within experimental scatter.

Again, this is an area where a kean ph.d. student will find ample experimental material and powerful theory that could be combined to a much more comprehensive picture than available here and now.

8.11 Third- and Higher-Order Moments (⋆)

8.11.1 Moments and Cumulants

Interest in higher moments of an energy-loss spectrum is mainly motivated theoretically. These quantities are increasingly governed by rare events with large energy transfers. Therefore, measurements of higher moments tend to be influenced by background noise. Conversely, calculations of higher moments get easier because of the gradual disappearance of a number of complications associated with the binding of target electrons. The main purpose of performing calculations of higher moments is to test the validity of approximate calculations of energy-loss profiles.

Although the method sketched in Sect. 2.2.4 to derive the connection between energy-loss straggling Ω^2 and second moment over the energy-loss cross section can be generalized to higher moments (problem 8.16), a more elegant method will emerge in Chapter 9. In either case, you obtain the following results for the *cumulants* of third to fifth order,

$$\left\langle (\Delta E - \langle \Delta E \rangle)^3 \right\rangle = NxQ_3, \tag{8.117a}$$

$$\left\langle (\Delta E - \langle \Delta E \rangle)^4 \right\rangle = NxQ_4 + 3\left(NxW\right)^2, \tag{8.117b}$$

$$\left\langle (\Delta E - \langle \Delta E \rangle)^5 \right\rangle = NxQ_5 + 10\left(NxW\right)\left(NxQ_3\right), \tag{8.117c}$$

where

$$Q_\nu = \sum_j T_j^\nu \sigma_j = \int T^\nu \, d\sigma(T) \tag{8.118}$$

in discrete or continuous notation. From this follows the skewness

$$\frac{\left\langle (\Delta E - \langle \Delta E \rangle)^3 \right\rangle}{\left\langle (\Delta E - \langle \Delta E \rangle)^2 \right\rangle^{3/2}} = \frac{NxQ_3}{(NxW)^{3/2}}, \tag{8.119}$$

which evidently decreases toward zero with increasing pathlength, and the curtosis

$$\frac{\left\langle (\Delta E - \langle \Delta E \rangle)^4 \right\rangle}{\left\langle (\Delta E - \langle \Delta E \rangle)^2 \right\rangle^2} = 3 + \frac{NxQ_4}{(NxW)^2}, \tag{8.120}$$

which approaches 3 with increasing pathlength. This happens to be the value characteristic of a gaussian profile. While this indicates that energy-loss profiles approach gaussian shape with increasing pathlength, nothing is said at this point about the minimum pathlength to reach this limit.

8.11.2 Free-Coulomb Scattering

For nonrelativistic free-Coulomb scattering,

$$d\sigma(T) = \frac{2\pi Z_1^2 Z_2 e^4}{mv^2} \frac{dT}{T^2}; \quad 0 < T \leq 2mv^2 \equiv T_{\max}, \tag{8.121}$$

you obtain

$$Q_\nu = \frac{4\pi Z_1^2 Z_2 e^4}{\nu - 1} T_{\max}^{\nu-2}. \tag{8.122}$$

From this follows the skewness

$$\frac{\left\langle (\Delta E - \langle \Delta E \rangle)^3 \right\rangle}{\left\langle (\Delta E - \langle \Delta E \rangle)^2 \right\rangle^{3/2}} = \frac{mv^2}{\sqrt{NxW_B}} \tag{8.123}$$

and the curtosis

$$\frac{\left\langle (\Delta E - \langle \Delta E \rangle)^4 \right\rangle}{\left\langle (\Delta E - \langle \Delta E \rangle)^2 \right\rangle^2} = 3\left[1 + \frac{T_{\max}^2}{9NxW_B}\right]. \tag{8.124}$$

Thus, a *necessary* condition for gaussian behavior must be

$$2\Omega_{\mathrm{Bohr}} \gg T_{\max} \tag{8.125}$$

to ensure small skewness, and

$$3\Omega_{\mathrm{Bohr}} \gg T_{\max} \tag{8.126}$$

to ensure a curtosis close to the gaussian value 3. At the same time, $\langle \Delta E \rangle$ should be well below E. We shall turn back to these relations in Chapter 9.

8.11.3 Bohr Theory

You may repeat the calculation leading to Fig. 8.2 for the skewness. Figure 8.27 shows the result. Comparison of the two graphs shows that the range of validity of the result found for straight Coulomb scattering is expanded down to lower values of ξ, i.e., lower projectile speeds. This must be a general trend that will be even more pronounced in higher moments: With increasing order ν of the moments Q_ν, high energy transfers T become increasingly important, causing effects due to atomic binding to become less and less significant on a relative scale, even if there is little change on an absolute scale.

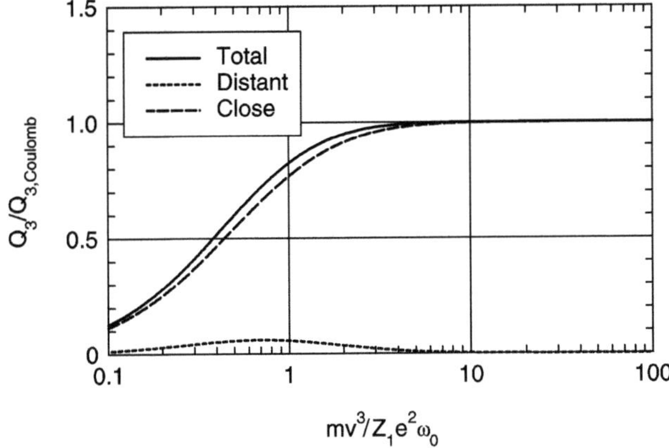

Fig. 8.27. Skewness in Bohr model. No shell correction

8.11.4 Born Approximation

Calculations within the Born approximation have been performed by Bichsel (1970) and by Mikkelsen et al. (1992). Fig. 8.28 shows numerical results for atomic hydrogen. While the qualitative behavior is similar to that of Fig. 8.27, a quantitative comparison makes little sense since Fig. 8.28, unlike Fig. 8.27, incorporates a shell correction.

The dashed curve in Fig. 8.28 refers to the quantal oscillator discussed in Sects. 4.5.2 and 8.4.1. The I-value has been chosen as 14.99 eV, i.e., the

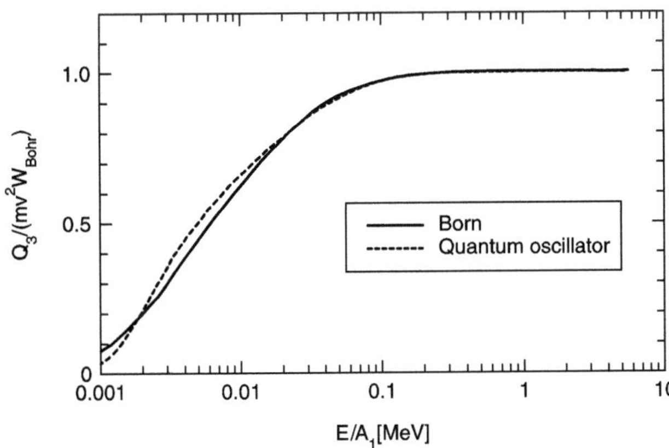

Fig. 8.28. Skewness evaluated in the first Born approximation. Solid line: Atomic hydrogen. Dashed line: Harmonic oscillator with the resonance frequency of atomic hydrogen. From Mikkelsen et al. (1992)

value for atomic hydrogen. The two curves show surprisingly close agreement. This was found to be a more general feature, predicted on theoretical grounds by Sigmund and Haagerup (1986) and confirmed for moments up to the fifth order by Mikkelsen et al. (1992) for the entire velocity range $v \gtrsim v_0/2$.

8.11.5 Relativistic Extension

The evaluation of higher moments within the relativistic Born approximation follows the scheme described in Sects. 5.6 and 8.4.3, the only new feature being the multiplication of the integrand by a factor $(\hbar\omega)^\nu$ instead of $\hbar\omega$ or $(\hbar\omega)^2$.

In the evaluation, sum rules are needed for the quantities

$$S_\nu = \sum_j (\epsilon_j - \epsilon_0)^{\nu-1} f_{j0}(Q). \tag{8.127}$$

Such sum rules have been evaluated by Fano and Turner (1964) and summarized by Inokuti (1971).

Here is a brief summary for the third moment Q_3:

- The longitudinal interaction delivers a low-Q contribution containing a Bethe-type logarithm multiplied by a factor const/v^4, where const lies in the nonrelativistic regime. For $v \sim c$, terms of order $1/c^4$ will be neglected.
- For intermediate Q, the longitudinal interaction delivers a term $(Q_1/2mv^2)^2$ as well as a term of the same order but with a logarithmic factor $\ln(Q_1/Q_0)$. Both terms are neglected for the same reason.
- The transverse interaction likewise delivers a term $\propto 1/v^4$ in the low-Q regime.
- In the intermediate-Q regime, the transverse interaction yields a contribution $\propto 1/c^6$ for $v \sim c$.

Thus, when terms $\propto 1/v^4$ and higher are neglected, the third moment is governed by the high-Q contribution, which is evaluated as indicated in the case of straggling, without splitting into longitudinal and transverse components. This reflects the increasing importance of large energy transfers in the integral and clearly holds for all higher moments.

Following the procedure sketched above, all integrations become elementary for all ν and yield

$$\frac{Q_\nu}{Q_{\nu,\text{Coulomb}}} = \frac{1 - \beta^2 \left(1 - 1/\nu\right)}{(1 - \beta^2)^{\nu-1}}. \tag{8.128}$$

This expression follows from (3.100) (problem 8.18). Indeed, for free-Coulomb scattering, (3.100) describes scattering in a reference frame moving with the ion, and the quantity called Z_2 in (3.100) is actually the atomic number of the ion, i.e., Z_1 in the present context.

8.11.6 Fermi Gas

Calculations for the Fermi gas along the scheme outlined in Sect. 8.5.3 have been performed by Sigmund and Johannessen (1985) and led to

$$\frac{Q_3}{Q_{3,\mathrm{Coulomb}}} = 1 + \left(\frac{\hbar\omega_\mathrm{P}}{mv^2}\right)^2 \left(\ln\frac{2mv^2}{\hbar\omega_\mathrm{P}} - \frac{1}{2}\right) - \frac{6}{5}\frac{v_\mathrm{F}^2}{v^2}\left(\frac{\hbar\omega_\mathrm{P}}{mv^2}\right)^2 \tag{8.129a}$$

$$\frac{Q_4}{Q_{4,\mathrm{Coulomb}}} = 1 + \frac{3}{8}\left(\frac{\hbar\omega_\mathrm{P}}{mv^2}\right)^2 + \frac{21}{80}\left(\frac{\hbar\omega_\mathrm{P}}{mv^2}\right)^2\frac{v_\mathrm{F}^2}{v^2}\left(\ln\frac{4mv^2}{\hbar\omega_\mathrm{P}} - \frac{5}{2}\right) \tag{8.129b}$$

$$\frac{Q_5}{Q_{5,\mathrm{Coulomb}}} = 1 + \frac{4}{15}\frac{v_\mathrm{F}^2}{v^2} + \frac{1}{2}\left(\frac{\hbar\omega_\mathrm{P}}{mv^2}\right)^4\left(\ln\frac{2mv^2}{\hbar\omega_\mathrm{P}} - \frac{3}{4}\right) \tag{8.129c}$$

The kinetic correction was evaluated only up to the quadratic term in v_F/v. Since the Fermi energy $mv_\mathrm{F}^2/2$ is typically of the order of the plasmon energy $\hbar\omega_\mathrm{P}$, terms $\propto v^{-2}$ need to be treated on equal footing. This implies that up to v^{-2},

$$\frac{Q_3}{Q_{3,\mathrm{Coulomb}}} = 1 + \mathcal{O}\left(\frac{1}{v^4}\right) \tag{8.130a}$$

$$\frac{Q_4}{Q_{4,\mathrm{Coulomb}}} = 1 + \mathcal{O}\left(\frac{1}{v^4}\right) \tag{8.130b}$$

$$\frac{Q_5}{Q_{5,\mathrm{Coulomb}}} = 1 + \frac{4}{15}\frac{v_\mathrm{F}^2}{v^2} + \mathcal{O}\left(\frac{1}{v^4}\right). \tag{8.130c}$$

This confirms the conclusion from section 8.11.3 that corrections due to binding – expressed by terms $\propto \hbar\omega_\mathrm{P}$ here – become insignificant for higher moments. However, while kinetic corrections do not contribute to the leading order in the third and fourth moment, they do come in again for the fifth moment. This was to be expected, cf. the discussion of the second moment in Sect. 8.5.3.

8.11.7 Kinetic Theory

Calculations of higher moments within kinetic theory have been performed by Sigmund and Johannessen (1985). First, a generalization of (8.50) to higher moments Q_ν was found. Unlike (8.50), Q_ν for $\nu \geq 3$ was found not to contain terms that diverge for straight Coulomb interaction. Therefore, electronic binding was neglected from the beginning.

The results were applied to the Fermi gas and compared with those mentioned in Sect. 8.11.6. The absence of kinematic terms to leading order in Q_3 and Q_4, (8.130a) and (8.130b) was confirmed, and Q_5 was found to be in quantitative agreement with (8.130c).

Comparisons were also reported of moments Q_ν up to $\nu = 10$ with results for hydrogen atoms evaluated in the Born approximation by Bichsel (1970). Surprisingly good agreement was found for $v^2/\langle v_e^2\rangle > 0.1$ and $\nu \leq 6$, while for higher moments, discrepancies were observed at the smallest beam velocities.

8.12 Discussion and Outlook

The first part of this chapter addressed straggling for independent target electrons. Here we learned that

- Plain Coulomb scattering produces a constant straggling parameter W_{B},
- Relativity causes an increase above the nonrelativistic Bohr value when v approaches the speed of light.
- Allowance for binding – disregarding higher-order-Z_1 effects – reduces straggling at low projectile speed,
- Allowance for orbital motion produces a shoulder in classical theory as well as in the Born approximation,
- Allowance for a Barkas-Andersen effect produces a shoulder for protons
- but a dip for antiprotons,
- The magnitude of the Barkas-Andersen effect is reduced by the shell correction,
- Quantum theory enters straggling mainly via orbital motion of target electrons.

Reliable estimates are possible of these effects. However, in the velocity range where shell and Barkas corrections are significant, also a bunching correction is necessary, which is much less studied in the literature but needs to be accounted for in comparisons with experiment. Best studied is the molecular correlation effect, which is always positive. The atomic bunching correction, while much less studied, is also positive and difficult to separate experimentally from other deviations from single-electron straggling. In solids, on the other hand, the correlation effect may have either sign, and direct experimental evidence has not been provided so far.

In the final part of this chapter, higher moments have been studied. Here, the dominating cause of deviations from binary-Coulomb interaction is the shell correction which becomes increasingly important with increasing order of the moment.

Problems

8.1. Derive (8.10) and (8.13) by following the procedure outlined in Sect. 4.5.1.

8.2. Derive (8.16) by following the procedure outlined in Sect. 4.5.2.

8.3. Derive (8.22) by the procedure outlined in Appendix A.4.4.

8.4. Generalize the calculation of problem 8.3 to a many-electron atom and identify terms that have to be neglected in order to achieve the final result (8.24).

8.5. Apply the procedure sketched in Sect. 5.3.2 to (8.34) to derive the expression (8.18) for straggling in the Bethe theory.

8.6. Derive (8.37), (8.38) and (8.39) following the procedure outlined in Sect. 5.7.4.

8.7. Find the position of the Bethe-Livingston shoulder by determining the maximum of the function given in (8.39).

8.8. Evaluate the integral (8.37) and try to reproduce Fig. 8.6.

8.9. Derive (8.47) following the procedure that led to the stopping cross section via (6.71).

8.10. Derive (8.51).

8.11. The expansion (8.54) can only be meaningful for $v^2 \gg \langle v_e^2 \rangle$. Use this, together with the Thomas-Fermi estimates $\hbar\omega \simeq Z_2\mathrm{R}$ and $v_e \simeq Z_2^{2/3}v_0$ to derive a condition for ξ.

8.12. Use a Thomas-Fermi argument to estimate the locations of the shoulders produced by the Barkas-Andersen effect and the shell correction, respectively.

8.13. Derive (8.92). Consult Appendix A.2.4 if necessary.

8.14. Derive the angular average (8.97), making use of spherical coordinates.

8.15. Somewhat schematic models to estimate the correlation effect may be based on an atomic energy-loss function

$$T(p) = Ce^{-p^2/a^2} \tag{8.131}$$

according to Sigmund (1991) or

$$T(p) = C'e^{-p/a'} \tag{8.132}$$

according to Grande and Schiwietz (1991) with constants C, a or C', a' that may be expressed by a stopping cross section S and a straggling parameter W. This type of *ansatz* allows analytical evaluation of the integral in (8.90) for a diatomic molecule as well as (8.106) for a solid. Perform the angular orientational averages *after* those integrations, and try to find estimates of the correlation terms ΔW in the two cases, if necessary by making use of numerical integration. [Hint: Make use of Fourier transform.]

8.16. Derive (8.117a) by the method described in Sect. 2.2.4.

8.17. Derive (8.129a).

8.18. Derive (8.128) from (3.100).

References

Besenbacher F. (1977): Stopping power og straggling for H og He ioner i gas targets. Master's thesis, Aarhus University

Besenbacher F., Andersen H.H., Hvelplund P. and Knudsen H. (1981): Straggling in energy loss of swift hydrogen and helium ions in gases. Mat Fys Medd Dan Vid Selsk **40 no. 9**, 1–42

Besenbacher F., Andersen J.U. and Bonderup E. (1980): Straggling in energy loss of energetic hydrogen and helium ions. Nucl Instrum Methods **168**, 1–15

Besenbacher F., Heinemeier J., Hvelplund P. and Knudsen H. (1977): Energy-loss straggling for protons and helium ions. Phys Lett A **61**, 75–77

Bichsel H. (1970): Straggling of heavy charged particles: Comparison of Born hydrogenic-wave-function approximation with free-electron approximation. Phys Rev B **1**, 2854–2862

Bohr N. (1913): On the theory of the decrease of velocity of moving electrified particles on passing through matter. Philos Mag **25**, 10–31

Bohr N. (1915): On the decrease of velocity of swiftly moving electrified particles in passing through matter. Philos Mag **30**, 581–612

Bonderup E. and Hvelplund P. (1971): Stopping power and energy straggling for swift protons. Phys Rev A **4**, 562–589

Chu W.K. (1976): Calculation of energy straggling for protons and helium ions. Phys Rev A **13**, 2057–2060

Cohen S.M. (2004): Aspects of relativistic sum rules. Adv Quantum Chem **46**, 241–265

Cohen S.M. (2005): Simple and accurate sum rules for highly relativistic systems. J Chem Phys **122**, 104105

Cohen S.M. and Leung P.T. (1998): General formulation of the semirelativistic approach to atomic sum rules. Phys Rev A **57**, 4994–4997

Dixmier M., L'Hoir A. and Amsel G. (1982): A statistical model for the energy loss fluctuations of energetic ions in condensed matter. Nucl Instrum Methods **197**, 537–544

Fano U. (1963): Penetration of protons, alpha particles, and mesons. Ann Rev Nucl Sci **13**, 1–66

Fano U. and Turner J.E. (1964): Contributions to the theory of shell corrections. In *Studies in penetration of charged particles in matter*, vol. 1133, 49–67. NAS-NRC, Washington

Grande P.L. and Schiwietz G. (1991): Impact-parameter dependence of electronic energy loss and straggling of incident bare ions on H and He atoms by using the coupled-channel method. Phys Rev A **44**, 2984–2992

Hansen F.Y. and Carneiro K. (1977): Procedure to obtain reliable pair distribution functions of non-crystalline materials from diffraction data. Nucl Instrum Methods **143**, 569–575

ICRU (2005): *Stopping of ions heavier than helium*, vol. 73 of *ICRU Report*. Oxford University Press, Oxford

Inokuti M. (1971): Inelastic collisions of fast charged particles with atoms and molecules – the Bethe theory revisited. Rev Mod Phys **43**, 297–347

Lindhard J. (1954): On the properties of a gas of charged particles. Mat Fys Medd Dan Vid Selsk **28 no. 8**, 1–57

Lindhard J. and Scharff M. (1953): Energy loss in matter by fast particles of low charge. Mat Fys Medd Dan Vid Selsk **27 no. 15**, 1–31

Lindhard J. and Sørensen A.H. (1996): On the relativistic theory of stopping of heavy ions. Phys Rev A **53**, 2443–2456

Livingston M.S. and Bethe H.A. (1937): Nuclear physics. C. Nuclear dynamics, experimental. Rev Mod Phys **9**, 245–390

Lombaard J., Conradie J. and Friedland E. (1983): Energy-loss and straggling of hydrogen and helium ions in silver. Nucl Instrum Methods **216**, 293–298

Malherbe J.B. and Albertz H.W. (1982a): Energy-loss straggling in C and Ge of p, D and alpha particles in the energy region 0.2 to 2.4 MeV. Nucl Instrum Methods **192**, 559–563

Malherbe J.B. and Albertz H.W. (1982b): Energy-loss straggling in lead of P, D and α-particles in the energy region 0.2 – 2.5 MeV. Nucl Instrum Methods **196**, 499

Mikkelsen H.H., Meibom A. and Sigmund P. (1992): Intercomparison of atomic models for computing stopping parameters from the Bethe theory - atomic hydrogen. Phys Rev A **46**, 7012–7018

Pitarke J.M., Ritchie R.H. and Echenique P.M. (1995): Quadratic response theory of the energy-loss of charged-particles in an electron-gas. Phys Rev B **52**, 13883–13902

Sigmund P. (1976a): Energy loss and angular spread of ions traversing matter. Ann Israel Phys Soc **1**, 69–120

Sigmund P. (1976b): Energy loss of charged particles to molecular gas targets. Phys Rev A **14**, 996

Sigmund P. (1978): Statistics of particle penetration. Mat Fys Medd Dan Vid Selsk **40 no. 5**, 1–36

Sigmund P. (1982): Kinetic theory of particle stopping in a medium with internal motion. Phys Rev A **26**, 2497–2517

Sigmund P. (1991): Statistics of charged-particle penetration. In A. Gras-Marti, H.M. Urbassek, N. Arista and F. Flores, editors, *Interaction of charged particles with solids and surfaces*, vol. 271 of *NATO ASI Series*, 73–144. Plenum Press, New York

Sigmund P. (2004): *Stopping of heavy ions*, vol. 204 of *Springer Tracts of Modern Physics*. Springer, Berlin

Sigmund P. and Fu D.J. (1982): Energy loss straggling of a point charge penetrating a free-electron gas. Phys Rev A **25**, 1450–1455

Sigmund P. and Haagerup U. (1986): Bethe stopping theory for a harmonic oscillator and Bohr's oscillator model of atomic stopping. Phys Rev A **34**, 892–910

Sigmund P. and Johannessen K. (1985): Higher moments of the energy loss spectrum of swift charged particles penetrating a thin layer of material. Nucl Instrum Methods B **6**, 486–495

Sigmund P. and Schinner A. (2002): Barkas effect, shell correction, screening and correlation in collisional energy-loss straggling of an ion beam. Europ Phys J D 201–209

Sternheimer R.M. (1960): Range straggling of charged particles in Be, C, Al, Cu, Pb, and air. Phys Rev **117**, 485–488

Titeica S. (1937): Sur les fluctuations de parcours des rayons corpusculaires. Bull Soc Roumaine Phys **38**, 81–100

Tofterup A.L. (1983): Relativistic binary-encounter and stopping theory: general expressions. J Phys B **16**, 2997–3003

Wang N.P. and Pitarke J.M. (1998): Nonlinear energy-loss straggling of protons and antiprotons in an electron gas. Phys Rev A **57**, 4053–4056

Yang Q., O'Connor D.J. and Wang Z.G. (1991): Empirical formulae for energy loss straggling of ions in matter. Nucl Instrum Methods B **61**, 149

9

Energy-Loss Spectra

9.1 Introductory Comments

This chapter addresses the important question of the shape of an energy-loss spectrum. All stopping theory described up to this point has addressed mean energy loss, variance and, to a much lesser extent, higher moments. It has already been mentioned that these quantities may be hard to measure. Experimentalists prefer to determine the peak position of an energy-loss profile and, perhaps, the half width. Measuring mean values and higher-order averages may lead to problems with the tails of a profile which may be affected by noise. However, with the exception of gaussian profiles, relating peak positions and half widths to mean values and standard deviations is not a trivial task.

This chapter focuses on statistical tools available to compute spectra and to extract expressions for peak energy loss and halfwidth[1]. Actual applications of these tools in the literature frequently focus on straight Coulomb interaction with as few modifications as possible. While this is a feasible strategy to demonstrate the reliability of statistical methods, it may not be adequate for comparison with measurements. Therefore, attention also needs to be given to how various complicating effects which have been discussed in Chapters 4–6 and 8 enter computed energy spectra as well as peak positions and halfwidths.

9.2 General Aspects

Based on considerations by Bohr (1915, 1948) it will be convenient in the following to distinguish between thin, moderately thick and very thick targets:

- In penetrating through a *thin target*, a projectile undergoes a very small number of interactions such that the energy-loss spectrum remains similar to the differential cross section for an individual encounter. That cross section resembles Rutherford's law, with a pronounced peak at fairly low

---

[1] Part of the material presented in this chapter has been extracted, revised and updated from a lecture series published previously (Sigmund, 1991).

energy transfers and a long tail extending to the maximum energy loss permitted by conservation laws. The main difference between a thin-target spectrum and the plain single-scattering cross section is that the former must be integrable: A properly calculated energy-loss spectrum has to be normalizable due to conservation of particles.

- In penetrating through a *moderately thick target*, a projectile undergoes a large number of encounters. This implies an energy-loss spectrum gradually approaching gaussian shape centered around the mean energy loss. However, the fractional energy loss is required to be small enough so that the dependence on beam energy of pertinent cross sections can be neglected.

- The latter restriction does not apply to a *very thick target*, where the relative energy loss is so large that cross sections may vary significantly over the trajectory. This breaks the trend toward gaussian shape of the energy-loss profile.

Hints on the range of validity of gaussian energy profiles were emerging from (8.125) and (8.126). Bohr (1915, 1948) argued that the requirement of the maximum energy loss in a single collision to be much smaller than the standard deviation of the profile,

$$T_{\max} \ll \Omega \tag{9.1}$$

with

$$\Omega^2 = NxW \tag{9.2}$$

actually must be a *sufficient* criterion for establishing a near-gaussian energy-loss spectrum. This point will be studied in considerable detail.

Here, assume for a moment a beam with a gaussian energy-loss profile at some path length x, characterized by a standard deviation Ω. Subsequent collisions may lead to energy transfers anywhere between some minimum value and $T_{\max}$. If (9.1) is fulfilled, the energy spectrum will shift and broaden slightly but remain gaussian, yet if the opposite relation

$$T_{\max} \gg \Omega \tag{9.3}$$

is fulfilled, the spectrum may undergo a dramatic change. Clearly, (9.3) does not go well along with a gaussian spectrum.

For an ion beam, where the maximum energy transfer to an electron is much smaller than the ion energy, (9.1) may well be fulfilled at not too small layer thicknesses x since $T_{\max} \ll E$, while penetrating electrons or positrons may lose all their energy in one single event. This implies that energy spectra of electrons and positrons tend to exhibit a significant inverse-power-like tail, even when the peak region is close to gaussian.

In the older literature (Williams, 1929, Landau, 1944, Symon, 1948), the straggling problem tends to be discussed in terms of the above extremes, (9.1) for ions and (9.3) for electrons. Subsequently, with increasing beam velocities and/or decreasing layer thicknesses, (9.3) has become increasingly relevant also for measurements with ion beams.

9.2.1 Bothe-Landau Formula

Go back for a moment to the rather general description outlined in Sect. 2.2 on page 28 where a collision was considered to be a detectable event involving the projectile and one or more target particles. Collisions were categorized by their outcome, such as excitation of the jth level of a target atom or molecule. Assume Poisson statistics to apply, i.e., let a dilute beam interact with a dilute and random target. The terminology utilized in the present discussion refers to electronic energy loss by target excitation, but the formalism will be extended in volume II to other processes such as energy loss accompanied by charge exchange as well as multiple scattering.

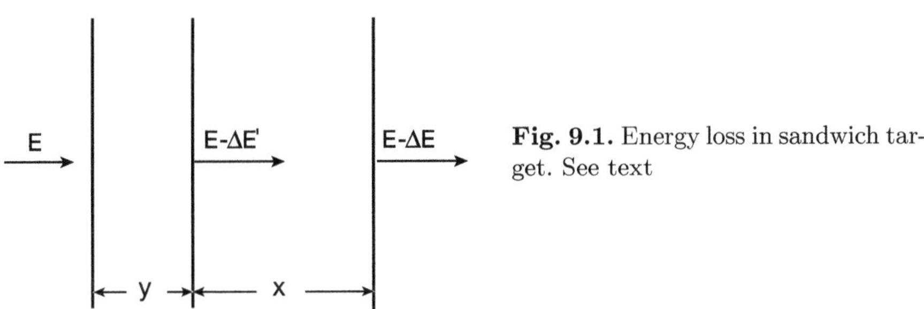

Fig. 9.1. Energy loss in sandwich target. See text

Let $F(\Delta E, x)\mathrm{d}(\Delta E)$ be the probability distribution in energy loss ΔE during a passage through a layer of thickness x, averaged over many projectiles. A central equation to be satisfied by $F(\Delta E, x)$ can be found by considering a sandwich target with thicknesses y and x in series (Fig. 9.1). Let the energy loss of a specific beam particle after passage through the first layer be $\Delta E'$. In order to exit with a total energy loss ΔE, the projectile must lose an energy $\Delta E - \Delta E'$ in the second layer. Because of the statistical independence of the collision events, the product of the two probabilities determines the joint probability. Since $\Delta E'$ is arbitrary, the resulting probabilities may be added up. Therefore,

$$F(\Delta E, x + y) = \int_0^{\Delta E} \mathrm{d}(\Delta E')\, F(\Delta E', y)\, F(\Delta E - \Delta E', x). \tag{9.4}$$

This is known as the Chapman-Kolmogorov equation in statistical physics (van Kampen, 1981).

The occurrence of a convolution suggests to go over to Fourier space[2],

$$F(\Delta E, x) = \frac{1}{2\pi} \int_{-\infty}^{\infty} dk \, e^{ik\Delta E} F(k, x) \tag{9.5}$$

with[3]

$$F(k, x) = \int_{-\infty}^{\infty} d(\Delta E) e^{-ik\Delta E} F(\Delta E, x). \tag{9.6}$$

Then (9.4) reduces to

$$F(k, x + y) = F(k, y) F(k, x) \tag{9.7}$$

with the exponential solution

$$F(k, x) = e^{xC(k)} \tag{9.8}$$

and an arbitrary function $C(k)$. You may give a proof by looking into problem 9.11.

The function $C(x)$ is governed by the individual events happening during the passage, which have not yet entered the description. Those are governed by the cross sections σ_j for energy loss T_j according to Sect. 2.4.2. If x is so small that the probability for two events is negligible, the energy-loss spectrum after passage through x is given by

$$F(\Delta E, x) = \left(1 - \sum_j \mathsf{P}_j\right) \delta(\Delta E) + \sum_j \mathsf{P}_j \, \delta(\Delta E - T_j), \tag{9.9}$$

where $\mathsf{P}_j = Nx\sigma_j$ is the probability for a j-event according to (2.15). Here the first term on the right-hand side expresses the probability that nothing happens, implying zero energy loss, while the second term collects all possibilities for exactly one collision and the associated energy loss.

Using the Fourier transform of the Dirac function (A.68) we may write (9.9) in Fourier space,

$$F(k, x) = 1 - Nx \sum_j \sigma_j \left(1 - e^{-ikT_j}\right). \tag{9.10}$$

---

[2] The reader is reminded of the adopted notation for Fourier transforms where $F(\Delta E, x)$ and $F(k, x)$ may have very different functional forms, cf. the remark on page 110.

[3] For convenience the factor $1/2\pi$ has been placed differently in (9.5) and (9.6) from the convention applied in Appendix A.2.2. This is justified and quite common, cf. the remark on page 383.

This may be compared with the small-x limit of (9.8),

$$F(k,x) = 1 + xC(k)\ldots,\tag{9.11}$$

where the unknown function $C(k)$ now can be identified as

$$C(k) = -N\sigma(k)\tag{9.12}$$

with the *transport cross section*

$$\sigma(k) = \sum_j \sigma_j \left(1 - e^{-ikT_j}\right).\tag{9.13}$$

From (9.5), (9.8), and (9.12) we finally obtain

$$F(\Delta E,x) = \frac{1}{2\pi} \int_{-\infty}^{\infty} dk \; e^{ik\Delta E - Nx\sigma(k)}.\tag{9.14}$$

This is the prototype of a *Bothe-Landau equation* which governs the statistics of cumulative events in particle penetration phenomena. For a continuous single-event spectrum, (9.13) needs to be rewritten in the form

$$\sigma(k) = \int d\sigma(T)\left(1 - e^{-ikT}\right),\tag{9.15}$$

and for a mixture of components $\mu = 1, 2\ldots$, the substitution

$$N\sigma(k) \rightarrow \sum_\mu N_\mu\sigma_\mu(k)\tag{9.16}$$

can be made, where N_μ is the number of μ-atoms per volume and

$$\sigma_\mu(k) = \sum_{j_\mu} \sigma_{j_\mu}\left(1 - e^{-ikT_{j_\mu}}\right).\tag{9.17}$$

Here, the running variable j_μ labels the states of an atom of type μ.

The central assumption entering here, apart from statistical independence of successive collision events (called *Markov assumption* in statistical physics), is translational invariance of the collision probabilities. In other words, the total energy loss needs to be small enough to justify the neglect of the variation of pertinent collision cross sections with beam energy. This means that the range of validity of (9.14)[4] is limited to thin and moderately thick targets.

The argument utilized in the derivation of (9.14) is quite general and applies to sums of many kinds of independent statistical variables (van Kampen, 1981). Bothe (1921) recognized several important application areas such as the

---

[4] *Correction to first edition:* Equation (9.14) in conjunction with (9.13) has occasionally been called the 'Vavilov formula'. This nomenclature is ambiguous, not so much with regard to the priority issue, but more seriously because it may give rise to a mix-up with Eq. (9.72) and its defining relationships which are due to Vavilov. As a matter of fact, the curves shown in Fig. 9.10 represent numerical evaluations of the Bothe-Landau equation (9.14) and not, as implied in Sect. 9.4.1, of the Vavilov equation (9.72).

statistical theory of errors, fluctuations in the local electric field in a dielectric, and multiple scattering of ions and electrons. Landau (1944) derived the specific formula for the energy-loss spectrum of a charged particle, even though his derivation had a more limited scope.

Equation (9.14) has been written as a Fourier transform rather than a Laplace transform, although the latter notation, introduced by Landau (1944), is more common in the literature. The Fourier notation allows for both positive and negative values of the energy loss. When the cross section only allows for positive energy loss, the Fourier and Laplace formulation are strictly equivalent. In other applications of the Bothe-Landau formula such as multiple scattering, Laplace transforms are of little use.

9.2.2 Bunching

The Bothe-Landau formula rests on Poisson's law, i.e., it assumes mutually independent interactions with the collision partners. We have seen in Sect. 8.9 that this assumption may be violated. However, violations may refer to different levels. Although we have the option to characterize bunching in atoms and molecules as deviations from stochastic behavior, we may alternatively consider the energy-loss spectrum to an atom or molecule as the basic quantity entering the Bothe-Landau formula. This, in fact, is the normal procedure.

However, energy loss in dense media such as crystalline or amorphous solids shows intrinsic deviations from random behavior which, when significant, are not contained in the Bothe-Landau formula. For a quantitative determination of an energy-loss spectrum one would then go back to Fig. 8.19 and the definition

$$F(\Delta E) = \frac{1}{A} \int d^2 p \, \delta \left(\Delta E - \sum_\nu T(p - p_\nu) \right) \tag{9.18}$$

in the notation of (8.103). Such a study has not been performed yet to the author's knowledge.

9.2.3 Moments and Cumulants to Arbitrary Order

It is useful to verify how the expressions for mean energy loss and straggling derived in Sect. 2.2.3 may be recovered from (9.14). Multiplication of (9.14) by some power of the energy loss and integration over ΔE yields

$$\langle (\Delta E)^n \rangle = \int d(\Delta E) \, \Delta E^n \, F(\Delta E, x)$$

$$= \frac{1}{2\pi} \int d(\Delta E) \int dk \, e^{-Nx\sigma(k)} (\Delta E)^n \, e^{ik\Delta E} \tag{9.19}$$

after rearrangement of the order of integrations and the factors making up the integrand. Making use of

$$\frac{\partial}{\partial k}e^{ik\Delta E} = i\Delta E e^{ik\Delta E} \tag{9.20}$$

we may replace ΔE^n by $(-i\partial/\partial k)^n$ and carry out the integration over ΔE. Repeated partial integration over k then yields

$$\int_{-\infty}^{\infty} \mathrm{d}(\Delta E)\,(\Delta E)^n\, F(\Delta E, x) = \left(i\frac{\mathrm{d}}{\mathrm{d}k}\right)^n e^{-Nx\sigma(k)}\Bigg|_{k=0}. \tag{9.21}$$

You may reproduce this by looking into problem 9.2. For $n = 0$ the right-hand side reduces to 1 by means of (9.15). This expresses the fact that $F(\Delta E, x)$ is normalized to 1 for all x. Obviously, the number of particles must be conserved during penetration.

For $n = 1$, differentiation of (9.15) and taking the limit $k = 0$ according to (9.21) yields the mean energy loss (2.19) on page 34. Similarly, for $n = 2$, (2.26) is recovered for the energy loss straggling Ω^2 (cf. problem 9.3).

You can derive higher cumulants (Symon, 1948) such as those listed in (8.117a), (8.117b) and (8.117c). It is an advantage, however, to start from a slightly rearranged form of (9.14),

$$F(\Delta E, x) = \frac{1}{2\pi} \int_{-\infty}^{\infty} \mathrm{d}k\, e^{ik(\Delta E - \langle\Delta E\rangle) - Nx\sigma_1(k)}, \tag{9.22}$$

where

$$\sigma_1(k) = \int \mathrm{d}\sigma(T)\big(1 - ikT - e^{-ikT}\big), \tag{9.23}$$

and to multiply by $(\Delta E - \langle\Delta E\rangle)^n$ directly instead of ΔE^n.

9.2.4 Diffusion Approximation

To provide a background for the discussion to follow, it is useful first to demonstrate the connection between the gaussian approximation to the energy-loss spectrum and the Bothe-Landau formula (9.14).

Eq. (2.26) predicts the energy loss spectrum to broaden with increasing target thickness. Therefore the Fourier integral should receive an increasing fraction of its contributions from small values of k. Expansion of (9.15) in powers of k yields

$$\sigma(k) = -\sum_{\nu=1}^{\infty} \frac{(-ik)^\nu}{\nu!} \int T^\nu \mathrm{d}\sigma(T) = ikS + \frac{1}{2}k^2 W - \frac{1}{6}ik^3 Q_3 \dots \tag{9.24}$$

where

$$Q_3 = \sum_j T_j^3 \sigma_j = \int T^3 \mathrm{d}\sigma(T) \tag{9.25}$$

is a skewness parameter discussed in Sect. 8.11.1.

If the series (9.24) is truncated after the term of second order in k, the integral (9.14) can be carried out and yields

$$F(\Delta E, x) = \frac{1}{\sqrt{2\pi N x W}} \exp\left(-\frac{(\Delta E - N x S)^2}{2N x W}\right), \tag{9.26}$$

i.e., a gaussian centered around the mean energy loss NxS with the standard deviation $\sqrt{NxW}$. The procedure sketched here is called diffusion approximation since the profile is of the type known from diffusion or heat conduction theory. One might expect that inclusion of higher-order terms could improve the accuracy of this distribution. This is only partially true: Truncation of the series (9.24) at any finite term beyond $\nu = 2$ was shown by Lindhard and Nielsen (1971) to lead to negative values of the probability density $F(\Delta E, x)$ at some interval of ΔE, regardless of the cross section. It is advisable, therefore, to base any improvements beyond the diffusion approximation upon the complete expression, which can be written in one more alternative form

$$\sigma(k) = ikS + \frac{1}{2}k^2W + \sigma_2(k), \tag{9.27}$$

with

$$\sigma_2(k) = \int d\sigma(T)\left(1 - ikT - \frac{1}{2}k^2T^2 - e^{-ikT}\right), \tag{9.28}$$

where the correction term $\sigma_2(k)$ can be suitably approximated if necessary.

9.2.5 An Integrable Energy-Loss Spectrum

It is useful to study a straight analytic evaluation of the Bothe-Landau formula. Lindhard and Nielsen (1971) have collected several cross sections allowing analytical solution, of which we here discuss

$$d\sigma(T) = \frac{C}{T^{3/2}}\, e^{-\alpha T} dT \text{ for } 0 < T < \infty, \tag{9.29}$$

where C and α are constants. This resembles Rutherford's law, (2.47) on page 42, although T^{-2} has been replaced by $T^{-3/2}$ and the cutoff at $T_{\max}$ has been replaced by an exponential with a parameter α which could be chosen to be $T_{\max}^{-1}$.

Eq. (9.15) and (9.29) yield the transport cross section

$$\sigma(k) = 2\sqrt{\pi}C\left(\sqrt{\alpha + ik} - \sqrt{\alpha}\right) \tag{9.30}$$

by means of the integral representation of the gamma function (see problem 9.4). Insertion into (9.14) and integration yields (cf. problem 9.5)

$$F(\Delta E, x) = \frac{NCx}{\Delta E^{3/2}} \exp\left[-\frac{\alpha}{\Delta E}(\Delta E - N x S)^2\right], \tag{9.31}$$

where S is the stopping cross section

$$S = \int_0^\infty T \,\mathrm{d}\sigma(T) = C\sqrt{\frac{\pi}{\alpha}} \qquad (9.32)$$

according to (2.29). You may easily derive from (2.30) that

$$\Omega^2 = \left\langle (\Delta E - \langle \Delta E \rangle)^2 \right\rangle = \frac{1}{2\alpha}\langle \Delta E \rangle. \qquad (9.33)$$

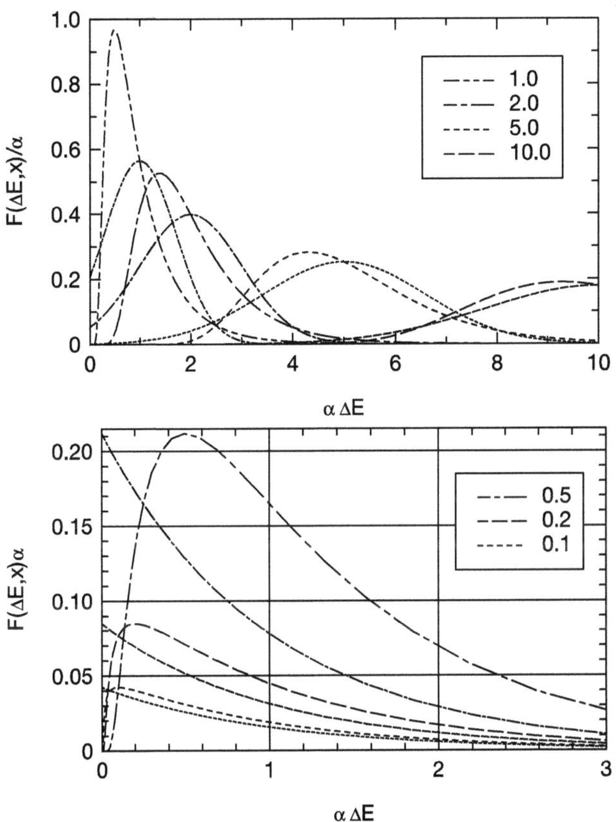

Fig. 9.2. Energy-loss spectra calculated for model cross section (9.29). Curves labelled by parameter $\langle \alpha \Delta E \rangle$, which is equivalent to $\langle \Delta E \rangle / T_{\mathrm{max}}$. Thick lines: Exact results, (9.31). Upper graph: Large thickness; thin lines: diffusion approximation, (9.26); Lower graph: Small thickness; thin lines: single-event probability defined by cross section 9.29

The upper graph in Fig. 9.2 shows calculated spectra for $\langle \alpha \Delta E \rangle \geq 1$, i.e., moderately thick targets. The position of the mean energy loss on the abscissa is identical with the label of the pertinent spectrum. While the spectra are seen to approach gaussian shape with increasing thickness, noticeable deviations are still present at $\alpha\Delta E = 10$, i.e., $\Omega^2 = 5T_{\mathrm{max}}^2$. The lower graph shows that for thinner targets the spectra get increasingly skew and approach the single-collision spectrum (9.29).

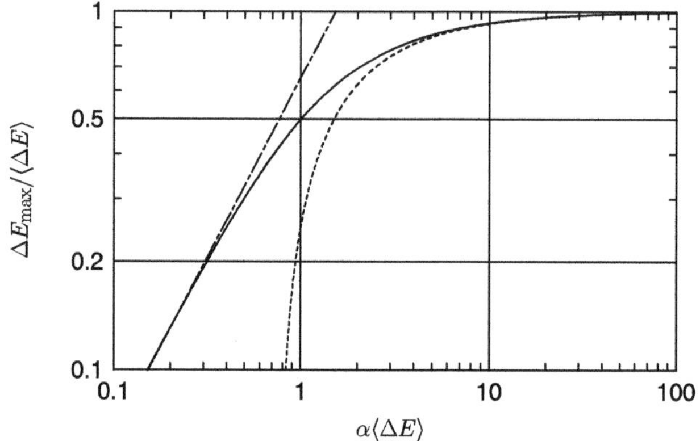

Fig. 9.3. Ratio of peak to mean energy loss, $\Delta E_{\mathrm{max}}/\langle\Delta E\rangle$ defined by (9.34) versus $\alpha\langle\Delta E\rangle$ which stands approximately for $\langle\Delta E\rangle/T_{\mathrm{max}}$. Solid line for cross section (9.29). Dotted line: Asymptotic relation (9.35). Dash-dotted line: Low-thickness limit $\Delta E_{\mathrm{max}} \propto x^2$, (9.36)

The spectrum (9.31) has its maximum at an energy loss ΔE_{max} given by

$$\Delta E_{\mathrm{max}} = \sqrt{\langle\Delta E\rangle^2 + \left(\frac{3}{4\alpha}\right)^2} - \frac{3}{4\alpha} \tag{9.34}$$

which, at large thicknesses, approaches

$$\Delta E_{\mathrm{max}} \simeq \langle\Delta E\rangle - \frac{3}{4\alpha}, \tag{9.35}$$

while at small thicknesses we find

$$\Delta E_{\mathrm{max}} \simeq \frac{2\alpha}{3}\langle\Delta E\rangle. \tag{9.36}$$

These relations are illustrated in Fig. 9.3. The x^2 dependence of ΔE_{max} for very thin targets shows up as a linear dependence of the peak-to-mean ratio $\Delta E_{\mathrm{max}}/\langle\Delta E\rangle$ for $x \ll 1$. The asymptotic relation – which will come up again more generally in the following – appears accurate as long as the deviation from unity does not exceed $\sim 10\,\%$.

9.3 Thin Targets

This section reviews several approaches to thin-target spectra. The Bohr-Williams approach, being applicable also to moderately thick targets, is analytically simple and quite flexible, but less quantitative than the others. Landau's solution is a common reference standard in particular for high-speed beams, but it has several limitations. Amongst several proposals to improve the Landau theory, attention will be paid primarily to the work of Lindhard (1985), whose approach allows for more realistic single-event spectra, and of Glazov (2000, 2002) who relaxed the limitations on allowed target thicknesses and in this way successfully bridged the gap between thin- and thick-target approaches.

9.3.1 Bohr-Williams Approach

The Bohr-Williams approach (Bohr, 1915, Williams, 1929, Bohr, 1948) represents an attempt to bridge the gap between eqs. (9.1) and (9.3) by the introduction of a limiting energy transfer T_1, such that the energy-loss spectrum is taken to approach the single-collision limit given by the differential cross section above some critical energy transfer T_1, and gaussian shape for $T \ll T_1$. The choice of the parameter T_1 – which must depend on target thickness – is crucial for the success of this approach.

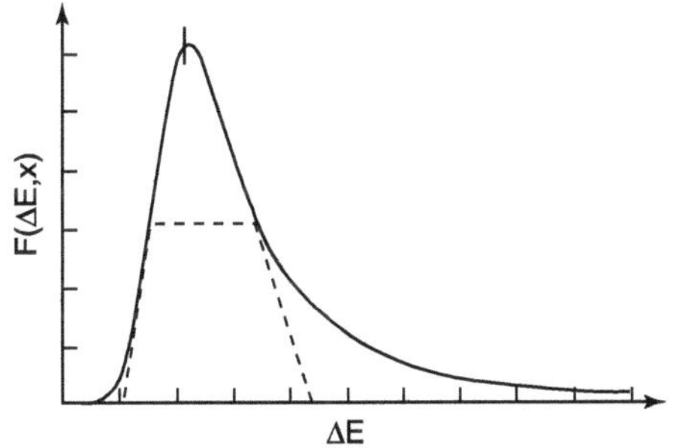

Fig. 9.4. Construction of an energy-loss spectrum according to Williams (1929)

Imagine the beam divided up into two groups of particles (Fig. 9.4 that have undergone an encounter with $T > T_1$ and those that have not. Let us assume T_1 to be so large that the probability for more than one event with $T > T_1$ is small. Then, the energy-loss spectrum for the first group can be approximated by the differential cross section,

$$F(\Delta E, x) \simeq N x \frac{\mathrm{d}\sigma(\Delta E)}{\mathrm{d}(\Delta E)}; \quad \Delta E > T_1. \tag{9.37}$$

Projectiles in the second group are likely to have undergone more than one interaction. One may be tempted to approximate that part of the spectrum

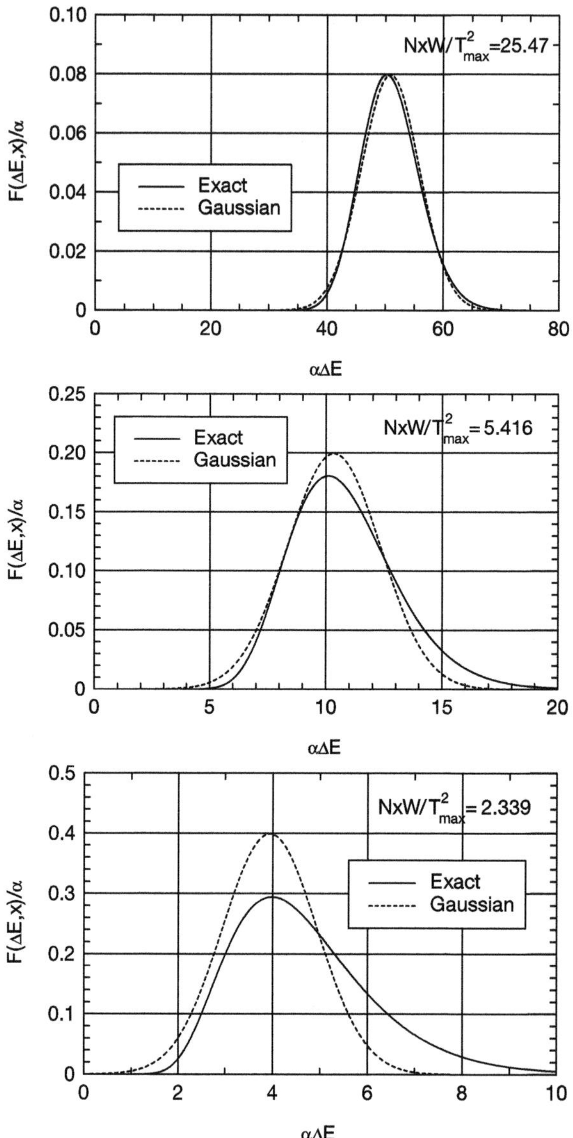

Fig. 9.5. Energy-loss spectra determined by the Bohr-Williams scheme with T_1 determined by (9.42) for model cross section (9.29)

by a gaussian,

$$F(\Delta E, \Delta x) = \frac{1}{\sqrt{2\pi N x W_1}} e^{-(\Delta E - N x S_1)^2 / 2 N x W_1} \tag{9.38}$$

centered around a mean value NxS_1 with

$$S_1 = \int_0^{T_1} T \mathrm{d}\sigma(T), \tag{9.39}$$

and a variance $\Omega_1^2 = NxW_1$ with

$$W_1 = \int_0^{T_1} T^2 \mathrm{d}\sigma(T), \tag{9.40}$$

following eqs. (2.29) and (2.30).

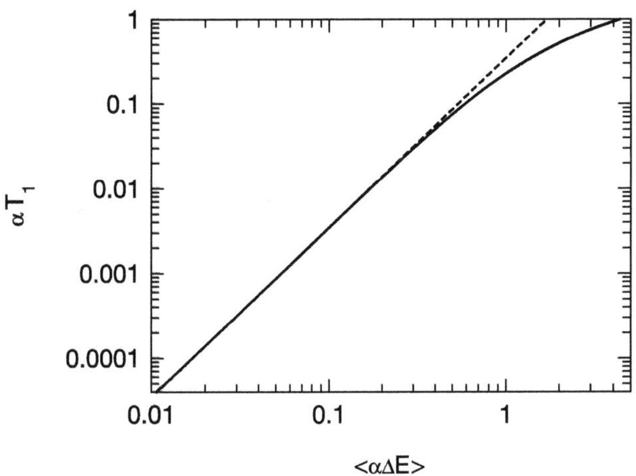

Fig. 9.6. Bohr-Williams parameter T_1 separating single- from multiple-encounter regime for cross section (9.29), determined by requiring the predicted peak energy loss to coincide with the exact value. Dashed line: quadratic dependence

A simple way to determine T_1 is the requirement that the probability for an event with $T > T_1$ be unity, i.e.,

$$Nx \int_{T \geq T_1} \mathrm{d}\sigma(T) = 1, \tag{9.41}$$

Another, quite successful procedure is to require that

$$T_1^2 = Nx \int_0^{T_1} T^2 \mathrm{d}\sigma(T) \tag{9.42}$$

according to Fastrup et al. (1966). Figure 9.5 shows three Bohr-Williams-type spectra determined in this manner, compared with the exact result for the model cross section (9.29). Only the multiple-collision portion of the Bohr-Williams spectrum is indicated, because inspection of the exact spectrum (9.31) reveals that the single-scattering limit is never reached for the relatively large pathlengths shown here. It is seen that for $\Omega^2/T^2_{\max}$ not too close to 1, the region around the peak energy loss is reasonably well described. Major discrepancies occur for thinner targets.

Another way of looking at the spectrum discussed in Sect. 9.2.5 is to fix the parameter T_1 by requiring the peak of the gaussian, NxS_1, to coincide with the exact value $\Delta E_{\max}$ (problem 9.6). You then find that

$$NxS_1 = NxS \operatorname{erf} \sqrt{\alpha T_1}, \tag{9.43}$$

and hence,

$$\frac{3}{2\alpha NxS} = \frac{1}{\operatorname{erf}\sqrt{\alpha T_1}} - \operatorname{erf}\sqrt{\alpha T_1} \tag{9.44}$$

This is illustrated in Fig. 9.6. It is seen that for thin targets, i.e. for $\alpha\langle\Delta E\rangle \simeq \langle\Delta E\rangle/T_{\max} < 1$, T_1 is roughly proportional to the square of the target thickness. Evidently, for thin targets, the overwhelming part of the spectrum follows the differential cross section. This behavior changes gradually for target thicknesses where the average energy loss NxS exceeds $T = 1/\alpha$.

9.3.2 Landau's Solution

The scheme of Landau (1944) is directed toward the thin-target limit. It operates with a cross section that is mainly Rutherford-like. The behavior at low energy transfers remains unspecified, but it is shown that it is insignificant for the *shape* of the spectrum, although not for its position on the energy-loss scale.

With this in mind we may, for clarity's sake, operate with a cross section

$$d\sigma(T) = C'\frac{dT}{T^2}; \quad T_{\min} \leq T \leq T_{\max}, \tag{9.45}$$

where

$$C' = \frac{W_B}{2mv^2} \text{ and } T_{\max} = 2mv^2, \tag{9.46}$$

and $W_B = 4\pi Z_1^2 Z_2 e^4$ according to (8.11). The Bethe stopping cross section (4.118) may be reproduced with the choice

$$T_{\min} = \frac{I^2}{2mv^2}. \tag{9.47}$$

With this, integration of (9.15) yields

$$\sigma(k) = \frac{C'}{T_{\min}}\left(1 - E_2(ikT_{\min})\right) - \frac{C'}{T_{\max}}\left(1 - E_2(T_{\max})\right), \tag{9.48}$$

where $E_2(z) = \int_1^\infty dt\, e^{-zt}/t^2$ is an exponential integral (Abramowitz and Stegun, 1964). If the main portion of the energy loss spectrum lies in the region where

$$T_{\min} \ll \Delta E \ll T_{\max}, \tag{9.49}$$

the integration over (9.14) is determined by the range of k-values for which $kT_{\min} \ll 1$ and $kT_{\max} \gg 1$. This allows Taylor expansions of the two terms on the right-hand side of (9.48) in terms of $kT_{\min}$ and $1/kT_{\max}$, respectively. The result is

$$\sigma(k) = C'ik\left[1 - \gamma - \ln(ikT_{\min})\right]\cdots, \tag{9.50}$$

where $\gamma = 0.5772$ is Euler's constant, and where all terms going to zero with $kT_{\min}$ or $1/kT_{\max}$ have been dropped.

With this, (9.14) reduces to

$$F(\Delta E, x)\mathrm{d}(\Delta E) = g_{\mathrm{L}}(\varLambda)\mathrm{d}\varLambda \tag{9.51}$$

with

$$g_{\mathrm{L}}(\varLambda) = \frac{1}{2\pi i}\int_{-i\infty}^{i\infty} \mathrm{d}u\, u^u\, e^{\varLambda u}, \tag{9.52}$$

where $u = ikNxC'$ and

$$\varLambda = \frac{\Delta E}{NxC'} - 1 + \gamma - \ln\frac{NxC'}{T_{\min}}. \tag{9.53}$$

Figure 9.7 shows a plot of the *Landau function* $g_{\mathrm{L}}(\varLambda)$, which has a maximum at $\varLambda_{\max} = -0.2258$, around which it has approximately gaussian shape. It falls off steeply below the maximum and slowly, Coulomb-like, at large values of $\varLambda$.

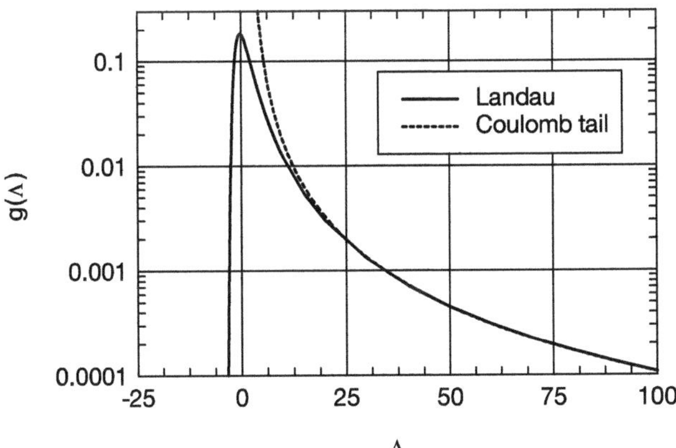

Fig. 9.7. Solid line: The Landau function $g(\varLambda)$ (9.52) tabulated by Börsch-Supan (1961). Dashed line: Single Coulomb scattering

Although the Landau function is properly normalized, as you may verify by direct integration or by applying (9.21), the mean energy loss and the straggling parameter diverge. This is caused by the complete neglect of $T_{\max}$ in (9.50). We shall see below that this is a severe limitation to the range of applicability of the Landau formula.

While the low-energy cutoff $T_{\min}$ is very small and hence barely significant for the spectrum, it does affect the position of the spectrum, i.e., the mean energy loss, according to (9.53). On the other hand, the assumption of straight Coulomb scattering down to energies far below atomic binding energies is a more serious drawback, as is the neglect of shell and Barkas corrections. Inclusion of these features requires a more general scheme which, at the same time, ought to take into account the maximum energy transfer and hence produce a distribution with the proper mean value and variance. Such schemes will be described in the following two sections.

Let us finally consider scaling properties. With the peak of the Landau function $g_{\mathrm{L}}(\Lambda)$ at $\Lambda = \Lambda_{\max}$ and the left and right half widths at $\Lambda_{-1/2}$ and $\Lambda_{1/2}$, respectively, and $T_{\min} = I^2/2mv^2$, (9.53) determines the most probable energy loss at

$$\Delta E_{\max} = \frac{4\pi Z_1^2 Z_2 e^4 N x}{mv^2} \ln \frac{\sqrt{4\pi Z_1^2 Z_2 e^4 N x C_W}}{I} \tag{9.54}$$

with

$$C_W = e^{\Lambda_{\max}+1-\gamma} = e^{0.2} \simeq 1.22, \tag{9.55}$$

while the right and the left halfwidth read

$$\left|\Delta E_{\pm 1/2} - \Delta E_{\max}\right| = \frac{4\pi Z_1^2 Z_2 e^4 N x}{mv^2} \left|\Lambda_{\pm 1/2} - \Lambda_{\max}\right| \tag{9.56}$$

with

$$\left|\Lambda_{-1/2} - \Lambda_{\max}\right| = 1.3637; \quad \left|\Lambda_{+1/2} - \Lambda_{\max}\right| = 2.655. \tag{9.57}$$

You may note that the factor in front of the logarithm in $\Delta E_{\max}$ is the one we know from the mean energy loss. Here, however, the logarithm is independent of the projectile speed and dependent on thickness, while the opposite is the case for the mean energy loss. Even though this is a striking difference, its quantitative consequences need not be pronounced because of the logarithmic dependence on either variable. Conversely, the fact that the halfwidth is proportional to the path length x, while the standard deviation is $\propto \sqrt{x}$ is dramatic.

9.3.3 Lindhard's Solution ($\star$)

Lindhard (1985) devised an elegant mathematical procedure to construct Landau-like energy-loss profiles, taking into account deviations of the differential cross section from the Landau spectrum (9.45). By introduction of suitable variables which incorporate the essential input from the cross section, a high degree of universality was achieved. The evaluation delivers two levels of accuracy, a first approximation offering a particularly simple scaling law that is adequate for gaining insight. The second approximation ensures better numerical accuracy but is less transparent.

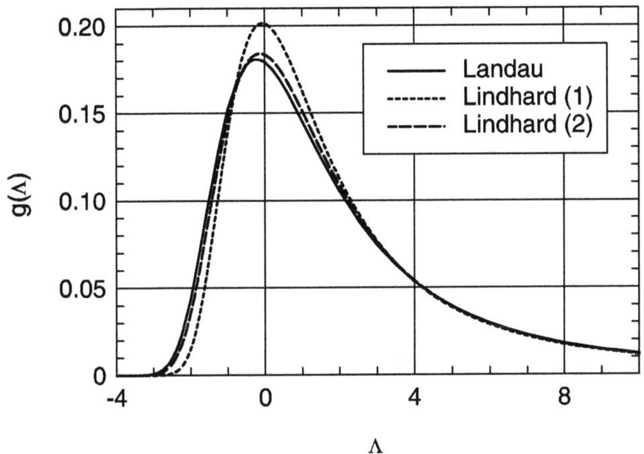

Fig. 9.8. Comparison of Landau function and the two approximations of Lindhard (1985)

By introduction of the functions

$$a(\Delta E, x) = \int_0^{\Delta E} F(\Delta E', x)\mathrm{d}(\Delta E'), \tag{9.58a}$$

$$\epsilon(\Delta E, x) = \int_0^{\Delta E} a(\Delta E', x)\mathrm{d}(\Delta E'), \tag{9.58b}$$

$$\eta(\Delta E, x) = \frac{\epsilon(\Delta E, x)}{a(\Delta E, x)}, \tag{9.58c}$$

$$\sigma(\eta) = \int_\eta^\infty \mathrm{d}\sigma(T), \tag{9.58d}$$

$$S(\eta) = \int_0^\eta T\,\mathrm{d}\sigma(T), \tag{9.58e}$$

and using η, (9.58c), as the independent energy-loss variable instead of ΔE, the function $a(\Delta E, x)$ was found to be given by

$$a(\eta, x) = \mathrm{e}^{-Nx\sigma(\eta)} \tag{9.59}$$

as a first approximation. With this, the spectrum may be determined by differentiation,

$$P(\Delta E, x) = \frac{\mathrm{d}a}{\mathrm{d}\eta} \frac{\mathrm{d}\eta}{\mathrm{d}(\Delta E)}, \tag{9.60}$$

and ΔE was found to be given by

$$\Delta E = \eta + NxS(\eta). \tag{9.61}$$

For the case of the Landau cross section the result is

$$g(\Lambda, x)\mathrm{d}\Lambda = \frac{\mathrm{d}\Lambda}{y(1+y)}\mathrm{e}^{-1/y}, \tag{9.62}$$

and

$$\Lambda = y + \ln y - 1 + \gamma. \tag{9.63}$$

Figure 9.8 shows surprisingly good agreement of this relation with a numerical tabulation. Near-perfect agreement is seen with a second approximation that involves a straggling parameter

$$W(\eta) = \int_0^\eta T^2 \, \mathrm{d}\sigma(T). \tag{9.64}$$

These relations become useful when deviations from the simple Coulomb-like scattering law (9.45) are estimated. In the first approximation, taking into account the maximum energy transfer $T_{\max}$ was found to imply a cutoff at high energy losses and multiplication with a normalizing factor, while modifications in the low-T behavior of the cross section have negligible influence.

In the second approximation, the low-T behavior of the cross section was found to have a substantial influence on the spectrum. However, that influence could be estimated from deviations of the straggling parameter W from the Bohr value W_{B}, i.e., effects discussed in Chapter 8.

9.3.4 Glazov's Solution

A very powerful and transparent approach has been developed by Glazov (2000, 2002). We have seen already that neglecting the kinematic limit in the Coulomb cross section is a significant drawback of Landau's solution. Hence, Glazov (2000) wrote the transport cross section in the form

$$\sigma(k) = \int_0^\infty \mathrm{d}\sigma(T)\left(1 - \mathrm{e}^{-ikT}\right) - \int_{T_{\max}}^\infty \mathrm{d}\sigma(T)\left(1 - \mathrm{e}^{-ikT}\right) \tag{9.65}$$

and rewrote the Bothe-Landau formula (9.14) in the form

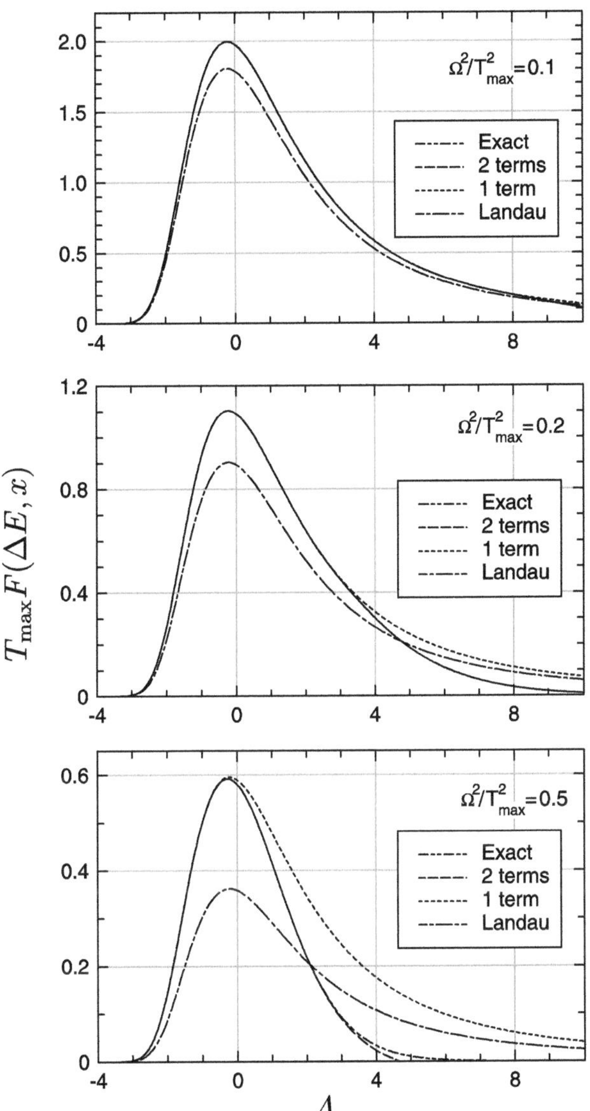

Fig. 9.9. Energy-loss spectra for cross section (9.45): Comparison between Landau spectrum, (9.66) taking into account 1 or 2 terms, and numerical evaluation of (9.14). Invisible curves coincide with the exact result. From Glazov (2000).

$$F(\Delta E, x) = e^{Nx\sigma_{\mathrm{ne}}} \left\{ F_\infty(\Delta E, x) - Nx \int_{T_{\max}}^\infty d\sigma(T) F_\infty(\Delta E - T, x) \right.$$

$$\left. + \frac{1}{2}(Nx)^2 \int_{T_{\max}}^\infty d\sigma(T) \int_{T_{\max}}^\infty d\sigma(T') \times F_\infty(\Delta E - T - T', x) \ldots \right\}, \quad (9.66)$$

where

$$\sigma_{\mathrm{ne}} = \int_{T_{\mathrm{max}}}^{\infty} \mathrm{d}\sigma(T) \tag{9.67}$$

is the *cross section for nonexisting events*, and $F_{\infty}(\Delta E, x)$ an energy spectrum calculated by ignoring the upper limit T_{max} in the transport cross section (cf. problem 9.10). The merit of the expansion (9.66) is twofold,

- All integrals go over large energy transfers, i.e., refer to rare events, thus ensuring rapid convergence for not too large pathlengths,
- Since energy spectra $F_{\infty}(\Delta E, x)$ vanish for $E < 0$, and since $T, T' \dots > T_{\mathrm{max}}$, the series reduces to a finite sum.

The scheme was applied to nonrelativistic Rutherford scattering, and Landau's solution (9.52) was introduced for $F_{\infty}(\Delta E, x)$. The accuracy of the resulting spectra was tested against accurate numerical evaluations of (9.14) for the same cross section. Remind that the low-T behavior of the cross section mainly affects the *position* of the spectrum on the energy-loss axis, while the *shape* is insensitive.

Figure 9.9 shows results calculated for three different target thicknesses, corresponding to values $\Omega^2/T_{\mathrm{max}}^2 = 0.1$, 0.2 and 0.5. Considering only the leading term in (9.66), implying renormalization of the Landau spectrum as mentioned in section 9.3.3, turns out to produce almost complete agreement with the numerical result for $\Omega^2/T_{\mathrm{max}}^2 = 0.1$. For higher values of $\Omega^2/T_{\mathrm{max}}^2$, up to ~ 0.3, two terms are found adequate, while three terms are needed up to $\Omega^2/T_{\mathrm{max}}^2 \sim 1$.

The main strength of this scheme appears to lie in the reduction of a complex spectrum characterized by at least two variables to a sum of very few one-parameter Landau-type functions, even at target thicknesses where a spectrum expressed by a single Landau function would be completely inadequate.

9.4 Moderately Thick Targets

This section reviews calculational schemes that have been utilized to determine energy-loss spectra in moderately thick targets. The reader is reminded that in the present terminology, a moderately thick target is still so thin that variation with beam energy of the differential cross section can be neglected, i.e., that the Bothe-Landau formula (9.14) may serve as a starting point.

9.4.1 Vavilov Scheme ($\star$)

The work of Vavilov (1957) represents an attempt to bridge the gap between the Landau spectrum and the gaussian limit. The starting point is the Bothe-Landau formula (9.14), but unlike Landau, Vavilov did not ignore the upper limit T_{max} in the evaluation of the transport cross section and thus ensured a finite stopping cross section and straggling parameter[5].

The evaluation starts at (9.22). In order to go beyond the diffusion approximation, Vavilov included a third-order term in the expansion of the transport cross section (9.23) but dropped all subsequent terms,

$$\sigma_1(k) \simeq -\frac{1}{2}(\mathrm{i}k)^2 W + \frac{1}{6}(\mathrm{i}k)^3 Q_3 + \mathcal{O}\left(k^4\right). \tag{9.68}$$

The exponent in the integral over $\mathrm{d}k$ can be simplified by means of the substitution

$$z = \mathrm{i}k - \frac{W}{Q_3}, \tag{9.69}$$

so that

$$F(\Delta E, x) = \exp\left(\frac{W}{Q_3}\eta - \frac{1}{6}NxQ_3\left(\frac{W}{Q_3}\right)^3\right)$$
$$\times \frac{1}{2\pi\mathrm{i}} \int_{c-\mathrm{i}\infty}^{c+\mathrm{i}\infty} \mathrm{d}z \, \exp\left(z\eta - \frac{1}{6}NxQ_3 z^3\right), \tag{9.70}$$

where

$$\eta = \Delta E - NxS + \frac{1}{2}NxQ_3\left(\frac{W}{Q_3}\right)^2. \tag{9.71}$$

The quantity c is nominally given by $c = -W/Q_3$ but can be chosen to be any real constant in view of the regularity of the exponential function in the complex plane.

The integral can be expressed by an Airy function so that[6]

$$F(\Delta E, x) = A \exp\left(\frac{W}{Q_3}\eta - \frac{1}{6}NxQ_3\left(\frac{W}{Q_3}\right)^3\right) \mathrm{Ai}(A\eta), \tag{9.72}$$

where

$$A = \left(\frac{2}{NxQ_3}\right)^{1/3} \tag{9.73}$$

and

$$\mathrm{Ai}(z) = \frac{1}{\pi}\sqrt{\frac{z}{3}} K_{1/3}\left(\frac{2}{3}z^{3/2}\right) \tag{9.74}$$

an Airy function in the notation of Abramowitz and Stegun (1964).

---

[5] The cross section utilized by Vavilov was equivalent to (9.45) except for an allowance for relativistic velocities in accordance with (3.92).

[6] Equation (9.72) corrects an error in the expression given in the author's lecture notes (Sigmund, 1991).

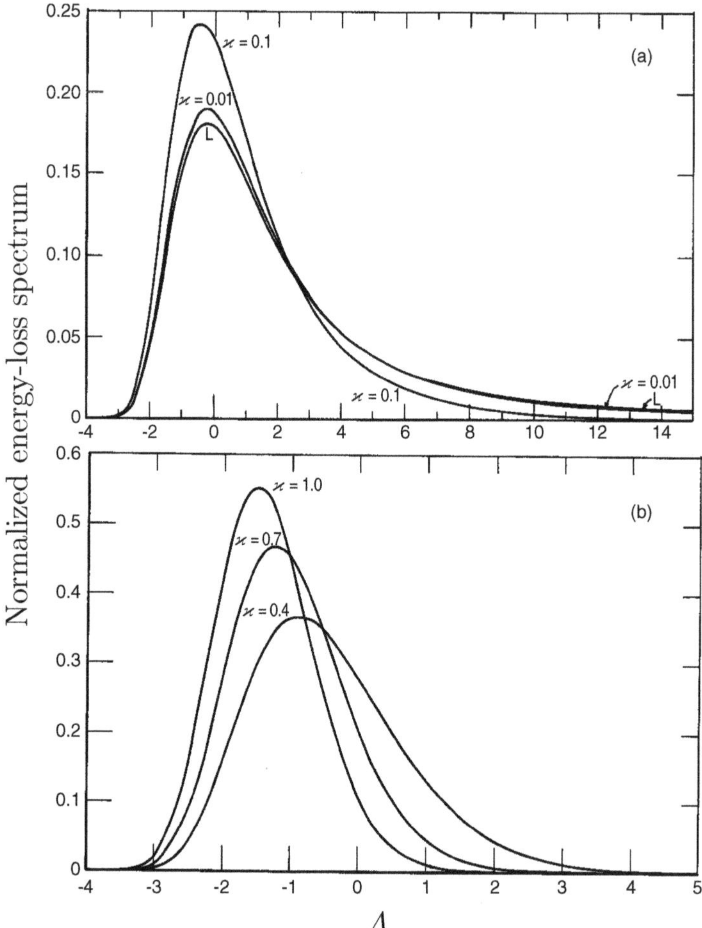

Normalized energy-loss spectrum

Fig. 9.10. The Vavilov[7] distribution for values of the parameter $\varkappa = NxS/T_{\mathrm{max}}$ according to Seltzer and Berger (1964). 'L' denotes the Landau spectrum

Writing the variable η in the form

$$\frac{\eta}{NxC'} = \frac{\Delta E}{NxC'} + 1 - 2L, \tag{9.75}$$

where $C' = W_{\mathrm{B}}/2mv^2$ is the constant in the Landau formula and L the stopping number, as well as

$$\frac{W}{Q_3}\eta = \frac{\Delta E}{mv^2} + \frac{NxW_{\mathrm{B}}}{(mv^2)^2}\left(\frac{1}{2} - L\right), \tag{9.76}$$

---

[7] See footnote on p. 331

indicates that η/NxC' is, apart from an additive constant, identical with the Landau energy-loss variable Λ, while $\eta W/Q_3$ approaches zero in the thin-target limit $NxW/(mv^2)^2 \ll 1$ and for energy losses near the peak where $\Delta E/mv^2 \ll 1$. Thus, the peak area of (9.72) comes close to Landau's formula in the limit of low thicknesses.

Fig. 9.10 shows plots of the Vavilov[8] distribution for several target thicknesses, expressed by the ratio $\langle \Delta E \rangle/T_{\max}$. A trend toward gaussian shape is seen with increasing thickness, but significant skewness is retained at the largest thickness.

Vavilov's *Ansatz* has weak points,

- The distribution results from a truncated series expansion in terms of a parameter $\sim kT_{\max}$ which is not necessarily small,
- Inspection of the properties of the Airy function shows that the spectrum (9.72) turns negative for

$$\Delta E - NxS < -\frac{NxW^2}{Q_3} - 2.338 \left(\frac{1}{2}mv^2 NxW\right)^{1/3}. \tag{9.77}$$

This general feature of diffusion approximations going beyond second order was already mentioned in Sect. 9.2.4. Inaccuracies will be observed not far outside the left halfwidth of the profile.
- The analytical form (9.70) is rather inconvenient for identifying scaling properties as well as determining peak energy loss and halfwidth,
- For the simple model cross section (9.29), where the exact profile (9.31) is available, the Vavilov solution is considerably more complex.

9.4.2 Method of Steepest Descent

The steepest-descent (or *saddle-point*) method is a wellknown mathematical procedure to evaluate the integral over functions with sharp maxima. It was applied to the Bothe-Landau formula by Moyal (1955). More recently, Sigmund and Winterbon (1985) showed that accurate results can be derived by this method and simple expressions can be extracted for peak position and half width of an energy-loss profile. Glazov (2000) demonstrated a considerable overlap between this and the expanded Landau scheme (9.66), so that combined use of the two methods allows to produce accurate energy-loss profiles for a very wide range of target thicknesses.

The method of steepest descent is of interest in the present context because contributions to the integral in (9.15) originate predominantly from the range around the maximum of the exponent. Expansion of the exponent around this maximum yields a parabola in the exponent and hence a gaussian which allows integration in closed form.

---

[8] See footnote on p. 331

This looks superficially like the diffusion approximation discussed in Sect. 9.2.4, but here the exponent is expanded around its maximum, whereas it is expanded around $k = 0$ in the diffusion approximation, irrespective of cross section and pathlength. In other words, the method of steepest descent will reproduce the diffusion approximation where appropriate but deliver more reliable results where not.

Now, write (9.14) in the form

$$F(\Delta E, x) = \frac{1}{2\pi} \int_{-\infty}^{\infty} \mathrm{d}k \, \mathrm{e}^{-f(k)} \tag{9.78}$$

with

$$-f(k) = \mathrm{i}k\Delta E - Nx\sigma(k). \tag{9.79}$$

Expansion around a point k_0 where $f(k)$ is stationary yields

$$f(k) = f_0 + \frac{1}{2}(k - k_0)^2 f_0'' \ldots \tag{9.80}$$

where

$$f_0' = \left. \frac{\mathrm{d}f(k)}{\mathrm{d}k} \right|_{k=k_0} = 0. \tag{9.81}$$

Dropping terms of higher than second order in $k - k_0$ allows you to carry out the integral and leads to

$$F(\Delta E, x) = \frac{\mathrm{e}^{-f_0}}{\sqrt{2\pi f_0''}}. \tag{9.82}$$

Insertion of (9.79) then yields

$$F(\Delta E, x) = \frac{\mathrm{e}^{Nx(k_0\sigma_0' - \sigma_0)}}{\sqrt{2\pi Nx\sigma_0''}}, \tag{9.83}$$

where $\sigma_0 = \sigma(k)|_{k=k_0}$ and similar for the derivatives, and k_0 is determined by

$$\Delta E = -\mathrm{i}Nx\sigma_0' \tag{9.84}$$

according to (9.81).

Equations (9.83) and (9.84) determine the spectrum for a given transport cross section as a parameter representation with k_0 as the independent variable. You may apply these relations to the cross section (9.45) (problem 9.11). In that particular case you recover the *exact* result.

We are now able to determine peak value and halfwidth of the profile as a function of k_0 and use (9.84) to find the corresponding values of ΔE. The peak value of (9.83) lies at a value k_{max} fulfilling the condition

$$Nx = \frac{\sigma_{\mathrm{max}}'''}{2k_{\mathrm{max}}\sigma_{\mathrm{max}}''^{\,2}}, \tag{9.85}$$

where

$$\sigma'_{\max} = \frac{\mathrm{d}}{\mathrm{d}k}\sigma(k)\bigg|_{k=k_{\max}} \tag{9.86}$$

and similarly for higher derivatives. This, together with (9.84), i.e.,

$$\Delta E_{\max} = -\mathrm{i}Nx\sigma'_{\max} \tag{9.87}$$

represents a parameter representation of $\Delta E_{\max}$ as a function of Nx with $k_{\max}$ as the independent variable.

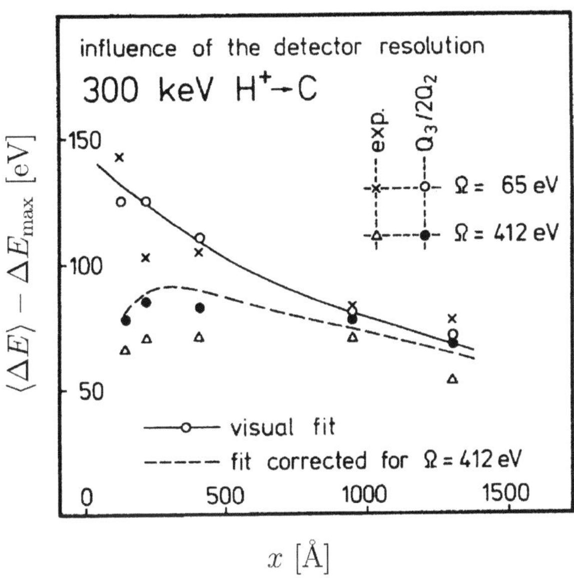

Fig. 9.11. The difference between mean and peak energy loss depends on detector resolution but is insensitive to foil thickness according to (9.94). From Mertens (1986)

Insertion of the transport cross section then leads to

$$Nx = \frac{Q_3(k_{\max})}{2\mathrm{i}k_{\max}[W(k_{\max})]^2}, \tag{9.88}$$

$$\Delta E_{\max} = NxS(k_{\max}), \tag{9.89}$$

where

$$S(k) = \int T\,\mathrm{d}\sigma(T)\mathrm{e}^{-\mathrm{i}kT} \tag{9.90}$$

$$W(k) = \int T^2\,\mathrm{d}\sigma(T)\mathrm{e}^{-\mathrm{i}kT} \tag{9.91}$$

$$Q_3(k) = \int T^3\,\mathrm{d}\sigma(T)\mathrm{e}^{-\mathrm{i}kT} \tag{9.92}$$

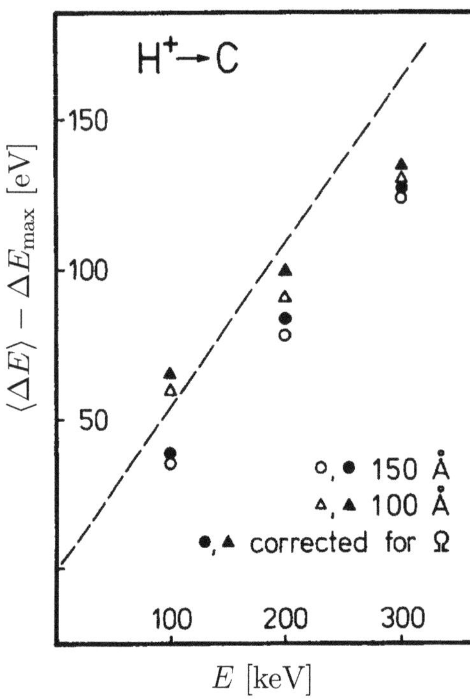

$\langle\Delta E\rangle - \Delta E_{\mathrm{max}}$ [eV]

E [keV]

Fig. 9.12. The difference between mean and peak energy loss is approximately proportional to the beam energy and insensitive to foil thickness according to (9.94). From Mertens (1986)

These relations are very suited for numerical evaluation. They also provide exact or approximate analytical solutions in a number of easily integrable cases such as the spectrum (9.45) (cf. problem 9.12).

Equation (9.89) predicts $\Delta E_{\mathrm{max}} = NxS = \langle\Delta E\rangle$ in the limit of $k_{\mathrm{max}} = 0$. For k_{max} small, (9.88) reduces to

$$\mathrm{i}k_{\mathrm{max}} = \frac{Q_3(0)}{2NxW(0)^2},\tag{9.93}$$

and expansion of (9.89) in powers of k_{max} yields

$$\Delta E_{\mathrm{max}} = NxS - \frac{Q_3(0)}{2W(0)}\ldots\tag{9.94}$$

This is the beginning of an asymptotic expansion in powers of the inverse thickness. It implies that a plot of the peak energy loss against thickness will approach a straight line at large velocities, but extrapolation toward zero thickness will show an intercept. This feature has been studied experimentally by Mertens (1986). Figure 9.11 shows the measured difference $\langle\Delta E\rangle - \Delta E_{\mathrm{max}}$ as a function of foil thickness for two experimental geometries, and Fig. 9.12 shows the same quantity as a function of beam energy for two target thicknesses. The dependence on thickness is weak at the high-energy end, as to be expected from (9.94), while a linear increase with beam energy is observed, as to be expected from (8.122).

For straight Coulomb scattering at nonrelativistic velocities, (9.94) reads

$$\Delta E_{\mathrm{max}} = NxS - \frac{mv^2}{2} \ldots \tag{9.95}$$

You may easily convince yourself that the term $Q_3/2W$ is identical with the expression $3/4\alpha$ in (9.35) for the particular case of the cross section (9.29).

A similar procedure determines the right and left halfwidth. Here only the result is given, while the reader is referred to Sigmund and Winterbon (1985) for details,

$$(\Delta E - \Delta E_{\mathrm{max}})_{\pm 1/2} = \pm\sqrt{2\ln 2\, NxW}\left(1 \pm \frac{Q_3}{6W}\sqrt{\frac{2\ln 2}{NxW}}\right). \tag{9.96}$$

Here, the leading term is the gaussian halfwidth. Since Q_3 is usually positive, the right halfwidth becomes greater, while the left halfwidth shrinks.

You may observe that changes in halfwidth due to skewness enter as a term of order $1/\sqrt{x}$, i.e., are more significant than the difference between mean and peak value in the high-thickness limit.

In the analysis of experiments, one might determine a peak value as the average between two half values. Going to one higher order in the expansion (9.96), Sigmund and Winterbon (1985) obtained a modified peak position

$$\Delta E'_{\mathrm{max}} = NxS - \frac{\ln 2}{3}\frac{Q_3(0)}{W(0)}\ldots \tag{9.97}$$

Evidently, a 'peak value' determined in this manner is less shifted from the mean value than the true peak value (9.94). The difference is significant: It is more than a factor of two!

Figure 9.13 shows a comparison of calculated peak energy losses and halfwidths for the cross section (9.45). It is seen that a combination of the asymptotic steepest-descent result (9.94) at large thicknesses and Glazov's solution (9.66) at small thicknesses provides an excellent basis for theoretical predictions as compared with exact numerical solutions.

A weak point of the method of steepest descent is that it does not preserve the normalization of the original distribution function. In cases of minor discrepancies this can be repaired by a normalizing factor. In more drastic cases the validity of the approximation may need an independent check.

9.4.3 Applications

This section reviews explicit calculations of energy-loss profiles for specific systems. The main thrust here is the effect of a realistic cross section as opposed to (9.45). Early attempts in this direction made use of two assumptions,

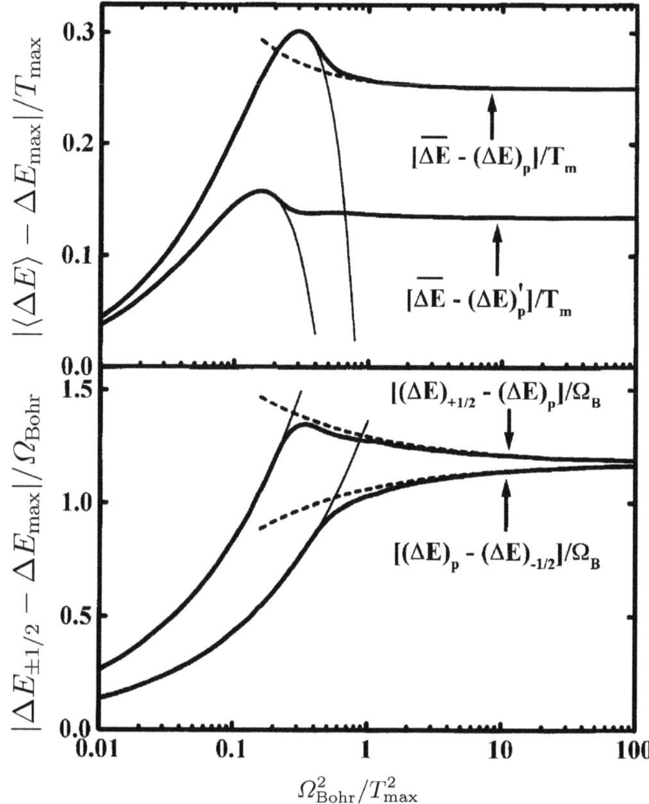

Fig. 9.13. Relation between peak and mean energy loss (upper graph) and between halfwidth and standard deviation (lower graph) for cross section (9.45). Thick solid lines: Numerical; thin solid lines: Glazov solution, 1-term approximation; dotted lines: Steepets descent, asymptotic. From Glazov (2002)

- Deviations of the cross section from (9.45) enter primarily via the second moment W, and
- The Vavilov expansion (9.68) is an adequate approximation to the transport cross section.

While the validity of the first assumption has been confirmed by Lindhard (1985), as mentioned already in Sect. 9.3.3, the second assumption is not universally valid. Therefore, early attempts to evaluate spectra for realistic cross sections are generally hampered by inaccuracies introduced by the Vavilov expansion (Blunck and Leisegang, 1950, Shulek et al., 1966, Bichsel, 1970).

9.4.4 Straight Convolution

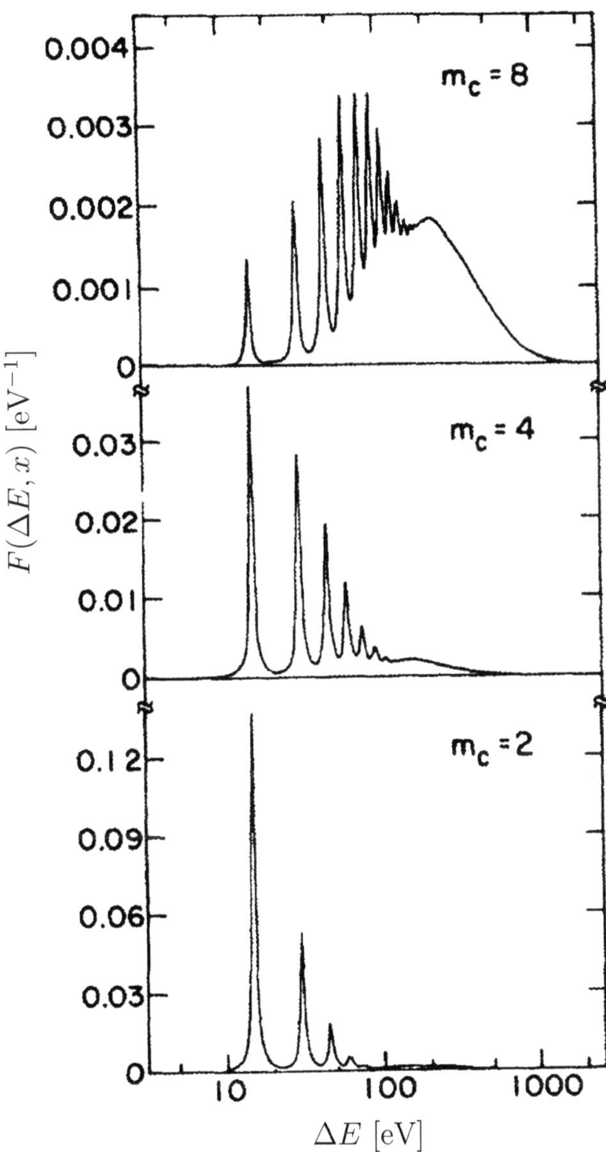

Fig. 9.14. Calculated energy-loss spectra for 20 MeV protons in aluminium found by straight convolution. $m_c = Nx\sigma_{tot}$ is the mean number of collisions. From Bichsel and Saxon (1975)

Bichsel and Saxon (1975) evaluated an energy-loss spectrum for protons in aluminium by straight numerical convolution of a single-collision spectrum. They operate with a finite total cross section $\sigma_{\mathrm{tot}} = \int_0^\infty d\sigma(T)$, so that $Nx\sigma_{\mathrm{tot}}$ is the mean number of collisions undergone by the projectile while penetrating the target. The procedure starts at (9.9), which is applied to a very thin foil, corresponding to $Nx\sigma_{\mathrm{tot}} = 1/1024$. Insertion of this expression into (9.4) for $y = x$, and numerical integration yields $F(\Delta E, 2x)$, i.e., a spectrum for $Nx\sigma_{\mathrm{tot}} = 1/512$. Doubling the thickness the necessary number of times provides a rapid procedure to construct the desired spectrum for an arbitrary thickness.

The cross section applied received contributions from K and L electrons, evaluated by the Born approximation, and from a model for M electrons allowing for a plasmon peak, but less sophisticated than the models discussed in Chapter 5.

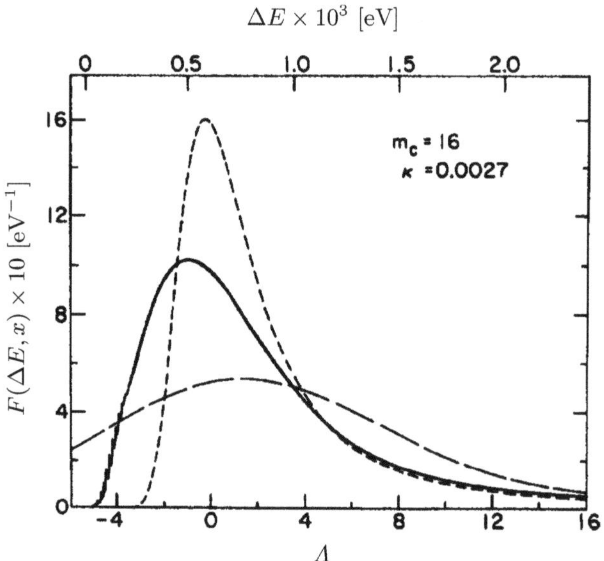

Fig. 9.15. Same as Fig. 9.14 for $m_c = 16$. Solid line: convolution; dotted line: Landau spectrum; dashed line: Vavilov spectrum. Absolute values refer to the upper abscissa scale. From Bichsel and Saxon (1975)

Figure 9.14 shows results calculated from this model for 20 MeV protons having undergone 2, 4 and 8 collisions, corresponding to 0.029, 0.058 and 0.116 μm foil thickness. For the lowest thickness, the spectrum is dominated by multiple plasmon peaks, while the continuum portion increases in significance with increasing thickness. Resolving energy losses at the 10 eV level for an ion beam in the 10 MeV range is experimentally difficult if not impossible. Figure 9.15 shows graphs evaluated for $Nx\sigma_{\mathrm{tot}} = 16$. Discrete features are no longer resolved here. More important, significant discrepancies are observed from Landau and Vavilov spectrum.

9.5 Transport Equations

Before proceeding to very thick targets we need to look at a very useful tool which has been used almost universally in the literature. Transport equations govern stopping in targets of all thicknesses. For thin and moderately thick targets, the Bothe-Landau equation (9.14) is a general solution, but for very thick targets that solution is not available.

9.5.1 Derivation by Two-Layer Argument

Let $F(E, E', x)\mathrm{d}E'$ be the probability that a projectile with initial energy E has an energy in the interval $(E', \mathrm{d}E')$ after a pathlength x. In this notation, the convolution equation (9.4) reads

$$F(E, E', x + y) = \int_0^E \mathrm{d}E'' F(E, E'', y) F(E'', E', x). \tag{9.98}$$

Note that the spectrum is now characterized by two energy variables E, E' rather than just the difference.

Now assume the thickness y of the first layer in Fig. 9.1 to be so small that the chance for more than one collision in that layer be negligible. Then the factor $F(E, E'', y)$ in (9.98) is determined by (9.9),

$$F(E, E'', y) = \left(1 - \sum_j N y \sigma_j(E)\right) \delta\left(E - E''\right)$$
$$+ \sum_j N y \sigma_j(E) \delta\left(E - E'' - T_j\right). \tag{9.99}$$

Insert this into (9.98) and apply Taylor expansion in y on both sides, keeping only terms up to first order. The terms of zero order in y then drop out, and the terms of first order yield

$$-\frac{\partial F(E, E', x)}{\partial x} = N \sum_j \sigma_j(E)\{F(E, E', x) - F(E - T_j, E', x)\}. \tag{9.100}$$

Although $F(E, E', x)$ has three variables, only two of them, E and x are active in (9.100) while E' is a dummy variable.

Eq. (9.100) is a differential equation of first order in x and hence requires an initial condition to be imposed. The proper condition here is

$$F(E, E', 0) = \delta(E - E') \tag{9.101}$$

for an initially monoenergetic beam.

With regard to the E-variable we deal with a difference equation. However, going over to continuum notation, i.e., substituting

$$\sigma_j(E) \to d\sigma(E, T) \equiv K(E, T)\,dT, \tag{9.102}$$

yields the following form for (9.100),

$$-\frac{\partial F(E, E', x)}{\partial x} = N \int dT\, K(E, T)\{F(E, E', x)$$
$$-F(E - T, E', x)\}, \tag{9.103}$$

which is an integro-differential equation.

Before going on let us do the same calculation once more, starting from (9.98) but assuming now that the thickness x of the second layer be small instead of y. Applying the same steps and afterwards renaming y to x we arrive at

$$-\frac{\partial F(E, E', x)}{\partial x} = N \sum_j \{F(E, E', x)\sigma_j(E')$$
$$-F(E, E' + T_j, x)\sigma_j(E' + T_j)\}, \tag{9.104}$$

which looks similar but by no means identical to (9.100). The most visible difference is in the active energy variable which was the initial energy E in (9.100) but now is the exit energy E' in (9.104).

In continuum notation, (9.104) reads

$$-\frac{\partial F(E, E', x)}{\partial x} = N \int dT\{K(E', T)F(E, E', x)$$
$$-K(E' + T, T)F(E, E' + T, x)\}. \tag{9.105}$$

These two kinds of integro-differential equations are well known in statistical physics. They go under names such as *kinetic equations*, *master equations*, *rate equations*, *linear transport equations* or even *Boltzmann equations*. Since Boltzmann's original equation is nonlinear in general, the latter nomenclature implies a linearized version.

9.5.2 Forward and Backward Equations

Rewrite (9.104) by dropping the initial energy E and changing sign,

$$\frac{\partial F(E', x)}{\partial x} = \sum_j \{-F(E', x)N\sigma_j(E') +$$
$$F(E' + T_j, x)N\sigma_j(E' + T_j)\}. \tag{9.106}$$

The physical content of this equation is readily grasped: The left-hand side indicates the local change in the energy spectrum of a particle beam per traveled pathlength. The first term on the right-hand side expresses the loss of projectiles from an energy interval around E' by collisions, and the second term represents a corresponding gain of projectiles scattered from an energy interval around $E' + T_j$ into E'. The equation is 'local' in the sense that only processes happening at depth x enter into the argument. Numerous rate equations in physics, chemistry and other branches of science follow this general scheme. In transport theory this equation is called a 'forward equation'.

Conversely, (9.100) or its continuum version is called a 'backward equation'. The physical difference may be appreciated by looking at Fig. 9.1 and realizing that the physical processes entering into the argument are collisions taking place in a thin layer of thickness y near the entrance surface, even though the quantity ultimately determined is the energy spectrum at the exit surface. This explains the alternative nomenclature of a *propagator equation.*

For homogeneous media the two formulations yield equivalent results. The forward form is more well-established in a general-physics context, but the backward form is more convenient when averages are taken over the exit energy. Numerous examples will be discussed in Volumes II and III. A very illuminating discussion of the entire complex in general terms was given by Lindhard and Nielsen (1971).

9.6 Very Thick Targets (⋆)

In the derivation of the Bothe-Landau formula it was assumed that the cross sections σ_j were constant throughout the penetrated layer. With increasing layer thickness the decrease in projectile speed may become too large for this assumption to be maintainable. Then, two major complications arise,

- The energy-loss spectrum $F(E, \Delta E, x)$ entering the convolution equation (9.4) now depends explicitly on energy. Therefore the convolution does not reduce to a product in Fourier space.
- Cross sectional input is needed for an extended range of beam energies.

Regarding the first point, the Chapman-Kolmogorov equation now reads

$$F(E, \Delta E, x + y)$$
$$= \int_0^{\Delta E} \mathrm{d}(\Delta E')\, F(E, \Delta E', y)\, F(E - \Delta E', \Delta E - \Delta E', x). \quad (9.107)$$

While closed analytical expressions for the energy spectrum are harder to obtain, this relation is still an appropriate starting point for numerical convolutions of the type described in Sect. 9.4.4. However, in view of the second point, the necessary computational effort may increase substantially.

9.6.1 Continuous Slowing-Down Approximation

A useful starting point is the *continuous-slowing-down approximation*, where all straggling is neglected. You may then operate with a unique projectile energy $E(x)$ as a function of pathlength x, which is connected to the stopping cross section $S(E)$ via the relation

$$\frac{\mathrm{d}E}{\mathrm{d}x} = -NS(E). \tag{9.108}$$

This has the general solution

$$x = \int_E^{E_0} \frac{\mathrm{d}E'}{NS(E')}, \tag{9.109}$$

which determines the inverse function $x = x(E)$ if E_0 is the energy at $x = 0$.

You may derive the continuous-slowing-down approximation from the backward transport equation by making the assumption that only very small energy transfers contribute to the cross section $K(E,T)$, so that straggling can be ignored. You may then try to expand

$$F(E - T, E', x) = F(E) - T\frac{\partial}{\partial E}F(E, E', x)\ldots \tag{9.110}$$

and truncate the series after the term $\propto T$. Equation (9.103) then reduces to

$$\left[\frac{\partial}{\partial x} + NS(E)\frac{\partial}{\partial E}\right] F(E, E', x) = 0, \tag{9.111}$$

with the solution

$$F(E, E', x) = \delta\left(E' - E(x)\right), \tag{9.112}$$

where $E(x)$ is given by (9.109).

9.6.2 Ionization Yield

The continuous-slowing-down approximation is a convenient tool to estimate radiation effects. Consider e.g., the number of ionizations generated by a projectile over a pathlength x. Let the ionization cross section be given by some function $\sigma_I(E)$. For a thin foil, the mean number n_I of ionizations will be given by

$$n_I = Nx\sigma_I(E), \tag{9.113}$$

but in case of substantial energy loss through the foil this expression cannot remain valid. Neglecting straggling you may write

$$n_I = \int_0^x N \mathrm{d}x'\, \sigma_I\big(E(x')\big), \tag{9.114}$$

and by introducing E' instead of x' as the independent variable you arrive at

$$n_I = \int_{E(x)}^{E_0} \mathrm{d}E' \frac{N\sigma_I(E')}{NS(E')}, \tag{9.115}$$

where E_0 and $E(x)$ are the entrance and exit energy, respectively.

Although relations of this type are extremely useful, they are not rigorous. More sophisticated estimates will be postponed to Volume III.

9.6.3 Stopping Measurement on a Thick Target

Equation (9.109) can be utilized in the analysis of stopping-force measurements on thick targets, if the fractional energy loss is substantial but straggling inappreciable. Write the relation in the form

$$x = \int_{E_1}^{E_0} \frac{\mathrm{d}E'}{NS(E')}, \tag{9.116}$$

where

$$E_1 = E_0 - \Delta E \tag{9.117}$$

is the exit energy and x the foil thickness.

As a first-order approximation take

$$E = \frac{E_0 + E_1}{2} \tag{9.118}$$

and get

$$NS(E) \simeq \frac{\Delta E}{x}. \tag{9.119}$$

This is a popular and usually quite good approximation. To improve this, introduce a new variable

$$\xi = E' - E, \tag{9.120}$$

so that (9.116) reads

$$x = \int_{-\Delta E/2}^{\Delta E/2} \frac{\mathrm{d}\xi}{NS(E + \xi)}. \tag{9.121}$$

Now, inserting the Taylor expansion

$$S(E + \xi) = S(E) + \xi S'(E) + \frac{1}{2}\xi^2 S''(E)\cdots \tag{9.122}$$

into (9.121), and expanding the integrand up to second order in ΔE,

$$x = \frac{1}{NS(E)} \int_{-\Delta E/2}^{\Delta E/2} d\xi \left[1 - \xi \frac{S'}{S} + \xi^2 \left(\left(\frac{S'}{S}\right)^2 - \frac{1}{2}\frac{S''}{S} \right) \cdots \right], \tag{9.123}$$

we can carry out the integral and obtain

$$x = \frac{\Delta E}{NS(E)} \left[1 + \frac{\Delta E^2}{12} \left(\left(\frac{S'}{S}\right)^2 - \frac{1}{2}\frac{S''}{S} \right) \cdots \right]. \tag{9.124}$$

This is exact up to the third order in ΔE. We may solve for $NS(E)$,

$$NS(E) = \frac{\Delta E}{x} \left[1 + \frac{(\Delta E)^2}{12} \left(\left(\frac{S'}{S}\right)^2 - \frac{1}{2}\frac{S''}{S} \right) \cdots \right]. \tag{9.125}$$

Since the correction term on the right-hand side is $\propto \Delta E^2$, we may, up to second order in ΔE, replace S and its derivatives on the right-hand side by their zero-order values. With the notation

$$f(E) \equiv \frac{\Delta E}{x}, \tag{9.126}$$

we may then rewrite (9.125) in the form

$$NS(E) = f(E) \left[1 + \frac{(\Delta E)^2}{12} \left(\left(\frac{f'(E)}{f(E)}\right)^2 - \frac{1}{2}\frac{f''(E)}{f(E)} \right) \cdots \right], \tag{9.127}$$

which contains only measurable quantities on the right-hand side, provided that ΔE is measured as a function of E accurately enough so that both the first and the second derivative with respect to E can be determined with acceptable accuracy. The factor $1/12$ indicates that the second-order correction needs to be considered in high-precision measurements.

9.6.4 Straggling According to Symon ($\star$)

You may be tempted to apply (9.115) to estimate straggling in a very thick target and write

$$\Omega^2 = \int_E^{E_0} dE' \frac{NW(E')}{NS(E')}. \tag{9.128}$$

If this were applied to Bohr straggling (8.14) – which is independent of energy – the solution $\Omega^2 = NxW_{\mathrm{B}}$ would hold even for very thick targets.

However, Symon (1948) demonstrated that this is not so. The calculation is nontrivial but an interesting application of the forward transport equation and hence will be reproduced explicitly here[9].

We start at the forward equation (9.106), with E as the instantaneous energy variable,

$$\frac{\partial F(E,x)}{\partial x} = \sum_j \left\{ -F(E,x)N\sigma_j(E) \right.$$
$$\left. +F(E+T_j,x)N\sigma_j(E+T_j) \right\}. \quad (9.129)$$

The occurrence of the product $F(E+T_j,x)\sigma_j(E+T_j)$, in connection with the fact that $T_j \ll E$, suggests a Taylor series in powers of T_j, leading to

$$\frac{\partial F(E,x)}{\partial x} = N \sum_{\nu=1}^{\infty} \frac{1}{\nu!} \left(\frac{\partial}{\partial E} \right)^\nu Q_\nu(E)F(E,x), \quad (9.130)$$

where $Q_\nu(E)$ is defined as

$$Q_\nu(E) = \sum_j T_j^\nu \sigma_j(E). \quad (9.131)$$

In particular, $Q_1(E) = S(E)$ is the stopping cross section and $Q_2(E) = W(E)$ the straggling parameter.

Now, introduce the mean energy at a given path length,

$$\epsilon(x) = \int dE\, E F(E,x) \quad (9.132)$$

and the variance

$$\Omega^2(x) = \int dE\, \left[E - \epsilon(E,x) \right]^2 F(E,x). \quad (9.133)$$

Equation (9.130) then delivers the following equations of motion,

$$\frac{d\epsilon(x)}{dx} = N \sum_{\nu=1}^{\infty} \int dE\, E \left(\frac{d}{dE} \right)^\nu Q_\nu(E)F(E,x) \quad (9.134)$$

$$\frac{d\Omega^2(x)}{dx} = N \sum_{\nu=1}^{\infty} \int dE\, \left[E - \epsilon(x) \right]^2 \left(\frac{d}{dE} \right)^\nu Q_\nu(E)F(E,x). \quad (9.135)$$

---

[9] Symon's thesis has unfortunately not been published in the open literature. The present calculation has been reproduced by Payne (1969).

Repeated partial integration shows that the two series break at $\nu = 1$ and 2, respectively, so that

$$\frac{\mathrm{d}\epsilon(x)}{\mathrm{d}x} = -N \int \mathrm{d}E \, S(E) F(E, x) \tag{9.136}$$

$$\frac{\mathrm{d}\Omega^2(x)}{\mathrm{d}x} = -2N \int \mathrm{d}E \big[E - \epsilon(x) \big] S(E) F(E, x)$$
$$+ N \int \mathrm{d}E \, W(E) F(E, x). \tag{9.137}$$

So far, these relations are exact. Now the assumption enters that the variation of the cross sections with energy is weak, so that

$$S(E) \equiv S\left(\epsilon(x) + E - \epsilon(x)\right) \simeq S\left(\epsilon(x)\right)$$
$$+ \left[E - \epsilon(x)\right] S'\left(\epsilon(x)\right) + \dots \tag{9.138}$$

and similarly for $\Omega^2(E)$, where the prime indicates the derivative with respect to energy. Now consider only the two leading terms on the right-hand side, insert into (9.136) and (9.137), respectively and make use of the definitions (9.132) and (9.133) as well as the normalization condition

$$\int \mathrm{d}E \, F(E, x) = 1. \tag{9.139}$$

Then you arrive at

$$\frac{\mathrm{d}\epsilon(x)}{\mathrm{d}x} = -N S\left(\epsilon(x)\right) \tag{9.140}$$

$$\frac{\mathrm{d}\Omega^2(x)}{\mathrm{d}x} = N W\left(\left(\epsilon(x)\right)\right) - 2N \Omega^2(x) S'\left(\epsilon(x)\right). \tag{9.141}$$

Here, (9.140) recovers the continuous-slowing-down approximation, while (9.141) demonstrates that (9.128) cannot in general be correct.

Equation (9.140) has the solution

$$x = \int_E^{E_0} \frac{\mathrm{d}\epsilon}{N S(\epsilon)} \tag{9.142}$$

which was already mentioned in (9.109). However, (9.109) was based on the continuous slowing-down approximation where straggling is ignored. The present derivation identifies the solution of (9.142) as the mean energy $\epsilon(x)$ at pathlength x and takes full account of straggling within the limits introduced by the approximation (9.138).

The solution of (9.141) may be written in x-variables,

$$\Omega^2(x) = \int_0^x \mathrm{d}x' \, N W\left(\epsilon(x')\right) e^{-2 \int_{x'}^x \mathrm{d}x'' N S'\left(\epsilon(x'')\right)}. \tag{9.143}$$

Now, going over to energy variables we find that the integral in the exponent reduces to

$$\int_{x'}^{x} dx'' \, NS' \left(\epsilon(x'') \right) = \int_{E}^{\epsilon'} \frac{d\epsilon''}{NS(\epsilon'')} \frac{d \left(NS(\epsilon'') \right)}{d\epsilon''} = \ln \frac{S(\epsilon')}{S(E)}. \tag{9.144}$$

Hence,

$$\Omega^2(x) = \int_{E}^{E_0} \frac{d\epsilon'}{NS(\epsilon')} NW(\epsilon') \left(\frac{S(E)}{S(\epsilon')} \right)^2 \tag{9.145}$$

where, again, E_0 is the incident and E the exit energy. This was first derived by Symon (1948).

The only approximative step that has been made in this derivation is the truncated Taylor expansion (9.138) which assumes slow variation of the stopping cross section. The corresponding relation for the straggling parameter W also enters but is less critical, at least in the velocity regime where the Bohr expression applies.

In principle, higher cumulants like skewness and curtosis could be evaluated similarly (Symon, 1948). However, this is of minor importance in view of the fact that gaussian spectra prevail in the range of thicknesses obeying the Bohr criterion $\Omega \gg T_{\text{max}}$.

9.6.5 Nonstochastic Broadening and Skewing

The Taylor expansion (9.138) sets a limit on the range of validity of Symon's approximation (9.145). Eventually, recourse has to be made to (9.136) and (9.137). Now, considering that the dependence of $W(E)$ on E is much weaker than that of $S(E)$, the first term on the right-hand side of (9.137) increases more rapidly than the second one, as the effective integration interval increases with pathlength. Note also that the term is positive, because the contribution of energies $E > \epsilon(x)$ to the integrand is less than that from $E < \epsilon(x)$ since $S(E)$ decreases with increasing E. For the same reason, the latter interval broadens more rapidly, i.e., the profile tends to skew toward low energies, i.e., high energy losses.

In other words, the energy-loss spectrum tends to develop a high-loss tail again, just as for thin targets, but this is not caused by statistical fluctuations – which enter through the second term in (9.137) – but by the energy dependence of the stopping cross section.

Figure 9.16 shows a simple model calculation to illustrate this effect (cf. problem 9.18). Let the energy distribution at depth $x = 0$ be distributed uniformly over a certain interval, and assume continuous slowing down according to

$$-\frac{dE}{dx} = \frac{NC}{E}, \tag{9.146}$$

with C being a constant. It is seen that the fact that the high-energy end of the initial spectrum suffers less energy loss than the low-energy end gives rise to a deformation of the initially rectangular spectrum and a high-loss tail.

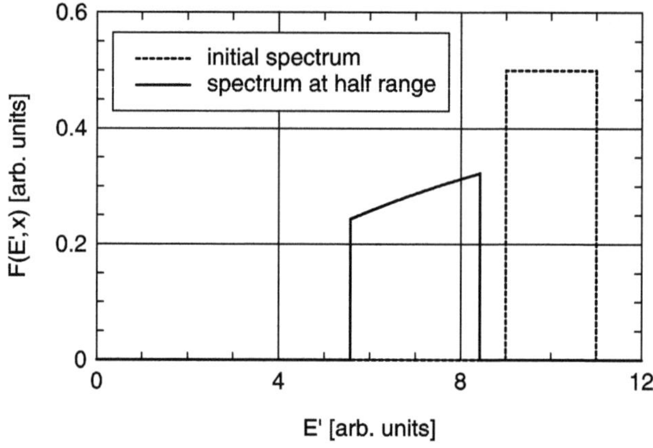

Fig. 9.16. Nonstochastic broadening of a rectangular profile. See text

Evidently, evaluation of the variance is insufficient to determine a profile, once skewing becomes efficient. Symon (1948) presented relations similar to (9.136) and (9.137) for higher cumulants, and Tschalär (1968) presented an elaborate numerical study of variance, skewness and curtosis for the cross section (9.45). Spectra were reconstructed from a Johnson distribution[10].

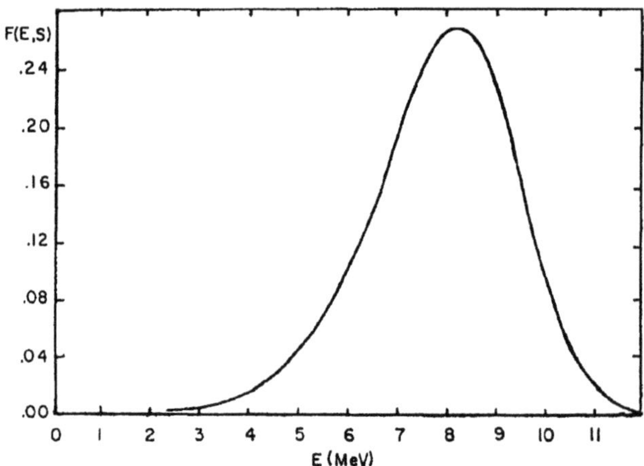

Fig. 9.17. Calculated energy spectrum of protons in Be for an initial energy of 50 MeV for a target thickness corresponding to a mean energy loss of 42 MeV. From Payne (1969)

---

[10] Johnson distributions are deformed gaussians found by a class of transformations (Johnson, 1949).

An alternative approach was taken by Payne (1969). Going back to the forward transport equation (9.105), and applying Taylor expansion in powers of T to the term $K(E' + T, T)F(E' + T, x)$ one obtains an infinite series

$$\frac{\partial F(E', x)}{\partial x} = N \sum_{\nu=1}^{\infty} \frac{1}{\nu!} \left(\frac{\partial}{\partial E'} \right)^{\nu} Q_\nu(E')F(E', x), \tag{9.147}$$

where $Q_\nu(E) = \int T^\nu \mathrm{d}\sigma(E, T)$. Truncation after the term of second order leads to the so-called Fokker-Planck equation. Payne also studied the case where the Q_2-term is neglected. This is relevant in the thickness regime where the term containing W in (9.137) becomes insignificant. A result is shown in figure 9.17, where the beam has lost 84 % of its initial energy.

9.6.6 Method of Moments

Once target thickness becomes so large that particles lose more than 90 % of their energy, the calculational tools described up to now become unreliable. Symon (1948) established a procedure to determining moments from the transport equation and proposed to reconstruct energy-loss profiles from moments by means of standard mathematic tools such as Gram-Charlier and Edgeworth expansions (Kendall and Stuart, 1963). This method has been explored by Tschalär (1968) and Payne (1969).

Reconstruction of a profile from moments is mathematically a much more problematic enterprise than approximating a function by a Taylor expansion. With the present availability of computer capacity, numerical tools such as straight convolution or Monte Carlo simulation are preferrable.

9.7 Simulation

9.7.1 Monte Carlo Schemes

The stochastic nature of the collision processes described so far makes particle penetration phenomena well suited for numerical treatment by Monte Carlo simulation. In fact, particle penetration has been one of the first major applications of computer simulation in physics. An early review was written by Berger (1963). In principle, a Monte Carlo simulation is equivalent with the solution of a transport equation for identical input. Differences appear in the practical implementation of either scheme.

Note that accurate solution of a transport equation normally requires reliable numerical procedures. Although typically requiring less computer capacity than Monte Carlo simulations, the necessary programming effort may be substantial. This makes Monte Carlo simulations a competitive tool.

The present section is devoted to fairly general aspects and avoids specific discussion of numerous available codes.

9.7.2 Procedure

Most Monte Carlo simulations make heavily use of the mean free path $\lambda(E)$ defined as

$$\lambda(E) = \frac{1}{N\sigma(E)}, \tag{9.148}$$

where

$$\sigma(E) = \sum_j \sigma_j(E) \tag{9.149}$$

is the *total cross section*, i.e., the sum of the cross sections for all possible events.

A typical Monte Carlo simulation starts with some particle that has been given a well-defined energy[11]. The particle is allowed to move uniformly until a collision is going to happen. That first collision takes place after a certain free-flight-path interval (x', dx') which is determined by means of a random-number generator from the probability

$$\frac{dx'}{\lambda(E)} e^{-x'/\lambda(E)}, \tag{9.150}$$

which follows from (2.7) for $n = 0$. The outcome of this event is then determined by another throw of the dice: The probability for event j to happen is given by

$$\frac{\sigma_j(E)}{\sigma(E)} = N\lambda\sigma_j(E). \tag{9.151}$$

For the problems discussed above, this implies that a quantum of energy T_j has been transmitted to the surroundings, and the projectile has gotten a new starting point for a certain free-flight distance with a new energy $E-T_j$. Angular deflection can, but need not be part of the bookkeeping. If so, a third throw of the dice is necessary to determine the new direction of motion: Even if there is a unique relation between energy loss and scattering angle, as is the case for elastic collisions, the azimuthal scattering angle is a random variable that needs to be fixed. Further random variables such as the state of the projectile after a collision may be of interest, as will be seen in Volume II.

After complete characterization of the projectile – and if necessary other particles emerging from a collision – the procedure is repeated with a properly revised distribution of pathlength and a new probability (9.151) reflecting the

---

[11] Monte Carlo is the preferred technique to compute slowing-down profiles of electrons, where angular deflection is an essential effect that cannot normally be neglected. It is straightforward to include this additional dimension in the simulation, but explicit reference will not be made here.

emerging projectile. The procedure may then be iterated as often as necessary until the projectile has slowed down below some threshold energy that is prescribed by the physical situation. This whole sequence of events is usually called a *trajectory*. Gaining statistically significant predictions requires simulation of a large number of trajectories. Dependent on the amount of detail required, a run may require simulation of millions of trajectories.

9.7.3 Equivalence with Transport Theory (⋆)

It is of interest to formulate the above procedure analytically. The probability for a free flight up to some pathlength x is $\exp(-x/\lambda(E))$, and the associated energy distribution is $\delta(E - E')$. Conversely, the probability for the first collision in $(x', \mathrm{d}x')$ is given by (9.150) and the type of event by (9.151). Since we are asking for the probability for energy $(E', \mathrm{d}E')$ at pathlength x, and since the particle has already travelled x' and lost an energy T_j, we now need the probability $F(E - T_j, E', x - x')$. Collecting these probabilities we arrive at

$$F(E, E', x) = \mathrm{e}^{-x/\lambda(E)}\delta(E' - E) + \int_0^x \frac{\mathrm{d}x'}{\lambda(E)}\,\mathrm{e}^{-x'/\lambda(E)}$$
$$\times \sum_j \frac{\sigma_j(E)}{\sigma(E)}\,F(E - T_j, E', x - x'). \quad (9.152)$$

Relations of the type of (9.152) are alternative forms of transport equations which require reasonably large mean free paths in order to be of practical use. They have been in use very much in neutron transport theory. You may demonstrate the equivalence with (9.100) in problem 9.16.

Note that unlike (9.100), (9.152) contains both a summation over j and an integration over x. This is a manifestation of the fact that the random number generator is consulted twice in each step, first to find the travelled path length and then to determine the type of collision. On the other hand, no differentiations occur. This implies that initial conditions have been incorporated into the integral equation and need not be formulated independently.

Simulation codes have also been developed that avoid a free-path distribution and, therefore, utilize the random number generator only once in each step, namely to determine the type of collision.

9.8 Discussion and Outlook

Energy-loss spectra of charged particles much heavier than electrons are conventionally classified into three regimes of target thickness,

- For thin targets, where the variance $\Omega^2 = NxW \ll T_{\mathrm{max}}^2$, the maximum energy loss in a single event, the spectrum resembles the single-event spectrum, i.e., , the differential cross section. In case of high energy resolution, structure of the latter may be visible. More frequently the spectrum resembles the Rutherford cross section.

- For moderately thick targets, when $\Omega^2 = NxW \gg T_{\max}^2$ but $\langle \Delta E \rangle \ll E$, the spectrum tends toward gaussian shape with the width determined by Ω. The approach to gaussian shape may be quite slow.

- For very thick targets, when $\langle \Delta E \rangle$ approaches the initial beam energy E, nonstochastic broadening becomes noticeable. Consequently, significant deviations from gaussian shape may be observed.

The most versatile theoretical tool is the kinetic equation, in forward or backward form, but for thin and moderately thick layers it has a general solution, the Bothe-Landau formula which can serve as a starting point for rigorous calculations. The exactly solvable model spectrum following from the Lindhard-Nielsen cross section (9.29) serves as a useful lesson if you tend to the opinion that the diffusion approximation is a universally-valid tool. The Bohr-Williams approach, handled with skill, is still powerful but has been superceded by more rigorous methods with the upcome of ever more capable computational tools.

The classic description for thin targets is the one by Landau (1944). Theoretical work on deviations from the Landau spectrum – which operates with classical Coulomb scattering – used to focus on the fine structure of the cross section. This may be significant for penetrating electrons but is rarely of interest for heavier projectiles. Even for classical Coulomb scattering, the range of validity of Landau's description is limited to rather small target thicknesses. Deviations used to be treated by the formalism of Vavilov. This can now safely be replaced by a combination of Glazov's scheme and the method of steepest descent.

Glazov's calculations indicate that the approach to gaussian behavior is usually quite slow, and gaussian spectra provide satisfactory descriptions mainly because experimental accuracy is rarely sufficient to pin down minor discrepancies with theory. Deviations from gaussian shape for moderately thick targets are well described by the scheme of Sigmund and Winterbon (1985).

Nuclear energy loss has been ignored in this as well as all previous chapters except Chapter 2. In the present context, this implies neglecting high-loss tails in the spectra.

Spectra for ions allowing to lose a sizeable fraction of their initial energy are most efficiently calculated by numerical solution of the kinetic equation or by the equivalent technique of Monte Carlo simulation. Although analytical solution of the transport equation in combination with the moments method is an efficient alternative, angular deflection may become important at the same time.

A major topic belonging into the present context is the influence of charge exchange on the energy-loss spectrum. This topic – which has a long history – will be discussed in volume II. The effect has two distinctly different implications: Energy lost in a charge-changing event may be substantial and hence give rise to splitting an energy-loss spectrum into a multiple-peak structure. Moreover, since the stopping force depends on the ion charge, fluctuating ion charge is a source of energy-loss fluctuation, called *charge-exchange straggling*.

This chapter dealt exclusively with Poissonian collision statistics. Charge exchange is not the only process that has been omitted for that reason. Also bunching, discussed in Sect. 8.9 has likewise been omitted, albeit for a different reason: Estimates of energy-loss spectra seem unavailable. Moreover, effects of crystal structure, in particular channeling, have been reserved for Volume II. In that connection, also simulation techniques going beyond the Monte Carlo method will come up for discussion.

Finally, phenomena connected with complete slowing down, such as ion ranges and radiation effects, have been reserved for Volumes II and III, respectively.

Problems

9.1. Derive (9.8) from (9.7). A simple procedure is by Taylor expansion in powers of the penetration depth.

9.2. Carry out the calculation leading from (9.19) to (9.21).

9.3. Derive (2.19) and (2.26) from (9.21).

9.4. Derive (9.30) from (9.15) and (9.29). You will need the integral representation of the gamma function $\Gamma(z+1) = \int_0^\infty dt\, t^z e^{-t}$ as well as $\Gamma(z+1) = z\Gamma(z)$ and $\Gamma(1/2) = \sqrt{\pi}$.

9.5. Derive (9.31) from (9.30). Hint: Introduce a variable $z = \alpha + ik$ and keep track of the proper path of integration. Then reduce the exponential to a simple gaussian by substituting $z = \pm t + A$ with a suitable constant A.

9.6. Determine T_1 defined in Sect. 9.3.1 for the spectrum (9.38) by requiring the approximate peak position NxS_1 to be identical with the exact value ΔE_p given by (9.34).

9.7. Use the Bothe-Landau formula to show that the quantity Q_3 defined by (9.25) fulfills the relation

$$\left\langle \left(\Delta E - \langle \Delta E\rangle\right)^3 \right\rangle = NxQ_3 \tag{9.153}$$

without specifying a cross section.

9.8. Derive (9.26) from (9.14) and (9.24).

9.9. Evaluate the energy-loss profile for the model spectrum (9.30) in the diffusion approximation.

9.10. Derive (9.66) from (9.14).

9.11. Apply the method of steepest descent to the cross section (9.29). You will arrive at the exact energy-loss spectrum (9.31).

9.12. Make a plot of peak energy loss ΔE_p versus mean energy loss NxS for the cross section (9.45) on the basis of (9.89).

9.13. Go explicitly through the derivation of (9.104).

9.14. An alternative derivation of (9.104) considers the change of the distribution $F(E, E', x)$ after passage through a small pathlength increment Δx. Express $F(E, E', x + \Delta x)$ by the loss of particles from $F(E, E', x)$ and the gain of particles from $F(E, E' + T_j, x)$ and take the limit $\Delta x \to 0$.

9.15. In this problem you will derive the transport equation (9.104) from the nonlinear Boltzmann equation.

1. Find the Boltzmann equation in a text on statistical mechanics or in the original reference (Boltzmann, 1875). If it is written in terms of velocity variables, try to rewrite it in terms of energy variables and eliminate any directional variables by orientational averaging. You will then end up with a nonlinear equation for an energy distribution $f(E', t)dE'$ as a function of time t. The central input quantity is a cross section $K(E, E'; E'', E''')dE''dE'''$ for two particles with energies E and E' to collide with exit energies (E'', dE'') and (E''', dE'''), respectively.
2. You will find several ways in the literature to linearize the Boltzmann equation. In the present context, the pertinent physical assumption is that a moving ion meets the scattering centers, i.e., the atoms of the target, at rest. If you have found the Boltzmann equation for a multicomponent medium (for example in Boltzmann's original paper), all you have to do is to assume that there is only one particle of the first species, and that all atoms of all other species are at rest.
3. If you have only found the Boltzmann equation for a monoatomic medium – which is most common in textbooks – you have to assume that the number of moving atoms is infinitesimally small, i.e.,

$$f(E', t) \simeq N\delta(E') + F(E', t) \tag{9.154}$$

   and apply Taylor expansion up to first order in $F(E', t)$.
4. Arrive at the final form by substituting $dx = vdt$.

9.16. Demonstrate the equivalence of (9.152) with (9.100). Replace $x' \to x - x'$ in the integral on the right-hand side, multiply the whole equation by $\exp(x/\lambda''(E))$, differentiate with respect to x and observe (9.148).

9.17. Show that (9.14) satisfies both (9.100) and (9.104), when $K_j(E)$ is independent of E.

9.18. Derive an expression for the energy spectrum shown in Fig. 9.16, assuming continuous slowing down according to the energy-loss function (9.146). Let the initial spectrum extend from energy 9 to 11 (arbitrary units) and assume $2NCx = 50$.

References

Abramowitz M. and Stegun I.A. (1964): *Handbook of mathematical functions.* Dover, New York

Berger M.J. (1963): Monte Carlo calculation of the penetration and diffusion of fast charged particles. Methods Comp Phys **1**, 135–215

Bichsel H. (1970): Straggling of heavy charged particles: Comparison of Born hydrogenic-wave-function approximation with free-electron approximation. Phys Rev B **1**, 2854–2862

Bichsel H. and Saxon R.P. (1975): Comparison of calculational methods for straggling in thin absorbers. Phys Rev A **11**, 1286–1296

Blunck O. and Leisegang S. (1950): Zum Energieverlust schneller Elektronen in dünnen Schichten. Z Physik **128**, 500–505

Bohr N. (1915): On the decrease of velocity of swiftly moving electrified particles in passing through matter. Philos Mag **30**, 581–612

Bohr N. (1948): The penetration of atomic particles through matter. Mat Fys Medd Dan Vid Selsk **18 no. 8**, 1–144

Boltzmann L. (1875): Über das Wärmegleichgewicht von Gasen, auf welche äussere Kräfte wirken. Sitzungsber Akad Wiss Wien, math-nat wiss Kl **72**, 427–457

Börsch-Supan W. (1961): On the evaluation of the function $\phi(\lambda) = (1/2\pi i) \int_{\sigma-i\infty}^{\sigma+i\infty} \exp(u \ln u + \lambda u)\, du$ for real values of λ. J Res Natl Bur Stand **65B**, 245–250

Bothe W. (1921): Die Gültigkeitsgrenzen des Gaußschen Fehlergesetzes für unabhängige Elementarfehlerquellen. Z Physik **4**, 161–177

Fastrup B., Hvelplund P. and Sautter C.A. (1966): Stopping cross section in carbon of 0.1-1.0 MeV atoms with $6 < Z_1 < 20$. Mat Fys Medd Dan Vid Selsk **35 no. 10**, 1–28

Glazov L.G. (2000): Energy-loss spectra of swift ions. Nucl Instrum Methods B **161**, 1–8

Glazov L.G. (2002): Energy-loss spectra of swift ions: Beyond the Landau approximation. Nucl Instrum Methods B **192**, 239–248

Johnson N.L. (1949): Systems of frequency curves generated by methods of translation. Biometrica **36**, 149–176

van Kampen N.G. (1981): *Stochastic processes in physics and chemistry.* North Holland, Amsterdam

Kendall M.G. and Stuart A. (1963): *The advanced theory of statistics*. Griffin, London

Landau L. (1944): On the energy loss of fast particles by ionization. J Phys USSR **8**, 201–205

Lindhard J. (1985): On the theory of energy loss distributions for swift charged particles. Phys Scr **32**, 72–80

Lindhard J. and Nielsen V. (1971): Studies in statistical dynamics. Mat Fys Medd Dan Vid Selsk **38 no. 9**, 1–42

Mertens P. (1986): Experiments on the difference between most probable and mean energy-loss for foil transmitted protons. Nucl Instrum Methods B **13**, 91–94

Moyal J.E. (1955): Theory of ionization fluctuations. Philos Mag **46**, 263–280

Payne M.G. (1969): Energy straggling of heavy charged particles in thick absorbers. Phys Rev **185**, 611–622

Seltzer S.M. and Berger M.J. (1964): Energy loss straggling of protons and mesons: Tabulation of the Vavilov distribution. In *Studies in penetration of charged particles in matter*, vol. 1133, 187–203. National Academy of Sciences – National Research Council, Washington

Shulek P., Golovin B.M., Kulyukina L.A., Medved S.V. and Pavlovich P. (1966): Fluctuations of ionization loss. Yadern Fiz **4**, 564–566. Engl. transl. Sov. J. Nucl. Phys. 4, 400-401 (1967)

Sigmund P. (1991): Statistics of charged-particle penetration. In A. Gras-Marti, H.M. Urbassek, N. Arista and F. Flores, editors, *Interaction of charged particles with solids and surfaces*, vol. 271 of *NATO ASI Series*, 73–144. Plenum Press, New York

Sigmund P. and Winterbon K.B. (1985): Energy loss spectrum of swift charged particles penetrating a layer of material. Nucl Instrum Methods B **12**, 1–16

Symon K. (1948): Fluctuations in energy lost by high energy charged particles in passing through matter. Ph.D. thesis, Harvard University

Tschalär C. (1968): Straggling distributions of large energy losses. Nucl Instrum Methods **61**, 141–156

Vavilov P.V. (1957): Ionization losses of high-energy heavy particles. Zh Eksp Teor Fiz **32**, 920–923. [English translation: Sov. Phys. JETP **5**, 749-751 (1957)]

Williams E.J. (1929): The straggling of beta-particles. Proc Roy Soc **125**, 420–445

Part IV

Appendices

A

Selected Tutorials

A.1 Units

A.1.1 Electromagnetic Units

Amongst a considerable number of systems of electromagnetic units that have been in use over the past century, mainly the official SI units and the Gaussian system of units seem to have survived. The latter is frequently used in the literature on particle penetration and collision theory, and in atomic physics in general. It is, therefore, of utmost importance that the reader be able to correctly evaluate the contents of an equation written in gaussian units.

Modern textbooks in electromagnetic theory employ SI units in which Maxwell's equations in vacuum take on the form

$$\boldsymbol{\nabla} \cdot \boldsymbol{E}' = \frac{\rho_e{}'}{\epsilon_0} \tag{A.1a}$$

$$\boldsymbol{\nabla} \cdot \boldsymbol{B}' = 0 \tag{A.1b}$$

$$\boldsymbol{\nabla} \times \boldsymbol{E}' = -\frac{\partial \boldsymbol{B}'}{\partial t} \tag{A.1c}$$

$$\boldsymbol{\nabla} \times \frac{\boldsymbol{B}'}{\mu_0} = \frac{\partial(\epsilon_0 \boldsymbol{E}')}{\partial t} + \boldsymbol{J}', \tag{A.1d}$$

where $\boldsymbol{E}'$ and $\boldsymbol{B}'$ denote the electric and the magnetic field and $\rho_e{}'$ and $\boldsymbol{J}_e{}'$ the charge and current density, respectively. The relation between the fields and the force $\boldsymbol{F}$ on a point charge q' is given by

$$\boldsymbol{F} = q'(\boldsymbol{E}' + \boldsymbol{v} \times \boldsymbol{B}'), \tag{A.2}$$

and the Coulomb force between two point charges q_1', q_2' has the magnitude

$$F_{\mathrm{Coul}} = \frac{q_1' q_2'}{4\pi\epsilon_0 r^2}. \tag{A.3}$$

Coulomb's law (A.3) takes a particularly simple appearance,

$$F_{\mathrm{Coul}} = \frac{q_1 q_2}{r^2}. \tag{A.4}$$

if the quantity

$$q = \frac{q'}{\sqrt{4\pi\epsilon_0}} \tag{A.5}$$

is taken as a measure of electric charge. This implies the following measures of charge and current density,

$$\rho_e = \frac{\rho_e'}{\sqrt{4\pi\epsilon_0}}; \ \ \boldsymbol{J} = \frac{\boldsymbol{J}'}{\sqrt{4\pi\epsilon_0}}. \tag{A.6}$$

It is appropriate, then, also to introduce alternative quantities $\boldsymbol{E}, \boldsymbol{B}$ for the electric and magnetic field.

For the electric field, convenience suggests to require that $q\boldsymbol{E} = q'\boldsymbol{E}'$. Then,

$$\boldsymbol{E} = \sqrt{4\pi\epsilon_0}\,\boldsymbol{E}', \tag{A.7}$$

and the force equation reads

$$\boldsymbol{F} = q\big(\boldsymbol{E} + \sqrt{4\pi\epsilon_0}\,\boldsymbol{v} \times \boldsymbol{B}'\big). \tag{A.8}$$

Amongst several feasible options for $\boldsymbol{B}$, the one adopted in the gaussian system of units is based on the convention that the electric and magnetic field have the same dimension. This is achieved by setting

$$\sqrt{4\pi\epsilon_0}\,\boldsymbol{v} \times \boldsymbol{B}' = \frac{\boldsymbol{v}}{c} \times \boldsymbol{B}, \tag{A.9}$$

where $c = 1/\sqrt{\epsilon_0\mu_0}$ is the speed of light in vacuum. This leads to

$$\boldsymbol{B} = \sqrt{\frac{4\pi}{\mu_0}}\,\boldsymbol{B}', \tag{A.10}$$

and Maxwell's equations take the form

$$\boldsymbol{\nabla} \cdot \boldsymbol{E} = 4\pi\rho_e \tag{A.11a}$$

$$\boldsymbol{\nabla} \cdot \boldsymbol{B} = 0 \tag{A.11b}$$

$$\boldsymbol{\nabla} \times \boldsymbol{E} = -\frac{1}{c}\frac{\partial \boldsymbol{B}}{\partial t} \tag{A.11c}$$

$$\boldsymbol{\nabla} \times \boldsymbol{B} = \frac{1}{c}\frac{\partial \boldsymbol{E}}{\partial t} + \frac{4\pi}{c}\boldsymbol{J}. \tag{A.11d}$$

For $q' = 1$ Coulomb (C) and $\epsilon_0 = 8.854 \times 10^{-12}$ C V^{-1} m^{-1} we find

$$q = 9.4803 \times 10^4 \ \text{kg}^{1/2}\text{m}^{3/2}\text{s}^{-1} = 3 \times 10^9 \ \text{g}^{1/2}\text{cm}^{3/2}\text{s}^{-1}$$

$$\equiv 3 \times 10^9 \ \text{esu}, \tag{A.12}$$

where 1 esu $= 1$ g$^{1/2}$cm$^{3/2}$s^{-1} is called the electrostatic unit of charge.

Gaussian units are used everywhere in this book except in Chapter 1 and the first parts of this appendix. From a practical point of view, the most visible feature of gaussian units is the absence of the factor $1/4\pi\epsilon_0$ in all equations containing e^2.

A.1.2 Atomic Units

Experimental parameters discussed in this book are most often expressed in terms of the elementary charge e, the electron mass m, the speed of light c, and Planck's constant $\hbar = h/2\pi$. The evaluation of expressions involving some of these parameters is an easy routine if atomic units are utilized. Table A.1 shows expressions for the most frequently used parameters both in SI and gaussian units along with their numerical values.

Table A.1. Basic atomic parameters

Quantity	Symbol	Expression (SI units)	Expression (gauss. units)	Value
Bohr radius	a_0	$4\pi\epsilon_0\hbar^2/me^2$	$\hbar^2/me^2$	0.0529177 nm
Rydberg energy	R	$e^2/8\pi\epsilon_0 a_0$	$e^2/2a_0$	13.60569 eV
				$=2.179872\times10^{-18}$ J
Fine structure constant	α	$e^2/4\pi\epsilon_0\hbar c$	$e^2/\hbar c$	1/137.0360
Bohr velocity	v_0	$e^2/4\pi\epsilon_0\hbar$	$e^2/\hbar$	2.187691×10^6 m/s
				$=\alpha c$
Electron radius	r_e	$e^2/4\pi\epsilon_0 mc^2$	e^2/mc^2	2.81894×10^{-15} m
				$=\alpha^2 a_0$

As an example, take the factor preceding the Bethe logarithm in the expression for the electronic stopping cross section (4.118). From the definition (2.20) of the stopping cross section we know that it has the dimension of energy $\times$ area. Therefore we express the factor in terms of Ra_0^2 which leads to

$$\frac{4\pi Z_1^2 Z_2 e^4}{mv^2} = \frac{Z_1^2 Z_2}{(v/v_0)^2}\frac{4\pi e^4}{mv_0^2} = \frac{Z_1^2 Z_2}{(v/v_0)^2}8\pi a_0^2 R$$

$$= 0.96\frac{Z_1^2 Z_2}{(v/v_0)^2}\ \mathrm{eV\,nm^2} \quad (A.13)$$

If SI were used, the same quantity would read

$$\frac{4\pi Z_1^2 Z_2 e^4}{(4\pi\epsilon_0)^2 mv^2} = \frac{Z_1^2 Z_2}{(v/v_0)^2}\frac{4\pi e^4}{(4\pi\epsilon_0)^2 mv_0^2} = \frac{Z_1^2 Z_2}{(v/v_0)^2}8\pi a_0^2 R$$

$$= 0.96\frac{Z_1^2 Z_2}{(v/v_0)^2}\ \mathrm{eV\,nm^2} \quad (A.14)$$

A.1.3 Length Measures

In the field of particle penetration, length measures such as $\mathrm{mg/cm^2}$ or $\mathrm{\mu g/cm^2}$ are frequently encountered. This denotes weight/area or density $\times$ thickness. There are three obvious points supporting this choice,

- A metal foil 100 nm thick with an area of 1 cm^2 weighs a few tens of micrograms. Area and weight are more readily accessible to accurate experimental determination than thickness,
- The density of a material may not be accurately known, and it may change during particle bombardment because of heating, defect formation, disordering, phase changes etc.,
- Measurable quantities depend typically on dimensionless variables like $Nx\sigma$, where N is the number of atoms per volume while x and σ denote pertinent lengths and cross sections, respectively.

The third item identifies weight/area as being closer to a 'natural' length measure in particle penetration than length proper, although that is strictly true only for the ideal case of an elemental, isotopically pure material.

Denoting the mass density by ρ_{m} we find the following expression for the weight per area

$$\text{weight/area} = \rho_{\mathrm{m}}x = \sum_{\ell} N_{\ell}M_{\ell}x \tag{A.15}$$

for a polyatomic material of thickness x containing atoms of mass M_{ℓ} or, more generally,

$$\text{weight/area} = \sum_{\ell} M_{\ell} \int \mathrm{d}x\, N_{\ell}(x) \tag{A.16}$$

for a material with a depth-dependent composition.

A simple conversion formula for homogeneous poly- or monoatomic materials may be found by noting that a weight per area of

$$\rho_{\mathrm{m}}x = 1\,\frac{\mu\mathrm{g}}{\mathrm{cm}^2} \tag{A.17}$$

represents a thickness of

$$x = \frac{100}{\rho_{\mathrm{m}}[\mathrm{g/cm}^3]}\,\text{Å}. \tag{A.18}$$

A.2 Calculus

A.2.1 Poisson Statistics

Consider an ideal gas with a density N [molecules/volume] contained in a very large volume. We are interested in the probability $\mathsf{P}_n(\mathsf{v})$ to find precisely n gas molecules within some well-defined volume v at a given time.

Assume first that v is so small that the chance to find more than one molecule within its boundaries is negligible. The average number of molecules Nv in v may then be expressed by

$$Nv = 0 \times P_0(v) + 1 \times P_1(v),$$

i. e.,

$$P_1(v) = Nv. \tag{A.19}$$

Moreover, since the probabilities must sum up to 1 we also have

$$P_0(v) = 1 - P_1(v) = 1 - Nv. \tag{A.20}$$

Next, consider two volumes v, v' of arbitrary size and ask for the probability $P_n(v + v')$ to find n molecules in $v + v'$. Evidently we must have

$$P_n(v + v') = \sum_{m=0}^{n} P_m(v)P_{n-m}(v') \tag{A.21}$$

from the laws of multiplication and addition of probabilities.

A solution of (A.21) can conveniently be found by introduction of a generating function

$$g(\zeta, v) = \sum_{n=0}^{\infty} P_n(v)\zeta^n, \tag{A.22}$$

where ζ is a dimensionless variable. Multiplying (A.21) by ζ^n and summing over n we find

$$g(\zeta, v + v') = g(\zeta, v)g(\zeta, v'). \tag{A.23}$$

From this follows that

$$g(\zeta, v) = e^{\alpha(\zeta)v} \tag{A.24}$$

with a yet undefined function $\alpha(\zeta)$.

For small v (A.22) reduces to

$$g(\zeta, v) \simeq P_0(v) + P_1(v)\zeta = [1 - Nv] + Nv\zeta \simeq 1 + \alpha(\zeta)v, \tag{A.25}$$

by means of (A.19) and (A.20), and hence

$$\alpha(\zeta) = (\zeta - 1)N. \tag{A.26}$$

This leads to

$$g(\zeta, v) = e^{(\zeta-1)Nv} = e^{-Nv} \sum_{0}^{\infty} \frac{(Nv\zeta)^n}{n!} \tag{A.27}$$

or, by comparison with (A.22),

$$\mathsf{P}_n(\mathsf{v}) = \mathrm{e}^{-N\mathsf{v}} \frac{(N\mathsf{v})^n}{n!}. \tag{A.28}$$

It is seen that $\sum_{n=0}^{\infty} \mathsf{P}_n(\mathsf{v}) = 1$ as it must be.

Eq. (A.28) represents the Poisson distribution. The essential assumption entering its derivation is that we deal with the molecules of an ideal gas, i. e., particles that are statistically independent. In particular it is implied that there is no upper limit on the number of molecules which, in principle, could find space within the volume v.

A.2.2 Fourier Transform

One Dimension

Fourier's theorem addresses *periodic* functions. Let $f(t)$ be some periodic function of t with a period τ,

$$f(t + \tau) = f(t) \qquad \text{for all } t. \tag{A.29}$$

$f(t)$ may be written in the form of a Fourier series

$$f(t) = \sum_{n=0}^{\infty} (a_n \cos n\omega_0 t + b_n \sin n\omega_0 t), \tag{A.30}$$

where $\omega_0 = 2\pi/\tau$ and a_n, b_n are constants. Eq. (A.30) can be recast into the more compact form

$$f(t) = \sum_{n'=-\infty}^{\infty} A_{n'} \mathrm{e}^{in'\omega_0 t}, \tag{A.31}$$

where the $A_{n'}$ are linear combinations of the a_n and b_n.

Multiplication of (A.31) by $\mathrm{e}^{-in\omega_0 t}$ with n being an arbitrary integer and integration over one period τ yields

$$A_n = \frac{1}{\tau} \int_{-\tau/2}^{\tau/2} \mathrm{d}t \, f(t) \mathrm{e}^{-in\omega_0 t}. \tag{A.32}$$

Fourier integrals deal with *aperiodic* functions. Fourier's theorem may be applied to such functions by the transition $\tau \to \infty$. For finite τ we may rewrite (A.31) in the form

$$f(t) = \sum_{\omega} \Delta\omega \, f_\omega \, \mathrm{e}^{i\omega t} \tag{A.33}$$

with $\omega = n'\omega_0$, $\Delta\omega = \omega_0$, and $f_\omega = A_{n'}/\omega_0$. When τ becomes large, $\Delta\omega = 2\pi/\tau$ becomes small and (A.33) turns into an integral,

$$f(t) = \int_{-\infty}^{\infty} d\omega\, \bar{f}(\omega) e^{i\omega t} \tag{A.34}$$

with

$$\bar{f}(\omega) = f_\omega = \lim_{\tau=\infty} \frac{1}{\omega_0\tau} \int_{-\tau/2}^{\tau/2} dt\, f(t) e^{-i\omega t} = \frac{1}{2\pi} \int_{-\infty}^{\infty} dt\, f(t) e^{-i\omega t}. \tag{A.35}$$

Eqs. (A.34) and (A.35) represent a Fourier transformation and its inverse transformation, respectively. The function $\bar{f}(\omega)$ is called the Fourier transform of $f(t)$. A condition for a function to have a Fourier transform is square integrability, i.e., the integral $\int_{-\infty}^{\infty} dt\, |f(t)|^2$ exists.

There is no fundamental difference between a Fourier transform and its inverse. Therefore the bar in $\bar{f}(\omega)$ is usually omitted, and it is the name of the variable that determines whether a function is considered in real or Fourier space. For the same reason the choice of the sign in the exponent is arbitrary, except that the sign in the Fourier transform has to be the opposite of that in the inverse transform. Finally there is some flexibility with regard to the arrangement of the factor $1/2\pi$. In quantum mechanics it has become customary to let the Fourier transform and its inverse each have a factor $(2\pi)^{-1/2}$. This notation has been adopted in appendix A.4.1.

In physics problems one frequently deals with the Fourier transform of an observable quantity. Let $f(t)$ be such a quantity which must then be a *real function*. Then (A.35) shows that

$$f^*(\omega) = \frac{1}{2\pi} \int_{-\infty}^{\infty} dt\, f(t) e^{i\omega t} = f(-\omega), \tag{A.36}$$

where the asterisk denotes the complex conjugate.

Higher Dimensions

The Fourier transform is not restricted to functions of one variable. In three-dimensional space it may be written in the form

$$f(\boldsymbol{r}) = \int d^3k\, f(\boldsymbol{k})\, e^{i\boldsymbol{k}\cdot\boldsymbol{r}} \tag{A.37a}$$

$$f(\boldsymbol{k}) = \frac{1}{(2\pi)^3} \int d^3r\, f(\boldsymbol{r})\, e^{-i\boldsymbol{k}\cdot\boldsymbol{r}}, \tag{A.37b}$$

and similarly in two or higher dimensions.

For the special case of a spherically-symmetric function $f(\boldsymbol{r}) = f(r)$ the directional variable drops out also from the Fourier transform, and

$$f(\boldsymbol{k}) = f(k) = \frac{1}{(2\pi)^3} \int 4\pi r^2 \, \mathrm{d}r \, f(r) \frac{\sin kr}{kr}$$

$$= \frac{1}{(2\pi)^3} \int 4\pi r^2 \, \mathrm{d}r \, f(r) j_0(kr) \quad (\mathrm{A}.38)$$

in terms of the spherical Bessel function $j_0(z) = \sin z/z$. Similarly, in two dimensions,

$$f(\boldsymbol{k}) = f(k) = \frac{1}{(2\pi)^2} \int 2\pi r \, \mathrm{d}r \, f(r) J_0(kr) \quad (\mathrm{A}.39)$$

in terms of the standard Bessel function $J_0(z)$ (Abramowitz and Stegun, 1964).

Coulomb Potential

As an example we evaluate the Fourier transform of the Coulomb potential, or

$$f(\boldsymbol{r}) = \frac{1}{r}. \quad (\mathrm{A}.40)$$

The integral

$$f(\boldsymbol{k}) = \frac{1}{(2\pi)^3} \int \mathrm{d}^3 r \, \frac{1}{r} \, \mathrm{e}^{-i\boldsymbol{k}\cdot\boldsymbol{r}} \quad (\mathrm{A}.41)$$

is readily evaluated by means of (A.38) and leads to

$$f(\boldsymbol{k}) = \frac{1}{(2\pi)^3} \frac{4\pi}{k} \int_0^\infty \mathrm{d}r \sin kr = \frac{1}{2\pi^2 k^2}. \quad (\mathrm{A}.42)$$

A reader who feels uneasy about the validity of the last step should first evaluate the Fourier transform of a screened-Coulomb potential function, $f(\boldsymbol{r}) = (1/r)\exp(-r/a)$, leading to

$$f(k) = \frac{1}{2\pi^2} \frac{1}{k^2 + 1/a^2} \quad (\mathrm{A}.43)$$

and subsequently let the screening radius a go toward infinity.

A.2.3 Spherical Harmonics and Legendre Polynomials

Spherical harmonics form a complete orthonormal set of functions on the unit sphere. This implies that a 'reasonably well-behaved' function $\Psi(\theta, \phi)$ of the polar angle θ and the azimuthal angle ϕ can be expanded in terms of spherical harmonics $Y_{\ell\mu}(\boldsymbol{\Omega})$,

$$\Psi(\theta, \phi) = \sum_{\ell=0}^{\infty} \sum_{\mu=-\ell}^{\ell} a_{\ell\mu} Y_{\ell\mu}(\boldsymbol{\Omega}) \tag{A.44}$$

with

$$a_{\ell\mu} = \int_{-1}^{1} d\cos\theta \int_{0}^{2\pi} d\phi\, \Psi(\theta, \phi) Y_{\ell\mu}^{*}(\boldsymbol{\Omega}), \tag{A.45}$$

where $\boldsymbol{\Omega}$ represents the unit vector

$$\boldsymbol{\Omega} = (\sin\theta\cos\phi, \sin\theta\sin\phi, \cos\theta). \tag{A.46}$$

(A.44) and (A.45) show strong similarities to the Fourier expansion, (A.30) and (A.31). It is helpful to keep this one-dimensional analog in mind when getting familiar with spherical harmonics.

While harmonic functions come out as solutions of the differential equation $d^2\psi(x)/dx^2 + k^2\psi(x) = 0$, spherical harmonics emerge as solutions of the angular part of the equation $\boldsymbol{\nabla}^2\psi(\boldsymbol{r}) + k^2\psi(\boldsymbol{r}) = 0$.

Just as in case of Fourier expansions there is some ambiguity which warrants precise definitions. This concerns both the functional form and various normalization factors. In the notation of Jackson (1975) we have

$$Y_{\ell\mu}(\boldsymbol{\Omega}) = N_{\ell\mu} P_{\ell}^{\mu}(\cos\theta)\, e^{i\mu\phi}, \tag{A.47}$$

where $P_{\ell}^{\mu}(\eta)$ represents an associated Legendre function,

$$P_{\ell}^{\mu}(\eta) = \frac{(-)^{\mu}}{2^{\ell}\ell!} (1 - \eta^2)^{\mu/2} \frac{d^{\ell+\mu}}{d\eta^{\ell+\mu}} \left(\eta^2 - 1\right)^{\ell} \tag{A.48}$$

and $N_{\ell\mu}$ a normalization constant

$$N_{\ell\mu} = \sqrt{\frac{2\ell + 1}{4\pi} \frac{(\ell - \mu)!}{(\ell + \mu)!}} \tag{A.49}$$

such that

$$\int_{-1}^{1} d\cos\theta \int_{0}^{2\pi} d\phi\, Y_{\ell\mu}^{*}(\boldsymbol{\Omega}) Y_{\ell'\mu'}(\boldsymbol{\Omega}) = \delta_{\ell\ell'}\delta_{\mu\mu'}. \tag{A.50}$$

Spherical harmonics could be (and are occasionally) defined via $\cos(\mu\phi)$ and $\sin(\mu\phi)$ instead of $\exp(i\mu\phi)$, and several legitimate versions exist for the factorization into $N_{\ell\mu}$ and $P_{\ell}^{\mu}(\cos\theta)$, as well as the definition of P_{ℓ}^{μ} for negative values of μ.

In the special but not infrequent case of vanishing azimuthal dependence all terms with $\mu \neq 0$ drop out. Then (A.50) reduces to

$$\int_{-1}^{1} d\cos\theta\, P_{\ell}(\cos\theta)\, P_{\ell'}(\cos\theta) = \frac{2}{2\ell + 1}\, \delta_{\ell\ell'}, \tag{A.51}$$

where

$$P_\ell(\eta) = P_\ell^0(\eta), \quad \ell = 0, 1, 2 \dots \tag{A.52}$$

are the Legendre polynomials, the definition of which seems to be unique in the more recent literature. They obey the differential equation

$$(1 - \eta^2)P_\ell'' - 2\eta P_\ell' + \ell(\ell+1)P_\ell = 0, \tag{A.53}$$

where the prime indicates a derivative.

In the following a few relationships are listed that are utilized in various parts of this book. Many more may be found in standard compilations like Abramowitz and Stegun (1964) as well as textbooks on quantum mechanics and electromagnetic theory.

The lowest-order Legendre polynomials read

$$P_0(\eta) = 1 \tag{A.54a}$$
$$P_1(\eta) = \eta \tag{A.54b}$$
$$P_2(\eta) = \frac{1}{2}(3\eta^2 - 1) \tag{A.54c}$$
$$P_3(\eta) = \frac{1}{2}(5\eta^3 - 3\eta) \tag{A.54d}$$
$$\dots$$

and the lowest-order spherical harmonics are

$$Y_{00} = \sqrt{\frac{1}{4\pi}} \tag{A.55a}$$
$$Y_{10} = \sqrt{\frac{3}{4\pi}} \cos\theta \tag{A.55b}$$
$$Y_{1\pm 1} = \mp\sqrt{\frac{3}{8\pi}} \sin\theta\, e^{\pm i\phi} \tag{A.55c}$$
$$Y_{20} = \sqrt{\frac{5}{4\pi}} \left(\frac{3}{2}\cos^2\theta - \frac{1}{2}\right) \tag{A.55d}$$
$$Y_{2\pm 1} = \mp\sqrt{\frac{15}{8\pi}} \sin\theta\,\cos\theta\, e^{\pm i\phi} \tag{A.55e}$$
$$Y_{2\pm 2} = \sqrt{\frac{15}{32\pi}} \sin^2\theta\, e^{\pm 2i\phi} \tag{A.55f}$$

Useful to know is the completeness relation

$$\sum_{\ell=0}^{\infty} \sum_{\mu=-\ell}^{\ell} Y_{\ell\mu}^*(\boldsymbol{\Omega}) Y_{\ell\mu}(\boldsymbol{\Omega}')$$
$$= \delta(\boldsymbol{\Omega} - \boldsymbol{\Omega}') \equiv \delta(\cos\theta - \cos\theta')\delta(\phi - \phi') \tag{A.56}$$

or, after integration over the azimuth,

$$\sum_{\ell=0}^{\infty} (\ell + \tfrac{1}{2}) P_\ell(\eta) P_\ell(\eta') = \delta(\eta - \eta').$$
(A.57)

Then there is the addition theorem

$$P_\ell(\boldsymbol{\Omega} \cdot \boldsymbol{\Omega}') = \frac{4\pi}{2\ell + 1} \sum_{\mu=-\ell}^{\ell} Y_{\ell\mu}^*(\boldsymbol{\Omega}) Y_{\ell\mu}(\boldsymbol{\Omega}').$$
(A.58)

There are numerous recurrence relations, of which only one is mentioned explicitly,

$$\eta P_\ell^\mu = \frac{1}{2\ell + 1}\left[(\ell - \mu + 1)P_{\ell+1}^\mu + (\ell + \mu)P_{\ell-1}^\mu\right].$$
(A.59)

Last but not least, mention is made of a generating function

$$\frac{1}{\sqrt{1 - 2\eta t + t^2}} = \sum_{\ell=0}^{\infty} P_\ell(\eta)\, t^\ell,$$
(A.60)

which is convenient to determine explicit expressions and matrix elements.

A.2.4 Dirac Function

One Dimension

The Dirac function is defined by the relations

$$\delta(x) = \begin{cases} 0 & x \neq 0 \\ & \text{for} \\ \infty & x = 0 \end{cases}$$
(A.61)

and

$$\int_{-\infty}^{\infty} dx\, \delta(x) = 1.$$
(A.62)

One may represent this function analytically by a limiting process such as a gaussian with a zero width,

$$\delta(x) = \lim_{a=0} \frac{1}{\sqrt{2\pi}\, a} e^{-x^2/2a^2},$$

but such representations are rarely needed in practical applications.

From either of the above definitions one deduces that

$$\delta(Ax) = \frac{\delta(x)}{|A|}$$
(A.63)

for an arbitrary nonvanishing constant A.

For an arbitrary function $f(x)$, continuous in $x = 0$, the above definition yields

$$\int_{-\infty}^{\infty} \mathrm{d}x\, f(x)\, \delta(x) = f(0) \int_{-\infty}^{\infty} \mathrm{d}x\, \delta(x) = f(0) \tag{A.64}$$

or, more generally,

$$\int_{-\infty}^{\infty} \mathrm{d}x'\, f(x')\, \delta(x - x') = f(x) \tag{A.65}$$

for any value of x where $f(x)$ is continuous.

The Fourier transform of $\delta(x)$ is defined by

$$\bar{\delta}(k) = \frac{1}{2\pi} \int_{-\infty}^{\infty} \mathrm{d}x\, \delta(x)\, \mathrm{e}^{-ikx} \tag{A.66}$$

according to (A.35). By (A.64) this reduces to

$$\bar{\delta}(k) = \frac{1}{2\pi}. \tag{A.67}$$

This yields the identity

$$\delta(x) = \frac{1}{2\pi} \int_{-\infty}^{\infty} \mathrm{d}k\, \mathrm{e}^{ikx} \tag{A.68}$$

which is a useful tool not the least in the handling of Fourier transforms.

In practical applications one frequently encounters Dirac functions of more complex arguments such as $\delta\big(f(x)\big)$, where $f(x)$ is some function of x. It is evident that $\delta\big(f(x)\big)$ is nonvanishing only at the zeros of $f(x)$. Let x_i be such a root so that $f(x_i) = 0$, and try to define an interval $\mathcal{C}_i$ around x_i within which $y = f(x)$ depends monotonically on x so that $f(x) \simeq (x - x_i)f'(x_i)$, where $f'(x) = \mathrm{d}f/\mathrm{d}x$. Then the contribution of the interval $\mathcal{C}_i$ to the integral $\int_{-\infty}^{\infty} \mathrm{d}x\, \delta\big(f(x)\big)$ reads

$$\int_{\mathcal{C}_i} \mathrm{d}x\, \delta\big(f(x)\big) = \int_{\mathcal{C}_i} \mathrm{d}x\, \delta\big((x - x_i)f'(x_i)\big)$$

$$= \frac{1}{|f'(x_i)|} \int_{\mathcal{C}_i} \mathrm{d}x\, \delta(x - x_i) = \frac{1}{|\mathrm{d}f/\mathrm{d}x|_{x=x_i}} \tag{A.69}$$

by use of (A.63). If there are several roots the integral reads

$$\int_{-\infty}^{\infty} \mathrm{d}x\, \delta\big(f(x)\big) = \sum_i \frac{1}{|\mathrm{d}f/\mathrm{d}x|_{x=x_i}}. \tag{A.70}$$

Example

As an example, evaluate the integral

$$\int_0^{2\pi} \mathrm{d}x\, \delta(\cos x) = \frac{1}{|\sin x|}\bigg|_{x=\pi/2} + \frac{1}{|\sin x|}\bigg|_{x=3\pi/2} = 2\,. \tag{A.71}$$

Three Dimensions

In three dimensions one defines

$$\delta(\boldsymbol{r}) = \delta(x)\delta(y)\delta(z). \tag{A.72}$$

Then (A.68) yields

$$\delta(\boldsymbol{r}) = \frac{1}{(2\pi)^3} \int_{-\infty}^{\infty} \mathrm{d}^3\boldsymbol{k}\, \mathrm{e}^{\mathrm{i}\boldsymbol{k}\cdot\boldsymbol{r}}. \tag{A.73}$$

It is useful to have expressions for multidimensional Dirac functions in non-cartesian coordinates, such as

$$\delta(\boldsymbol{r} - \boldsymbol{r}') = \frac{\delta(r - r')}{r}\delta(\phi - \phi')\delta(z - z') \tag{A.74}$$

in cylindrical coordinates r, ϕ, z, and

$$\delta(\boldsymbol{r} - \boldsymbol{r}') = \frac{\delta(r - r')}{r^2}\delta(\cos\theta - \cos\theta')\delta(\phi - \phi') \tag{A.75}$$

in spherical coordinates $r, \cos\theta, \phi$.

Point Charge

We may express the charge density $\rho(\boldsymbol{r})$ of a point charge e placed in the origin by a Dirac function,

$$\rho(\boldsymbol{r}) = e\delta(\boldsymbol{r}). \tag{A.76}$$

The electrostatic potential $V(r) = e/r$ is known to satisfy Poisson's equation

$$\boldsymbol{\nabla}^2 V(r) = -4\pi\rho(r) \tag{A.77}$$

in gaussian units. In Fourier space,

$$V(r) = \int \mathrm{d}^3\boldsymbol{k}\, V(\boldsymbol{k})\mathrm{e}^{\mathrm{i}\boldsymbol{k}\cdot\boldsymbol{r}} \tag{A.78}$$

$$\rho(r) = \int \mathrm{d}^3\boldsymbol{k}\, \rho(\boldsymbol{k})\mathrm{e}^{\mathrm{i}\boldsymbol{k}\cdot\boldsymbol{r}},$$

Poisson's equation reduces to

$$k^2 V(\boldsymbol{k}) = 4\pi \rho(\boldsymbol{k}). \tag{A.79}$$

From (A.73) we deduce that

$$\rho(\boldsymbol{k}) = \frac{e}{(2\pi)^3} \tag{A.80}$$

and hence

$$V(\boldsymbol{k}) = \frac{e}{2\pi^2 k^2} \tag{A.81}$$

in accordance with (A.42)

A.2.5 Green Functions

General

Green functions serve to provide explicit or implicit solutions to a wide class of differential or integral equations. As a simple prototype consider

$$L_x \psi(x) = f(x), \tag{A.82}$$

where $f(x)$ is a known function of some variable x, L_x a linear (differential and/or integral) operator acting in x-space, and $\psi(x)$ an unknown solution. Recall that

$$L_x \Big(\psi_1(x) + \psi_2(x) \Big) = L_x \psi_1(x) + L_x \psi_2(x) \tag{A.83}$$

for arbitrary ψ_1, ψ_2 if L_x is linear.

Try to write $\psi(x)$ as an integral

$$\psi(x) = \int \mathrm{d}x' \, G(x, x') f(x'), \tag{A.84}$$

where the 'Green function' $G(x, x')$ is required to satisfy the relation

$$L_x G(x, x') = \delta(x - x'), \tag{A.85}$$

and $\delta(x - x')$ represents the Dirac function.

Then, evaluating $L_x \psi(x)$ by inserting $\psi(x)$ from (A.84) you find

$$L_x \psi(x) = \int \mathrm{d}x' \, L_x G(x, x') f(x')$$
$$= \int \mathrm{d}x' \, \delta(x - x') f(x') = f(x) \tag{A.86}$$

by use of (A.64), in agreement with (A.82). Interchanging the order of L_x and integration over dx' is justified since L_x is linear according to (A.83) and acting only on the x variable.

A more general solution of (A.82) can then be written in the form

$$\psi(x) = \int dx' \, G(x, x') f(x') + \psi_0(x), \tag{A.87}$$

where $\psi_0(x)$ is an arbitrary solution of the homogeneous equation,

$$L_x \psi_0(x) = 0. \tag{A.88}$$

Clearly this scheme is not restricted to functions of one variable. Indeed, x may symbolize a set of arbitrarily many continuous variables, and an extension to discrete variables is straightforward. This allows the scheme to be applied to a variety of operators.

One evident advantage is that solving (A.85) once for all provides solutions in closed form to a whole family of equations of the type of (A.84) for different $f(x)$. This is particularly useful if boundary conditions to be satisfied by $\psi(x)$ can be incorporated already into $G(x, x')$.

A very important application, used in this monograph as well as many branches of theoretical physics, emerges when the function $f(x)$ is replaced by some functional containing the unknown function, i. e.,

$$f(x) \rightarrow f\{x, \psi(x)\}. \tag{A.89}$$

While the derivation above still remains valid, (A.84) or (A.87) then do not any longer constitute explicit solutions but, instead, integral equations from which the unknown function may be determined as a second step. This may be an advantage since important clues about the analytic behavior of the complete solution may emerge readily. Moreover, (A.87) may form a more suitable starting point for a perturbation expansion, an iterative scheme, or a straight numerical solution than the original equation. When utilized in this manner, Green functions may be useful tools in the solution of both linear and nonlinear equations, provided that it is possible to single out a linear operator L_x characterizing some essentials of the system under consideration.

Harmonic Oscillator

As a first example consider the differential equation of a forced harmonic oscillator

$$\frac{d^2\psi}{dt^2} + \Gamma \frac{d\psi}{dt} + \omega_0^2 \psi = \frac{1}{m} f(t), \tag{A.90}$$

where $\psi(t)$ is the displacement *versus* time, m the mass, ω_0 the resonance frequency, Γ an infinitesimal damping constant and $f(t)$ some external force. The solution (A.87) may be written in the form

$$\psi(t) = \frac{1}{m} \int dt'\, G(t, t') f(t') + \psi_0(t),$$ (A.91)

where $\psi_0(t)$ describes a free, damped oscillation and $G(t, t')$ must obey

$$\frac{\partial^2 G(t, t')}{\partial t^2} + \Gamma \frac{\partial G(t, t')}{\partial t} + \omega_0^2 G(t, t') = \delta(t - t').$$ (A.92)

Two procedures will be used to solve this differential equation. The first is more elegant but the second is more powerful in general.

Procedure A

Equation (A.92) represents a free, damped oscillation initiated by a short pulse given at time t'. For weak damping ($\Gamma \to 0$) we may write

$$G(t, t') = A \cos \omega_0 (t - t') + B \sin \omega_0 (t - t') \text{ for } t > t'$$ (A.93)

with constants A, B to be determined from the initial conditions at $t = t'$. Integration of (A.92) over t over the interval $(t' - 0, t' + 0)$ yields

$$\left(\frac{\partial G(t, t')}{\partial t} + \Gamma G(t, t') \right)_{t=t'-0}^{t'+0} = 1,$$ (A.94)

where a third term on the left-hand side dropped out since the integral of $G(t, t')$ over an interval of vanishing size vanishes for any finite $G(t, t')$. Now assume the oscillator to be at rest so long as there is no force, $G(t, t') = 0$ for $t < t'$. Then (A.94) is satisfied by the initial conditions $G(t', t') = 0$ and $\partial G(t, t')/\partial t|_{t=t'} = 1$. This yields $A = 0$ and $B\omega_0 = 1$ in (A.93) and, therefore,

$$G(t, t') = \begin{cases} \dfrac{\sin \omega_0 (t - t')}{\omega_0} & t > t' \\ 0 & t < t' \end{cases} \quad \text{for}$$ (A.95)

Procedure B

This paragraph requires some knowledge of integration in the complex plane, especially Cauchy's theorem and the concept of a residue.

Write (A.92) in Fourier space,

$$G(t, t') = \int d\omega\, G(\omega) e^{i\omega(t-t')},$$ (A.96)

so that

$$\left(-\omega^2 + i\omega\Gamma + \omega_0^2 \right) G(\omega) = \frac{1}{2\pi}.$$ (A.97)

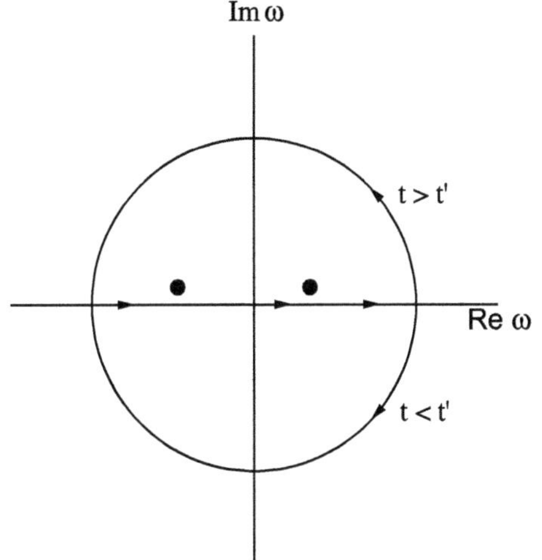

Fig. A.1. Paths of integration in the complex ω plane (see text). Dots indicate poles at $\pm\omega_0 + i\Gamma/2$

Solving for $G(\omega)$ and insertion into (A.96) lead to

$$G(t,t') = \frac{1}{2\pi} \int_{-\infty}^{\infty} d\omega \, \frac{e^{i\omega(t-t')}}{\omega_0^2 - \omega^2 + i\omega\Gamma}. \tag{A.98}$$

For $t - t' > 0$ the integral can be closed in the complex ω-plane by means of a semicircle in the *upper half-plane* (Fig. A.1) which does not contribute to the integral in the limit of infinite radius. The latter can be seen by setting $\omega = \omega_1 + i\omega_2$ and noticing that the imaginary part ω_2 is positive in the upper half-plane. Therefore the real part of the product $i\omega(t - t')$ becomes a large negative number for $t - t' > 0$ and $|\omega|$ approaching infinity. Also the integration path becomes large, but the denominator more than compensates and the exponential dominates since $\exp[i\omega(t-t')]$ goes even more rapidly to zero in that limit.

The value of the closed integral is determined by the sum of residues lying in the upper half-plane. In the context of this monograph we are interested mostly in the case where the damping constant Γ is infinitesimally small. Then, poles of the denominator are located in $\omega \simeq \pm\omega_0 + i\Gamma/2$, i.e., in the upper half-plane close to the real axis. This yields

$$G(t,t') = 2\pi i \sum \text{Res} \left\{ \frac{e^{i\omega(t-t')}}{2\pi(\omega_0 - \omega)(\omega_0 + \omega)} \right\}, \tag{A.99}$$

where 'Res' indicates the residues. Evaluation of the latter reproduces the upper part of (A.95).

For $t - t' < 0$ the integral in the complex ω-plane can be closed by a semicircle in the *lower half-plane* ($\omega_2 < 0$). Again the value of the integral reduces to the sum of residues. Since no poles are located in the lower half-plane the integral vanishes. This is in accordance with the lower part of (A.95).

Thus, from both procedures one deduces that the general solution for the forced oscillator with infinitesimal damping can be written in the form

$$\psi(t) = \frac{1}{m} \int_{-\infty}^{t} \mathrm{d}t' \, \frac{\sin \omega_0(t - t')}{\omega_0} f(t') + \psi_0(t). \tag{A.100}$$

This form is consistent with the law of causality: Only the force acting up to time t determines the displacement at t.

One of several alternative solutions would be

$$\psi(t) = -\frac{1}{m} \int_{t}^{\infty} \mathrm{d}t' \, \frac{\sin \omega_0(t - t')}{\omega_0} f(t') + \psi_1(t), \tag{A.101}$$

which is mathematically equivalent with (A.100) for a suitably chosen free oscillation $\psi_1(t)$. Physically this solution reflects a situation in which the roles of cause and effect have been interchanged. An analogous case is known from electromagnetic theory where both 'retarded' and 'advanced' potentials are in use.

Wave Equation

The Green function of the wave equation is central to the quantum theory of elastic and inelastic scattering. Writing the stationary Schrödinger equation for a single particle in a potential $\mathcal{V}(\boldsymbol{r})$,

$$\left(-\frac{\hbar^2}{2m} \boldsymbol{\nabla}^2 + \mathcal{V}(\boldsymbol{r}) - E \right) \psi(\boldsymbol{r}) = 0 \tag{A.102}$$

in the form

$$\left(-\frac{\hbar^2}{2m} \boldsymbol{\nabla}^2 - E \right) \psi(\boldsymbol{r}) = -\mathcal{V}(\boldsymbol{r})\psi(\boldsymbol{r}), \tag{A.103}$$

we may obtain a formal solution

$$\psi(\boldsymbol{r}) = \psi_0(\boldsymbol{r}) - \int \mathrm{d}^3 r' \, G(\boldsymbol{r}, \boldsymbol{r}')\mathcal{V}(\boldsymbol{r}')\psi(\boldsymbol{r}') \tag{A.104}$$

where $\psi_0(\boldsymbol{r})$ is a solution of the force-free wave equation

$$\left(-\frac{\hbar^2}{2m} \boldsymbol{\nabla}^2 - E \right) \psi_0(\boldsymbol{r}) = 0, \tag{A.105}$$

and the Green function $G(\boldsymbol{r}, \boldsymbol{r}')$ obeys

$$\left(-\frac{\hbar^2}{2m} \boldsymbol{\nabla}^2 - E \right) G(\boldsymbol{r}, \boldsymbol{r}') = \delta(\boldsymbol{r} - \boldsymbol{r}'). \tag{A.106}$$

You may solve (A.106) in Fourier space along the same line as was followed in case of the forced oscillator. However, a more direct approach is again possible. Set $r - r' = R$ and assume G to depend on R only. Then,

$$(\nabla^2 + k^2)G(R) = -\frac{2m}{\hbar^2}\delta(R) \tag{A.107}$$

with $k^2 = 2mE/\hbar^2$. Since the Dirac function shows spherical symmetry we must be able to find a spherically symmetric solution of (A.107). Then, outside $R = 0$, we have two independent solutions,

$$G(R) = \text{const}\,\frac{e^{\pm ikR}}{R} \qquad \text{for } R \neq 0. \tag{A.108}$$

The value of the constant is determined by the singular behavior near $R = 0$ which is insensitive to the value of k. For $k = 0$ (A.107) reduces to Poisson's equation for the potential of a point charge of magnitude $2m/4\pi\hbar^2$, i. e.,

$$G(R) = \frac{m}{2\pi\hbar^2}\frac{1}{R} \text{ for } R \simeq 0. \tag{A.109}$$

This determines the constant in (A.108), and hence

$$G(R) = \frac{m}{2\pi\hbar^2}\frac{e^{\pm ikR}}{R} \tag{A.110}$$

or

$$G(r, r') = \frac{m}{2\pi\hbar^2}\frac{e^{\pm ik|r-r'|}}{|r - r'|}. \tag{A.111}$$

Solutions of this type are familiar from electromagnetic theory (Jackson, 1975).

The difference between the two solutions contained in (A.110) is readily identified by adding the time-dependent factor to the wave function. Then the two phases read

$$\pm ikR - iEt/\hbar = \pm i(kR \mp Et/\hbar). \tag{A.112}$$

The function with the upper sign describes an outgoing spherical wave, i.e., a wave moving radially away from the center of force. The lower sign, on the other hand, describes an incoming spherical wave.

A.3 Mechanics

A.3.1 Classical Perturbation Theory

This section serves to derive results reported without proof in Sect. 3.3.6.

Scattering angle

Consider first the scattering integral

$$\theta = \pi - 2p \int_{r_m}^{\infty} \frac{\mathrm{d}r}{r^2} \frac{1}{\sqrt{1 - \mathcal{V}(r)/E_r - p^2/r^2}} \tag{A.113}$$

according to (3.35) and (3.34). A perturbation expansion is synonymous with Taylor expansion in powers of the interaction potential $\mathcal{V}(r)$. Straight expansion of the square root, however, would lead to increasingly strong singularities at $r = p$. A way to circumvent this problem has been designed by Lehmann and Leibfried (1963).

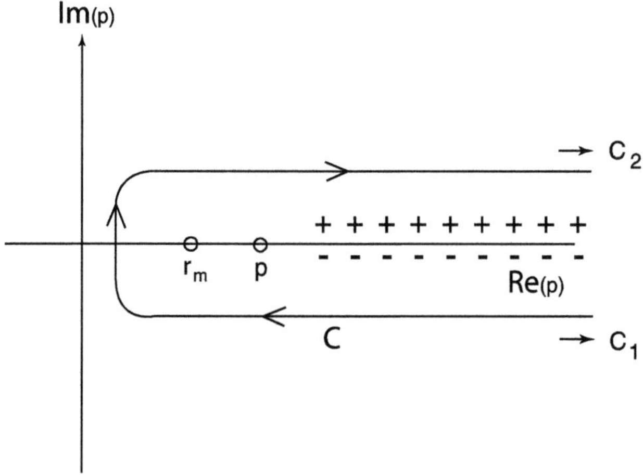

Fig. A.2. Integration path $\mathcal{C}$ for scattering integral. Attractive interaction assumed

We first note that in the complex r-plane, the integrand is a meromorphous (multi-valued) function which, however, can be made unique by making a cut on the real axis for $r > r_m$, such that the positive value of the square root is approached from the positive-imaginary half-plane, and vice versa for the negative square root. This means that (A.113) can be rewritten in the form

$$\theta = \pi - p \int_{\mathcal{C}} \frac{\mathrm{d}r}{r^2} \frac{1}{\sqrt{1 - \mathcal{V}(r)/E_r - p^2/r^2}}, \tag{A.114}$$

where the integration path $\mathcal{C}$ indicated in Fig. A.2 circumvents the cut. This is allowed by Cauchy's theorem since the integrand is regular outside the cut.

Taylor expansion now yields

$$\theta = \pi - p \int_C \frac{\mathrm{d}r}{r^2} \frac{1}{\sqrt{1 - p^2/r^2}}$$

$$\times \left[1 + \frac{1}{2} \frac{\mathcal{V}(r)/E_r}{1 - p^2/r^2} + \frac{3}{8} \frac{[\mathcal{V}(r)/E_r]^2}{(1 - p^2/r^2)^2} \cdots \right]. \tag{A.115}$$

Consider now the first two terms,

$$\theta^{(0)} = \pi - 2p \int_p^\infty \frac{\mathrm{d}r}{r^2} \frac{1}{\sqrt{1 - p^2/r^2}}, \tag{A.116}$$

where the integration path has been pulled back to the real axis as close as possible. Substituting p/r as the integration variable you immediately see that $\theta^{(0)}$ vanishes, as it has to in the absence of an interaction.

Consider now the first-order term

$$\theta^{(1)} = -\frac{p}{2} \int_C \frac{\mathrm{d}r}{r^2} \frac{\mathcal{V}(r)/E_r}{(1 - p^2/r^2)^{3/2}}. \tag{A.117}$$

You may rewrite this in the form

$$\theta^{(1)} = \frac{1}{2pE_r} \int_C \mathrm{d}r [r\mathcal{V}(r)] \frac{\mathrm{d}}{\mathrm{d}r} \frac{1}{\sqrt{1 - p^2/r^2}}, \tag{A.118}$$

which is ready for integration by parts. The integrated term vanishes since *both* end points C_1 and C_2 of C lie at infinity (Fig. A.2), where $r\mathcal{V}(r)$ will vanish for a screened Coulomb potential[1]. The remaining integral has an integrable singularity at $r = p$. We may then pull the integration path back on the real axis and obtain

$$\theta^{(1)} = -\frac{1}{pE_r} \int_p^\infty \frac{\mathrm{d}r}{\sqrt{1 - p^2/r^2}} \frac{\mathrm{d}}{\mathrm{d}r} [r\mathcal{V}(r)]. \tag{A.119}$$

This well-known standard expression can also be written in the alternative form

$$\theta^{(1)} = -\frac{p}{E_r} \int_p^\infty \frac{\mathrm{d}r}{\sqrt{r^2 - p^2}} \frac{\mathrm{d}\mathcal{V}(r)}{\mathrm{d}r}, \tag{A.120}$$

as can be seen by differentiating the product $r\mathcal{V}(r)$, substituting $r^2 = r^2 - p^2 + p^2$ and partial integration (in the complex plane).

The second-order term

$$\theta^{(2)} = -\frac{3p}{8} \int_C \frac{\mathrm{d}r}{r^2} \frac{[\mathcal{V}(r)/E_r]^2}{(1 - p^2/r^2)^{5/2}} \tag{A.121}$$

---

[1] Lehmann and Leibfried (1963) also provided an alternative procedure which only requires $\mathcal{V}(r)$ to vanish at infinity.

can be evaluated similarly. Indeed,

$$\theta^{(2)} = -\frac{p}{8E_r^2} \int_C \mathrm{d}r \left[r^2 \mathcal{V}(r)\right]^2 \frac{\mathrm{d}}{\mathrm{d}r} \frac{1}{r} \frac{\mathrm{d}}{\mathrm{d}r} \frac{1}{\sqrt{r^2 - p^2}} \tag{A.122}$$

reduces after repeated partial integration and subsequent deformation of the integration path to

$$\theta^{(2)} = -\frac{p}{4E_r^2} \int_p^\infty \frac{\mathrm{d}r}{\sqrt{r^2 - p^2}} \frac{\mathrm{d}}{\mathrm{d}r} \frac{1}{r} \frac{\mathrm{d}}{\mathrm{d}r} \left[r\mathcal{V}(r)\right]^2 . \tag{A.123}$$

An alternative form, which can be found by substituting $\mathcal{V} \Rightarrow (1/r)(\mathrm{d}/\mathrm{d}r)r^2\mathcal{V}^2$ in the two expressions for $\theta^{(1)}$ above, reads

$$\theta^{(2)} = -\frac{1}{4pE_r^2} \int_p^\infty \frac{r\mathrm{d}r}{\sqrt{r^2 - p^2}} \frac{\mathrm{d}^2}{\mathrm{d}r^2} r^2 \mathcal{V}(r)^2 \tag{A.124}$$

Now, both (A.119) and (A.123) have integrable singularities at $r = p$. As an example, take the Yukawa potential

$$\mathcal{V}(r) = \frac{e_1 e_2}{r} \mathrm{e}^{-r/a}. \tag{A.125}$$

Then,

$$\theta^{(1)} = \frac{e_1 e_2}{aE_r} K_1 \left(\frac{p}{a}\right) \tag{A.126}$$

and

$$\theta^{(2)} = -\left(\frac{e_1 e_2}{aE_r}\right)^2 K_1 \left(\frac{2p}{a}\right) . \tag{A.127}$$

The ratio of the two leading contributions,

$$\frac{\theta^{(2)}}{\theta^{(1)}} = -\frac{e_1 e_2}{aE_r} \frac{K_1(2p/a)}{K_1(p/a)}, \tag{A.128}$$

is seen to be governed both by the energy E_r and the impact parameter. The latter dependence is illustrated in Fig. A.3.

Time Integral

The time integral (3.61) has been evaluated in the same manner by Sigmund and Schinner (2000). Indeed, writing it in the form

$$\tau(p, v) = \int_p^\infty \frac{\mathrm{d}r}{\sqrt{1 - p^2/r^2}} - \int_{r_m}^\infty \frac{\mathrm{d}r}{\sqrt{1 - \mathcal{V}(r)/E_r - p^2/r^2}}, \tag{A.129}$$

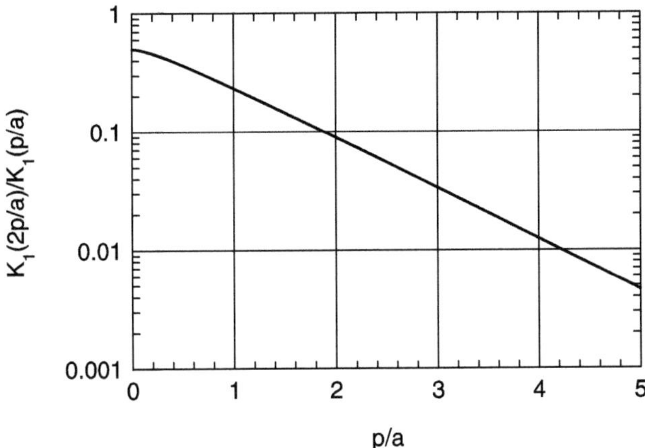

Fig. A.3. The ratio $K_1\left(2p/a\right)/K_1\left(p/a\right)$ governing the relative significance of the second-order perturbation for Yukawa interaction

we again pull the path of integration into the complex plane in accordance with Fig. A.2. This leads to

$$\tau(p,v) = \frac{1}{2} \int_{\mathcal{C}} \mathrm{d}r \left(\frac{1}{\sqrt{1 - p^2/r^2}} - \frac{1}{\sqrt{1 - \mathcal{V}(r)/E_r - p^2/r^2}} \right), \qquad \text{(A.130)}$$

where a common integration path $\mathcal{C}$ has been chosen by making use of Cauchy's theorem.

Taylor expansion up to the first order in $\mathcal{V}(r)$ then yields

$$\tau(p,v) = -\frac{1}{4E_r} \int_{\mathcal{C}} \mathrm{d}r \frac{\mathcal{V}(r)}{(1 - p^2/r^2)^{3/2}} \qquad \text{(A.131)}$$

or, after partial integration,

$$\tau(p,v) = -\frac{1}{4E_r} \int_{\mathcal{C}} \frac{1}{\sqrt{r^2 - p^2}} \frac{\mathrm{d}}{\mathrm{d}r} \left[r^2 \mathcal{V}(r) \right]. \qquad \text{(A.132)}$$

For Yukawa interaction we then obtain

$$\begin{aligned} \tau(p,v) &= -\frac{e_1 e_2}{2E_r} \int_p^{\infty} \frac{\mathrm{d}r}{\sqrt{r^2 - p^2}} \left(1 - \frac{r}{a} \right) \mathrm{e}^{-r/a} \\ &= \frac{e_1 e_2}{2E_r} \left[K_1 \left(\frac{p}{a} \right) - \frac{p}{a} K_0 \left(\frac{p}{a} \right) \right]. \end{aligned} \qquad \text{(A.133)}$$

A.3.2 Relativity

This is just a brief recapitulation of pertinent results from special relativity. The reader who needs a genuine introduction into the field is referred to Landau and Lifshitz (1971) or Jackson (1975).

Kinematics

The *Lorentz transformation* describes coordinate transformations between two reference frames S and S' in *uniform* motion relative to each other. For simplicity of notation, we may place the x axis of S and the x' axis of S' in the direction of the relative velocity $\boldsymbol{v}$. With a suitable choice of y and z axes as well as time $t = 0$ we may write the transformation equations in the form

$$x' = \gamma_v \left(x - vt \right) \tag{A.134a}$$
$$y' = y \tag{A.134b}$$
$$z' = z \tag{A.134c}$$
$$t' = \gamma_v \left(t - \frac{vx}{c^2} \right), \tag{A.134d}$$

where c is the speed of light and

$$\gamma_v = \frac{1}{\sqrt{1 - v^2/c^2}}. \tag{A.135}$$

The so-called Galilei transformation – which is valid in Newtonian mechanics – emerges from this by letting c go to infinity.

Direct consequences of this are *Lorentz contraction* and *time dilatation*. From the first equation follows

$$x_1' - x_2' = \gamma_v \left[x_1 - x_2 - v(t_1 - t_2) \right]. \tag{A.136}$$

Now, consider a ruler with length $\Delta \ell' = x_1' - x_2'$ in S', where it is at rest. In S, at time $t_1 = t_2$ we see a length

$$\Delta \ell = x_1 - x_2 = \frac{\Delta \ell'}{\gamma_v}, \tag{A.137}$$

i.e., the ruler looks shorter.

Conversely, for a person moving uniformly with the speed v relative to S, we have $x = vt$ and hence, from the last equation,

$$t' = \frac{1}{\gamma_v} t, \tag{A.138}$$

i.e., time runs more slowly in the moving reference frame.

The set of variables x, y, z, ct is said to form a *four-vector*. You may define a four-vector as a set of four variables that satisfies the same transformation equations as x, y, z, ct.

Consider now a particle moving with a velocity $\boldsymbol{u} = (u_x, u_y, u_z)$ in the frame S. We want to know the connection of $\boldsymbol{u}$ to the corresponding velocity $\boldsymbol{u}' = (u_x', u_y', u_z')$ in S'.

From (A.134d) we find

$$\mathrm{d}x' = \gamma_v \left(\mathrm{d}x - v\mathrm{d}t\right) \tag{A.139a}$$

$$\mathrm{d}y' = \mathrm{d}y \tag{A.139b}$$

$$\mathrm{d}z' = \mathrm{d}z \tag{A.139c}$$

$$\mathrm{d}t' = \gamma_v \left(\mathrm{d}t - \frac{v}{c^2}\mathrm{d}x\right), \tag{A.139d}$$

and hence,

$$u'_x = \frac{\mathrm{d}x'}{\mathrm{d}t'} = \frac{u_x - v}{1 - vu_x/c^2} \tag{A.140a}$$

$$u'_y = \frac{\mathrm{d}y'}{\mathrm{d}t'} = \frac{u_y}{\gamma_v(1 - vu_x/c^2)} \tag{A.140b}$$

$$u'_z = \frac{\mathrm{d}z'}{\mathrm{d}t'} = \frac{u_z}{\gamma_v(1 - vu_x/c^2)} \tag{A.140c}$$

This is one of several ways to express the *addition theorem of velocities*. Note that the velocity so defined is not a four-vector.

Dynamics

The momentum $\boldsymbol{P}$ of a particle with the *rest mass m*, moving with a velocity $\boldsymbol{v}$, is given by

$$\boldsymbol{P} = m\gamma_v\boldsymbol{v}. \tag{A.141}$$

It forms a four-vector together with the quantity

$$E = \sqrt{(mc^2)^2 + P^2c^2}, \tag{A.142}$$

which defines the energy of the particle,

$$P'_x = \gamma_v \left(P_x - \frac{vE}{c^2}\right) \tag{A.143a}$$

$$P'_y = P_y \tag{A.143b}$$

$$P'_z = P_z \tag{A.143c}$$

$$E' = \gamma_v \left(E - P_x v\right), \tag{A.143d}$$

where, as in the previous section, the x-axis is chosen to point in the direction of the relative velocity $\boldsymbol{v}$.

For $\boldsymbol{P} = 0$, (A.142) represents the *rest energy mc^2*, and the difference denotes the *kinetic energy*

$$E_{\mathrm{kin}} = \sqrt{(mc^2)^2 + P^2c^2} - mc^2 \equiv mc^2(\gamma_v - 1). \tag{A.144}$$

Newton's second law remains valid if written in the form

$$\frac{\mathrm{d}\boldsymbol{P}}{\mathrm{d}t} = \boldsymbol{F}, \tag{A.145}$$

where $\boldsymbol{F}$ is the force acting on the particle.

The angular momentum, defined by

$$\boldsymbol{L} = \boldsymbol{r} \times \boldsymbol{P}, \tag{A.146}$$

is easily seen to be conserved for a central force, as in nonrelativistic dynamics.

Electromagnetic fields

The Lorentz transformation for electromagnetic fields is most easily written in terms of the potentials $\boldsymbol{A}, \Phi$ discussed in Sect. 5.2 since they form a four-vector. In terms of the electric and the magnetic field the transformation equations read

$$E'_x = E_x \tag{A.147a}$$

$$E'_y = \gamma_v \left(E_y - \frac{v}{c} B_z \right) \tag{A.147b}$$

$$E'_z = \gamma_v \left(E_z + \frac{v}{c} B_y \right) \tag{A.147c}$$

$$B'_x = B_x \tag{A.147d}$$

$$B'_y = \gamma_v \left(B_y + \frac{v}{c} E_z \right) \tag{A.147e}$$

$$B'_z = \gamma_v \left(B_z - \frac{v}{c} E_y \right). \tag{A.147f}$$

A.4 Quantum Mechanics

A.4.1 Gaussian Wave Packets

The wave function

$$\psi(x, t) = \frac{1}{\sqrt{2\pi}} \int_{-\infty}^{\infty} \mathrm{d}k \, A(k) \mathrm{e}^{\mathrm{i}(kx - \omega_k t)} \tag{A.148}$$

with $\omega_k = \hbar k^2 / 2m$ is a linear superposition of plane waves, a 'wave packet'. It is a solution of Schrödinger's equation for a free particle regardless of the choice of weight function $A(k)$ since each of the partial waves $\exp\left(i(kx - \omega_k t)\right)$ satisfies that equation. This follows from the linearity of Schrödinger's equation.

Comparison of (A.148) with the one-dimensional analog of (A.37a) shows that $A(k)\exp(-\omega_k t)$ may be read as a Fourier transform of $\psi(x, t)$ with respect to the spatial variable, apart from the assignment of the factor $(2\pi)^{-1}$ which, in accordance with common practice in quantum mechanics, has been chosen differently here.

Making use of the inverse transform one easily verifies that

$$\int_{-\infty}^{\infty} \mathrm{d}x\, \psi^{\star}(x,t)\psi(x,t) = \int_{-\infty}^{\infty} \mathrm{d}k\, A^{*}(k)A(k). \tag{A.149}$$

Hence, if $A(k)$ is normalized to 1, $\psi(x,t)$ is normalized to 1 at all times t.

Now choose a weight function $A(k)$ having a narrow peak around some value k_0. The resulting wave function $\psi(x,t)$ consists of plane waves which all travel with a velocity $v \simeq \hbar k_0/m$. Therefore, it represents a free particle moving with this velocity but with the most notable difference to a plane wave that the wave packet is localized in space. In accordance with Heisenberg's uncertainty principle, some spread is assigned to both position and momentum.

The expectation value of the position operator x is given by

$$\langle x \rangle = \int_{-\infty}^{\infty} \mathrm{d}x\, \psi^{\star}(x,t)\, x\, \psi(x,t) \tag{A.150}$$

as a function of time. Insertion of (A.148) and integration over $\mathrm{d}x$ leads to

$$\langle x \rangle = \mathrm{i} \int_{-\infty}^{\infty} \mathrm{d}k\, A^{\star}(k) \left(\frac{\mathrm{d}A(k)}{\mathrm{d}k} - \mathrm{i}\,\frac{\hbar k t}{m} A(k) \right). \tag{A.151}$$

Similarly one finds

$$\langle x^2 \rangle = \int_{-\infty}^{\infty} \mathrm{d}x\, \psi^{\star}(x,t)x^2\psi(x,t)$$

$$= \int_{-\infty}^{\infty} \mathrm{d}k\, \left| \frac{\mathrm{d}A(k)}{\mathrm{d}k} - \mathrm{i}\,\frac{\hbar k t}{m} A(k) \right|^2. \tag{A.152}$$

Now, assume a gaussian weight function

$$A(k) = \frac{1}{(2\pi\Delta k^2)^{1/4}}\, e^{-(k-k_0)^2/4\Delta k^2}. \tag{A.153}$$

Here, form and constants have been chosen such that the probability density

$$|A(k)|^2 = \frac{1}{\sqrt{2\pi\Delta k^2}}\, e^{-(k-k_0)^2/2\Delta k^2} \tag{A.154}$$

has the common gaussian form, normalized to 1 with the mean value

$$\langle k \rangle = \int_{-\infty}^{\infty} \mathrm{d}k\, k \left|A(k)\right|^2 = k_0 \tag{A.155}$$

and the variance

$$\left\langle (k-k_0)^2 \right\rangle = \int_{-\infty}^{\infty} \mathrm{d}k\, (k-k_0)^2 \left|A(k)\right|^2 = \Delta k^2. \tag{A.156}$$

The spread in wave number Δk is a key parameter that can be chosen freely.

For this choice of weight function the spatial extent of the wave packet is characterized by the average values

$$\langle x \rangle = \frac{\hbar k_0}{m} t = vt \tag{A.157}$$

which follows by insertion of (A.153) into (A.150), and

$$\Delta x^2 = \left\langle \left(x - \langle x \rangle \right)^2 \right\rangle = \frac{1}{4\Delta k^2} + \left(\frac{2\hbar \Delta k}{m} \right)^2 t^2 . \tag{A.158}$$

It is seen from (A.158) that the spatial extent Δx of a wave packet increases as time goes on.

After replacement of the wave number k by the momentum variable $P = \hbar k$, (A.158) reduces to

$$\Delta P^2 \Delta x^2 = \left(\frac{\hbar}{2} \right)^2 + \left(\frac{2\Delta P^2}{m} t \right)^2 . \tag{A.159}$$

It is seen that the uncertainty product $\Delta P \Delta x$ has its minimum value $\hbar/2$ at time $t = 0$ whereas it is larger at any other instant. It can be shown that this minimum value of the uncertainty product is a unique feature of a gaussian weight function (Schiff, 1981).

A.4.2 Time-Dependent Perturbation Theory

This appendix serves to recapitulate standard time-dependent perturbation theory (Schiff, 1981, Merzbacher, 1970) up to the point where it is needed in this monograph. The notation is kept as close as possible to the applications discussed in Chapters 4–6.

Consider a system governed by a hamiltonian H with eigenstates $|j\rangle$ and energies ϵ_j, perturbed by an interaction $\mathcal{V}$ which may depend explicitly on time, $\mathcal{V} = \mathcal{V}(\boldsymbol{r}, t)$.

Note first that according to Schrödinger's equation (4.33),

$$\frac{\partial}{\partial t} \Psi^* \Psi = \frac{1}{i\hbar} \left\{ \Psi^* \left[H + \mathcal{V}(t) \right] \Psi - \Psi \left[H + \mathcal{V}(t) \right] \Psi^* \right\}. \tag{A.160}$$

After integration over all spatial variables one finds that

$$\frac{\mathrm{d}}{\mathrm{d}t} \langle \Psi | \Psi \rangle = 0 \tag{A.161}$$

because of the hermiticity of both H and $\mathcal{V}$. In other words, if the wave function Ψ is normalized at one time it will remain normalized at all times.

A distinguishing feature of scattering problems is an interaction $\mathcal{V}(t)$ which vanishes at times $t = \pm\infty$ and goes through a maximum at some intermediate time which, frequently, will be chosen around $t = 0$. The system may be assumed to occupy an eigenstate $|0\rangle$ at $t = -\infty$. Since eigenfunctions form a complete basis set, we may expand the wave function according to

$$\Psi = \sum_j c_j(t)\mathrm{e}^{-\mathrm{i}\omega_j t}|j\rangle, \tag{A.162}$$

where $\omega_j = \epsilon_j/\hbar$. The coefficients $c_j(t)$ have to be determined from Schrödinger's equation, (4.33) which reduces to

$$\sum_j \mathcal{V}(t)c_j(t)\mathrm{e}^{-\mathrm{i}\omega_j t}|j\rangle = \mathrm{i}\hbar \sum_j \frac{\mathrm{d}c_j(t)}{\mathrm{d}t}\mathrm{e}^{-\mathrm{i}\omega_j t}|j\rangle. \tag{A.163}$$

In the absence of an interaction, $c_j(t)$ would be constant and identical with the value at $t = -\infty$, $c_j(t) \equiv c_j^{(0)} = \delta_{j0}$. Multiplication of (A.163) with $\langle \ell|$ yields

$$\sum_j \mathcal{V}_{\ell j}(t)c_j(t)\mathrm{e}^{-\mathrm{i}\omega_j t} = \mathrm{i}\hbar \frac{\mathrm{d}c_\ell(t)}{\mathrm{d}t}\mathrm{e}^{-\mathrm{i}\omega_\ell t} \tag{A.164}$$

or, after integration,

$$c_\ell(t) = \delta_{\ell 0} + \frac{1}{\mathrm{i}\hbar}\sum_j \int_{-\infty}^{t} \mathrm{d}t'\, \mathcal{V}_{\ell j}(t')\mathrm{e}^{\mathrm{i}\omega_{\ell j} t'}c_j(t'), \tag{A.165}$$

where $\omega_{\ell j} = \omega_\ell - \omega_j$ and $\mathcal{V}_{\ell j}(t) = \langle \ell|\mathcal{V}(t)|j\rangle$.

We may try to expand the time-dependent coefficients $c_j(t)$ in powers of the interaction $\mathcal{V}$,

$$c_j(t) = c_j^{(0)}(t) + c_j^{(1)}(t) + c_j^{(2)}(t)\dots. \tag{A.166}$$

Then (A.165) reduces to

$$c_j^{(\nu)}(t) = \frac{1}{\mathrm{i}\hbar}\sum_\ell \int_{-\infty}^{t} \mathrm{d}t'\, \mathcal{V}_{j\ell}(t')\mathrm{e}^{\mathrm{i}\omega_{j\ell} t'}c_\ell^{(\nu-1)}(t') \tag{A.167}$$

for $\nu \geq 1$, by collection of terms of equal order in $\mathcal{V}$. This yields

$$c_j^{(0)}(t) = \delta_{j0} \tag{A.168a}$$

$$c_j^{(1)}(t) = \frac{1}{\mathrm{i}\hbar}\int_{-\infty}^{t} \mathrm{d}t'\, \mathcal{V}_{j0}(t')\mathrm{e}^{\mathrm{i}\omega_{j0} t'} \tag{A.168b}$$

$$c_j^{(2)}(t) = \frac{1}{(\mathrm{i}\hbar)^2}\sum_\ell \int_{-\infty}^{t} \mathrm{d}t' \int_{-\infty}^{t'} \mathrm{d}t''\, \mathcal{V}_{j\ell}(t')\mathcal{V}_{\ell 0}(t'')\mathrm{e}^{\mathrm{i}\omega_{j\ell} t'}\mathrm{e}^{\mathrm{i}\omega_{\ell 0} t''} \tag{A.168c}$$

$$\dots$$

These expressions are applied as they stand in dispersion theory, Appendix A.5 and Chapter 5. In Chapters 4 and 6 mainly the total energy absorbed by the system due to the interaction $\mathcal{V}$ is of interest. Since the interaction is assumed to vanish at $t = \infty$ the mean energy change is given by

$$T = \langle \Psi(t) | H | \Psi(t) \rangle_{t=\infty} - \langle \Psi(t) | H | \Psi(t) \rangle_{-\infty} . \tag{A.169}$$

Insertion of the expansion (A.166) and observation of the normalization (A.161), or $\sum_j |c_j|^2 = 1$, yields

$$T = \sum_j \hbar\omega_{j0} \left(\left[c_j^{(1)*} c_j^{(1)} \right] + \left[c_j^{(1)*} c_j^{(2)} + c_j^{(2)*} c_j^{(1)} \right] \right.$$

$$\left. + \left[c_j^{(1)*} c_j^{(3)} + c_j^{(2)*} c_j^{(2)} + c_j^{(3)*} c_j^{(1)} \right] + \dots \right)_{t=\infty} \tag{A.170}$$

up to terms of fourth order in $\mathcal{V}$.

A.4.3 Generalized Oscillator Strengths for the Harmonic Oscillator

This appendix serves to illustrate one of several powerful methods of evaluating matrix elements. Use is made of generating functions. An example of a generating function has been mentioned in (A.27), Appendix A.2.1. Several examples of the utilization of generating functions in the determination of matrix elements may be found in ref. Schiff (1981).

A linear harmonic oscillator has eigenfunctions

$$|n\rangle = N_n H_n(\xi) e^{-\xi^2/2}, \; n = 0, 1, 2 \dots \tag{A.171}$$

where

$$\xi = \alpha x; \; \alpha = \sqrt{m\omega_0/\hbar}, \tag{A.172}$$

$H_n(\xi)$ are Hermite polynomials and N_n a set of normalizing constants.

A generating function for the H_n is given by Abramowitz and Stegun (1964),

$$S(\xi, s) = e^{-s^2 + 2s\xi} = \sum_{n=0}^{\infty} \frac{s^n}{n!} H_n(\xi), \tag{A.173}$$

where s is a dimensionless variable. Write this relation once more with another dimensionless variable t,

$$S(\xi, t) = e^{-t^2 + 2t\xi} = \sum_{m=0}^{\infty} \frac{t^m}{m!} H_m(\xi), \tag{A.174}$$

take the product between the two relations, multiply by $\exp(iqx - \xi^2)$, and integrate both sides over x. This yields

$$\sum_{n=0}^{\infty}\sum_{m=0}^{\infty}\frac{s^n t^m}{n!m!}\int_{-\infty}^{\infty}\mathrm{d}x\, H_n(\xi)H_m(\xi)\mathrm{e}^{-\xi^2/2+iqx}$$

$$= \frac{\sqrt{\pi}}{\alpha}\mathrm{e}^{2st+iq(s+t)/\alpha-q^2/4\alpha^2}. \qquad (\mathrm{A}.175)$$

Information about matrix elements is then found by Taylor expansion of the right-hand side in powers of s and t. Instead of the general expansion we consider two special cases.

Set first $q = 0$, so that

$$\sum_{n=0}^{\infty}\sum_{m=0}^{\infty}\frac{s^n t^m}{n!m!}\int_{-\infty}^{\infty}\mathrm{d}x\, H_n(\xi)H_m(\xi)\mathrm{e}^{-\xi^2/2} = \frac{\sqrt{\pi}}{\alpha}\mathrm{e}^{2st}$$

$$\equiv \frac{\sqrt{\pi}}{\alpha}\sum_{n=0}^{\infty}\frac{(2st)^n}{n!}. \qquad (\mathrm{A}.176)$$

Comparison of equal powers of s and t leads to

$$\int_{-\infty}^{\infty}\mathrm{d}x\, H_n(\xi)H_m(\xi)\mathrm{e}^{-\xi^2} = \delta_{nm}\frac{\sqrt{\pi}}{\alpha}2^n n!, \qquad (\mathrm{A}.177)$$

from which N_n is determined to

$$N_n = \sqrt{\frac{\alpha}{\sqrt{\pi}2^n n!}}. \qquad (\mathrm{A}.178)$$

Next consider q arbitrary but $t = 0$,

$$\sum_{n=0}^{\infty}\frac{s^n}{n!}\int_{-\infty}^{\infty}\mathrm{d}x\, H_n(\xi)H_0(\xi)\mathrm{e}^{-\xi^2/2+iqx}$$

$$= \frac{\sqrt{\pi}}{\alpha}\mathrm{e}^{iqs/\alpha-q^2/4\alpha^2} \equiv \frac{\sqrt{\pi}}{\alpha}\mathrm{e}^{-q^2/4\alpha^2}\sum_{n=0}^{\infty}\frac{s^n}{n!}\left(\frac{iq}{\alpha}\right)^n. \qquad (\mathrm{A}.179)$$

Comparison of equal powers of s yields

$$\int_{-\infty}^{\infty}\mathrm{d}x\, H_n(\xi)H_0(\xi)\mathrm{e}^{-\xi^2+iqx} = \frac{\sqrt{\pi}}{\alpha}\left(\frac{iq}{\alpha}\right)^n\mathrm{e}^{-q^2/4\alpha^2} \qquad (\mathrm{A}.180)$$

or, after multiplication with $N_n N_0$,

$$\langle n|\mathrm{e}^{iqx}|0\rangle = \frac{1}{\sqrt{2^n n!}}\left(\frac{iq}{\alpha}\right)^n\mathrm{e}^{-q^2/4\alpha^2}, \qquad (\mathrm{A}.181)$$

from which (4.109) emerges by means of (4.107).

A.4.4 Sum Rules

This appendix presents proofs of sum rules that have been utilized in various chapters. The simplest and most well-known sum rule is the one involving dipole oscillator strengths (4.48). Since this follows as a limiting case of Bethe's sum rule for generalized oscillator strengths, the latter will be derived first.

Bethe's Sum Rule

According to (4.114) the generalized oscillator strengths defined by (4.107),

$$f_{j0}(Q) = \frac{1}{Z_2}\frac{\epsilon_j - \epsilon_0}{Q}\,|F_{j0}(\boldsymbol{q})|^2\,, \tag{A.182}$$

satisfy the sum rule

$$\sum_j f_{j0}(Q) = 1. \tag{A.183}$$

For a proof consider first the case of a one-electron atom. Then,

$$\sum_j f_{j0}(Q) = \frac{1}{Q}\sum_j (\epsilon_j - \epsilon_0)\left\langle 0\left|e^{-iqx}\right|j\right\rangle\left\langle j\left|e^{iqx}\right|0\right\rangle, \tag{A.184}$$

where it has been assumed that the vector $\boldsymbol{q}$ points into the x direction.
Now,

$$(\epsilon_j - \epsilon_0)\left\langle j\left|e^{iqx}\right|0\right\rangle = \left\langle j\left|He^{iqx} - e^{iqx}H\right|0\right\rangle \tag{A.185}$$

because of $H|j\rangle = \epsilon_j|j\rangle$.
The commutator reduces to

$$He^{iqx} - e^{iqx}H \equiv [H, e^{iqx}] = \left[\frac{p_x^2}{2m}, e^{iqx}\right]$$

$$= \frac{\hbar q}{2m}e^{iqx}(\hbar q + 2p_x), \tag{A.186}$$

where $p_x = -i\hbar\partial/\partial x$ is a momentum operator.
With this we find

$$\sum_j f_{j0}(Q) = \frac{\hbar q}{2mQ}\sum_j\left\langle 0\left|e^{-iqx}\right|j\right\rangle\left\langle j\left|e^{iqx}(\hbar q + 2p_x)\right|0\right\rangle$$

$$= \frac{1}{\hbar q}\left\langle 0\left|(\hbar q + 2p_x)\right|0\right\rangle. \tag{A.187}$$

Here the completeness relation $\sum_j|j\rangle\langle j| = 1$ has been applied.

For an atom at rest we must have

$$\langle 0 | p_x | 0 \rangle = 0. \tag{A.188}$$

With this, (A.187) reduces to (A.183).

For a many-electron atom the same procedure leads to

$$\sum_j f_{j0}(Q) = \frac{1}{Z_2 \hbar q} \sum_{\mu,\nu} \left\langle 0 \left| e^{iq(x_\nu - x_\mu)}(\hbar q + 2p_{x\nu}) \right| 0 \right\rangle. \tag{A.189}$$

Here the terms for $\mu = \nu$ make up a sum of Z_2 identical contributions corresponding to the single-electron case. Hence the sum rule is proven if it can be shown that the sum over $\mu \neq \nu$ vanishes. To this end, write the contribution to (A.189) containing the momentum operator in the form

$$\frac{1}{Z_2 \hbar q} \sum_{\mu \neq \nu} \left\langle 0 \left| e^{iq(x_\nu - x_\mu)} 2p_{x\nu} \right| 0 \right\rangle = \frac{1}{Z_2 \hbar q} \sum_{\mu \neq \nu} \left\langle 0 \left| e^{iq(x_\nu - x_\mu)} p_{x\nu} \right| 0 \right\rangle$$
$$+ \frac{1}{Z_2 \hbar q} \sum_{\mu \neq \nu} \left\langle 0 \left| p_{x\nu} e^{-iq(x_\nu - x_\mu)} \right| 0 \right\rangle, \tag{A.190}$$

which must be true since the sum is real.

Now,

$$p_{x\nu} e^{-iq(x_\nu - x_\mu)} |0\rangle = -p_{x\mu} e^{-iq(x_\nu - x_\mu)} |0\rangle \tag{A.191}$$

for $\mu \neq \nu$. This invokes the antisymmetry of the ground-state wave function $|0\rangle$ in the electron coordinates. Then,

$$\frac{1}{Z_2 \hbar q} \sum_{\mu \neq \nu} \left\langle 0 \left| e^{iq(x_\nu - x_\mu)} 2p_{x\nu} \right| 0 \right\rangle = \frac{1}{Z_2 \hbar q} \sum_{\mu \neq \nu} \left\langle 0 \left| \left[e^{iq(x_\nu - x_\mu)}, p_{x\nu} \right] \right| 0 \right\rangle$$
$$= \frac{1}{Z_2 \hbar q} \sum_{\mu \neq \nu} \left\langle 0 \left| -\hbar q e^{iq(x_\nu - x_\mu)} \right| 0 \right\rangle, \tag{A.192}$$

which cancels the remaining term in (A.189).

Dipole Limit

In view of (A.182), generalized oscillator strengths vanish for degenerate states, i.e., for $\epsilon_j = \epsilon_0$. Therefore the constant term in the Taylor expansion for small q,

$$\left\langle j \left| e^{iqx} \right| 0 \right\rangle = \langle j | (1 + iqx + \ldots) | 0 \rangle \tag{A.193}$$

vanishes because of orthogonality, $\langle j | 0 \rangle = 0$. In the limit of $q = 0$, (A.182) reduces to

$$\lim_{Q=0} f_{j0}(Q) = \frac{1}{Z_2} \frac{2m}{\hbar^2} (\epsilon_j - \epsilon_0) \left| \left\langle j \left| \sum_\nu x_\nu \right| 0 \right\rangle \right|^2, \tag{A.194}$$

in agreement with the definition of the dipole oscillator strength, (4.47). Since (A.183) has been shown to be valid for arbitrary real values of q it must also remain valid in the limit of $q = 0$. This proves (4.48).

A.4.5 Dirac Equation

The Dirac equation is a relativistic extension of the Schrödinger equation. In its rigorous form it refers to the motion of a *single electron* in an electromagnetic field. Apart from minor differences in notation and units, we follow the presentation of Bransden and Joachain (2000).

Fundamentals

Consider first a free electron. We want to describe its motion by a an equation of the type

$$H\Psi = i\hbar \frac{\mathrm{d}\Psi}{\mathrm{d}t}, \tag{A.195}$$

just as in nonrelativistic quantum mechanics, and we want to keep the momentum operator

$$\boldsymbol{P} = -i\hbar\boldsymbol{\nabla}. \tag{A.196}$$

The energy operator H needs to be constructed from (A.142),

$$H = \sqrt{\boldsymbol{P} \cdot \boldsymbol{P} c^2 + (mc^2)^2}, \tag{A.197}$$

which we try to write in the form

$$H = \sum_{i=1}^{3} \alpha_i P_i + \beta mc^2 \tag{A.198}$$

with operators α, β to be determined so that (A.197) is fulfilled. Squaring (A.198) and equating the result with the square of (A.197) leads to the following conditions,

$$\alpha_i \alpha_j + \alpha_j \alpha_i = 0; \quad i \neq j \tag{A.199a}$$
$$\alpha\beta + \beta\alpha = 0; \tag{A.199b}$$
$$\alpha_i^2 = 1 \tag{A.199c}$$
$$\beta^2 = 1 \tag{A.199d}$$

These conditions are satisfied with the following 4×4 matrices,

$$\alpha_i = \begin{pmatrix} 0 & \sigma_i \\ \sigma_i & 0 \end{pmatrix}; \quad \beta = \begin{pmatrix} \mathbf{I} & 0 \\ 0 & -\mathbf{I} \end{pmatrix} \tag{A.200}$$

where

$$\mathbf{I} = \begin{pmatrix} 1 & 0 \\ 0 & 1 \end{pmatrix} \tag{A.201}$$

and σ_i are the *Pauli matrices*

$$\sigma_x = \begin{pmatrix} 0 & 1 \\ 1 & 0 \end{pmatrix}; \quad \sigma_y = \begin{pmatrix} 0 & -\mathrm{i} \\ \mathrm{i} & 0 \end{pmatrix}; \quad \sigma_z = \begin{pmatrix} 1 & 0 \\ 0 & -1 \end{pmatrix}. \tag{A.202}$$

In analogy with classical hamiltonian or nonrelativistic quantum mechanics, we may then incorporate an electromagnetic field by means of the substitutions

$$H \to H - q\Phi; \quad \mathbf{P} \to \mathbf{P} - \frac{q}{c}\mathbf{A}, \tag{A.203}$$

where $q = -e$ is the electron charge and Φ and $\mathbf{A}$ the electromagnetic scalar and vector potential, respectively.

Plane Waves

In order to describe the dynamics of a free electron, we look for plane-wave solutions of the Dirac equations in the form

$$\Psi(\mathbf{r}, t) = A\mathbf{u}\,\mathrm{e}^{\mathrm{i}(\mathbf{k}\cdot\mathbf{r} - Et/\hbar)}. \tag{A.204}$$

Here, A is a normalization constant, while $\mathbf{u}$ is a *spinor*, a four-dimensional array defined by the eigenvalue equation

$$\left(\hbar c \boldsymbol{\alpha} \cdot \mathbf{k} + mc^2 \beta\right) \mathbf{u} = E\mathbf{u}. \tag{A.205}$$

In the *Dirac representation* (A.200), the four linearly independent solutions are

$$\mathbf{u}^{(1)} = N \begin{pmatrix} 1 \\ 0 \\ \dfrac{\hbar c k_z}{E_+ + mc^2} \\ \dfrac{\hbar c k_+}{E_+ + mc^2} \end{pmatrix}; \quad \mathbf{u}^{(2)} = N \begin{pmatrix} 0 \\ 1 \\ \dfrac{\hbar c k_-}{E_+ + mc^2} \\ -\dfrac{\hbar c k_z}{E_+ + mc^2} \end{pmatrix} \tag{A.206}$$

and

$$\mathbf{u}^{(3)} = N \begin{pmatrix} -\dfrac{\hbar c k_z}{-E_- + mc^2} \\ -\dfrac{\hbar c k_-}{-E_- + mc^2} \\ 1 \\ 0 \end{pmatrix}; \quad \mathbf{u}^{(4)} = N \begin{pmatrix} -\dfrac{\hbar c k_-}{-E_- + mc^2} \\ \dfrac{\hbar c k_z}{-E_- + mc^2} \\ 0 \\ 1 \end{pmatrix}, \tag{A.207}$$

where

$$k_+ = k_x + ik_y; \quad k_- = k_x - ik_y \tag{A.208}$$

and

$$E_+ = \sqrt{(\hbar ck)^2 + (mc^2)^2}; \quad E_- = -\sqrt{(\hbar ck)^2 + (mc^2)^2} \tag{A.209}$$

and

$$N = \sqrt{\frac{E_+ + mc^2}{2E_+}}. \tag{A.210}$$

With this last relation, $\mathbf{u}$ is normalized according to

$$\mathbf{u}^+\mathbf{u} \equiv \sum_{\nu=1}^{4} u_\nu^2 = 1 \tag{A.211}$$

Unlike in nonrelativistic quantum mechanics, electron spin is an intrinsic part of the theory. Indeed, $\mathbf{u}^{(1)}$ and $\mathbf{u}^{(2)}$ represent electrons with two spin orientations 'up' and 'down', respectively. In addition, there are two more solutions, $\mathbf{u}^{(3)}$ and $\mathbf{u}^{(4)}$, representing negative-energy states E_-.

A.5 Dispersion and Absorption

A.5.1 Drude Theory for a Dilute Gas

The Drude theory considers the interaction between an electromagnetic wave and an atom as that between a harmonically bound electron and an electric field that may be considered constant in space over the dimensions of the atom but oscillating in time with a frequency ω. With the field pointing in the z direction the equation of motion

$$m\frac{d^2 z(t)}{dt^2} = -m\omega_0^2 z(t) - m\Gamma\frac{dz(t)}{dt} - e\mathrm{Re}E_0 e^{-i\omega t} \tag{A.212}$$

has the solution

$$z(t) = -\frac{e}{m}\mathrm{Re}\frac{E_0 e^{-i\omega t}}{\omega_0^2 - \omega^2 - i\Gamma\omega}, \tag{A.213}$$

where ω_0 is the resonance frequency of the oscillator, Γ a damping constant, and E_0 the amplitude of the field which may contain a phase and thus be complex.

Contrary to common usage the symbol Re denoting the real part has been inserted explicitly in the above equations. This will also be done in the following in order to leave no doubt about the occurrence of real and complex quantities, respectively.

The displacement $z(t)$ of the electron from its equilibrium position generates a dipole moment $-ez(t)$. For a local density of N oscillators per volume this implies a dielectric polarization $P = -Nez(t)$ and, hence, an electric displacement

$$D = E + 4\pi P = \mathrm{Re}\left(1 + \frac{4\pi Ne^2}{m}\,\frac{1}{\omega_0^2 - \omega^2 - \mathrm{i}\Gamma\omega}\right)E_0\mathrm{e}^{-\mathrm{i}\omega t} \qquad (A.214)$$

or a dielectric function

$$\epsilon = \frac{D}{E} = 1 + \frac{\omega_\mathrm{P}^2}{\omega_0^2 - \omega^2 - \mathrm{i}\Gamma\omega}, \qquad (A.215)$$

where

$$\omega_\mathrm{P} = \sqrt{\frac{4\pi Ne^2}{m}} \qquad (A.216)$$

is called the plasma frequency. Here as well as in the following, the function of the damping constant Γ is to ensure a physically acceptable behavior of $\epsilon(\omega)$ around zeros and singularities. It is sufficient, from this point of view, to assume Γ to be infinitesimally small. This implies that any known and important physical damping mechanism has to be included separately into the general description. Note that a specific damping force is not necessarily proportional to the velocity of the moving electron. This tends to complicate the analysis.

It is immaterial whether damping is introduced as a property of the unperturbed oscillator – as has been implied here – or as a factor turning on and off the field adiabatically such that $E(t) = E_0 \cos\omega t \exp(-\Gamma|t|)$.

A.5.2 Quantum Theory for a Dilute Gas

In an equivalent quantal description we may consider a one-electron atom for simplicity. Extension to a multi-electron atom is strictly analogous to the case considered in Sect. 4.3. We need to solve Schrödinger's equation

$$\left[H + \mathcal{V}(t)\right]\psi(\boldsymbol{r}, t) = \mathrm{i}\hbar\,\frac{\partial\psi(\boldsymbol{r}, t)}{\partial t}, \qquad (A.217)$$

where H is the hamiltonian of the unperturbed atom and $\mathcal{V}(t) = -e\Phi$ with $\Phi = -E_0 z \cos\omega t$ the perturbation induced by the field.

The equation is solved via perturbation theory to first order. In the notation of Chapter 4 the wave function may be written in the form

$$\psi(\boldsymbol{r},t) = \psi^{(0)}(\boldsymbol{r},t) + \psi^{(1)}(\boldsymbol{r},t)\dots$$

$$= \mathrm{e}^{-\mathrm{i}\omega_0 t}|0\rangle + \sum_j c_j^{(1)}(t)\mathrm{e}^{-\mathrm{i}\omega_j t}|j\rangle \dots . \tag{A.218}$$

From (A.168d) we find

$$c_j^{(1)}(t) = \frac{1}{\mathrm{i}\hbar}\int_{-\infty}^{t} \mathrm{d}t'\, \mathrm{e}^{\mathrm{i}\omega_{j0} t'}\langle j|\mathcal{V}(t')|0\rangle$$

$$= -\frac{eE_0}{2\hbar}\mathrm{e}^{\mathrm{i}\omega_{j0} t}\langle j|z|0\rangle \left(\frac{\mathrm{e}^{\mathrm{i}\omega t}}{\omega_{j0}+\omega-\mathrm{i}\Gamma} + \frac{\mathrm{e}^{-\mathrm{i}\omega t}}{\omega_{j0}-\omega-\mathrm{i}\Gamma}\right), \tag{A.219}$$

where again an infinitesimal damping constant Γ has been introduced, this time in the field as mentioned in the last paragraph of Sect. A.5.1.

The induced dipole moment is found as an expectation value

$$\langle -ez\rangle = \int \mathrm{d}^3\boldsymbol{r}\, \psi^*(\boldsymbol{r},t)(-ez)\psi(\boldsymbol{r},t)$$

$$= \int \mathrm{d}^3\boldsymbol{r}\, [\psi^{(0)}(\boldsymbol{r},t)]^*(-ez)\left[\psi^{(1)}(\boldsymbol{r},t)\right] + \text{ conj. compl.}$$

$$= \operatorname{Re}\frac{2e^2 E_0}{\hbar}\sum_j \omega_{j0}\langle 0|z|j\rangle\langle j|z|0\rangle \frac{\mathrm{e}^{-\mathrm{i}\omega t}}{\omega_{j0}^2 - \omega^2 - 2\mathrm{i}\Gamma\omega}. \tag{A.220}$$

Comparison with $-ez(t)$ given by (A.213) shows that the quantal result emerges from the classical one by the substitution

$$\frac{1}{\omega_0^2 - \omega^2 - \mathrm{i}\Gamma\omega} \to \frac{2m}{\hbar}\sum_j \frac{\omega_{j0}\langle 0|z|j\rangle\langle j|z|0\rangle}{\omega_{j0}^2 - \omega^2 - \mathrm{i}\Gamma\omega}. \tag{A.221}$$

Therefore, the quantal result for the dielectric constant may be written in the form

$$\epsilon(\omega) = 1 + \omega_\mathrm{P}^2 \sum_j \frac{f_{j0}}{\omega_{j0}^2 - \omega^2}, \tag{A.222}$$

with the dipole oscillator strengths

$$f_{j0} = \frac{2m}{\hbar}\omega_{j0}\langle 0|z|j\rangle\langle j|z|0\rangle = \frac{2m}{\hbar^2}(\epsilon_j - \epsilon_0)|\langle j|z|0\rangle|^2, \tag{A.223}$$

in complete agreement with (4.48).

Eq. (A.222) demonstrates that the classical theory of dispersion translates into quantum mechanics with the main modification that the electron can be viewed as an ensemble of classical oscillators characterized by the transition frequencies $\omega_{j0} = (\epsilon_j - \epsilon_0)/\hbar$, weighted according to the dipole oscillator strengths f_{j0}, (A.223).

A.5.3 Dense Media

In dense media the field acting upon an electron in a medium is affected by the induced fields due to the other electrons. In textbooks of classical electrodynamics (Jackson, 1975), this effect is taken into account in the Clausius-Mosotti scheme. In brief, the polarization field acting on a given electron in a medium is split into two parts, one from dipoles in the near vicinity and another one from those outside a sphere of a given radius. The contribution from the latter reduces to $4\pi \boldsymbol{P}/3$, while the former can be shown to vanish under specific assumptions on symmetry.

If those symmetry requirements are fulfilled, $E = E_0 \cos \omega t$ in (A.213) needs to be replaced by $E + 4\pi P/3$, and this results in the Lorentz-Lorenz relation

$$3\,\frac{\epsilon - 1}{\epsilon + 2} = \frac{\omega_{\mathrm{P}}^2}{\omega_0^2 - \omega^2 - \mathrm{i}\omega\Gamma}, \tag{A.224}$$

which approaches (A.215) in the limit of weak polarization, $\epsilon \sim 1$.

One major weakness of this scheme is the use of the dipole approximation even for interactions at very short distances. This is circumvented in the dielectric theory described in Chapter 5.

A.5.4 Lindhard Function of the Fermi Gas

This section contains details on calculations outlined in Chapter 5. In contrast to other appendices and previous sections, the present one is heavily dependent on the main text. It is recommended that the reader return to Chapter 5 and consult the present section only after proper reference.

We start at (5.150) and transform the sum over $\boldsymbol{k}_0$ into an integral according to (5.145). We then need to carry out the integral

$$\epsilon(k,\omega) = 1 + \frac{\omega_{\mathrm{P}}^2}{2\omega_k}\,\frac{3}{4\pi k_{\mathrm{F}}^3} \int_{k_0 < k_{\mathrm{F}}} \mathrm{d}^3 k_0 \big(g(k,\omega') + g(k,-\omega')\big), \tag{A.225}$$

with

$$g(k,\omega') = \frac{1}{\omega_k + \hbar k k_0 \eta/m - \omega'}, \tag{A.226}$$

where $\eta = \cos\theta$ and θ the angle between $\boldsymbol{k}_0$ and $\boldsymbol{k}$, and $\omega' = \omega + \mathrm{i}\Gamma$.

After integration over $\mathrm{d}^3 k_0$ this reads

$$\epsilon(k,\omega) = 1$$

$$+ \frac{3m\omega_{\mathrm{P}}^2}{4\omega_k k_{\mathrm{F}}^3 \hbar k}\Big(h(k,\omega') - h(-k,\omega') + h(k,-\omega') - h(-k,-\omega')\Big)\Big|_{k_0=0}^{k_{\mathrm{F}}}, \tag{A.227}$$

with

$$h(k,\omega) = \frac{1}{2}\left(k_0^2 - \frac{b^2}{a}\right)\ln(ak_0 + b) - \frac{1}{4}\left(k_0 - \frac{b}{a}\right)^2,\tag{A.228}$$

where $a = \hbar k/m$ and $b = \omega_k - \omega'$. After collecting terms with $\pm k$ and introduction of dimensionless variables

$$k = 2k_\mathrm{F}z; \quad \omega' = kv_\mathrm{F}u'\tag{A.229}$$

this reads

$$\epsilon = 1 + \frac{3\omega_\mathrm{P}^2}{k^2v_\mathrm{F}^2}\left\{\frac{1}{2} + \frac{1}{8z}[1 - (z - u')^2]\ln\frac{1 + z - u'}{-1 + z - u'}\right.$$
$$\left. + \frac{1}{8z}[1 - (z + u')^2]\ln\frac{1 + z + u'}{-1 + z + u'}\right\}.\tag{A.230}$$

Inserting $u' = u + i\Gamma'$ we find

$$\ln\frac{1 + z - u'}{-1 + z - u'} = \ln\frac{(z - u)^2 - (1 - i\Gamma')^2}{(-1 + z - u)^2 + \Gamma'^2}\tag{A.231}$$

or, by the standard expression for the logarithm of a complex number $z = |z|\exp(i\varphi)$, i.e., $\ln z = \ln|z| + i\varphi$,

$$\ln\frac{1 + z - u'}{-1 + z - u'} = \ln\left|\frac{1 + z - u}{-1 + z - u}\right| + \begin{cases} i\pi & |z - u| < 1 \\ & \text{for} \\ 0 & |z - u| > 1 \end{cases},\tag{A.232}$$

and similarly

$$\ln\frac{1 + z + u'}{-1 + z + u'} = \ln\left|\frac{1 + z + u}{-1 + z + u}\right| + \begin{cases} -i\pi & |z + u| < 1 \\ & \text{for} \\ 0 & |z + u| > 1 \end{cases}.\tag{A.233}$$

This yields (5.155).

References

Abramowitz M. and Stegun I.A. (1964): *Handbook of mathematical functions.* Dover, New York

Bransden B.H. and Joachain C.J. (2000): *Quantum Mechanics.* Prentice Hall, Harlow, 2 edn.

Jackson J.D. (1975): *Classical electrodynamics.* John Wiley & Sons, New York

Landau L. and Lifshitz E.M. (1971): *The classical theory of fields*, vol. 2 of *Course of theoretical physics.* Pergamon Press, Oxford

Lehmann C. and Leibfried G. (1963): Higher order momentum approximations in classical collision theory. Z Physik **172**, 465–487

Merzbacher E. (1970): *Quantum mechanics*. Wiley, New York

Schiff L.I. (1981): *Quantum mechanics*. McGraw-Hill, Auckland

Sigmund P. and Schinner A. (2000): Binary stopping theory for swift heavy ions. Europ Phys J D **12**, 425–434

B

Books and Reviews

Although I am unaware of a textbook or monograph with a similar scope, numerous major and minor summaries have been written in which you may find parts of the material presented in this book. Here I give a list of books and reviews which I can recommend to the reader for further and/or alternative study, with brief comments about their respective merits.

Textbooks

Landau and Lifshitz (1960a,b,c)

Short chapters on central topics presented in a highly original and precise manner.

Jackson (1975)

Excellent presentation of central aspects of classical penetration theory.

Bonderup (1981)

Well-written transcript of a lecture series inspired by J. Lindhard.

Bethe and Jackiw (1986)

Concise presentation of Bethe stopping theory.

Monographs

Bohr (1948)

The classic in the field.

Kumakhov and Komarov (1981)

Useful reference.

ICRU (1984, 1993, 2005)

Official documentation of the International Commision on Radiation Units and Measurements.

Sigmund (2004)

Recent monograph emphasizing penetration of swift heavy ions.

Reviews

Bethe (1933), Livingston and Bethe (1937), Bethe and Ashkin (1953)

Classic reviews.

Uehling (1954), Whaling (1958), Birkhoff (1958)

Early reviews, not only of historic interest.

Fano (1963)

This paper raised the state of the art when it appeared and is still a standard reference.

Inokuti (1971)

A thorough study of quantitative aspects and implications of Bethe theory with numerous instructive illustrations.

Bichsel (1972)

Handbook article with extensive tables. Superceded by ICRU (1984) and ICRU (1993).

Sigmund (1975)

Emphasis on low-velocity stopping, a bit out of date.

Ahlen (1980)

Emphasis on relativistic heavy-ion stopping, a bit out of date.

Ziegler et al. (1985)

You will need to look into this book if you want to know what is behind the popular SRIM code.

Article collections

Uehling (1960), Fano (1964)

Several articles still relevant.

Gras-Marti et al. (1991)

Excellent summaries of selected aspects presented at a NATO summer institute.

Cabrera-Trujillo and Sabin (2004a,b)

Collection of articles of varying quality, but there are a few nice summaries of recent work.

Applications

Mayer et al. (1970)

Introduction to collision physics underlying ion implantation.

Feldman and Mayer (1986)

Introduction to physical principles of ion beam analysis.

Turner (1995)

Applications in health physics.

Nastasi et al. (1996)

Modern introduction to ion implantation.

Wieszczycka and Scharf (2001)

Ion beam therapy.

Conference series

Atomic collisions in solids (ICACS)

Biennial, starting 1965. Published in Nucl. Instrum. Methods B since 1980.

Symposia on stopping of charged particles

- Nucl. Instrum. Methods B **12**, 1-191 (1985)
- Nucl. Instrum. Methods B **27**, 249-353 (1987)
- Nucl. Instrum. Methods B **69**, 1-166 (1992)
- Nucl. Instrum. Methods B **93**, 113-226 (1994)
- Nucl. Instrum. Methods B **195**, 1-231 (2002).

Ion Beam Analysis (IBA)

Biennial, starting 1973. Published in Nucl. Instrum. Methods B since 1985

Radiation Research Congress

Every four years. Proceedings published in book form.

References

Ahlen S.P. (1980): Theoretical and experimental aspects of the energy loss of relativistic heavily ionizing particles. Rev Mod Phys **52**, 121–173

Bethe H. (1933): Stosstheorie. In *Handbuch der Physik*, vol. 24, 273. Springer Verlag, Berlin

Bethe H.A. and Ashkin J. (1953): The passage of heavy particles through matter. In *Experimental Nuclear Physics*, vol. 1, 166–201. Wiley, New York

Bethe H.A. and Jackiw R.W. (1986): *Intermediate quantum mechanics*. Benjamin/Cummings, Menlo Park, California

Bichsel H. (1972): Passage of charged particles through matter. In *American Institute of Physics Handbook*, 8.142–8.189. McGraw-Hill, New York

Birkhoff R.D. (1958): The passage of fast electrons through matter. In *Encyclopedia of Physics*, vol. 34, 53–138. Springer, Berlin

Bohr N. (1948): The penetration of atomic particles through matter. Mat Fys Medd Dan Vid Selsk **18 no. 8**, 1–144

Bonderup E. (1981): *Interaction of charged particles with matter*. Institute of Physics, Aarhus

Cabrera-Trujillo R. and Sabin J.R., editors (2004a): *Theory of the interaction of swift ions with matter. Part 1*, vol. 45 of *Adv. Quantum Chem*. Elsevier, Amsterdam

Cabrera-Trujillo R. and Sabin J.R., editors (2004b): *Theory of the interaction of swift ions with matter. Part 2*, vol. 46 of *Adv. Quantum Chem*. Elsevier, Amsterdam

Fano U. (1963): Penetration of protons, alpha particles, and mesons. Ann Rev Nucl Sci **13**, 1–66

Fano U., editor (1964): *Studies in penetration of charged particles in matter*, vol. 1133 of *Nucl. Sci. Ser*. Natl. Acad. Sci. - Natl. Res. Council, Washington

Feldman L.C. and Mayer J.W. (1986): *Fundamentals of surface and thin film analysis*. North Holland, New York

Gras-Marti A., Urbassek H.M., Arista N.R. and Flores F., editors (1991): *Interaction of charged particles with solids and surfaces*, vol. B 271 of *NATO ASI Series*. Plenum Press, New York

ICRU (1984): *Stopping powers for electrons and positrons*, vol. 37 of *ICRU Report*. International Commission of Radiation Units and Measurements, Bethesda, Maryland

ICRU (1993): *Stopping powers and ranges for protons and alpha particles*, vol. 49 of *ICRU Report*. International Commission of Radiation Units and Measurements, Bethesda, Maryland

ICRU (2005): *Stopping of ions heavier than helium*, vol. 73 of *ICRU Report*. Oxford University Press, Oxford

Inokuti M. (1971): Inelastic collisions of fast charged particles with atoms and molecules – the Bethe theory revisited. Rev Mod Phys **43**, 297–347

Jackson J.D. (1975): *Classical electrodynamics*. John Wiley & Sons, New York

Kumakhov M.A. and Komarov F.F. (1981): *Energy loss and ion ranges in solids*. Gordon and Breach, New York

Landau L. and Lifshitz E.M. (1960a): *Mechanics*, vol. 1 of *Course of theoretical physics*. Pergamon Press, Oxford

Landau L.D. and Lifshitz E.M. (1960b): *Electrodynamics of continuous media*, vol. 8 of *Course of theoretical physics*. Pergamon Press, Oxford

Landau L.D. and Lifshitz E.M. (1960c): *Quantum mechanics. Non-relativistic theory*, vol. 3 of *Course of theoretical physics*. Pergamon Press, Oxford

Livingston M.S. and Bethe H.A. (1937): Nuclear physics. C. Nuclear dynamics, experimental. Rev Mod Phys **9**, 245–390

Mayer J.W., Eriksson L. and Davies J.A. (1970): *Ion implantation in semiconductors. Silicon and germanium*. Academic Press, New York

Nastasi M., Hirvonen J.K. and Mayer J.W. (1996): *Ion-solid interactions: Fundamentals and applications*. Cambridge University Press, Cambridge

Sigmund P. (1975): Energy loss of charged particles in solids. In C.H.S. Dupuy, editor, *Radiation damage processes in materials*, NATO Advanced Study Institutes Series, 3–117. Noordhoff, Leyden

Sigmund P. (2004): *Stopping of heavy ions*, vol. 204 of *Springer Tracts of Modern Physics*. Springer, Berlin

Turner J.E. (1995): *Atoms, radiation, and radiation protection*. Wiley, New York, 2 edn.

Uehling E.A. (1954): Penetration of heavy charged particles in matter. Ann Rev Nucl Sci **4**, 315–350

Uehling E.A., editor (1960): *Penetration of charged particles in matter*, vol. 752 of *Nuclear Science Series*. Natl. Acad. Sci. - Natl. Res. Council, Washington

Whaling W. (1958): The energy loss of charged particles in matter. In *Encyclopedia of Physics*, vol. 34, 193–217. Springer, Berlin

Wieszczycka W. and Scharf W.H. (2001): *Proton radiotherapy accelerators*. World Scientific, New Jersey

Ziegler J.F., Biersack J.P. and Littmark U. (1985): The stopping and range of ions in solids. 1–319. Pergamon, New York

Author Index

Subject Index

Springer Series in
SOLID-STATE SCIENCES

Series Editors:

M. Cardona   P. Fulde   K. von Klitzing   R. Merlin   H.-J. Queisser   H. Störmer

MIX
Papier aus verantwortungsvollen Quellen
Paper from responsible sources
FSC® C105338

If you have any concerns about our products,
you can contact us on
ProductSafety@springernature.com

In case Publisher is established outside the EU,
the EU authorized representative is:
Springer Nature Customer Service Center GmbH
Europaplatz 3, 69115 Heidelberg, Germany

Printed by Libri Plureos GmbH
in Hamburg, Germany